A TEXTBOOK OF MEDICINAL CHEMISTRY

A TEXTBOOK OF MEDICINAL CHEMISTRY

Second Edition

MUNENDRA MOHAN VARSHNEY
Professor
Department of Pharmaceutical Sciences
Raj Kumar Goel Intitute of Technology
Ghaziabad

ASIF HUSAIN
Professor
Department of Pharmaceutical Chemistry
Jamia Hamdard University
New Delhi

TECHSAR™
NEW DELHI

Published by
I.K. International Pvt. Ltd.
4435-36/7, Ansari Road, Daryaganj
New Delhi–110 002 (India)
E-mail: tech_info@techsarworld.com
Website: www.techsarworld.com

ISBN 978-93-90620-24-1

Published/Printed by Krishan Makhijani for I.K. International Pvt. Ltd., 4435-36/7, Ansari Road, Daryaganj, New Delhi–110 002

Preface to the Second Edition

Medicinal chemists play vital roles in pharmaceutical research to synthesize, purify and analyse novel bioactive medicinal compounds which may be easily exploited. The actual phase of identification and optimum bioactive position on the basic skeleton of lead molecule is the important aspect for biological activity. Our main concern is to provide basic information to the students.

This book is the outcome of teaching of medicinal chemistry to the students of undergraduate and postgraduate in the field of pharmaceutical sciences.

Advancements in recent developments in medicinal chemistry have been taking place rapidly. We have tried to enrich with latest developments in the book. This book is meant for undergraduate students of pharmacy and those who opt for medicinal chemistry as a part of course curriculum. This book emphasizes the fundamentals of medicinal chemistry aspects, QSAR, various types of classification, drug synthesis, drug action and their pharmacological uses with recent developments of various classes of drugs. Thus, it would be beneficial to the readers to extend knowledge in respect of several known classes of drug molecules.

Suggestions for further improvement are welcome and we are responsible for any deficiencies or any kind of errors that might have crept in and we would be grateful to readers to call them for our attention.

Munendra Mohan Varshney

Asif Husain

Preface to the First Edition

Medicinal chemistry plays a vital role in pharmaceutical research to synthesize and analyse novel bioactive medicinal compounds. The actual phase of identification and optimum bioactive position on the basic skeleton of lead molecule is the important aspect for biological activity and for further research.

The advancement of recent developments in medicinal chemistry has been taking place rapidly. We have tried to enrich with latest developments in the book. This book is meant for undergraduate students of Pharmacy and those who offer medicinal chemistry as part of course curriculum. This book emphasizes the fundamentals of medicinal chemistry aspects, various classifications, drug synthesis, drug action and their pharmacological uses with recent developments of various classes of drug action and their pharmacological uses with recent developments of various classes of drugs. Thus, it would be beneficial to the readers to extend knowledge in respect of several known classes of drug molecules.

Suggestions and criticism for further improvement are welcome and we are responsible for all kinds of errors that might have crept in and we will be grateful if the readers bring them to our attention.

Munendra Mohan Varshney

Asif Husain

Contents

1

Physical Chemical Factors and Biological Properties

1.1 INTRODUCTION

In the history of drug, act has been of primary concern to scientists in the field of Biochemistry, Pharmacology and Medicinal Chemistry, etc. Since the early 20th century, efforts have been devoted for establishing a relation explanation for drug action. The drug or medicinal agents effect in the biological system is limited only by our ability to correlate the observed physiological aspects with a reasonable hypothesis or concept.

The concept of macromolecular protein/receptor has gained prominence. Nowadays new theories in receptor research, however, have provided the evidence that drug receptors exist in distinct moeities by the different methodologies such as isolation and molecular characterization of receptors. Drug-receptor concept emphasises the importance of physical and chemical properties of drug and relation of such distinguish properties to the biological action. Physical and chemical properties of the drug molecule may have effect on the biological platform. Properties include solubility, partition coefficient, degree of ionization, surface tension, isosterism, intermolecular forces, oxidation reduction potential and stereochemistry influence the absorption, distribution, metabolism and excretion of molecule, etc.

1.2 TYPES OF DRUG ACTION THEORIES

1.2.1 The Ferguson Principle

According to Ferguson, equilibrium condition must exist between the drug in the molecular biophase and the external phase (exobiophase) that exist in extracellular fluids. Present theory explains the tendency for the drug to escape from each phase is the same, although the concentration of each phase is different. This tendency is called thermodynamic activity which is equivalent to degree of saturation of each phase. Thermodynamic activity depends on vapour pressure (P) and molar concentration (S) ratio at the experimental temperature.

1.2.2 Structure and Activity

On the bases of pharmacological action, drugs may be divided into two main classes.

(i) Structurally Nonspecific Drugs

These are those drugs in which pharmacological action is not directly subordinated to the chemical structure except to the extent that structure affects physicochemical properties. Among these properties they may be adsorption solubility, PKa, oxidation and reduction potential.

(ii) Structurally Specific Drugs

Structurally specific drugs are those drugs whose biological action is a consequenes from their specific chemical structure which can be understood by its three dimensional structure bind to the receptors macromolecule at the specific target site. In these drugs chemical reactivity, shape, size stereochemical arrangement of the molecule and distribution of functional groups as well as resonance, Inductive effects, electronic distribution and possible binding with receptors play decisive roles.

Table 1.1: Minor structural change results in biological action change

Structure	Biological Response	Pharmacological classification
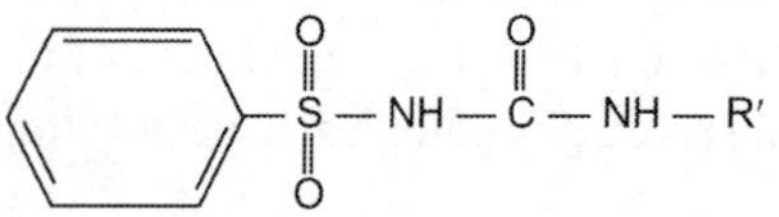	Short Acting	Hypoglycemic
R = CH_3, R' = C_4H_9 (Tolbutamide) R = Cl, R' = C_3H_7 (Chlorpropamide)	Long Acting	
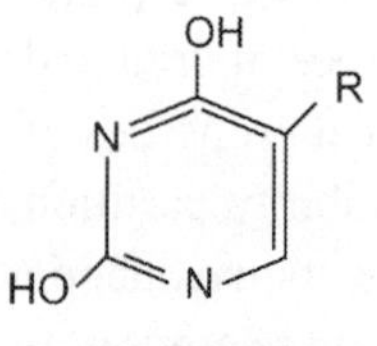	Metabolite	Metabolite
R = CH_3 (Thymine) R = F (5-Fluoro uracil)	Antimetabolite	
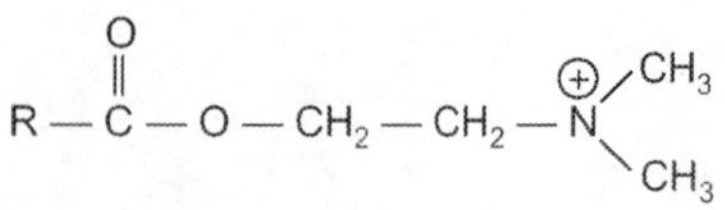	Short acting	Cholinergic
R = CH_3 (Acetylcholine) R = NH_2 (Carbachol)	Long acting	

1.2.3 Relationship between Molecular Structure and Biological Action

Concept of chemical and biological activity was introduced by Crum Brown and Fraser in 1869. They revealed that biological action of a drug is related to the molecular shape of the molecule, which includes the position of functional groups and stereochemistry. For illustration morphin, strychnine and nicotine compounds have different biological actions. Morphine shows analgesic activity while N-methylated morphin is a muscle relaxant. On the other hand, nicotine is a insecticide while N-methyl nicotine is a muscle relaxant. Conclusion is that after methylation in the lead compound there is change the biologic activity.

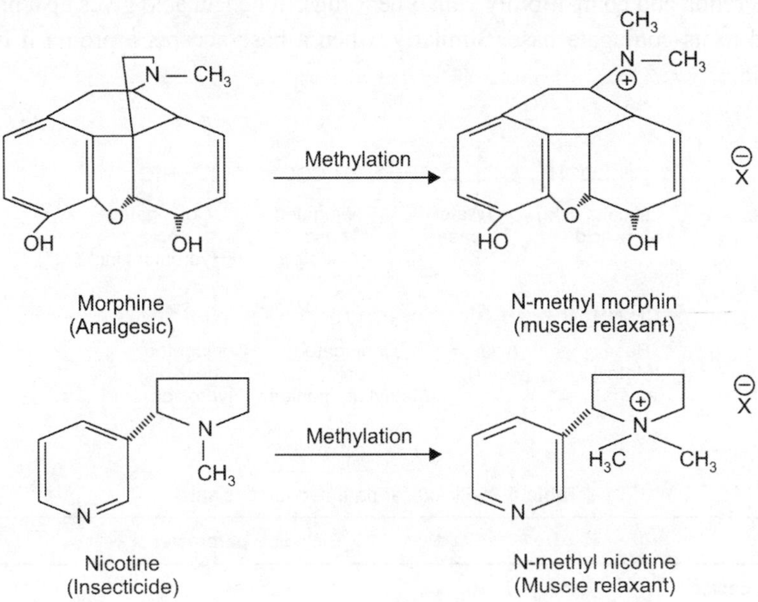

Fig. 1.1

1.3 PHYSICOCHEMICAL PARAMETERS AND PHARMACOLOGICAL ACTIVITY

1.3.1 Introduction

In early 1870, Crum-Brown and Fraser postulated the biological activity concerned with function of chemical structure of drug. This theory was the tool for establishing the structural activity relationship for the more specifically active and more potent drugs. Several mathematical equations have been approached which correlate chemical structural and pharmacological activity. The folowing aspects are directly related to biological action.

1.3.2 Solubility Parameters

Solubility parameter is also called lipophilic or hydrophobic parameter. It measures the degree of attraction of drugs by lipophilic membrane regions of macromolecules. They are related to the transport of the drugs from extracellular environment to receptor compartment with possible attraction and interaction between lipophilic portion of the drugs and receptors. The main parameters of this class are partition coefficient, hydrophobicity constant, solubility, molar mass, chromatographic parameter, substitution constant, surface tension, etc.

1.3.2.1 Acid Base Properties

According to Bronsted-Lowry theory, acids are donate H^+ (ions) and bases accept the same. Human body has 70–75% of water component (about 5.5 litres). The medicinal agents or drugs adopt the physicochemical environment in body dilute solution thus their acid/basic properties directly affect absorption, excretion and compatibility with other drugs. When an acid gives up a proton to a base, it is converted to its conjugate base. Similarly, when a base accepts a proton it is converted to conjugated acid.

Equations:

$$CH_3COOH + H_3O \rightleftharpoons CH_3COO^{\ominus} + H_3O^{\oplus}$$

CH_3COOH	H_3O	$CH_3COO^{\ominus}$	$H_3O^{\oplus}$
(Acetic acid) acid	(Water) base	Conjugated base (Acetate)	Conjugated acid (Hydronium ion)

$$CH_3NH_2 + H_2O \quad CH_3NH_3^{\oplus} + OH^{\ominus}$$

CH_3NH_2	H_2O	$CH_3NH_3^{\oplus}$	$OH^{\ominus}$
Base (Methyl amine)	Acid (water)	Conjugated acid (Methyl ammonium)	Conjugated base (Hydroxide)

Table 1.2: Biological parameters/constants

Source of activity		Biological parameter
1.	Isolated receptor	
	Rate constant	log k, log k_{cat}
	Michaelis-Menten constant	log 1/Km
	Inhibition constant	log 1/KI
	Affinity data	Data $pA_2 : pA_1$
2.	Cellular system	
	Inhibition constant	log 1/IC50
	Cross-resistance	log CR
	In vitro biological data	log 1/C
	Mutagenicity state	log TA_{98}
3.	In vivo system	
	Biocentration factor	log BCF
	In vivo reaction rates	log I (Induction)
	Pharmacodynamic rates	log T (Total clearance)

Table 1.3

1. Some acidic organic functional groups/conjugated base	
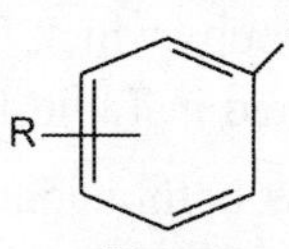 Phenol	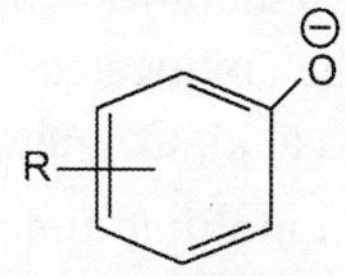 Phenolate
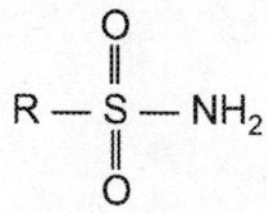 Sulphonamide	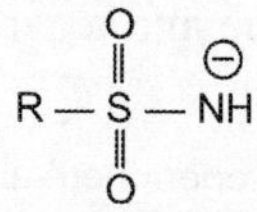 Sulphonamidate
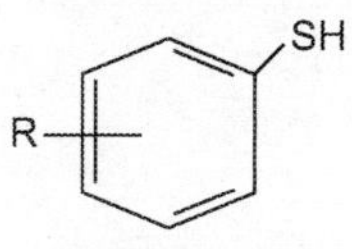 Thiophenol	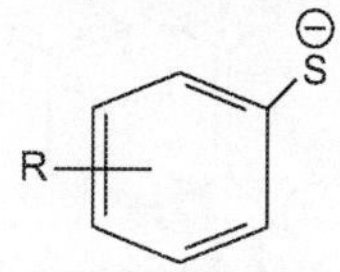 Thiophenolate
$R-SH$ Alkylthio	$R-S^{\ominus}$ Thiolate
2. Some basic organic functional groups/conjugated acid	
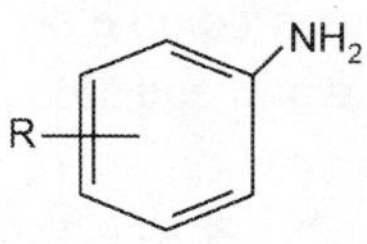 Arylamine	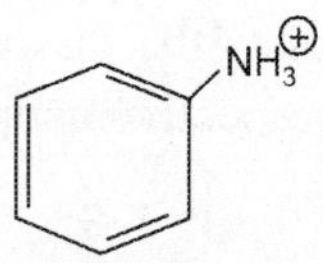 Arylammonium
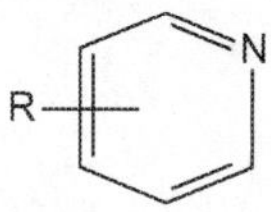	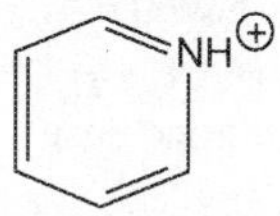 Aromatic ammonium
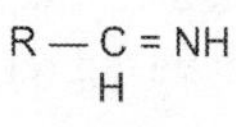 Imine	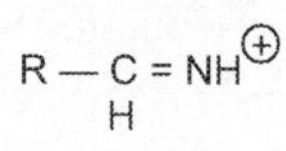 Iminium
3. Some organic functional groups are neutral at psychlogical condition	
$R-CH_2-OH$ Alkyl alcohol	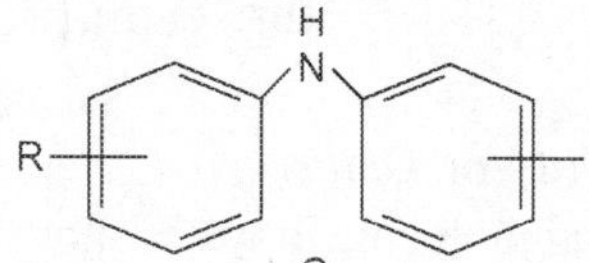 Diarylamine
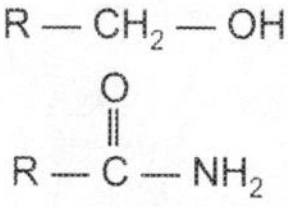 Amide	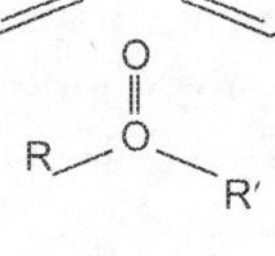 Ketone/Aldehyde
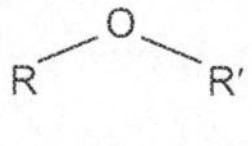 Ether	$R-C\equiv N$ Nitrile

When an acid loses its proton, it is left with an extra lone pair of electrons that are no longer neutralized by the proton. This is called the ionized form of acid and is more water soluble due to having charge. It also often undergoes "dissociation". Many different organic functional groups behave like acid and form the conjugated base and some of functional groups which behave like base and generate conjugated acid. Conjugated acid as well as base are listed in Table 1.3.

A molecule may contain multiple functional groups therefore possess both acidic as well as basic properties, for example, ciprofloxacin (fluoroquinolone antibiotic) contains a secondary alkyl group, two tertiary alkylamine and a carboxylic acid.

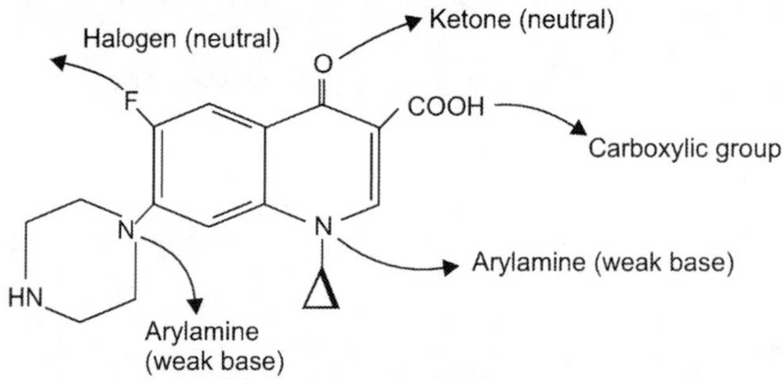

Fig. 1.2: Chemical structure of ciprofloxacin showing various functional groups.

1.3.2.2 *Relative Acid Strength Theory*

Relative acid strength depicts the dissociation of strong acids and bases completely in aqueous solution to produce their respective conjugated bases and acids as shown in equation.

Example

$$\underset{\text{(Strong acid)}}{HCl + H_2O} \rightleftharpoons Cl^{\ominus} + H_3O^{\oplus}$$

$$\underset{\text{(Strong base)}}{NaOH + H_2O} \rightleftharpoons Na^{\oplus} + OH^{\ominus} + H_2O$$

Degree of ionization prediction is based on if a molecule is going to be ionized or unionized at the given pH. This factor is associated with the functional groups attached to the molecule are acidic or basic. The Henderson-Hassalbach equation can be used to calculate the degree of ionization at given pH.

$$\text{PKa} = \text{PH} + \log\frac{[\text{acidic form}]}{[\text{base form}]}$$

1.3.2.3 *Water Solubility of Drugs*

The solubility of any drug molecule in water shows the pronounced effects on the respective routes of administration as well as its absorption, distribution and elimination. The two phenomena to be kept in mind when considering the water/fat solubility of a molecule are the hydrogen bond-forming potential of the various functional groups which are present in the molecule and ionization of functional groups.

1.3.3 Steric Parameters

Steric parameters reveal the factors which represent the form and the size of substituents introduced in the parent compound. They measure the intramolecular steric effect. Parameters are the following:

(a) Taft substituents constant
(b) Hancock's constant
(c) vander waals radii
(d) Steric constant
(e) Molecular connectivity
(f) Verloop's Sterimol parameters.

1.4 STEREOCHEMISTRY CONSIDERATION

Isomers are compounds that have the same molecular formula. These are named depending on the difference which may be structural or stereoisomeric represented in scheme:

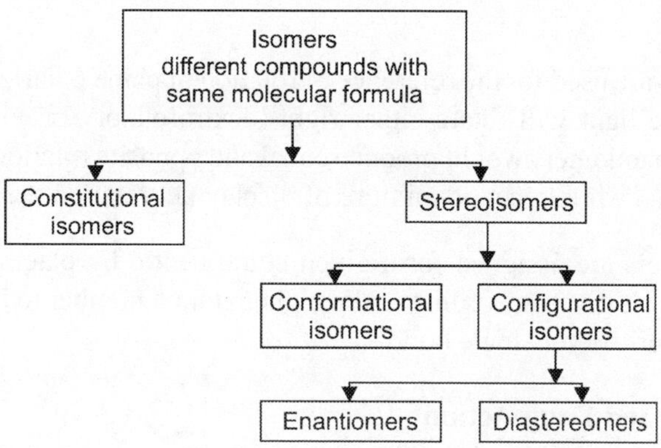

Fig. 1.3

1.4.1 Chirality

Concept of chirality was approached by J.H. Van't Hojj (first recipient of Nobel Prize in Chemistry in 1901) and J.A. Le Bel in the favour of four bonds of carbon atom in methane. They are arranged in such a way so as to point towards the corner of a regular tetrahedron (shown in Fig. 1.4) if the carbon atom was placed at its centre. Regular tetrahedron is highly symmetrical with four equivalent vertices. When these are occupied by four different achiral atoms or groups and this type of arrangement is non-superimposable on its mirror image. The tetracovalent carbon (the fifth point)

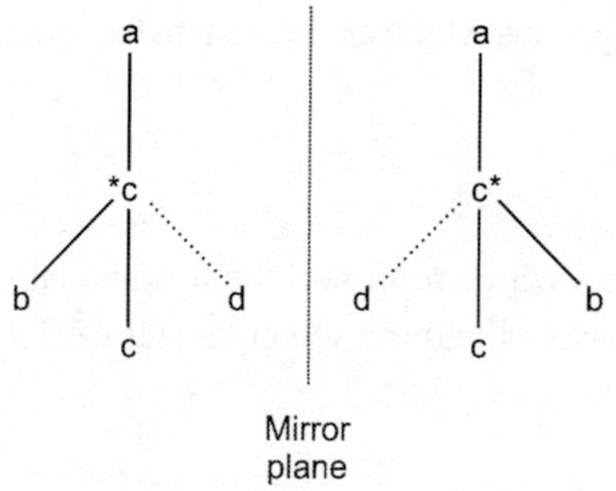

*c represents a stereocenter of chiral

Fig. 1.4

placed at the centre of the tetrahedron is known as stereocenter whose presence generally generates molecular chirality.

There are some major ways to express the chirality.

1. **D and L:** Means (dextro and levo) comes from the Latin words for 'Right' and 'Left'.

   For example: D-L- Glyceraldehyde.

   This is done by the most oxidized group prefer at the bottom and comparing the Fisher projection of the two molecules.

2. **d and l**: These letters are used for the reference of rotation of plane polarized light which passes through the solution. The light will rotate either right (dextrorotatory or +) or left (levorotatory or –). The pure form of enantiomers would produce equal and opposite rotation (50% of dextroform +, and 50% of levoform –) while recemic mixture of display net rotation zero.

3. **R and S:** These letters are assigned for a given chiral centre by placing the lowest priority group in the back. Priority is given according to the higher atomic number to lower atomic number. Double bond is given priority over single bond.

1.4.2 Stereochemistry and Drug Action

According to stereochemistry aspects, every function group is cited in such a way that has more spatial arrangement respectively. Drug molecules though come into the human body which is an asymmetric environment due to proteins or macromolecules which are asymmetric in nature and are also called receptors. When a particular drug molecule binds to macromolecule or receptor protein, this drug-receptor ligand is determined by three-dimensional orientation of functional groups. On other hand, if proper spatial arrangement of various functional groups surrounding is required, the macromolecules do not occur, then the drug-receptor bonding would not be possible. Although if respective functional groups are in proper three-dimensional orientation, this drug may produce very strong interaction with receptors and show biological activity.

1.4.3 Isomers Types

1.4.3.1 Constitutional Isomers (Positional Isomers)

Constitutional isomers have different sequence of bonding of their atoms as with the molecular formula, for example, C_2H_6O which may represent ethanol CH_3CH_2OH or dimethyl ether CH_3OCH_3.

Example of C_3H_7

$$CH_3-CH_2-CH_2-CH_3 \qquad CH_3-CH(CH_3)-CH_3$$

Butane Isobutane

Other examples are glucose and fructose.

Some other examples of constitutional isomers include:

(i) Tautomers: Isomerization involves atoms or group ingration
(ii) Proton tautomers: Where Keto-Enol isomerization occurs.
(iii) Valence isomers: Are interconvertible by reorganization of some bonding electrons.

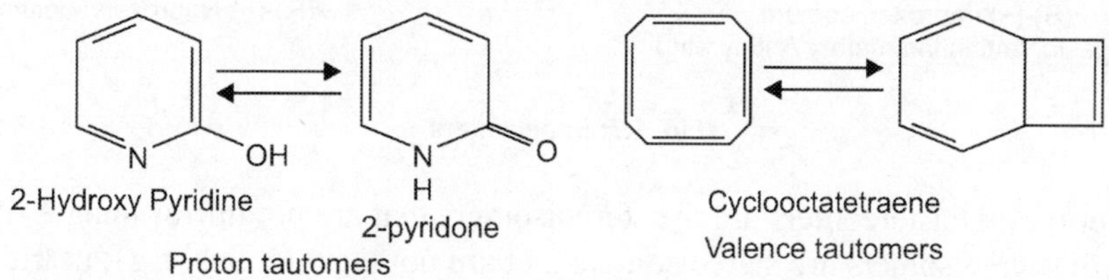

Fig. 1.5

1.4.3.2 Stereo Isomers

Isomers have the same sequence of covalent bonds but differ in the relative disposition of the atoms in space. Examples are:

(i) Enantiomers
(ii) Diastereomers (Epimers, Anomers)
(iii) Conformational Isomers

(i) Enantiomers—Optical Isomers: Enantiomers are also referred to as chiral compounds/ antipodes or enantiomorphs. Consider a simple molecule, e.g., a compound with an sp^3 hybridized carbon with four different substitutents as in 2-butanol. The molecule cannot be superimposed on its mirror images. Such molecules are said to be chiral (or handad) and the pair of butanol molecule are referred to as "enantiomers" (from the Greek word "enantio" means opposite) which are defined as pairs of molecules related to non-superimposable images. Enantiomers have identical physical properties except ability to rotate plane of polarized light with equal magnitude as well.

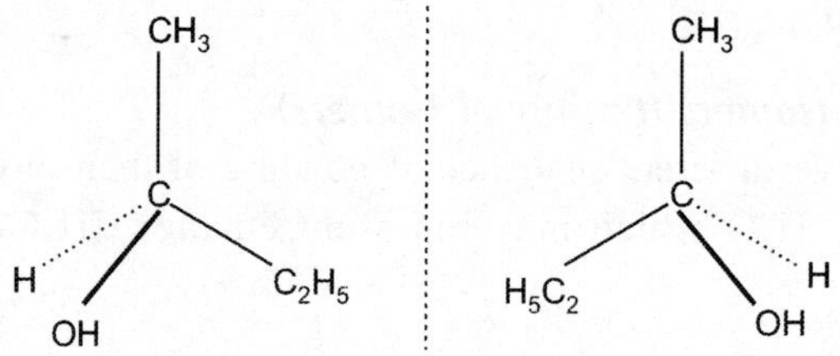

(S)-(+)-2-butanol R-(–)-2-butanol

Fig. 1.6

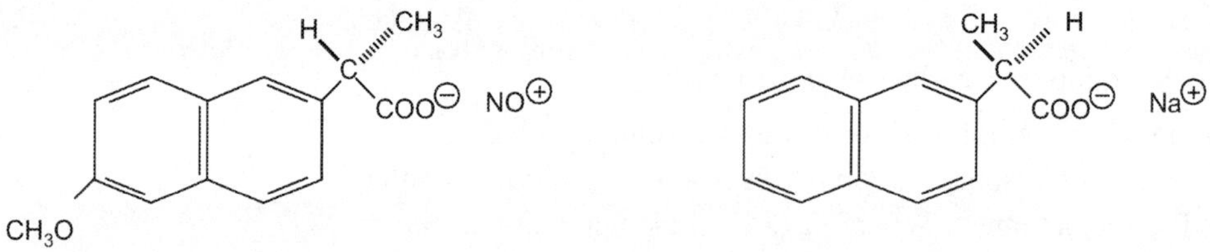

(Active) (Inactive)

(S)-(+) naproxen sodium
(Analgesic, Anti-inflammatory Antipyretic)

(R)-(–)-Naproxen sodium

Fig. 1.7: Enantiomers.

(ii) Diastereomers: Diastereomers are the stereoisomers that are not mirror images of each other. It means that diastereoisomers are stereoisomers that are not enantiomers. e.g. tartaric acid, meso-tartaric acid is a diastereomer of each of the enantiomers.

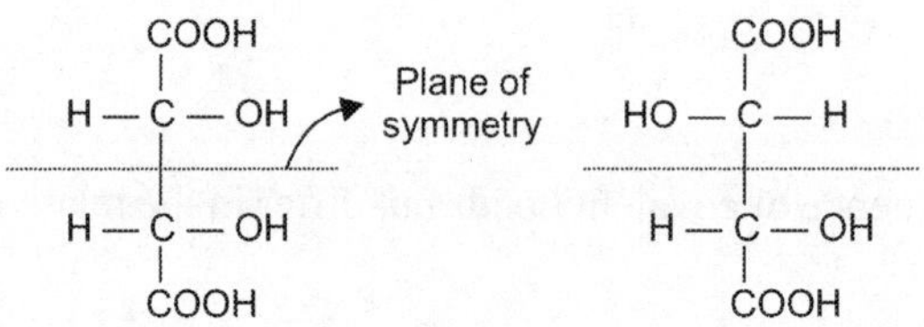

(MESO) (2s, 3s-tartaric acid)

Fig. 1.8: Tartaric acid.

Diastereoisomers display different physical and chemical properties such as chromatographic behaviours, solubility, melting point and boiling point.

Several illustrations of other compounds that would contains two or more than chiral centres and therefore are diastereoisomers. Carbon-carbon double bonds containing system such as imines (C = N) cause restriction of bond and therefore produce stereoisomers. They are also referred to as geometric isomers. substituents can be oriented on the same side or opposite site. Generally referred to as cis or Z-isomers [Greek word Zusammen (together)] and trans-isomers or E-isomers [Greek word Entgegen (opposite)] as well.

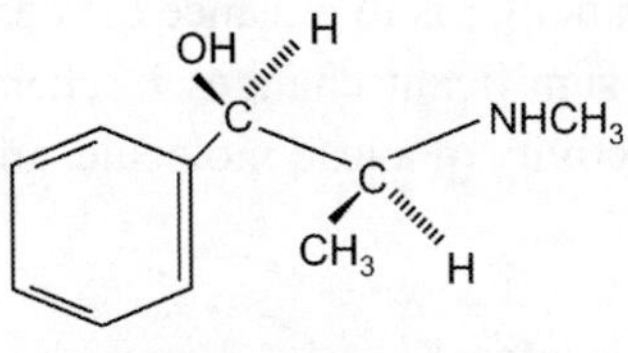

(–)-Ephedrin (1R, 2S)
(Active)
(CNS-stimulent)

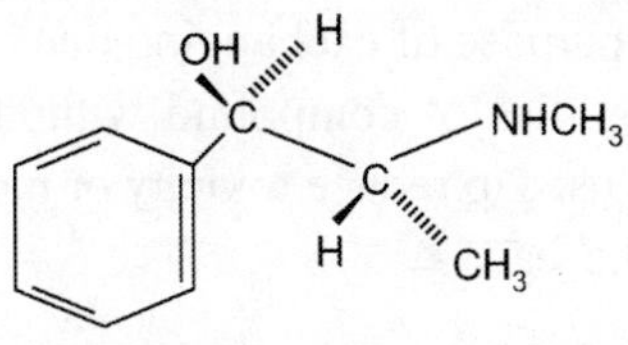

(–) Pseudoephedrine
(1R, 2R)
(Inactive)

Fig. 1.9: Diastereoisomers.

For Example: 2 Butane

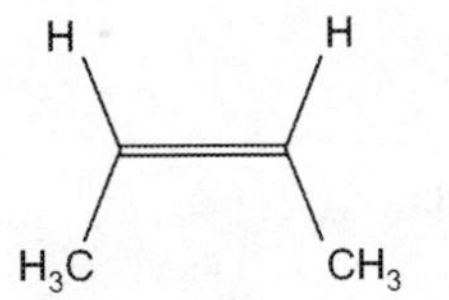

Cis or Z-isomer

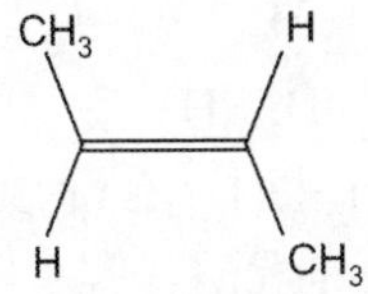

Trans or E-isomer

Fig. 1.10

1.5 ISOSTERISM AND BIOISOSTERISM

1.5.1 Pharmacophore or Lead

Pharmacophore of a drug molecule is that portion of the molecule which contains the essential functional groups and directly interact with receptor active site and therefore confers on the molecule the biological activity of interest. This phenomenon is very important and makes drug-receptor interaction or ligand. Alteration of any organic functional group in the specific site of drug results in change in biological activity. It may be in alteration of alkyl chains; chain length, chain branchning and ring alteration.

1.5.2 Isosteres

The isostere concept was formulated by Lrving Langmuir in 1919 and later modified by Grimm. After that Friedrich Erlenmeyer extended the concept to biological system in 1932. Biological-active compound containing an isosteres is called a "Bioisostere". Classical isosteres are defined as being atoms, ions or molecules that had identical outer shell of electrons. The definition has now been broadened to include groups that produce compounds that can sometimes have similar biological activities.

1.5.3 Bioisosteres

In medicinal chemistry, bioisosteres are substituents or groups with similar physical or chemical properties which formed broadly similar pharmacological activity to the chemical compound. In drug design the purpose of exchanging one biostere to another is to enhance the desired biological or physical properties of compound without making significant changes in chemical structure. Bioisosterism is used to reduce toxicity or modify the activity of a lead molecule, and may alter the metabolism of the lead.

1.5.3.1 Classical Bioisoster and Non-classical Bioisosters

Burger's definition significantly broadened this concept. Now denominating those atoms or molecular subunits or functional groups of the same valence and ring equivalents are the classical bioisosters. While non-classical bioisosters were those which practically did not fit the definition of the first class concept.

Classical Bioisosters

Monovalent

F, H

OH, NH

F, OH, NH or CH_3 for H

SH, OH

Cl, Br, CF_3

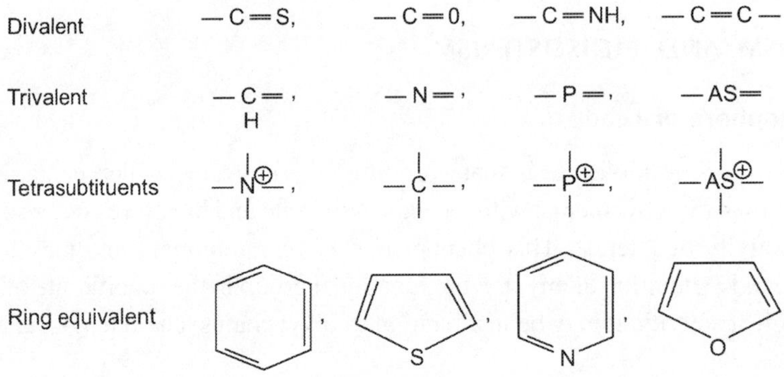

Fig. 1.11

Example

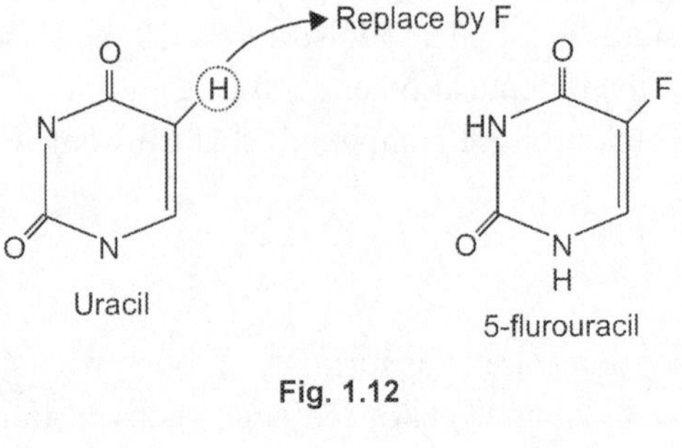

Fig. 1.12

Isosteric replacement of hydrogen by fluorin atom is an example of classical Bioisosters.

1.5.3.2 Non-classical Bioisosteres

(i) Cyclic or noncyclic
(ii) Functional groups
(iii) Retroisosterism

Some of the functional groups through mimic spacial arrangement, electronic properties or some other physico-chemical properties of the molecule or functional groups come under non-classical bioisosteres.

For Example:

Use of double bond to position essential functional groups into a particular spatial configuration may be critical for biological activity. Trans-estradiol is the synthetic analogue of estradiol (Steroidal Analogue) is equipotent as parent compound because hydroxy group mimic the correct orientation to drug-receptor interaction. On the other hand, cis form having 1/14th activity of the parent one.

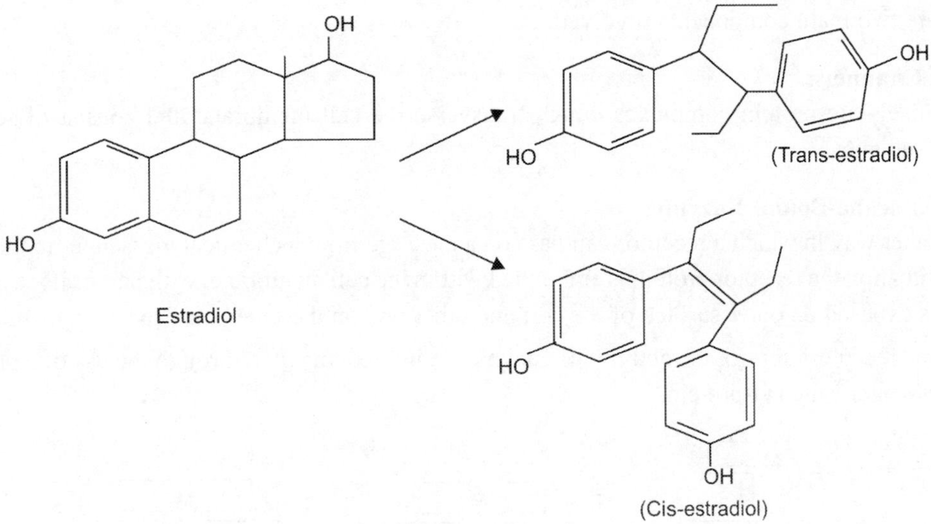

Fig. 1.13

1.6 DRUG-RECEPTOR INTERACTION

1.6.1 Receptor

A receptor is known to be a protein molecule embedded within cell membrane with part of its structure facing the outside of the cell. Receptor surface is a complicated shape containing hollows, ravines and ridges and somewhere amidst the complicated orientation there is an area which has the correct shape to accept the incoming messenger. This area is known as binding site and is analogue to the active site of an enzyme. When a chemical messenger fits into the site, it switches on the receptor molecule and the message is received. Although the main difference between enzyme and receptor is that the chemical messenger does not undergo a chemical reaction. It fits into the binding site of the receptor protein with distinguished chemical bonding, passes on its message and then leaves unchanged.

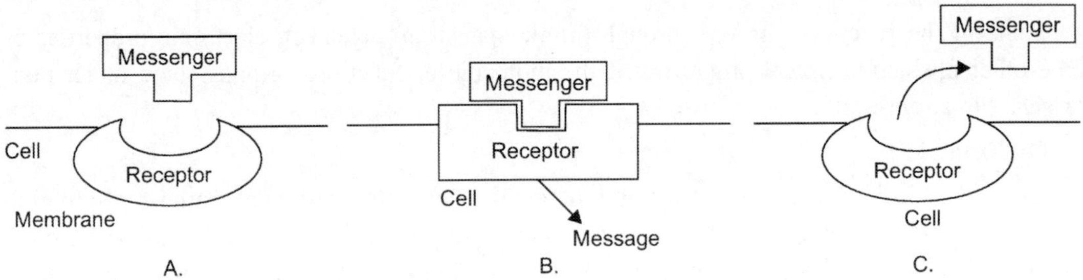

Fig. 1.14

There are two main components involved.

(i) Ion Channels:

Ion channels are protein complexes which transverse the cell membrane and consist of several protein subunits.

(ii) Membrane-Bound Enzymes

It is another way in which a receptor can pass on a message from a chemical messenger to the cell. In this situation the receptor protein is embedded within the cell membrane, with part of its structure which is exposed on outer surface of the cell and other one on the inner surface of the membrane.

Some receptors control the activity of enzymes in indirect manner, through the use of a protein unit messenger called G-protein.

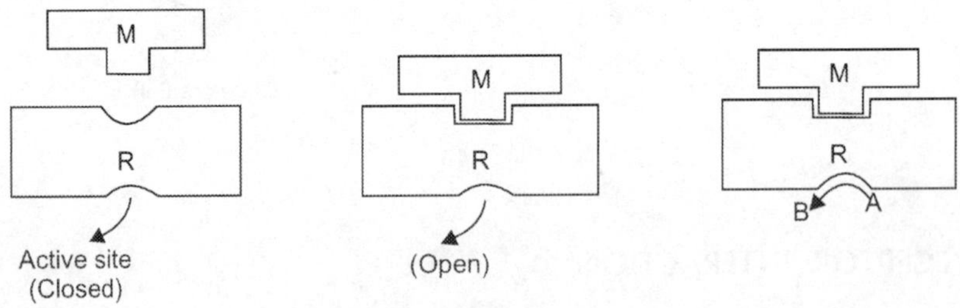

Fig. 1.15: Enzyme activation.

M – Messenger
R – Receptor

1.6.2 Requirement for Agonist Activity

The drug molecules can be designed to interact with receptor with the following requirements:

(i) The drug must have the correct binding functional group.
(ii) The drug must have these binding groups in correct position.
(iii) The drug must be of the right size for the binding site.

1.6.3 Receptor Families

There is a large number of different membrane bound receptors. On the basis of their structure and function, they are grouped into three "superfamilies":

(a) Ion channel receptors
  (i) 2-TM receptors
  (ii) 3-TM receptors
  (iii) 4-TM receptors

(b) G-Protein coupled receptors (GPCR) (7-TM receptors)

(c) Kinese linked receptors (1-TIM receptors)

1.6.4 Role of Chemical Bonding

Drug-receptor interaction is the ability of a drug to produce an effect which must result from specific chemical interaction between drug and specific site. When a drug interacts with receptor, it is believed that a number of chemical attractive forces are involved in the initial interaction. The compounds that are attracted to a receptor or macromolecular structure are said to have affinity for that receptor. The physical characteristics of a compound interact with receptor depend on its efficacy of chemical bonds types. The overall strengths of these bonds will vary and will determine the degree of affinity.

The chemical bonds are the following.

(i) Covalent bond: Strongest bond in which two atoms, one from the ligand and one from the receptor, share a pair of electron. Strength of covalent bond is (50-150 Kcal/mol)
Example:

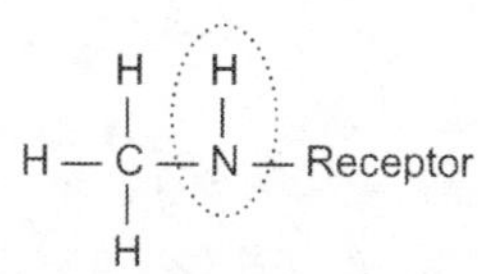

Fig. 1.16: Covalent bond.

(ii) Ionic Bond: Ionic bond is generated when two ions of opposite site/charge are attracted to each other through electrostatic forces. The strength of the ionic bond is 5–10 kcal/Mol.

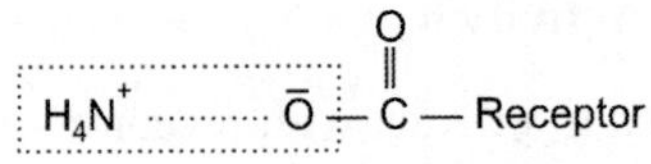

Fig. 1.17: Ionic bond.

(iii) Hydrogen Bond: Hydrogen bond is linked via a covalent bond to a strongly electronegative atom (O, N, S) which develops a relative positive charge and will be attracted to another which possesses a relative negative charge. This type of relationship is called 'hydrogen bond'. For example, a water molecule behaving as an electronic dipole can easily bond to other water molecule via hydrogen bonding. Strength of hydrogen bond is 2 to 5 kcal/Mol.

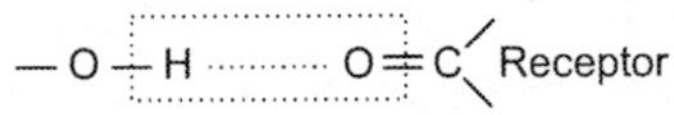

Fig. 1.18: Hydrogen bond.

(iv) Hydrophobic Interaction: It is established between non-polar organic molecules, which contribute to the binding forces that attract a ligand to its receptor. It is a very weak force and strength of hydrophobic interaction is 0.5–1 kcal/Mol.

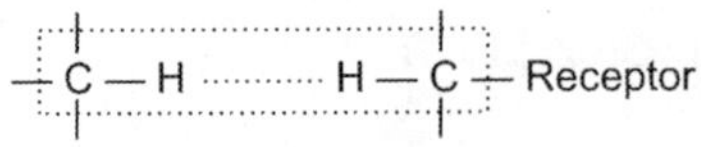

Fig. 1.19: Hydrophobic bond.

1.6.5 Types of Receptors

1.6.5.1 Ligand Gated Ion Channel Receptor

Those macromolecules which control ion channels are the part of the five-protein ion channel structure, and are called ligand-gated ion channel receptors. These types of receptors are based on the fact that glycoproteins which transfer the cell membrane and that the receptor is a part of the ion channel that bind to chemical messenger which leads to a rapid response. When the receptors binds to ligand, it can change the shape of receptor and this makes a knock-on effect on the protein complex which has the ability to cause the ion channel to opening. This mechanism is called "gating". There are cationic ion channel for Na^+, K^+ and Ca^{++} ions. With the effect of depolarization, these channels are open, for examples, nicotinic receptors, $5HT_3$, ATP receptors, on the other side anionic ions channels for chloride ion, are controlled by $GABA_A$ and glycine receptors.

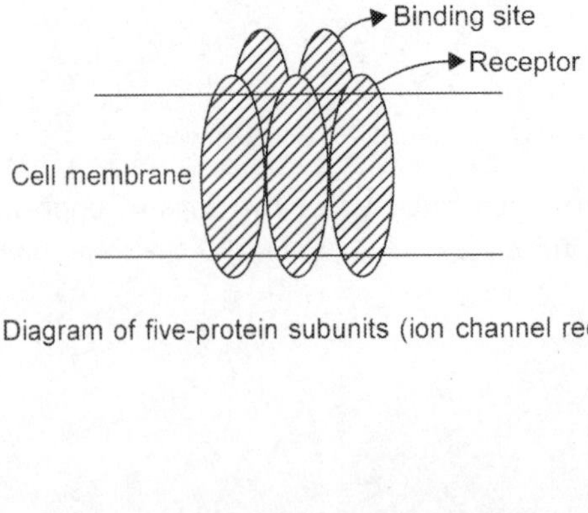

Fig. 1.20: Diagram of five-protein subunits (ion channel receptors).

1.6.5.2 G-Protein-Coupled Receptors (7-TM)

G-protein-coupled receptors are such proteins which are embedded in the cell membrane and which have regions exposed both to the outside and inside of the cell. The protein chain winds back and forth through the cell membrane sevenfolds, hence, the term (7-TM) transmembrane. Each of the seven transmembrane sections is lipophilic and helical in shape and thus usual to assign these helices with Roman numerals (I, II, etc.) starting from N-terminal end of the protein. The process makes three extracellular and intracellular loops that bind to the specific receptor. The N-terminal chain of extracellular loops is to the exterior of the cell and variable in length depending upon the receptor while the C-terminal chain is to interior.

There are a large number of different G-protein coupled receptors. Each one interacts with different neurotransmitter or hormone.

The chemical messengers are the following:

- Monoamines (Dopamine, Histamine, Serotonin, etc.)
- Nucleotides
- Calcium ions
- Lipids
- Glycoprotein hormone
- Glutamate

Signal Transduction: Signal transduction by a G-protein coupled receptor (GPCR) initiated with an inactive G-protein to the receptor. It exists as a heterotriner consisting of G_α, G_β, and G_γ. Once GPCR recognizes a ligand, the conformation of the receptor changes to activate G-protein, causing G_α to bind a molecule of GTP and dissociate from the other two-G-protein subunits. The dissociation exposes sites on the subunits that can interact with other molecules. The activated G-protein subunits release from the receptor and initiate signal from many downstream effectors. Protein such as 'phospholipases' and ion channels therefore release second messenger molecules. The total strength of signal by receptors is determined by the lifetimes of the ligand-receptor complex and receptor-effector protein complex and the deactivation time of activated effectors through intrinsic enzymatic activity.

1.7 PRODRUGS

1.7.1 Introduction

Prodrugs are the compounds which are inactive in themselves but which are converted in the body to the active moiety. Prodrugs are designed to improve the permeability and oral absorption of the parent drug. They are more lipid soluble than the parent one and should be rapidly converted to the parent compound during absorption from the gut wall, liver or site of action. They have been useful in tackling the problems such as acid sensitivity, poor membrane permeability, drug toxicity, unlikely taste and short duration of action. It is very important of designing prodrugs to ensure the prodrug should effectively converted to the active one and after absorb into the blood supply any groups cleaved from the molecule would be non-toxic.

1.7.2 Concept of Prodrug

Albert and coworkers established the concept of prodrug approach in 1950. They depicted prodrug as pharmacologically inactive chemical derivatives that have ability to change the physico-chemical properties of drug in a temporary manner. Such derivatives are also called "latentiated drugs', or bioreversible derivatives or congeners.

Prodrug can also be defined as a drug derivative that undergoes biotransformation, enzymatically. non-enzymtically, inside the body before exhibiting its therapeutic effect and followed by rapid elimination of released derivatizing group without causing side effects in the process.

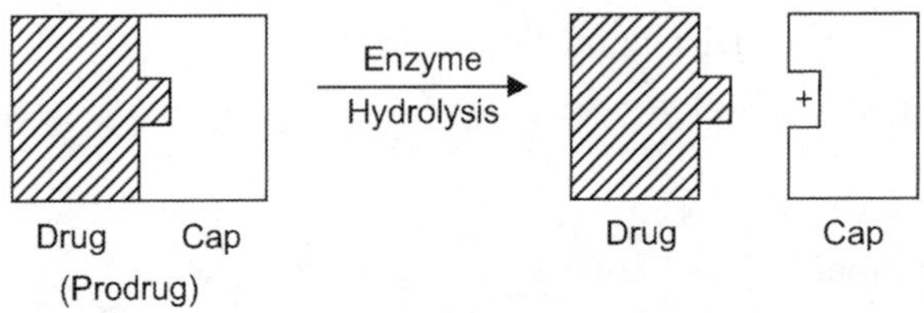

Fig. 1.21: Concept of prodrug.

1.7.3 Prodrugs are Designed for the Following Purposes

1. To improve patient acceptability
2. To improve absorption
3. Alter biodistribution
4. Alter metabolism
5. Reduce toxicity or adverse effect
6. To increase chemical stability
7. Prolong or shorten the pharmacological action.

1.7.4 Classification of Prodrugs

1. Carrier Linked Prodrugs
2. Bioprecursors

1.7.4.1 Carrier Linked Prodrugs

These pertain to a temporary linkage of active molecule with transport moiety which is hydrophobic in nature. A simple hydrolytic reaction cleaves this transport moiety. Such type of prodrugs are less active than the parent one or even are inactive.

Table 1.4: Chemical agents prodrugs (some examples)

1. Carrier linked prodrugs

Examples:

1. (R — OH, Ar — OH, —C(=O)— OH)

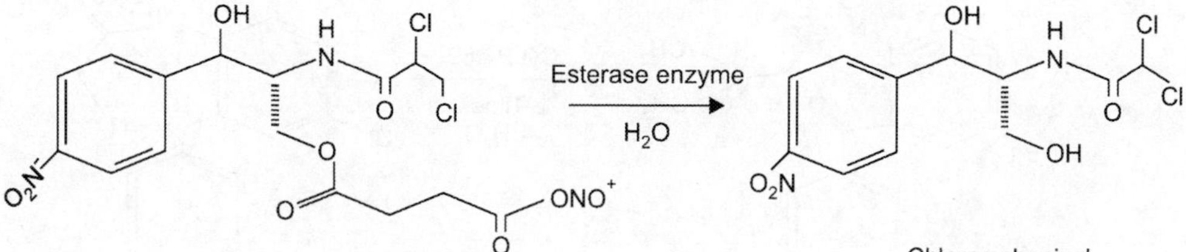

Chloramphenicol succinate

Chloramphenicol

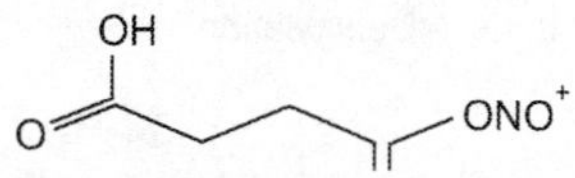

Sodium succinate

2. (Amide)

(—C(=O)— NH_2)

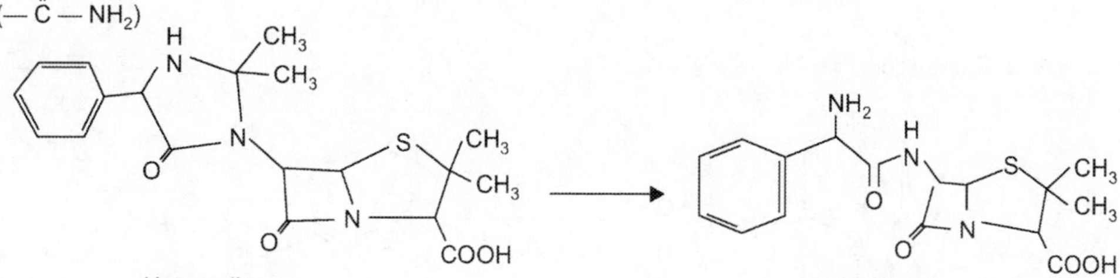

Hetacycline

Ampicillin

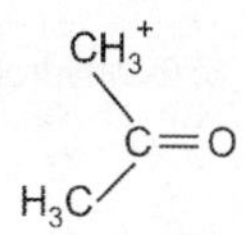

3. Increase water solubility

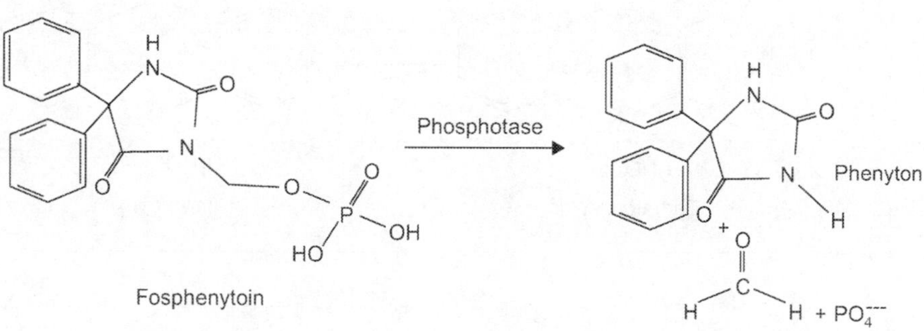

Fosphenytoin

1. Bioprecursor prodrugs

Examples:

1. N-Dealkylation

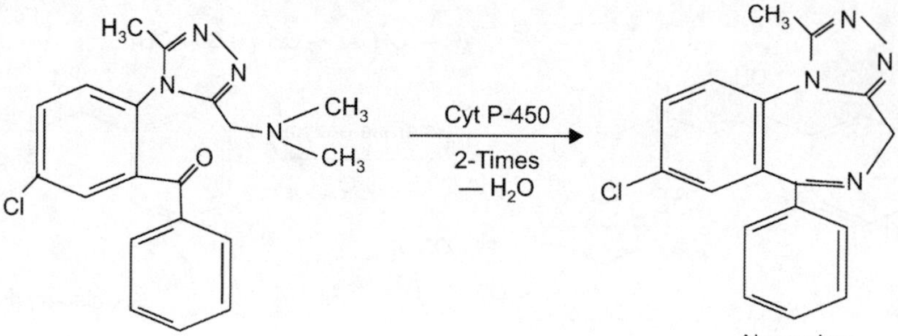

Alprazolam

2. O — Dealkylation

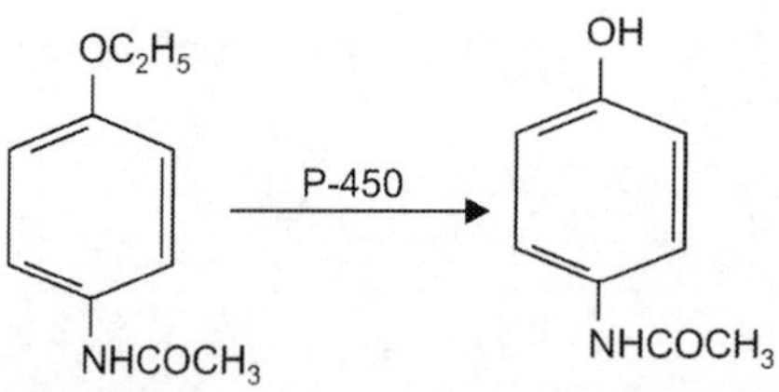

Phenacetine Paracetamol

3. N — Oxidation

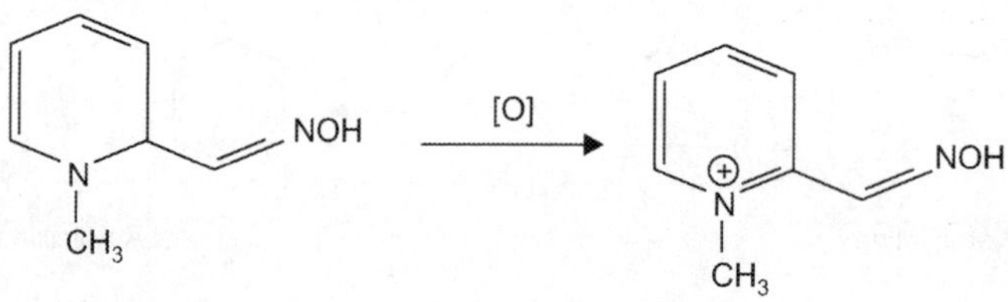

5, 6-Dihydropyridine Pralidoxime

4. Epoxidation

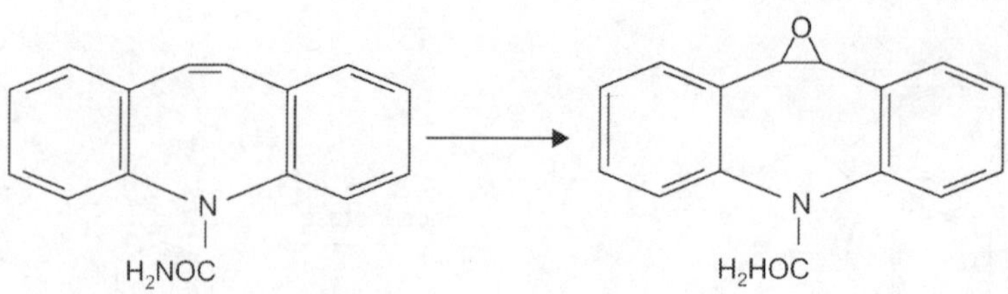

Carbamazepine Active epoxide

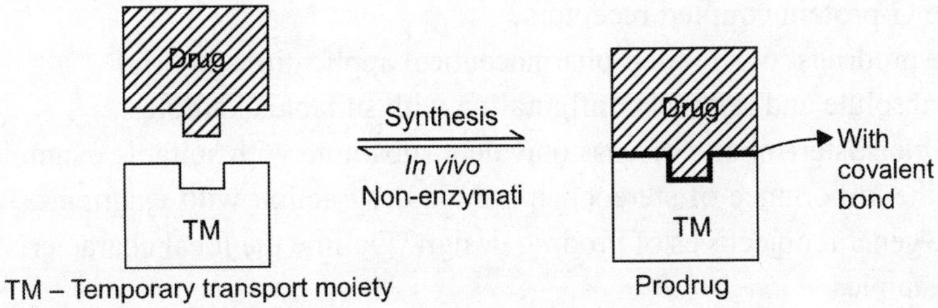

Fig. 1.22: Schematic diagram of carrier linked prodrug.

Further classification of carrier linked prodrugs:

- **Bipartate:** Composed of one carrier group attached to the drug.
- **Tripartate:** Carrier group attached via linker to drug
- **Mutual prodrugs:** Two drugs are linked together

1.7.4.2 Bioprecursor Prodrugs

These prodrugs do not imply the linkage to a carrier group but result from molecular modification of the active principle itself. This molecular modification generates a new compound which may be transformed metabolically or chemically. The final compound would be an active one:

Metabolic activation takes place by the following methods:

1. Oxidative Activation
   - N and O dealkylation
   - Oxidative deamination
   - N-oxidation
   - Epoxidation
2. Reductive Activation
   - Azo reduction
   - Sulfoxide reduction
   - Disulfide reduction
3. Nucleotide activation
4. Phosphorylation activation
5. Cyclization

EXERCISE

1. Explain optical isomerism of drug molecule and biological action.
2. Explain drug-receptor interaction.
3. Write a note on the concept of prodrug.
4. Explain how biological activities are influenced by optical isomers. Illustrate with suitable examples.

5. What are G-protein coupled receptors?
6. What are prodrugs? Write their pharmaceutical applications in brief.
7. Discuss absolute and relative configuration with suitable examples.
8. Classify bioisosterism and discuss univalent substitute with suitable examples.
9. Discuss the importance of stereochemistry in drug action with examples.
10. What are general objectives of prodrug design? Outline the ideal characteristics of prodrugs giving examples.
11. Enumerate the different types of receptor family and describe in detail about drug action mediated by ion channel.
12. What are the factors which govern the ability of drugs to reach to its active site.
13. What are the factors which govern drug action?
14. Write the concept of prodrugs.

2

Autonomic Nervous System Cholinergic Drugs and Adrenergic Agents

AUTONOMIC NERVOUS SYSTEM

2.1 INTRODUCTION

Cholinergic nerves are also called parasympathetic nerves which occur in the peripheral and central nervous system. Autonomic nervous system works as 'fight or flight' system inside the human body. Autonomic Nervous System (ANS) is also responsible for housekeeping mechanism of the body. Posterior/lateral nuclei of hypothalamus release the neurotransmitters that make the sympathetic nervous system while anterior or middle nuclei of hypothalamus release the neurotransmitters which make parasympathetic nervous system. Acetylcholine (Ach) serves as a neurotransmitter at both sympathetic and parasympathetic nervous system at the postganglionic nerve ending. Postganglionic nerve fibre in the parasympathetic division and some of the postganglionic fibres in the sympathetic division (e.g., salivary and sweat gland) of ANS regulate the activity of smooth muscles (respiratory circulation, digestion, body temperature and metabolism) and glandular secretion. Acetylcholine is prepared at the nerve ending by the transfer of Acetyl group from Acetyl-CoA to choline. This reaction is catalyzed by enzyme acetyltransferase. Ach is released from the synaptic vesicles by the action potential. The major site of Ach in periphery are postganglionic parasympathetic receptors, autonomic ganglia and skeletal neuromuscular junction. On the other hand adrenal medulla in CNS release the neurotransmitter adrenaline and nor-adrenaline. Adrenaline reinforces the action of nor-adrenaline therefore makes fight or flight process inside the body. Nor-adrenaline affects cardiac smooth muscles, smooth muscles, glandular secretion.

The peripheral nervous system is made up of afferent and efferent neurons of the automatic and somatic nervous system. Drugs acting at synaptic and neuroeffector junctional sites can either stimulate them or block them. The autonomic nervous system is also called vegetative (helps to control), visceral, automatic or involuntary nervous system.

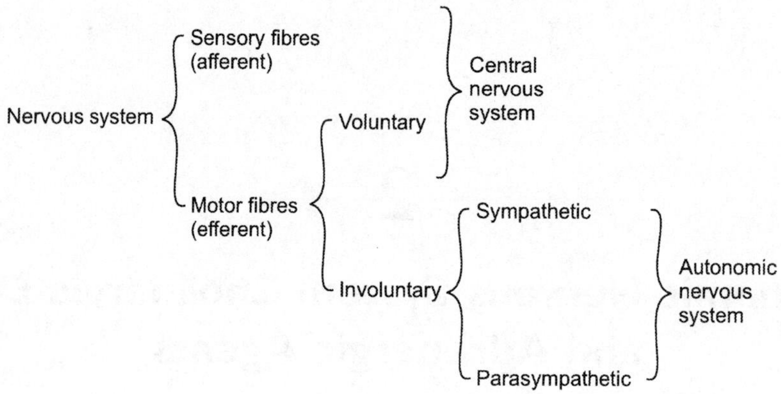

Fig. 2.1: Anatomy of nervous system.

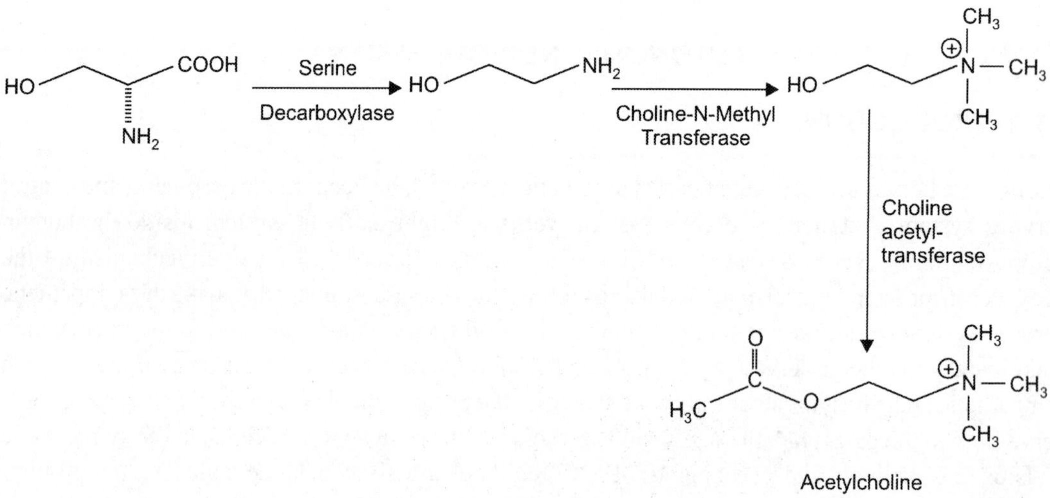

Fig. 2.2`: Biosynthesis of acetylcholine.

2.2 CHOLINERGIC DRUGS

2.2.1 Definition

The drugs and the chemical agents that act on cholinergic nerves or on the tissue innervate to either mimic or block the action of Acetylcholine (Ach). These are known as cholinergic drugs or agents. They mimic the action of Ach either by direct action on cholinergic receptors or by indirect action (inhibiting Acetylcholinersterage Enzyme) at the nerve terminals. The term is known as sympathomimetic (mimic the action of Ach), cholinomimetic parasympathominetics, (on parasympathetic division) whereas blocking the action is called parasympatholytic (on parasympathetic division) and sympatholytic (on sympathetic division).

Table 2.1: Action of ANS on various effectors

	Effector	Sympathetic	Parasympathetic
1.	**Eyes**		
	Pupil	Dilation	Constriction
	Ciliary Muscles	Constriction	Constriction
2.	**Glands**		
	Sweat	Secretion	—
	Salivary	Secretion	Secretion
	Bronchioles	Relaxation	Constraction
3.	**Heart**		
	Sinus nodal rhythm	Acceleration	Slowing
	Atrial contraction force	Increase	Decrease
	Ventricular contraction force	Increase	Decrease
4.	**Blood Vessels Muscle**	Dilation	—
	Salivary gland	Constriction	Dilation
5.	**Gastrointestinal muscles tract**	Relaxation	Constriction
6.	**Urinary Bladder**	Relaxation	Contraction
7.	**Liver**	Glycogenolysis	—

2.2.2 Cholinergic Receptors

On the basis of bound by naturally occurring alkaloids nicotine and muscarine.

(a) **Nicotinic Receptors (N):** (Ligand gated ion channels)

(b) **Muscarinic Receptors (M):** (G-protein coupled receptors)

Nicotinic receptors coupled directly to ion channels, the major role of nicotinic receptors is to control hypertension. Whereas muscarinic receptors play an essential role in stimulating secretion of salivary gland, sweat gland, contraction of gut, relaxation of smooth muscles and heart muscles.

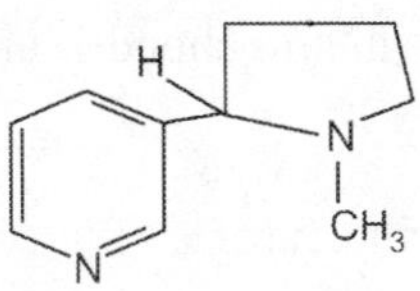

Nicotine

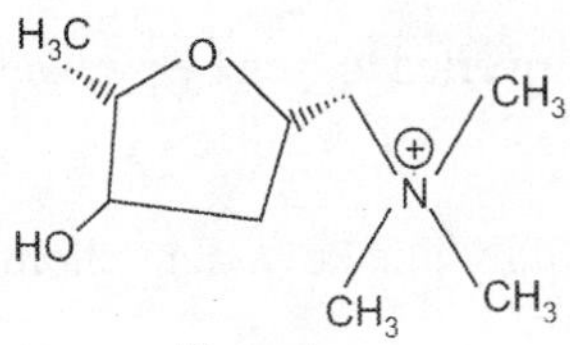

Muscarine

Fig. 2.3

2.2.3 Biosynthesis of Ach

Hypothetical model of synthesis, storage and release of Ach.

Step 1: Ach released from the granules under the influence of action potential and in the presence of Ca^{++}.

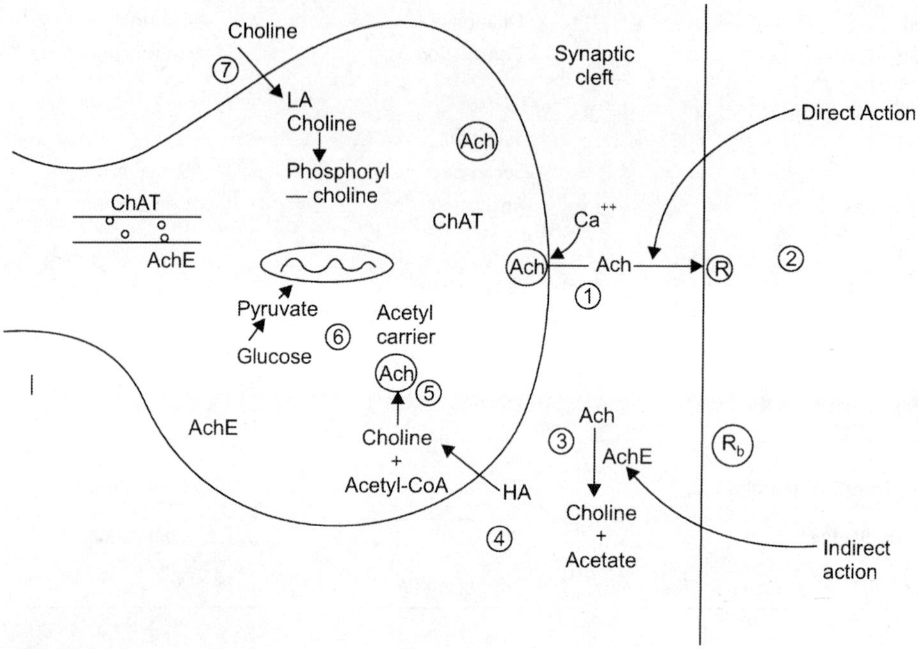

Fig. 2.4

Step 2: Ach acts on postsynaptic cholinergic receptors.

Step 3: Hydrolysis of Ach by AcetylcoE (AchE) in synaptic cleft.

Step 4: A high affinity uptake system returns choline to cytosole.

Step 5: ChAT synthesizes Ach in cytosol and stores in granules

Step 6: Glucose is converted to Ach by pyruvate intermediate in mitochondria and release by acetyl carrier.

Step 7: With low affinity uptake system choline is taken up in neurons.

2.2.4 Classification

1. **Cholinergic Receptor Agonist:** These agents have direct or indirect effects on muscarinic receptors on the organ system (smooth muscles, heart)

1. Direct acting cholinomimetics

(a) Acetylcholine chloride

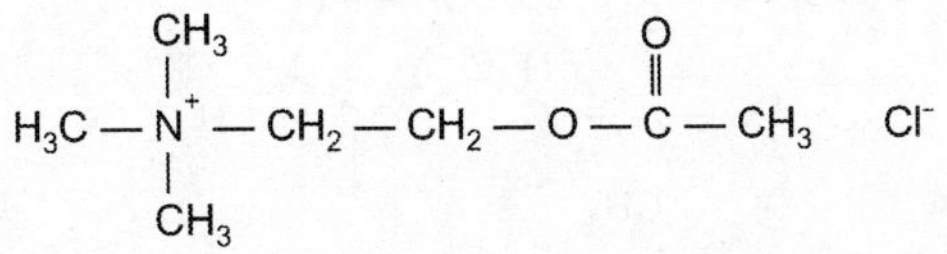

2-acetoxy – N, N, N - trimethylethyl ammonium chloride

Fig. 2.5

(b) Methacholine chloride

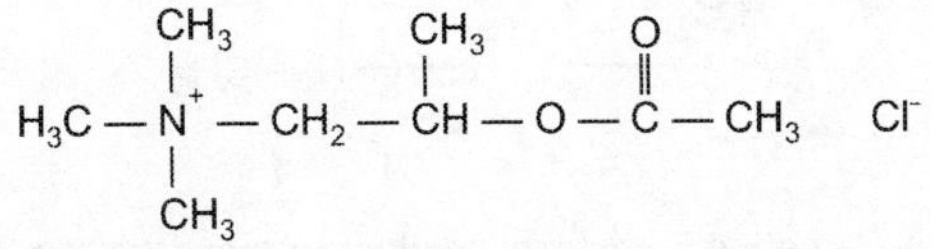

1-acetoxy-2-(N, N, N-trimethyl) propyl ammonium chloride

Fig. 2.6

(c) Carbachol chloride

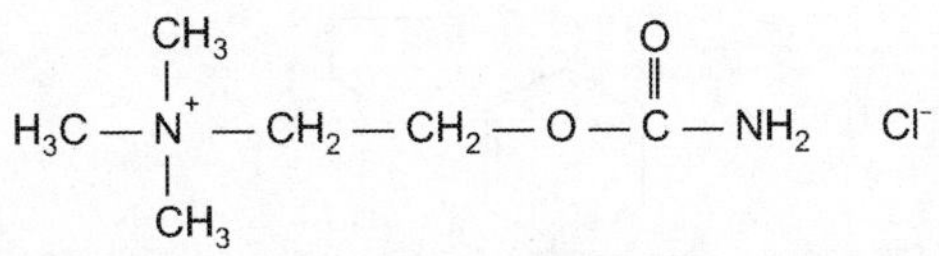

2-carbamoyloxy-N, N, N-trimethylethyl ammonium chloride

Fig. 2.7

(d) Batanechol

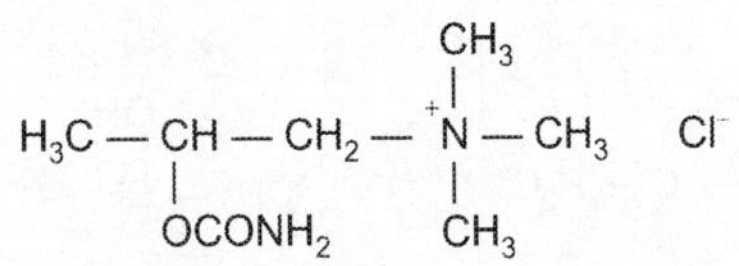

2-carbamoyloxy-1-(N, N, N-trimethyl) propyl-ammonium chloride

Fig. 2.8

(e) Carbachol

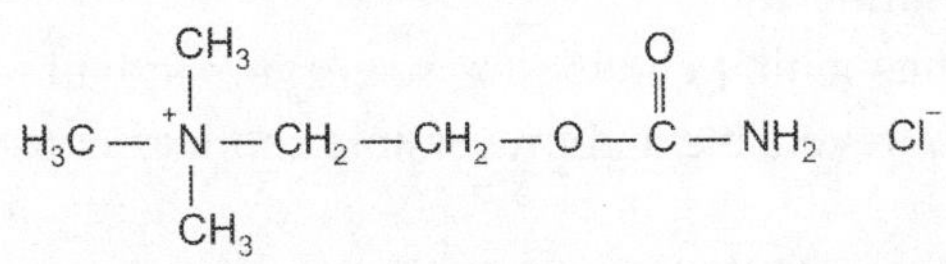

2-carbamoyloxy-N, N, N-trimethylethylammonium chloride.

Fig. 2.9

(f) Naturally occurring muscarinic alkaloids

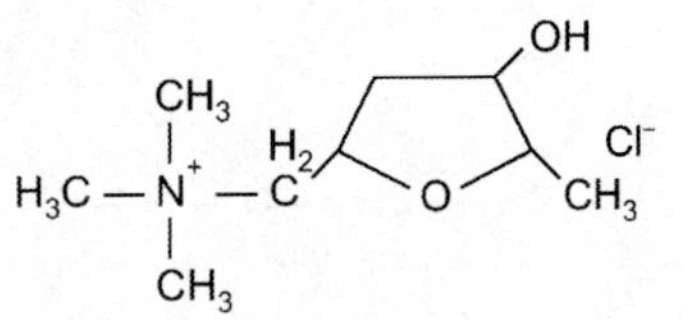

Fig. 2.10

(a) Muscarine

(b) Pilocarpine

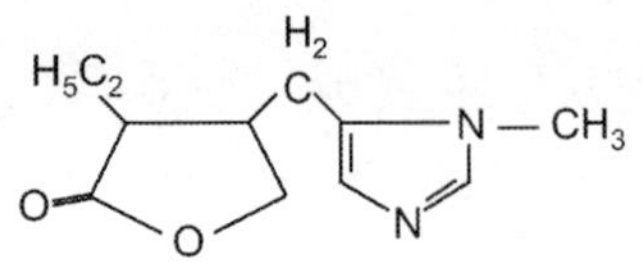

3-ethyl-4-(1-methyl-5-imidazolymethyl)-tetrahydrofuran-2-one.

Fig. 2.11

(g) Naturally occurring nicotinic alkaloids

(a) Nicotine

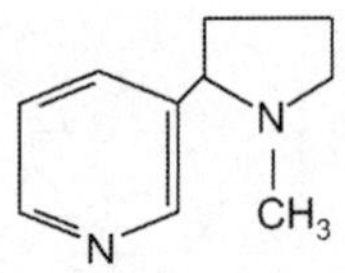

1-methyl-2-(3-pyridyl)-pirrolidine

Fig. 2.12

(b) Lobeline

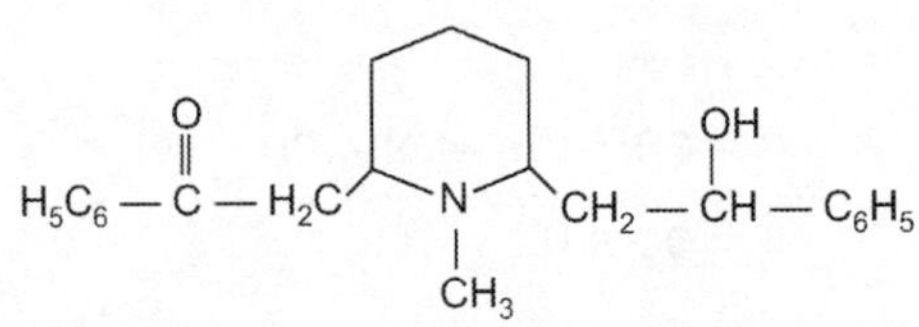

1-methyl-2-(β-hydroxy-β-phenylethyl)-6-phenacylpiperidine

Fig. 2.13

2. Indirect Acting Cholinomimetics

These drugs or chemical agents inhibit cholinesterase enzyme to build-up acetylcholine in neuro-effector compounds. They are reversible and irreversible cholinesterase inhibitors.

1. Reversible cholinesterase inhibitors

These inhibitors are directed to eyes, skeletal muscles, neuromuscular junction. GI-tract and urinary tract, heart.

(A) Carbamates

(a) Physostigmine

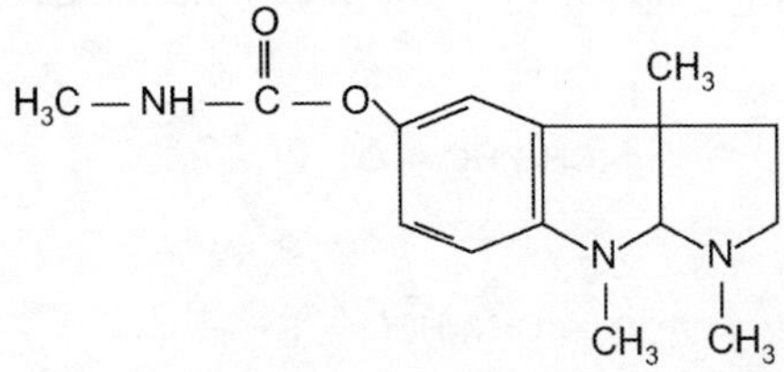

1, 3a, 8-trimethyl-2, 3, 3a, 8a-tetrahydropyrroloindol-5-yl-N-methylcarbamate.

Fig. 2.14

(b) Neostigmine

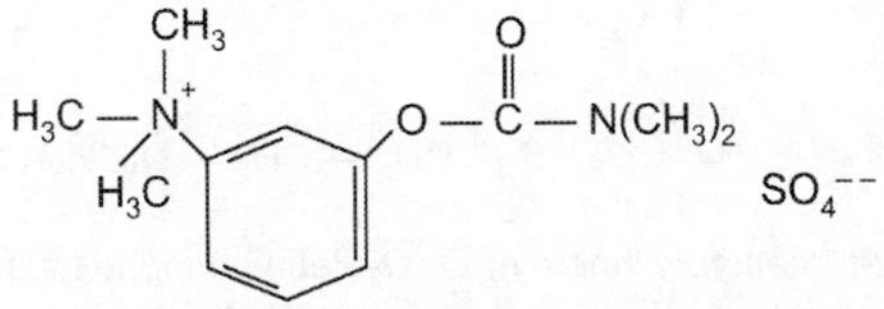

N, N, N-trimethyl-meta-(dimethylcarbomoyloxy)-phenylammonium methyl sulfonate.

Fig. 2.15

(B) Quaternary Amines

(a) Edrophonium

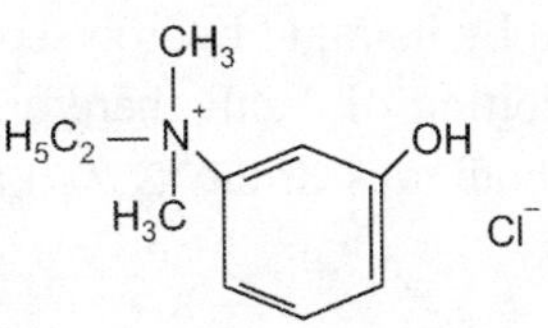

Ethyl-(3-hydroxyphenyl)-dimethylammonium chloride.

Fig. 2.16

(b) Demecarium

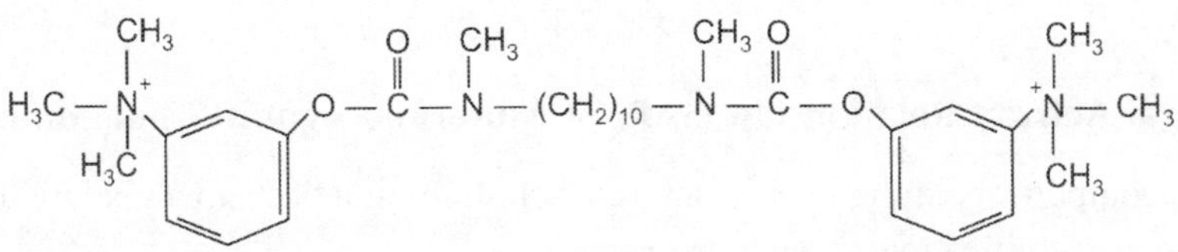

N, N'-Decamethylene-bis-[meta-(N-methylcarbamoyloxy)-Phenyl-trimethylammonium] hydroxide.

Fig. 2.17

2. Irreversible Cholinesterase Inhibitors

These are basically organophosphates and are made up of organophosphorus compounds. They act by forming a complex with hydroxyl group of serine of cholinesterage enzyme with covalent bonding.

(a) Isofluorophosphate

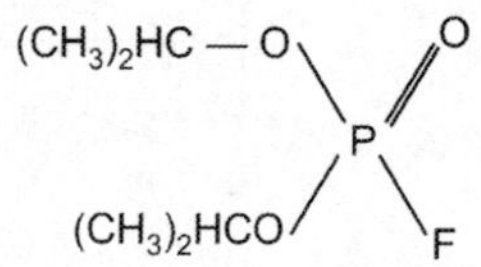

(di-iso-propylester of fluorophosphoric acid)

Fig. 2.18

(b) Echothiophate

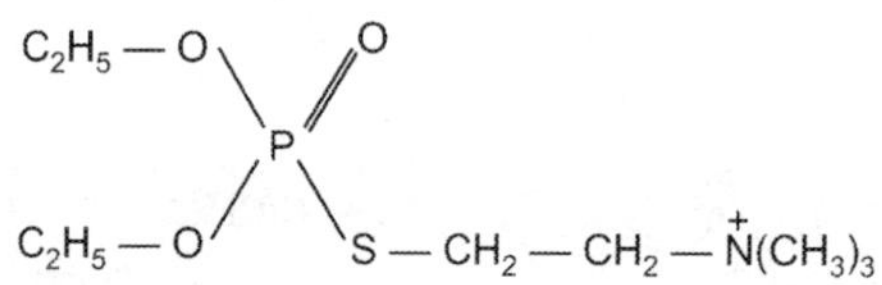

S-(2-trimethylaminoethyl) O, O-diethyl thiophosphate.

Fig. 2.19

2.2.5. Action of Cholinesterase Inhibitors

It is another means to produce an autonomic response to interfere with the synthesis of acetylcholine by which action of the neurotransmitter is terminated in the parasympathetic nervous system. However, action of Ach is terminated by its rapid hydrolysis by acetylcholine esterase enzyme to acetic acid and choline moiety. Inhibition of AchE therefore increases the conc. of Ach in the synapse and results in generation of both muscarinic as well as nicotinic responses.

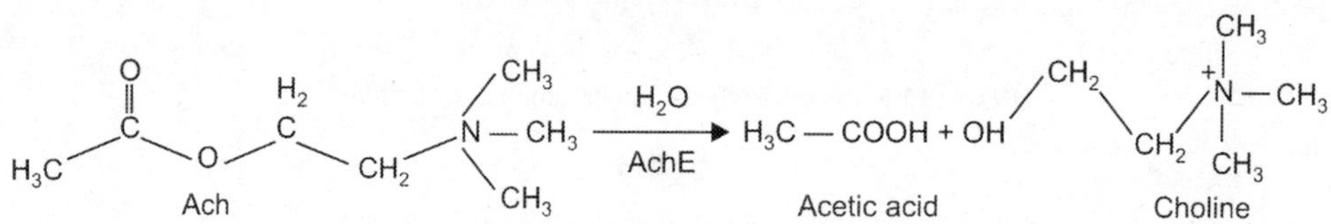

Fig. 2.20

2.2.6 Structural Activity Relationship (SAR) of Musarinic Agonist/Cholinomimetics

There are three components in the lead structure of cholinomimetics shown below to examine the effects of chemical modification of each group.

(A) Modification of Acyloxy Group

(i) It has been observed that introduction of higher homologus of methyl groups like propionyl or butyryl groups are less active.

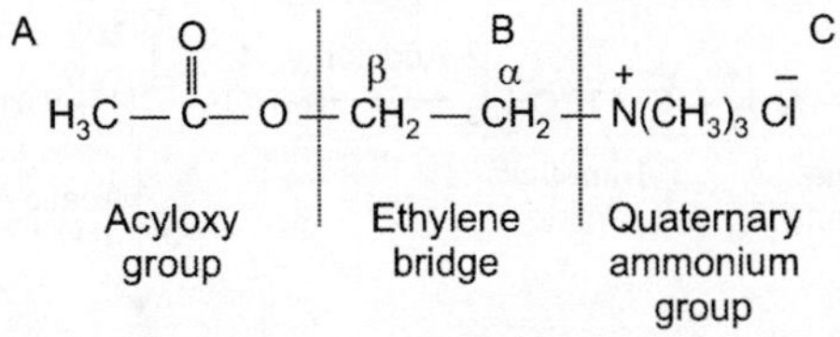

Fig. 2.21: Lead structure.

(ii) Esterification of aryl acids shows cholinergic antagonist activity.

(iii) Methyl esters of acyloxy group are rapidly hydrolysed by cholinesterase to choline and acetic acid. To avoid from hydrolysis, carbamate esters of choline (carbachol) were synthesized and found to be more stable than esters that can be administered orally.

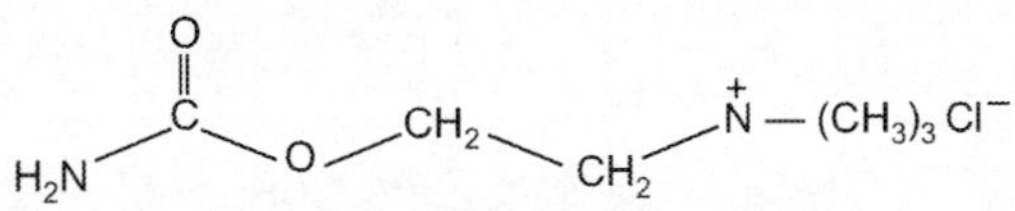

Carbachol

Fig. 2.22

(iv) Replacement of ester group with ether or keton generate chemically stable and potent compound.

(B) Modification of Ethylene Bridge

(i) It has been observed that increase in the chain length results in decreasing the muscarinic activity. No more than five atoms are required between nitrogen atom and terminal hydrogen for maximal muscarinic activity.

(ii) Introduction of a methyl group on carbon beta to quaternary nitrogen atom afford methacholine (acetyl-b-methylcholine), which is equipotent to Ach. While at a carbon to quaternary nitrogen atom, reduces the activity at nicotinic receptor as well as muscarinic receptor.

(C) Modifiction of Quaternary Ammonium Group

(i) It is essential for activity.

(ii) If nitrogen is replaced by arylring, phosphorus and sulfur it results in less active compounds.

(iii) Replacement of methyl group by ethyl or large alkyl group produces inactive compounds.

2.2.7 Cholinomimetic Agents

(1) **Methacholine (Provocholine®):** Methacholine is marketed as recemic mixture. It is selective muscarinic receptor agonist its S-(+)-enantiomers is 240 times more potent than R-(-) isomers at muscarinic receptor.

Uses: Methacholine is used in the treatment of asthma as bronchodilators.

Synthesis:

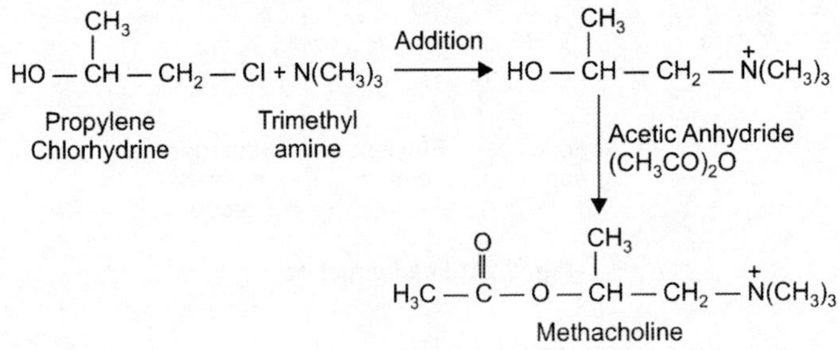

Fig. 2.23

(2) **Carbachol (Isoptocarachol)®:** Carbachol is 2-carbamoyloxy-N, N, N-trimethyl ethylammonium chloride. It increases hydrolytic stability compared to Ach due to having carbamates which are more stable than esters.

Synthesis:

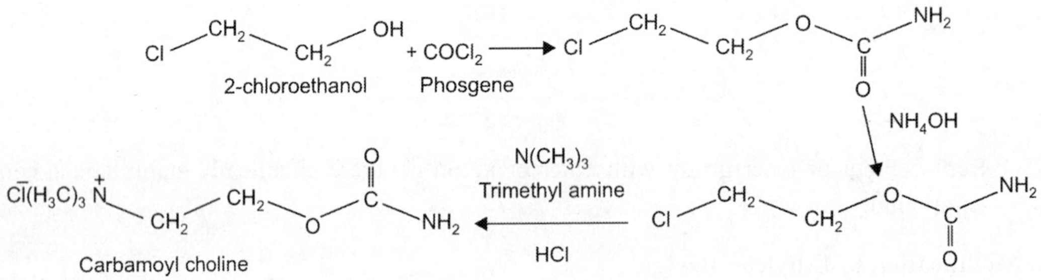

Fig. 2.24

Carbamol is used in the treatment of glucoma.

(3) **Pilocarpine (Isoptocarpine)®:** Pilocarpine is a natural product obtained from the leaves of *pilocarpus jaborandi*. It is muscarinic agonist. It is marketed as an ophthalmic preparation like solution, gel, tablet and ocusert delivery system. It is chemically unstable to alkali (hydrolysis) and due to epimerization in the lactone ring to isopilocarpine.

(4) **Physostigmine:** Physostigmine is assumed to be a classical AchE inhibitor. It is an alkaloid obtained from the seed of calabar beans, *Phystostigma venenosum*.

It acts as an antidote for atropine poisoning and in the treatment of glucoma.

Synthesis:

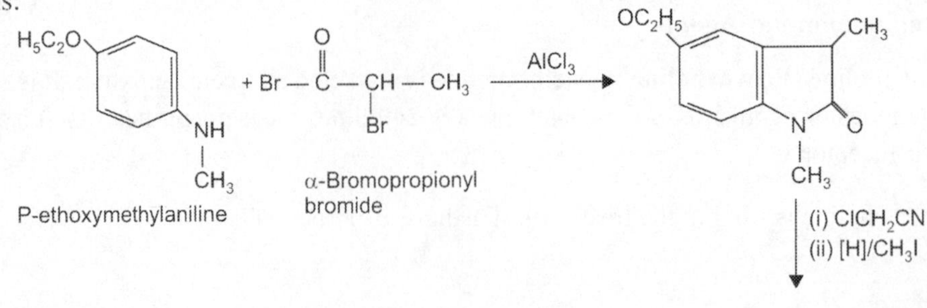

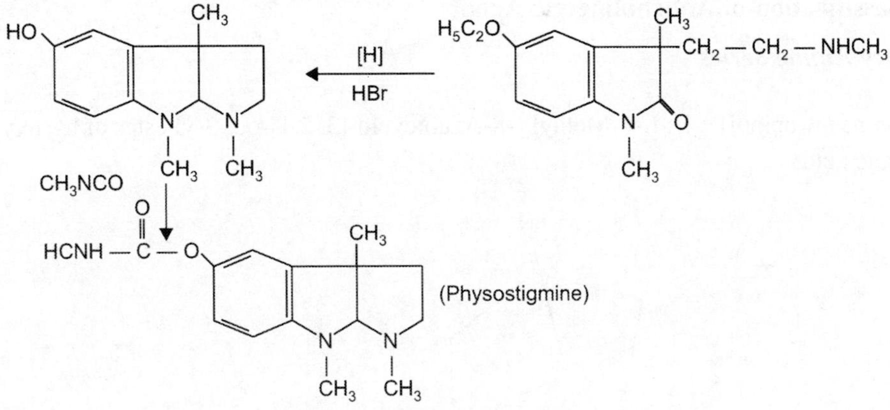

Fig. 2.25

(5) **Neostigmine (Prostigmine)®:** It is also an AchE inhibitor like physostigmine. Neostigmine is indicated for prophylaxis of post-operative abdominal distension, urinary retention and neuromuscular disorders.

Synthesis:

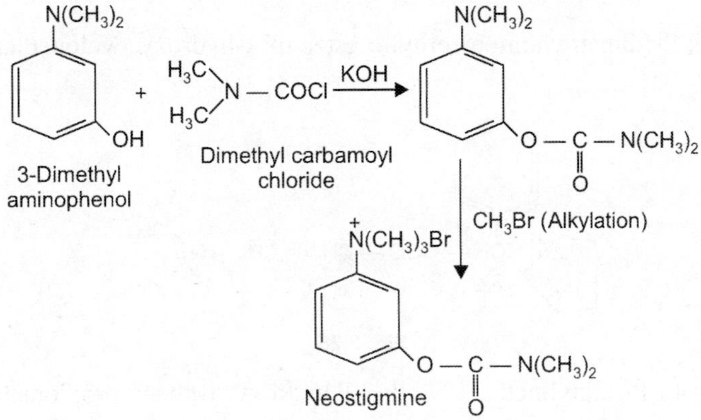

Fig. 2.26

2.2.8 Anticholinergic/Cholinergic Blocking Agents/Parasympatholytics/Antispasmodics Agents/Muscarinic Receptors Antagonist

Definition

Muscarinic antagonists are those compounds which have high binding affinity for muscarinic receptors but have no intrinsic activity at the smooth muscles, secretory cells, and certain central synapses.

Mechanism of Action

There are two approaches to treat the condition characterized by overstimulation of cholinergic nerves, one is to inhibit the synthesis or release of Ach and second to block the Ach from with the receptors competitively. These agents cause decreased gastric secretion, dry mouth and mucous membrane, mydriasis, loss of accommodation, urinary-retention, bronchial and biliary dilation.

2.2.9 Classification of Anticholinergic Agents

(1) Tertiary Amine Series

(A) Atropine (Atopinol)®: D, L-8-Methyl (-8-Azabicyclo-[3, 2.1]-oct-3-yl ester of hyroxy methyl phenylacetic acid.

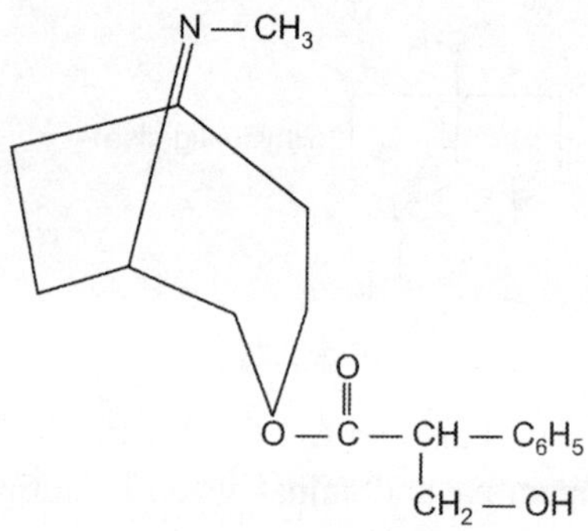

Fig. 2.27

(B) Cyclopentolate: 2-(dimethylamino) ethylic ester of 1-hydroxy cyclopentane-α-phenylacetic acid.

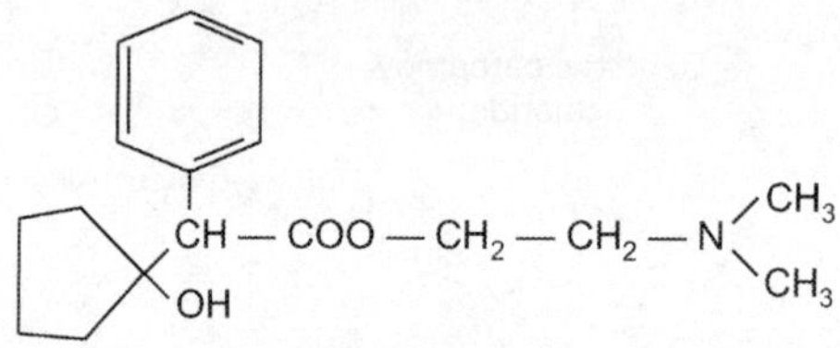

Fig. 2.28

(C) Tropicamide: N-(4-Piridinylmethyl)-N-ethyl-β hydroxy-α-phenylpropionamide.

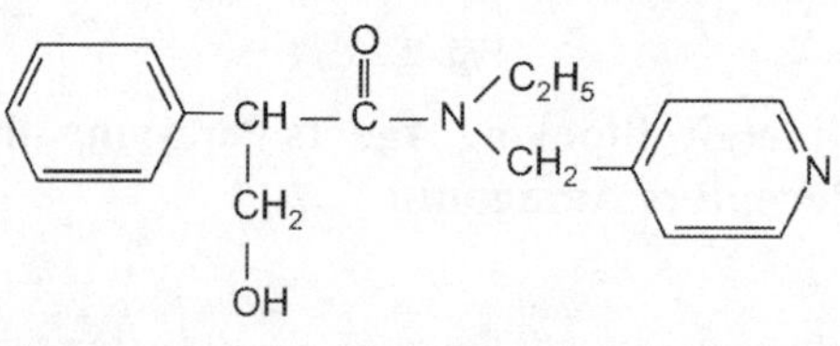

Fig. 2.29

(D) Dicyclomine: Diethylaminoethyl ester of 1-cyclohexyl-cyclohexane carboxylic acid.

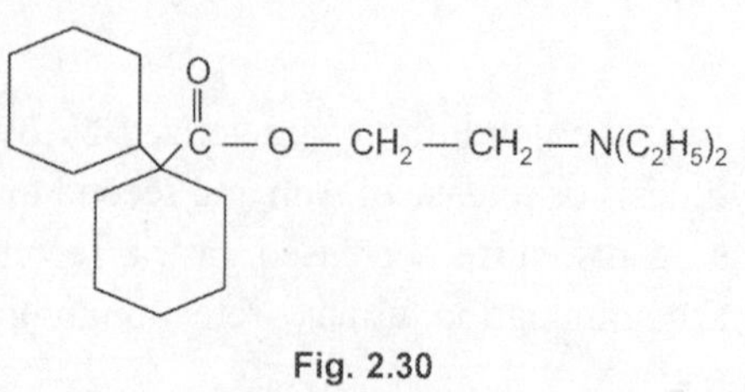

Fig. 2.30

2. Quaternary Amine Series

(A) Methscopolamine: 7-(3-hydroxy-1-oxo-2-phenyl propoxy)-9, 9-dimethyl-3-oxa-9-azoniticyclo nonane nitrate.

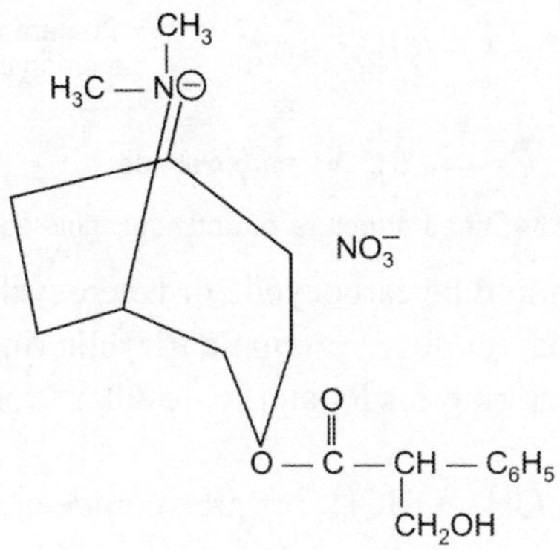

Fig. 2.31

(B) Propnantheline: N-methyl-N-(1-methylethyl)-N-[2-[(9H-Xanthen-9-yl carbonyl) oxy]ethyl]-2-Propanaminium bromide

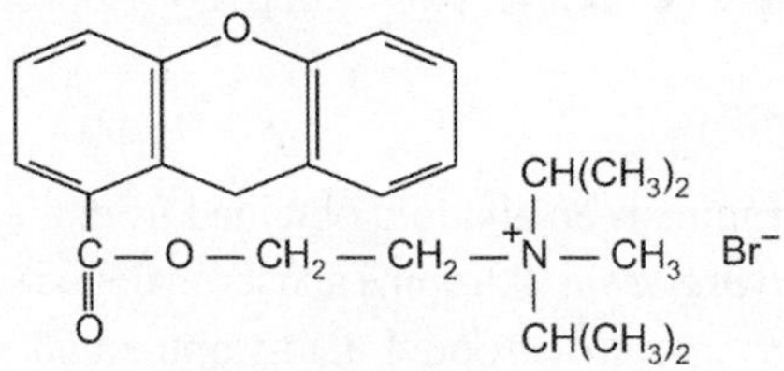

Fig. 2.32

(C) Isopropamide: (3-carbamoyl-3.3-diphenyl propyl)-di-iso propylmethyl ammonium iodide.

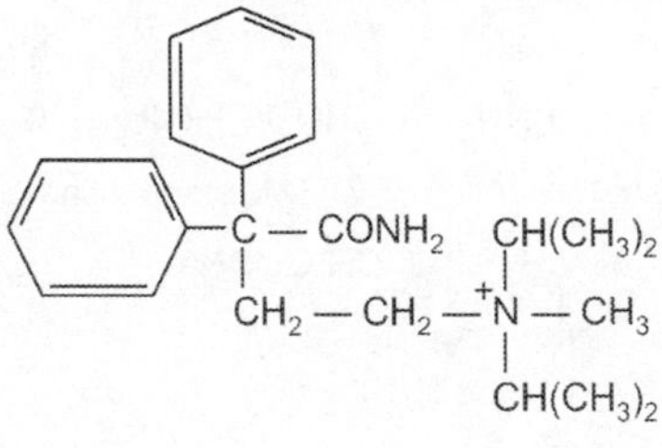

Fig. 2.33

2.2.10 Structural Activity Relationship (SAR) of Anticholinergic Agent

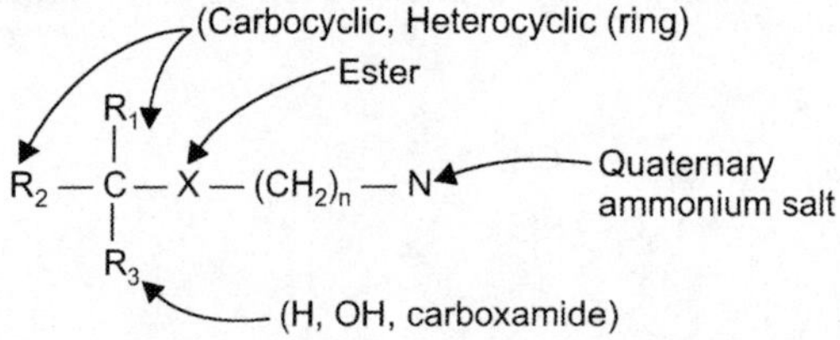

Fig. 2.34: Lead structure of antimuscarinic agents.

(i) Substituents R_1 and R_2 should be carbocyclic or heterocyclic rings for maximal antagonist activity. However, they may be fused aromatic tricyclic ring system.

(ii) Substitution of naphthalene ring for R_1 and R_2 resultant compounds would inactive due to steric hindrance.

(iii) R_3 substituents may be H, OH, $-OCH_3$ or carboxamide makes the compound more potent. However, –OH group increases however the potency of binding to receptor.

(iv) X substituents in the most anticholinergic is an ester. It may also be substituted by ethereal oxygen.

(v) N substituents is a quaternary ammonium salt. Presumably binding to the receptor in cationic form.

(vi) The alkyl substituents may be methyl, ethyl, propyl or isopropy for anticholinergic activity.

2.2.11 Anticholinergic Agent

(A) Atropine (Atropinol®): Atropine is an alkaloid obtained from the solanaceous plant belladonna (*A tropa belladonna*) Dubosia, *Hyoscyamus*. Atropine and its analogue scopolamine are most important antimuscarinic drugs. These are esters of tropoyl acid-tropine and scopine respectively. Scopine differs from tropine in that it has an expoxide ring bridge between carbon atoms C_6 and C_7 of tropine. Atropine is used during anesthesia in surgery, in ophthalmic preparation, for the treatment of Parkinsonism. Atropine is also used as an antidote with combination of other drugs for poisonous anticholinesterase agents like organophosphorus insecticides.

Synthesis:

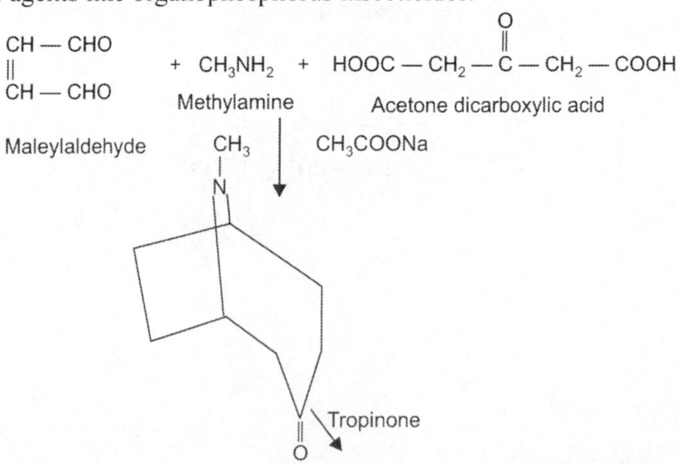

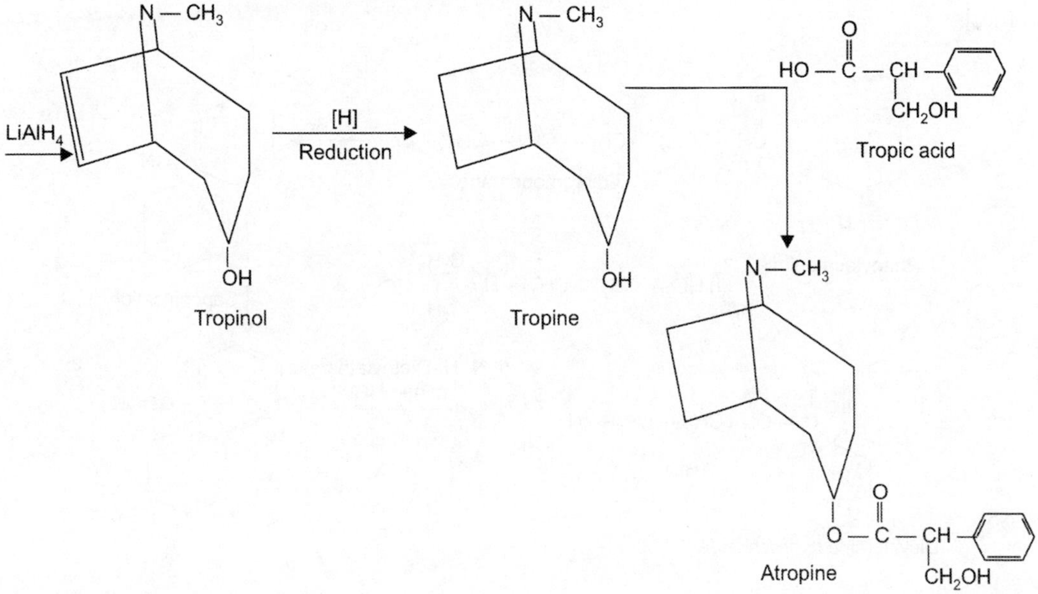

Fig. 2.35

(B) Scopolamine: Scopolamine is used for the treatment of Parkinsonism symptoms, stomach ulcer, spasm of urinary tract and bronchial asthma.

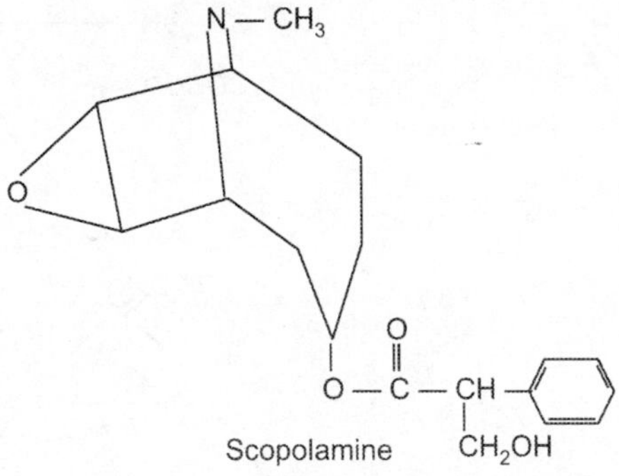

Fig. 2.36

(C) Glycopyrrolate (Robinul®): glycopyrrolate prevents the secretion of digestive juices and restores normal stomach functions. It is used for the treatment of stomach ulcer, inflammed intestine.
(D) Dicyclomine (Bentyl®): Dicyclomine is an antispasmodic drug which exhibits direct antispastic action on smooth muscles and inhibits the action of acetylcholine on smooth muscles. It is used in the treatment of stomach ulcer, colic in children and for treating irritable bowel syndrome.

Synthesis:

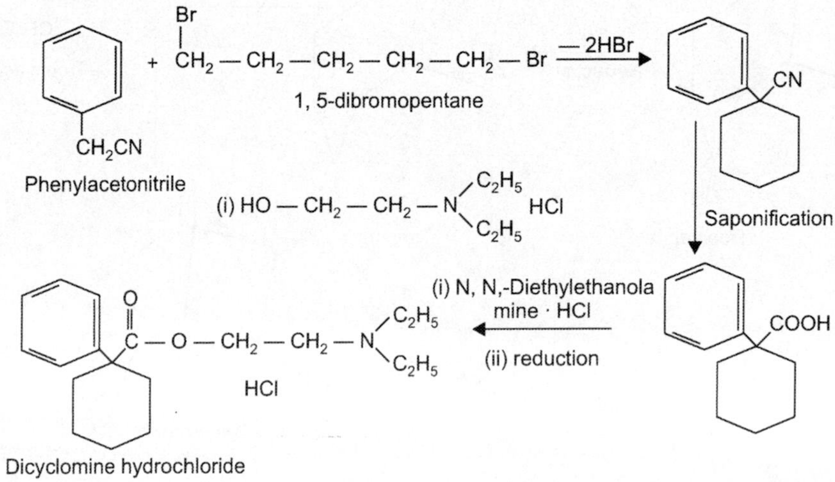

Fig. 2.37

Synthesis: (Glycopyrrolate)

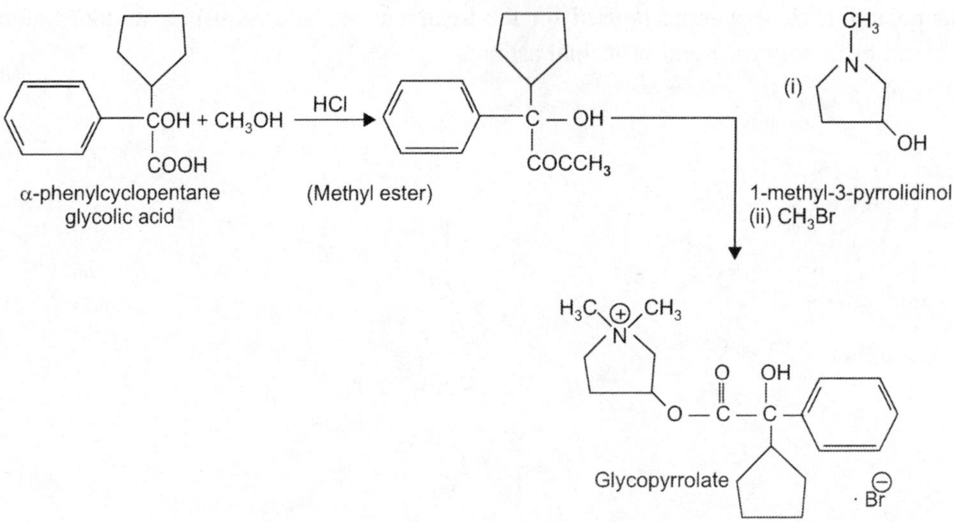

Fig. 2.38

2.2.12 Newer Drug

1. Donepezil (Aricept)®: 2-[(1-benzyl-4-piperidyl) methyl]-5, 6-dimethoxy-2, 3-dihydroinden-1-one

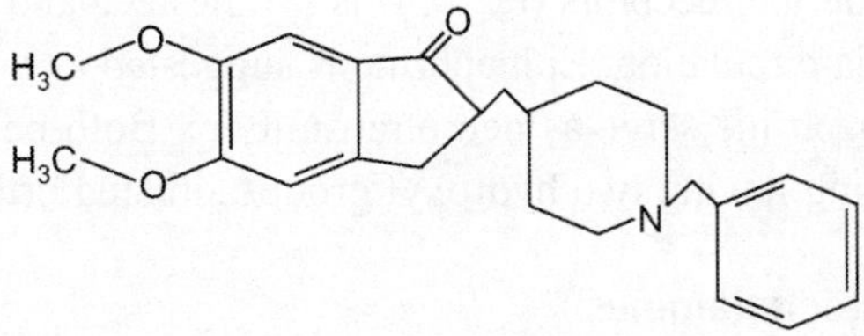

Fig. 2.39: Donepezil.

Donepezil is non-classic centrally acting, reversible non-competitive AchE inhibitor. It is clinically useful in the treatment of mild to moderate AD and dementia.

2. Rivastigmine (Exelon)®: (S)-N-Ethyl-N-methyl-3-[1-(dimethylamino)ethyl]-phenyl carbamate.

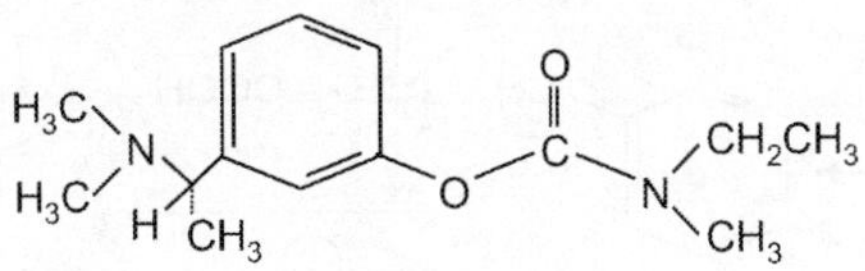

Fig. 2.40: Rivastigmine.

Rivastigmine is centrally acting arylcarbonate AchE inhibitor. It is also suggested for the treatment of Alzheimer's disease (AD).

3. Galantamine (Razadyne)®: Galantamine was obtained from the plant of family amaryllidaceae. It is an alkaloid. Galantamine is an reversible AchE inhibitor. It is used in the treatment of mild to moderate Alzheimer's disease.

2.3 ADRENERGIC AGENTS

2.3.1 Introduction

Adrenergic agents or drugs are exhibit their pharmacological action by their enhancing or reducing the activity of the various components of sympathetic division of autonomic nervous systems. The chemical agents that produce the similar effects to stimulation of sympathetic system are known as sympathomimetics or adrenergic stimulants. On the other hand, these decrease the effect called sympatholytics or antiadrenergics or adrenergic blocking agents.

2.3.2 Mechanism of Action

Norepinephrine is a neurotransmitter of the post-ganglionic sympathetic neurons. It is released from the sympathetic nerve ending into the synaptic cleft, cause sympathetic nerve stimulation. It interacts with specific adrenergic receptors (α, β). It is synthesized and stored in Adrenal medulla from which it is released via circulation. Epinephrine is suggested as a neuro-hormone. In certain area of CNS, NE and Epinephrine serve as neurotransmitters. Both belong to the chemical class catecholamines (aromatic ring having two hydroxyl groups situated ortho position to each other).

2.3.3 Biosynthesis of Catecholamines

Biosynthesis of catecholamines takes place in adrenergic and dopaminergic neurons, in sympathetic neurons in ANS and in adrenal medulla. Enzymes which are responsible for synthesis of catecholamines, are synthesized in cell bodies at neurons and are transported to the axons to their nerve terminals. The amino acid L-tyrosin is the main precursor for the catecholamines biosynthesis which is given below.

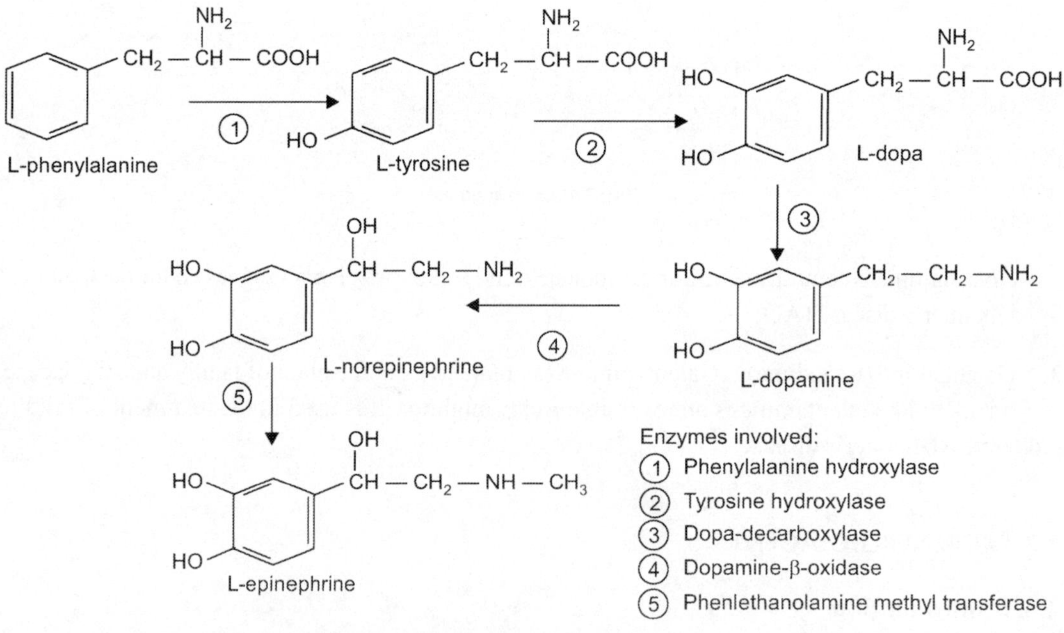

Fig. 2.41: Biosynthesis of catecholamines.

2.3.4 Pharmacological Action

(i) Adrenergic agents exhibit excitatory effects on smooth muscles of blood vessels, salivary glands and sweat glands.

(ii) Also show inhibitory response to smooth muscles of GI tract, bronchial tract and skeletal muscles.

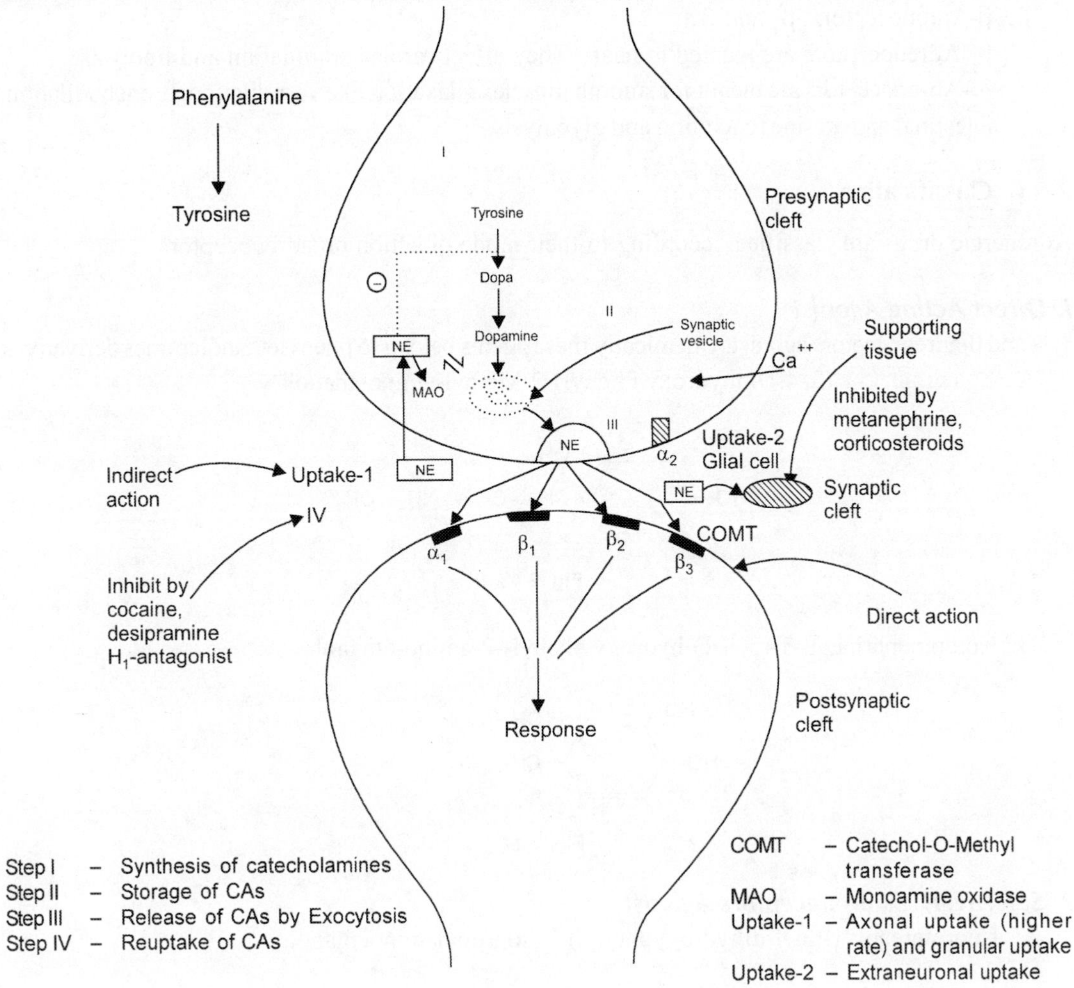

Fig. 2.42: Schematic projection of adremergic neurotransmission.

(iii) These increase the force of contraction (+ ionotropic effect)

(iv) Increase level of catecholamines in CNS leads to respiratory stimulation and decreased appetite.

(v) Catecholamines promote glycogenolysis in liver as well as in muscles.

2.3.5 Adrenergic

Receptors: There are two main adrenergic receptors:

A. α-Adrenoceptors: α_1 and α_2 in which α_1 is responsible for smooth muscles stimulation, sweating and salivation. It is located in post-functional of effector organs while α_2 inhibits the presynaptic release of nor-epinephrine and postsynaptic activation of adenylate cyclase.

B. β-Adrenoceptors: β_1 and β_2

β_1-Adrenoceptors are located in heart. They affect cardiac stimulation and lipolysis.

β_2-Adrenoceptors are meant for smooth muscles relaxation like vasodilation, bronchodilation, intestinal and uterine relaxation and glycolysis.

2.3.6 Classification

Adrenergic drugs are classified according to their mode of action on adrenoceptors

I. Direct Acting Agonist

1. α and β-adrenoceptors agonist: Chemically these agents belong to phenylethanolamines derivatives

A. Epinerine: L-1-(3, 4-Dinhydroxy Phenyl)-2-Methyl-aminoethanol

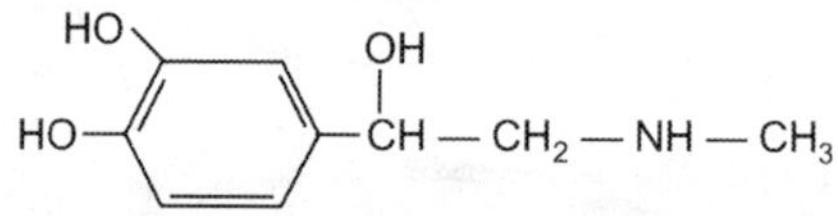

Fig. 2.43

B. Norepinephrine: L-1-(3, 4-Dihydroxy phenyl)-2-amino-ethanol.

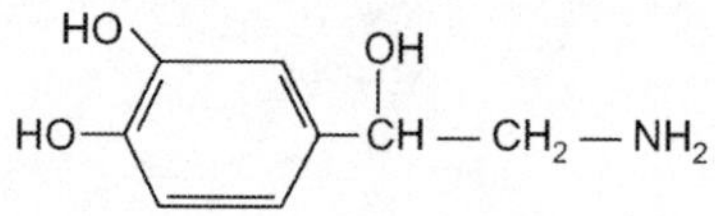

Fig. 2.44

2. Selectively β-adrenoceptors agonist

A. Isoproterenol: 1-(3, 4-dihydroxyphenyl) 2-isopropylaminoethanol

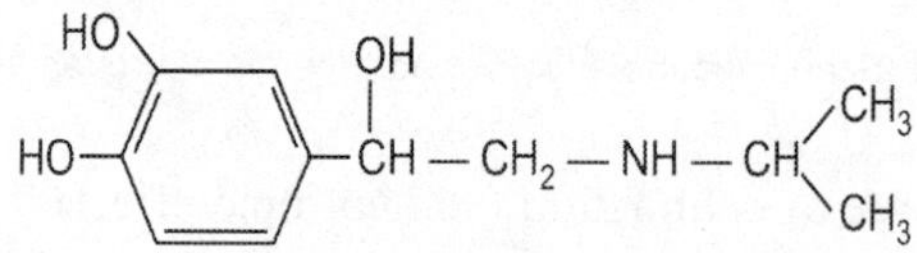

Fig. 2.45

B. Isoetharine: 3, 4-dihydroxy-α-[1-(iso-propylamino) propyl] benzylic alcohol.

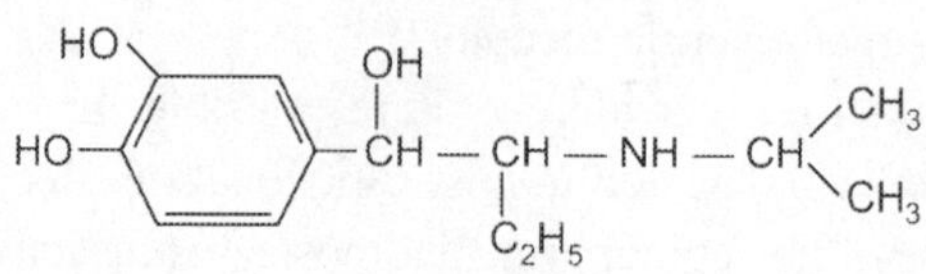

Fig. 2.46

C. Terbutaline: α-[(tert-butylamino) methyl]-3, 5-dihydroxy-benzylic alcohol.

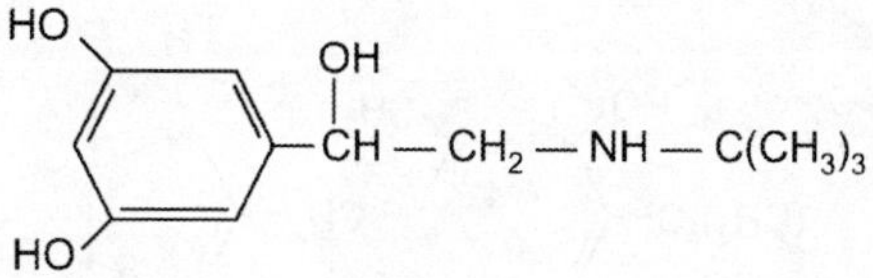

Fig. 2.47

D. Metaproterenol: α[(isopropylamino) methyl]-3, 5- dihydroxy benzylic alcohol.

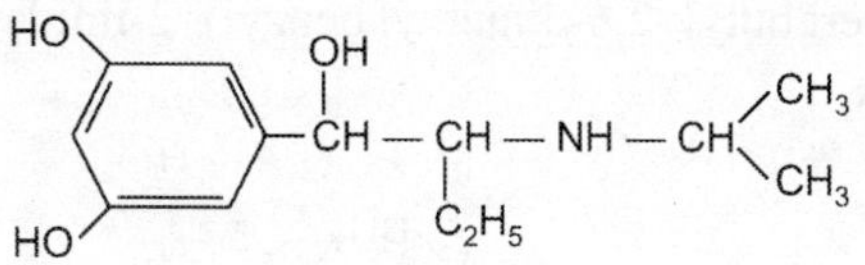

Fig. 2.48

E. Salbutamol or albuterol: 2-tert-butylamino-1-(4-hydroxymethylphenyl) ethanol

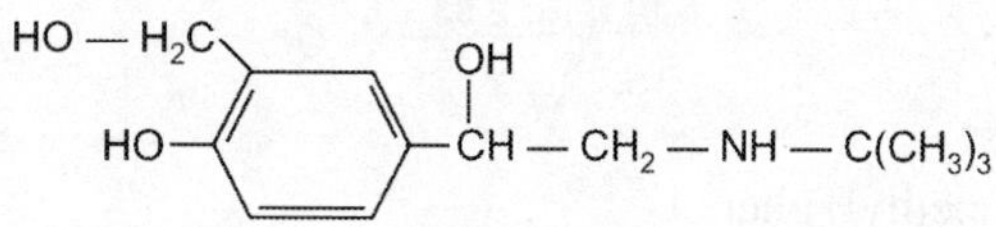

Fig. 2.49

3. *Selectively α Adrenoceptors agonist:*

A. Clonidine: 2-(2, 6-diochlorophenylamino) Imidazoline

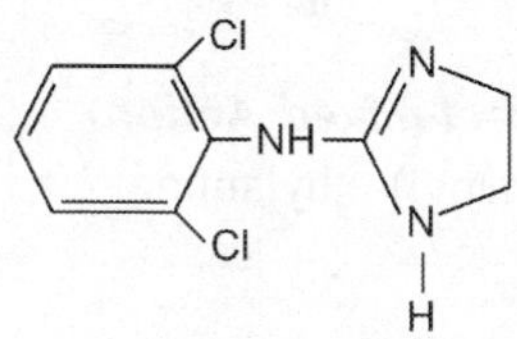

Fig. 2.50

4. Antiedema Sympathomimetics: (Imidazoline Derivative)

A. Naphazoline: 2-(1-Naphthylmethyl)-2-Imidazoline

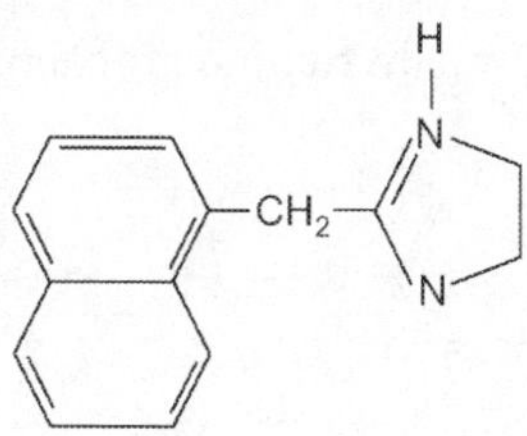

Fig. 2.51

B. Oxymetazoline: 6-tert butyl-3-(2-Imidazolin-2-yl(methyl) -2, 4-dimethylphenol

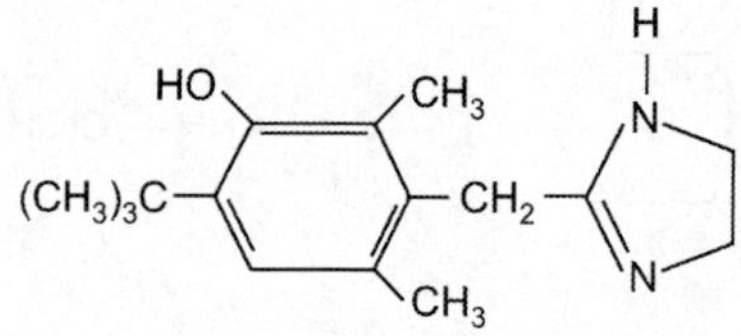

Fig. 2.52

C. Xylometazoline: 2-(4-tert butyl -2,6-dimethyl benzyl) -2-imidazoline

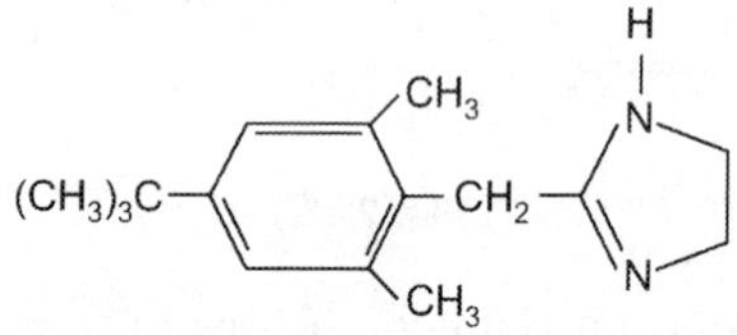

Fig. 2.53

II. Indirect Acting Agonist

A. Tyramine: 4-(2-aminoethyl) phenol

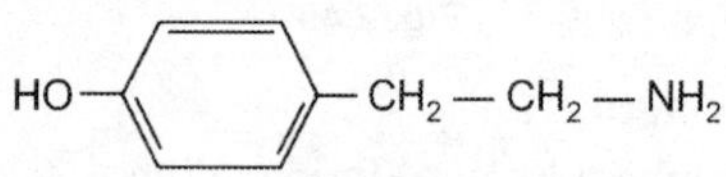

Fig. 2.54

III. α and β adrenomimetics (Indirect-mixed Action)

A. Dopamine: 2-(3, 4-dihydroxyphenyl)-ethylamine

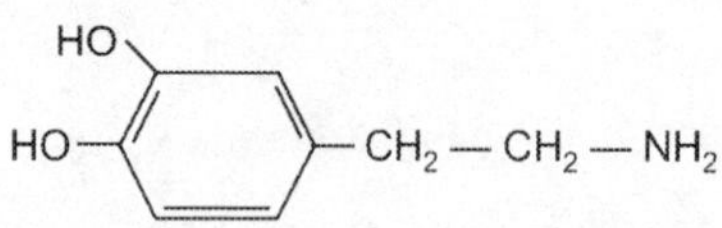

Fig. 2.55

B. Epherine: L-erythro-1-phenyl-2-methyl amino propanol

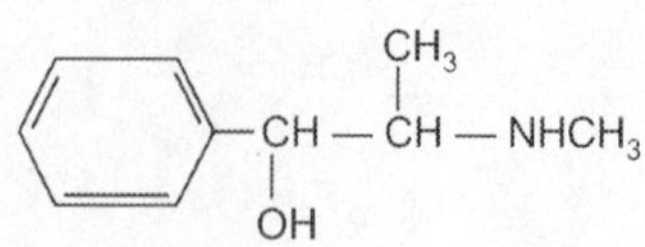

Fig. 2.56

C. Amphetamine: It is racemate of dextroamphetamine and methamphetamine

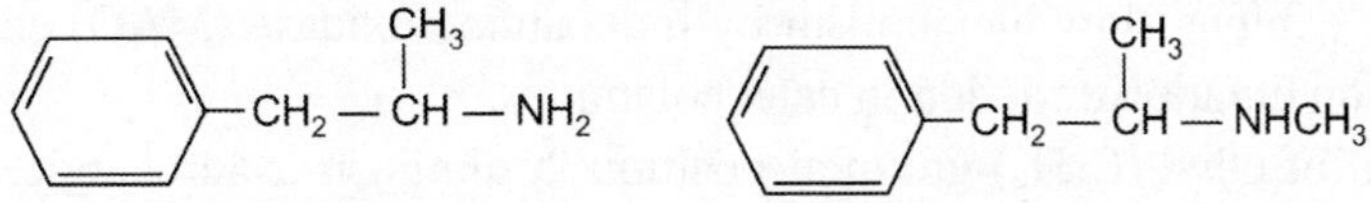

Dextroamphetamine Methamphetamine

Fig. 2.57

2.3.7 SAR of Phenylethanolamine Derivatives/Direct Acting Agonist

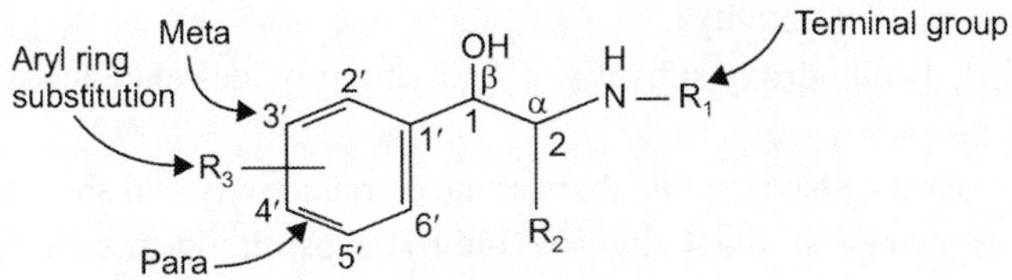

Fig. 2.58

Substitution are the following

	Derivatives	R_1	R_2	R_3	Adrenergic receptor activity
1.	Nor-epinephrine	H	H	3', 4'-diol	$\alpha_1\ \alpha_2\ \beta_1$
2.	Epinephrine	CH_3	H	3', 4'-diol	$\beta_1 > \alpha_1\ \alpha_2\ \beta_1$
3.	Isoproterenol	$—CH(CH_3)_2$	H	3', 4'-diol	$\beta_1 + \beta_2$
4.	Terbutaline	$—C(CH_3)_3$	H	3', 5'-diol	$\beta_2 > \beta_1$
5.	Ephedrine	CH_3	CH_3	H	$\alpha_1, \alpha_2, \beta_2$
6.	Salbutamol	$—C(CH_3)_3$	H	3'-methylhydroxy 4'-methylhydroxy	β_2
7.	Phenylephrine	$—CH_3$	H	3'-OH	α

1. R_1 substitution on terminal nitrogen group

(i) Terminal nitrogen group substitution plays a vital role in the adrenergic receptor activity. It has been shown that increase in the size of R_1 results in the decrease of the selective activity at α receptors followed by increase in the β-adrenergic activity.

(ii) The activity at both α and β receptors shown to be maximal when R_1 is methyl substitution, e.g., epinephrime.

(iii) It has been shown that if R_1 becomes larger than butyl group affinity for α_1-receptors returns but does not exhibit Intrinsic activity.

2. R_2-substitution

(i) Presence of small alkyl groups such as methyl or ethyl at α to the basic nitrogen atom. Such substitution exhibits slow metabolism by monoamine oxidase (MAO) enzyme, but shows little effect on duration of action in catecholamines.

(ii) Introduction of ethyl (C_2H_5) group at position 2, diminish α-adrenergic activity far more than β-selectivity.

(iii) Carbon atom at position No. 2 and its substitution plays as asymmetric center into this molecule which produces a pair of diastereomers that exhibit significantly distinct biological as well as chemical properties.

3. Substitution on Aromatic Ring (R_3)

(i) Aryl ring is essential for activity.

(ii) The nature of 3', 4'-dihydroxy substituted Aryl ring provides excellent receptor activity for both α and β receptors.

(iii) Other substitutions are also possible that enhance oral activity and show β_2 selective activity such as 3'-hydroxymethyl and 4'-hydroxymethyl substitution in salbutamol. Such type of substitution also resist COMT mechanism.

(iv) At least one of the groups must be capable of forming hydrogen bond at the receptor site and retain the β-adrenergic activity e.g. Ritodrine.

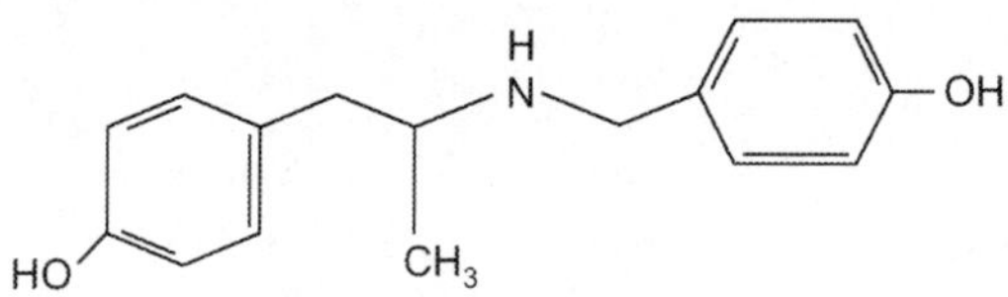

Fig. 2.59: Ritodrine.

(v) Introduction of 3'-hydroxy group only, however, the activity is reduced at α-site and almost eliminated at β-site thus affording selective α-adrenergic agonist for example pheylephrine and metaraminol.

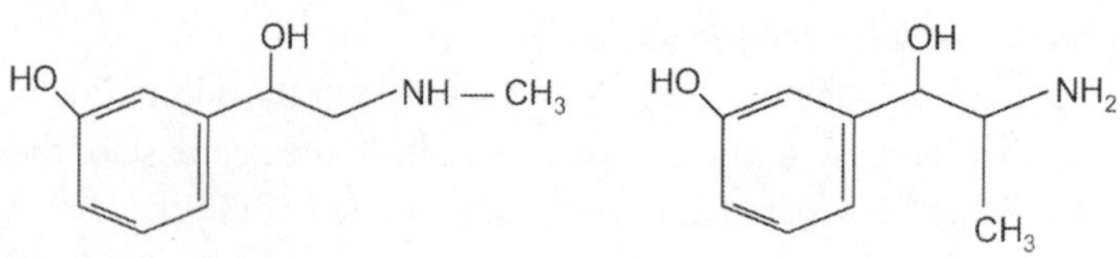

Fig. 2.60

2.3.8 SAR of Indirect Acting Agonist/Arylethylamine Derivatives

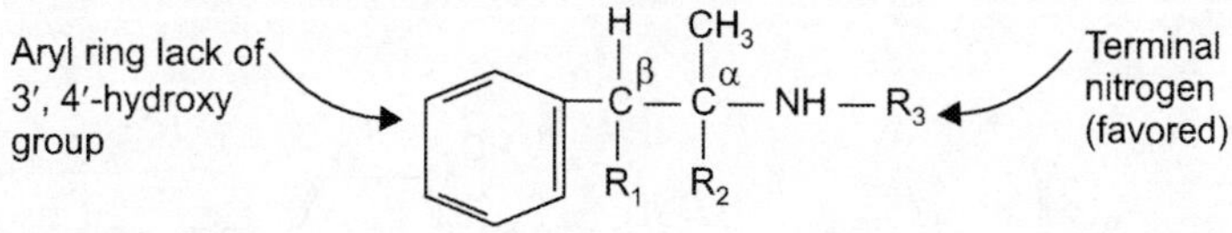

Fig. 2.61

Substitutions are the following:

	Derivatives	R_1	R_2	R_3
1.	Phenyl propanolamine	OH	H	H
2.	Amphetamine	H	H	H
3.	Metamphetamine	H	H	CH_3
4.	Phentermine	H	CH_3	H

(i) The above mentioned figure of pharmacophore indicates that indirect acting adrenergic agonist do not contain the phenolic hydroxyl groups at 3', 4' position. Lack of these ionic groups potentiate the oral absorption and increase the penetration in CNS.

(ii) The phenolic hydroxyl groups may or may not be present in indirect acting compounds. Those compound without this alcoholic hydroxyl groups are less polar and pass more readily through BBB (blood brain barrier) and exhibit greater CNS stimulation.

(iii) The phenylic group may be replaced by other groups, i.e., cycloalkyl or alkyl groups which results in change of the activity.

(iv) Propylhexedrine is an analogue of amphetamine in which aryl ring is replaced by cyclohexane ring. This effect still possesses the adrenergic activity. This drug is used for vasoconstriction and decongestant effect.

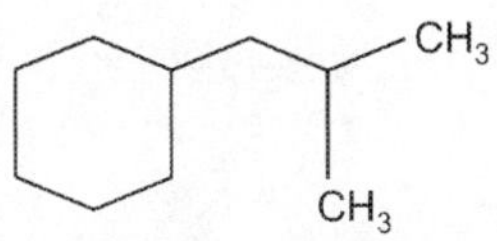

Fig. 2.62: Propylhexedrine.

2.3.9 Adrenergic Agents

1. Epinephrine (Adrenaline) : Epinephrine is also known as adrenaline. It has potent stimulatory effect on both α and β receptors. It is a light sensitive drug and for avoiding the oxidation it is kept with the addition of reducing agent sodium bisulphite. Ephedrine is used in the treatment of open

angle glaucoma, in nasal congestion, and also for enhancing the activity of local anaesthetics. Ephedrine is obtained from adrenal gland tissue of livestock as well as from synthetic route.

Synthesis:

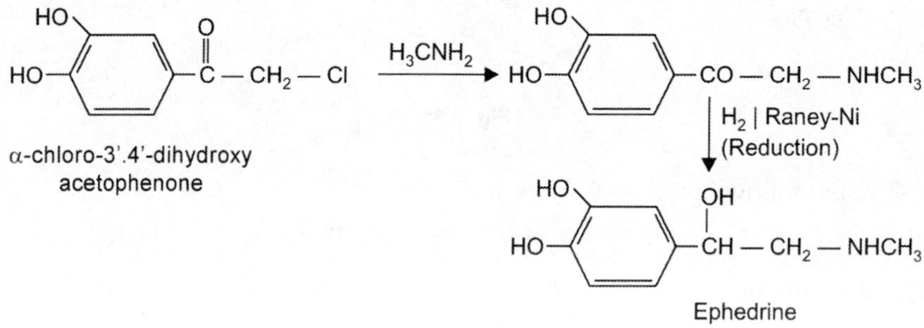

Fig. 2.63

2. Nor-Epinephrine: Nor-epinephrine is also known as nor-adrenaline. It is a primary neurotransmitter and produces and releases by adrenergic neuron. Nor-epinephrine stimulate α-adrenoceptors and activate of β_1-adrenoceptors to produce cardiostimulatory effects.

3. Isoproterenol (Isoprenaline)®: Isoproterenol is highly selective to β-adrenoceptors due to presence of bulky iso-propyl or tert butyl group at the terminal nitrogen atom of phenylethanolamine skeleton. Isoproterenol is used in bronchospasma, asthma, cardiac block and in shock.

Synthesis:

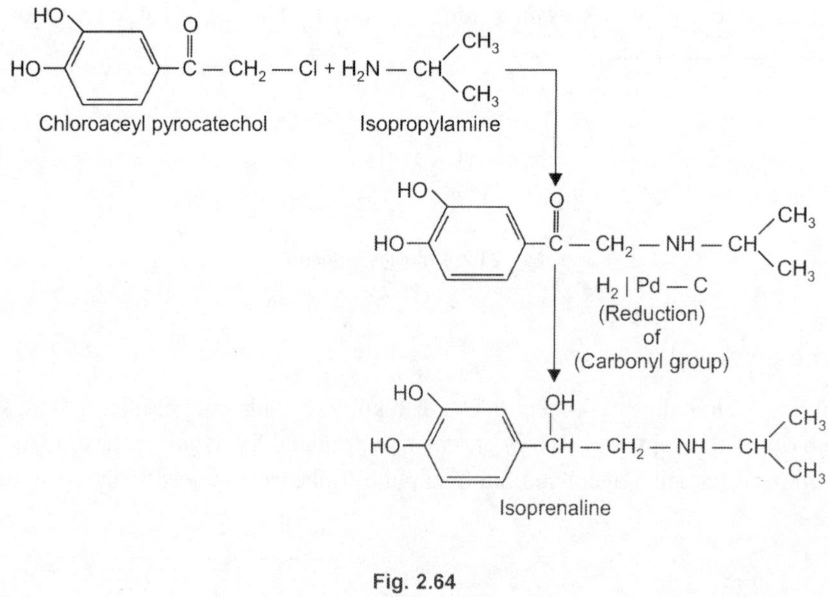

Fig. 2.64

4. Terbutaline (Bricanyl)®: Terbutaline is an important and selective direct acting β_2-adrenoceptors agonist. It stimulates smooth musles of bronchi β_2- receptors and less effect on β_1-receptors of the heart. It is used in the treatment of bronchial asthma, chronic bronchitis, and other pulmonary disease.

5. Salbutamol (Albuterol)®: Salbutamol is also known as Albuterol. It is a non-catechol selective β_2-adrenergic receptor agonist. Its R/S enantiomers have been shown that the (s) enantiomer may be proinflammatory activity and is responsible for enhancing bronchial muscles contraction thus show opposite bronchodilating effects of the (R)-enantiomer levoalbuterol. The recemic mixture of albuterol has expressed broncholytic effects and increases the vital capacity of lungs. It is widely used in the treatment of bronchial asthma and other illness of the respiratory tract.

Synthesis:

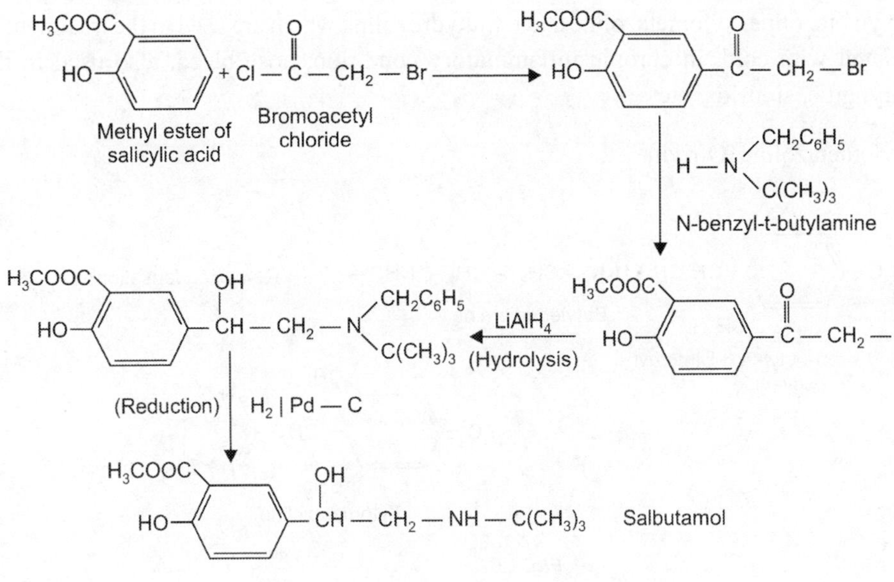

Fig. 2.65

6. Ephedrine: Ephedrine is an alkaloid obtained from the plants *extract ephedra sinica* and *ephedra-equisetina.* In the structure of ephedrine, two asymmetric atoms are present thus form four isomeric forms. Pseudoepinephrine (d-isoephrine) is the stereoisomer of ephedrine. It is typical a non-catecholamine sympathomimetics of mixed action. It stimulates both α- and β-adrenoceptors. Its vasoconstructive properties are approximately 100-folds weaker than epinephrine but duration of action is 10 times longer. Ephedrine is widely used in the treatment of bronchial asthma and in allergic rhinitis condition. It is also used locally in ophthalmology for dilation of pupils.

Synthesis:

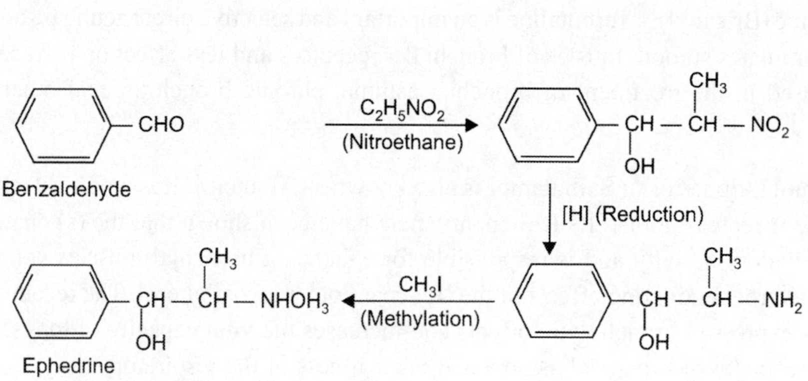

Fig. 2.66

7. Imidazoline Sympathomimetics Derivatives: Imidazoline ring containing sympathomimetics: naphazoline, oxymetazoline, xylometazoline and tetrahydrozoline which are used in the treatment of rhinitics associated with cold, in chronic inflammatory condition, nose bleed, also used in the treatment of laryngitis, sinusitis, etc.

Synthesis of Xylometazoline [Otrivin®]:

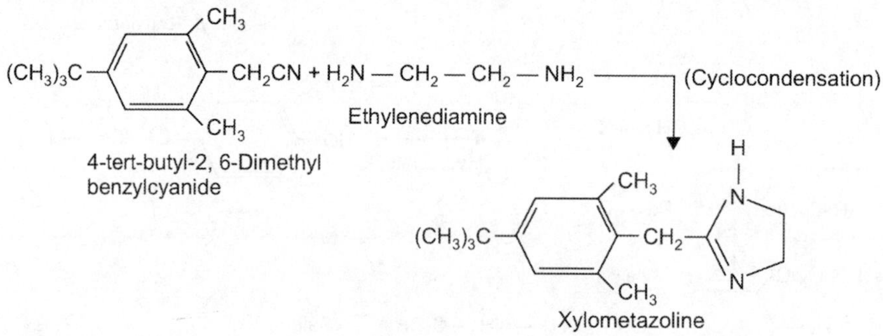

Fig. 2.67

8. Amphetamine (Paredrine)®: Amphetamine and metamphetamine are the derivatives of phenylisopropylamines. Both lack ring substituents, a side chain hydroxyl groups and thus behave sufficiently lipophilic to cross blood brain barrier readily which causes dramatic CNS stimulation. Clinically useful (S)-(+) amphetamine shows central appetite suppressant effects. Another function of amphetamines is their psychostimulatory effects. Larger dose of amphetamines causes hallucination and related mental condition similar to the paranoid schizophrenia.

Synthesis: (Leucart Reaction)

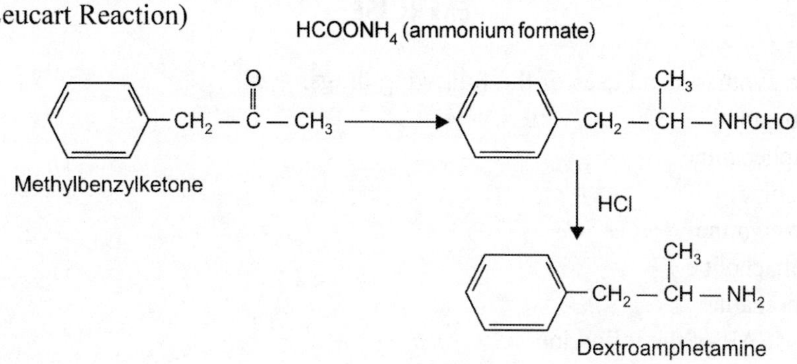

Fig. 2.68

2.3.10 Newer Drugs

1. Salmeterol (Servent)®: 2-(hydroxymethyl)-4-{1-hydroxy-2-[6-(4-phenyl-butoxy) hexylamino] ethyl} phenol

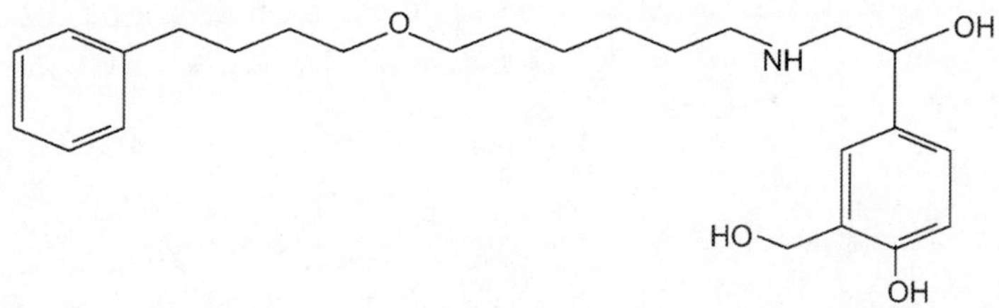

Fig. 2.69

Salmeterol serves as a long acting β_2-adrenergic receptor agonist. It is suggested for the treatment of asthma and (COPD) chronic obstructive pulmonary disease. It is 50 times more potent than albuterol for β_2-receptor activity.

2. Formoterol (Foradil)®: N-[2-hydroxy-5-[1-hydroxy-2-[1-(4-methoxy phenyl)-propane-2-ylamino]ethyl]phenyl] formamide

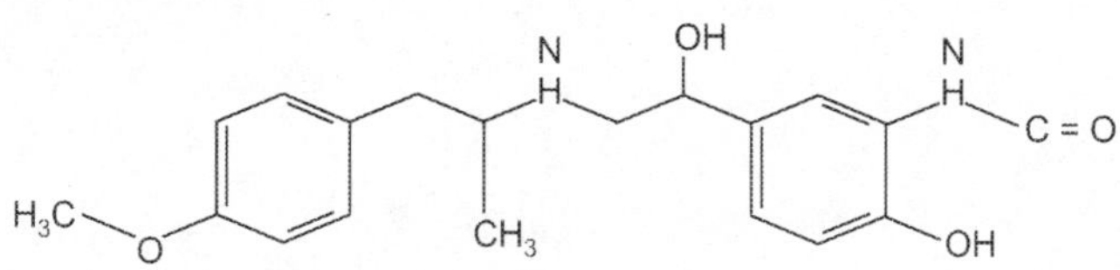

Fig. 2.70: Formoterol.

Formoterol is long acting β_2-receptor agonist like salmeterol. It is a mixture of (R.R.) (–) and (S.S.)—(+)-stereoisomers. (R.R. (–) isomers are 1000 times more potent than (S,S) isomer. It has faster onset of action. It is clinically prescribed for the treatment of asthma and COPD.

EXERCISE

1. Give the synthesis and uses of the following drugs.
   a. Salbutamol
   b. Amphetamine
   c. Atropine
   d. Neostigmine
   e. Methacholine
   f. Isoprenaline
2. Write the SAR of the following
   a. Antimuscarinic agents
   b. Adrenomimetics
3. Give the structures, nomenclature of any four antimuscarinic agents.
4. Give the general structure of muscarinic agonists and discuss their SAR.
5. Write a note on medicinal uses of anticholinesterage.

3

General Anaesthetics

3.1 INTRODUCTION

General anaesthetic agents are those agents which are used to produce unconsciousness and loss of perceptions in painful surgical operations. General anaesthetics is a class of CNS depressant drugs which produce a partial or total loss of sense of pain with a controlled and reversible action of central nervous system in order to perform more complicated surgical operations. A number of chemical agents or drugs are clinically used as general anaesthetics such as inhalation anaesthetics, alcohols, phenol derivatives, barbiturates, benzodiazepines and neurosteroids are acted on the allosteric binding sites of $GABA_A$ receptors.

3.2 IDEAL CHARACTERISTICS OF GENERAL ANAESTHETICS

1. General anaesthetis are non-specific in action and do not interact with specific receptors.
2. They are used in high concentration and have access to all areas of the biological system.
3. They must be rapid in action and pleasant in behaviour induction of surgical anaesthesia.
4. After performing biological action, they should withdraw rapidly.
5. They must have adequate relaxation properties for skeletal muscles.
6. They must be potent enough to allow adequate oxygen supply in the mixture.
7. They should have wide margin of safety as well as nontoxicity.
8. They should have absence of adverse effects.
9. They must be non-inflammable and non-explosive in nature.
10. They should be economical.

3.3 IMPORTANT PHYSIOLOGICAL PROPERTIES RESPONSIBLE FOR ENTRY OF COMPOUNDS INTO BRAIN

1. The degree of ionization plays an important role into transport of drugs to cross blood brain barrier at physiological condition (pH). It has been observed that non-ionized compounds are more prone to penetrate easily.
2. Molecular size of the molecule must be less than 400 to increase penetration.
3. Lipophilic character of the molecule should be ($\log Po = 2 \pm 0.31$). Ideal for penetration into BBB (blood brain barrier)

3.4 CONCEPTS OF MECHANISM OF ANAESTHESIA

1. Meyer-Overton Theory

This theory was concluded by Hans Meyer and Charles Overton in 1900. According to this theory, potency of the compounds is directly related to its lipid solubility and oil-gas partition coefficient with the reference of olive oil, octanol and other membranes like lipids.

According to this concept:
(i) All neutral lipid-soluble substances have depressant properties on neurons.
(ii) Anaesthetic properties are more pronounced in lipid rich cells.
(iii) Anaesthetic effects are gradually increased with increase in the partition coefficient.

Arguing against this theory however is the finding that all highly lipid soluble substances are not capable of producing anaesthesia and moreover the protein drug interaction was the major problem.

2. Ion Channel Theory

General anaesthetics block ion channels by interacting with cellular membranes and reducing the flow of Na^+ ions and enhancing the K^+ ions flow into the cells which leads to development of anaesthesia.

3.5 MINIMUM ALVEOLAR CONCENTRATION (MAC)

It is defined as the concentration at 1 atmosphere in anaesthetic agents in the alveoli, which is required to produce immobility in 50% of the patient subjected to surgical operation. Further increase, 1 to 1.3 MAC, will cause immobility in 99% of patients. At equilibrium the concentration of anesthetic agents in the alveoli is equal to that in brain and it is this concentration in the brain at the site responsible for pharmacological action.

3.6 STAGES OF ANAESTHESIA

Stage-I (Analgesia): Features are loss of consciousness, pain is progressively abolished, patient remains conscious, i.e., he can hear, see and feel, skeletal muscles relaxed and respiration normal.

Stage-II (Delirium): Stage of delirium are, loss of consciousness, apparent excitement, patient may shout, struggle and hold his breath, muscle tone increases, jaws are tightly closed, breathing jerky, vomiting and involuntary micturition.

Stage-III (Surgical Anaesthesia): In this stage the following planes show the activities:

- Plane-I – Roving eyeballs
- Plane-II – Loss of corneal and laryngeal reflexes
- Plane-III – Pupils start dilation.
- Plane-IV – Intercostal paralysis, shallow, abdominal respiration, pupils dilated, muscles tone decreases, B.P. decreases and decrease in respiration.

Stage-IV (Medullary Paralysis): This stage consists of cessation of breathing to failure of circulation and further pupil widely dilated, muscles are totally flabby, pulse thready and ultimately death.

3.7 CLASSIFICATION OF GENERAL ANAESTHESIA

I. Inhalation Anaesthetics

A. Liquid Anaesthetics

(i) Enflurane: 2-chloro-1, 1, 2-trifluoroethyldifluoro-methyl ether

$$CHClF—CF_2—OCHF_2$$

Fig. 3.1

(ii) Isoflurane: 1-chloro-2, 2, 2-trifluoroethyl-di-fluoromethyl ether.

$$CF_3CHCl—O—CHF_2$$

Fig. 3.2

(iii) Methoxyflurane: 2, 2-dichloro-1, 1-difluoroethyl-methylether

$$CHCl_2—CF_2—OCH_3$$

Fig. 3.3

(iv) Halothane: 2-bromo-2-chloro-1,1,1-trifluoroethane

$$CF_3CHClBr$$

Fig. 3.4

(v) Ether: Diethyl ether

$$C_2H_5—O—C_2H_5$$

Fig. 3.5

B. Gaseous Anaesthetics

(i) Cyclopropane:

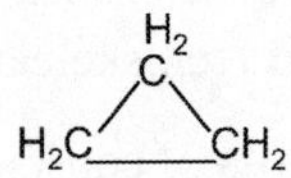

Fig. 3.6

(ii) Nitrous oxide

N_2O

Fig. 3.7

II. Intravenous Anaesthetics

A. Ultra Short Acting Barbiturates

(i) Thiopental sodium: 5-Ethyl-5-(1-methyl butyl)-2-thiobarbiturate

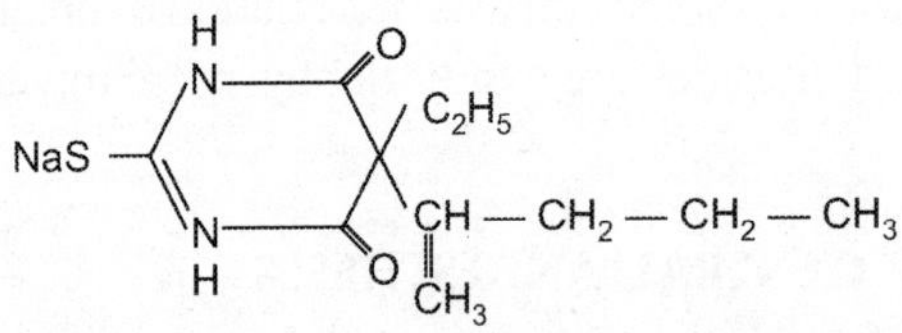

Fig. 3.8

(ii) Methohexital Sodium: 5-allyl-1-methyl-5-(1-methyl-2-pentynyl) barbiturate

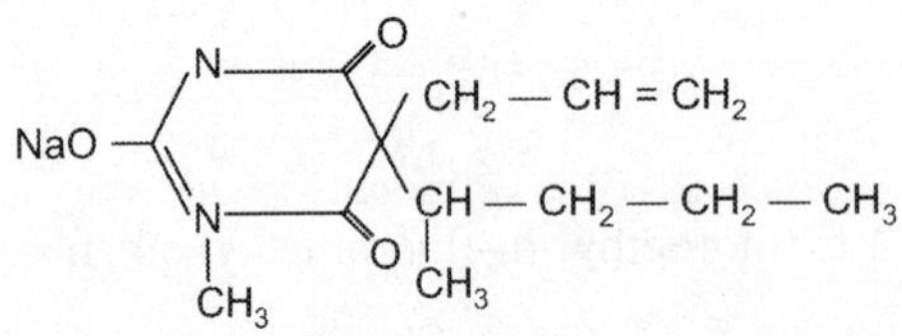

Fig. 3.9

(iii) Thiamylal Sodium: 5-allyl-5-(1-methyl butyl)-2-thiobarbiturate

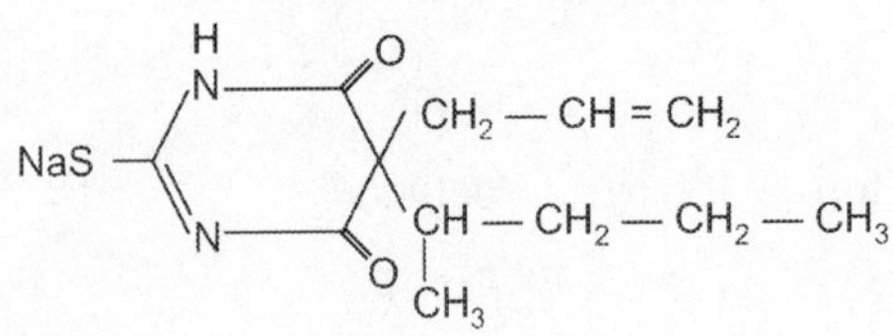

Fig. 3.10

B. Arylcyclo Hexylamine Derivatives

(i) Ketamine: 2-(2'-chlorophenyl)-2-(methylamino)-cyclohexanone

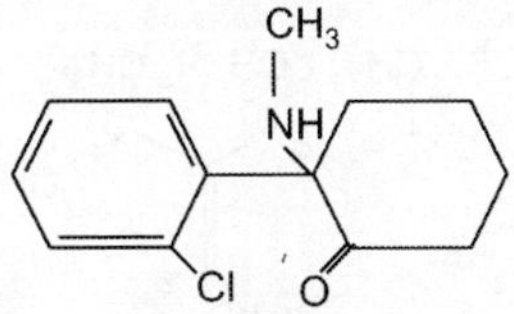

Fig. 3.11

C. Benzodiazepine

(i) Midazolam: 8-chloro-6-(2'-fluorophenyl)-2-methyl-imidazo benzodiazepine.

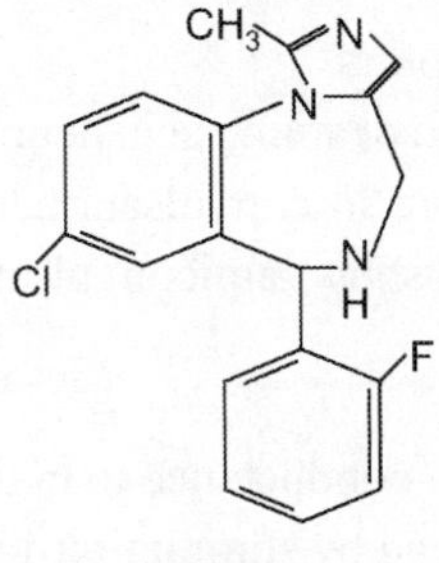

Fig. 3.12

D. Narcotic Analgesics

(i) Fentanyl Citrate: N-(1-phenylethyl-4-piperidinyl)-Propionalide citrate

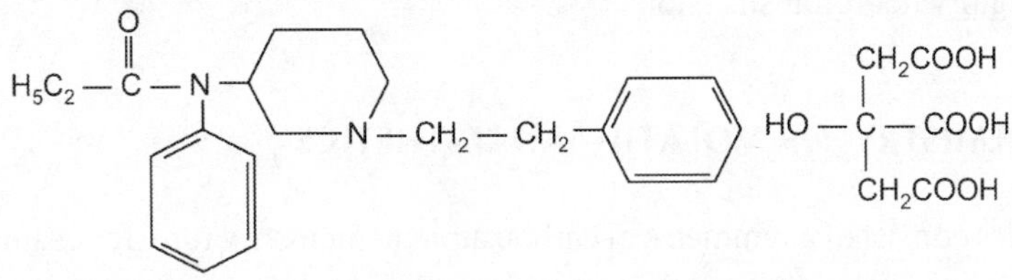

Fig. 3.13

(ii) Alfentanil: N-[1-[2-(4-ethyl-4, 5-dihydro-5-oxo-lH-Tetrazol-1-yl)ethyl]-4-(methoxymethyl)-4-piperidinyl]-N-phenyl propionamide

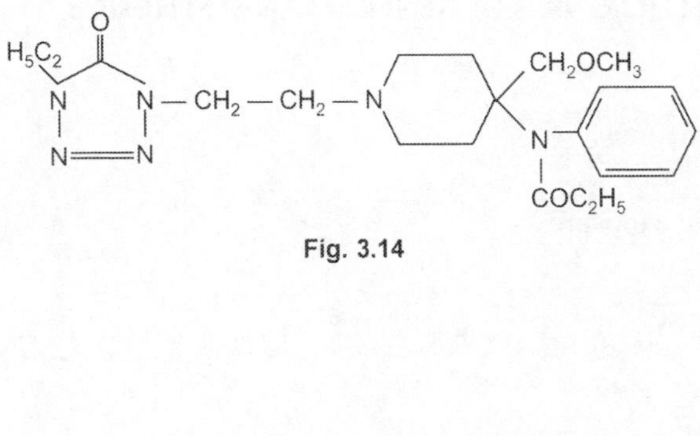

Fig. 3.14

E. Miscellaneous Drugs

(i) Propofol: 2,6-disopropyl Phenol

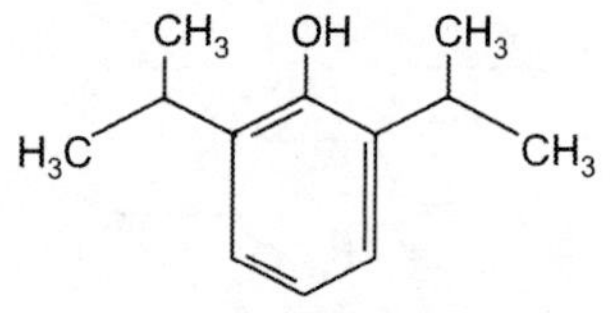

Fig. 3.15

3.8 RECEPTORS INVOLVED IN THE MECHANISM

1. NMDA (N-methyl-D-asparate) Receptors

These receptors are activated by excitation of amino acid neurotransmission and glutamate to increase the conductance of sodium ions resulting in convulsant action. Ketamine hydrochloride exhibits antagonist action on this receptor site and shows anticonvulsants, sedative and anaesthetic properties.

2. G-Protein Coupled Receptors

G-protein coupled receptors acted via K^+ conductance to maintain polarized state of neurons and to assist in repolarization of neurons followed by stimulation-induced depolarization.

3. GABA (γ-amino butyric Acid) Receptors

It is initiated by chloride ion channel and normally mediated, inhibitory response to central nervous system. Some agents like halothan, isoflurane and other general anaesthetics are capable of increasing GABA-ergic inhibitory neurotransmission.

3.9 STEREOCHEMISTRY OF VOLATILE ANAESTHETICS

Volatile anaesthetics consist of asymmetrical carbon atom in their structure, for examples, isoflurane, desflurane, enflurane and halothane and thus can exist in their both (+) or (–) enantiomers. Anaesthetic properties and MAC value also depend on the specific enantiomer of anaesthetic for illustration (+) if isoflurane enantiomer is 50% more potent than (–) enantiomer.

3.10 TOXICITY OF FLUORINATED GENERAL ANAESTHETICS

(i) Hepatotoxicity
(ii) Malignant hyperthermia
(iii) Nephrotoxicity
(iv) Low level chronic exposure

3.11 GENERAL ANAESTHETICS AGENTS

1. Fluorinated Hydrocarbons

During anaesthesia formation of toxic degradants is a major concern with used carbon dioxide absorbants. This reaction converts to secondary chemical structure from primary structure. For example, Halothane converts into -2-bromo-2-chloro-1, 1-difluoroethylene and sevoflurane converts into 2-(fluoromethoxy)-1, 1, 3, 3, 3-pentafluoro-1-propene. Fluorinated hydrocarbon anaesthetics are potent anaesthetics because of the presence of fluorine and halogen atoms in the structures which makes the compound more potent and non-inflammable as well as intermediate blood solubility. It has been observed that during short-term infrequent administration shows nephrotoxicity, hepatotoxicity by halothane and methoxy flurane.

Synthesis:

Enflurance (Ethrane)®

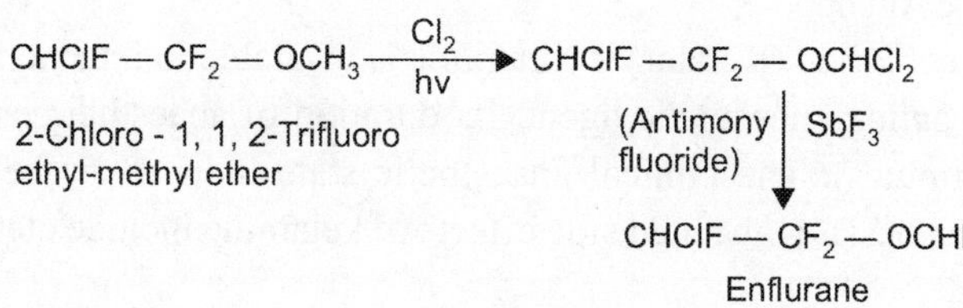

Fig. 3.16

Isoflurane (Forane)®

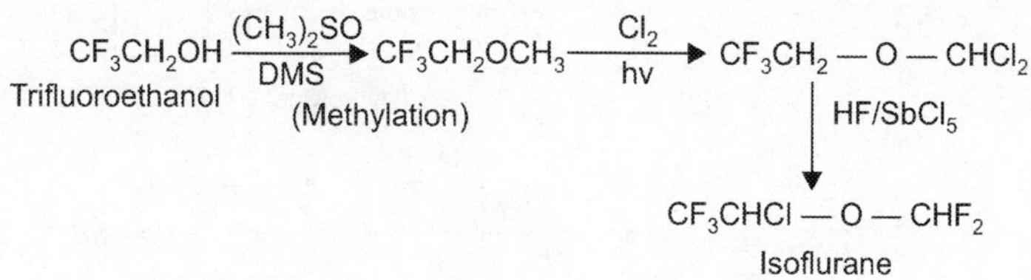

Fig. 3.17

Methoxyflurane: (Penthrane)®

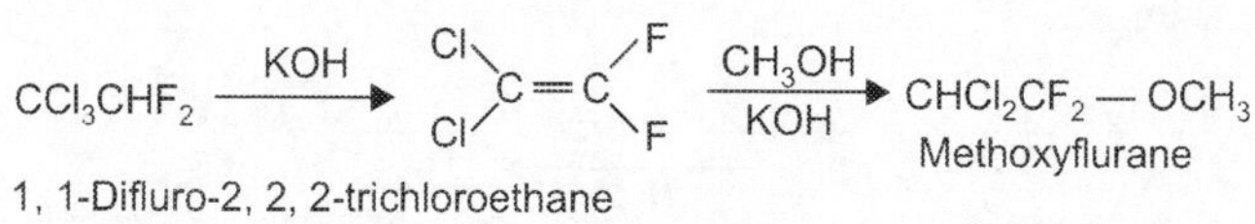

Fig. 3.18

2. Ultrashort Acting Barbiturates

(i) Thiopental Sodium (Pentothal)®: Thiopental sodium is clinically used intravenously. It produces rapid unconsciousness for surgical and basal anaesthesia within 30–40 seconds. These agents are used initially to induce anaesthesia. Unwanted reactions like respiratory depression, myocardial depression, cardiac arythmia have been seen during anaesthesia.

Synthesis:

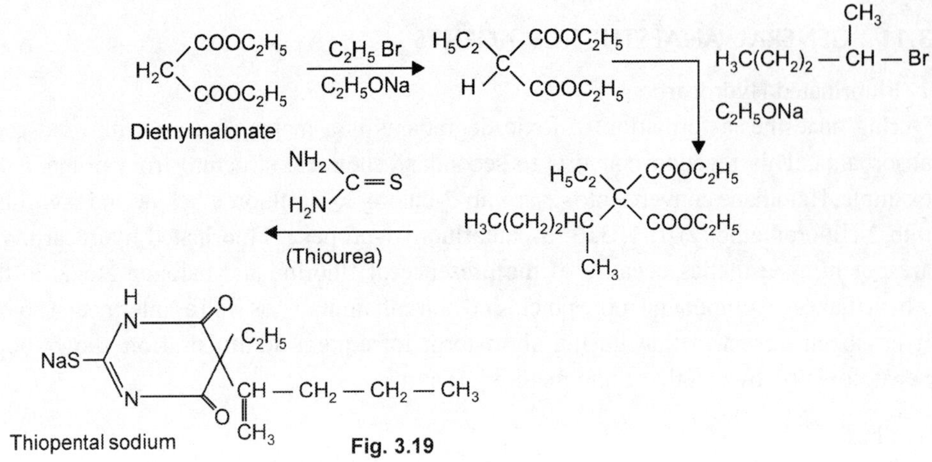

Fig. 3.19

3. Arylcyclo Hexylamine Agents

(i) Ketamine: Hydrochloride (Ketalar)®: Ketamine is available in injactable form. Ketamine is meant for rapid acting anaesthetic agent. Its duration of anaestheic property is very short about (10–25 minutes). The clinical anaesthetic state induced by ketamine is known as dissociative anaesthesia. Unwanted side effects of ketamine include elevated blood pressure and pulse rate.

Synthesis:

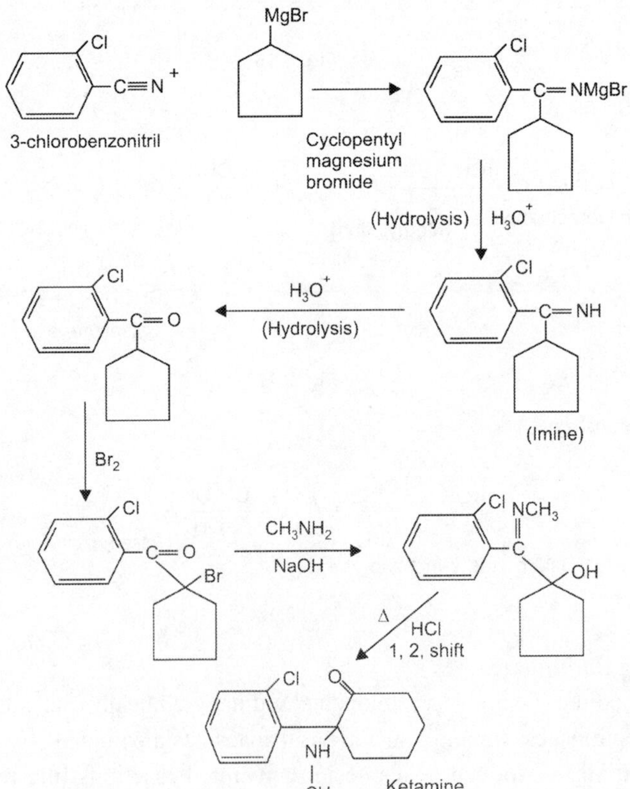

Fig. 3.20

4. Narcotic Anaesthesia

(i) Propofol (Diprivin)®: Propofol is the most common parenteral anaesthetic used in the United States. It is supposed that propofol acts via enhancing Gabargic neurotransmissions within the central nervous system. It is a poor water soluble agent thus it is formulated as 1 to 2% emulsion with soyabean oil, egg lecithin and glycerol. Propofol is used as an induction agent and for short anaesthtic procedure.

Synthesis:

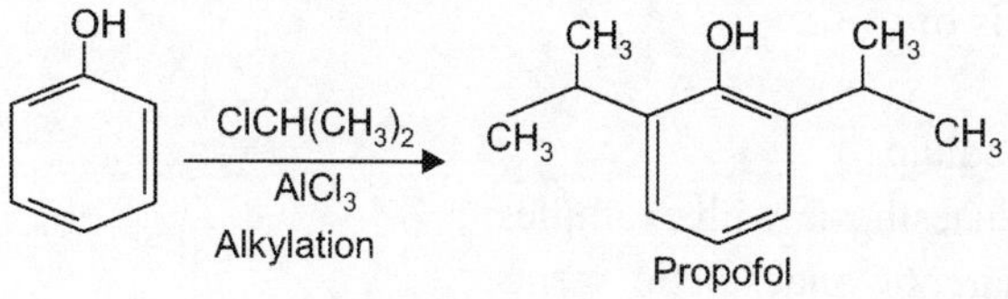

Fig. 3.21

3.12 NEWER AGENTS

(i) Desflurane: 2-(difluoromethoxy)-1, 1, 1, 2-tetrafluoroethane.

Disflurane contains a number of fluorine atoms, used for maintenance of general anaesthesia with sevoflurance. It has rapid onset of action and low solubility in blood.

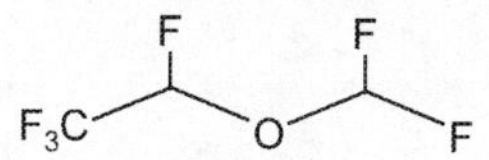

Fig. 3.22

(ii) Sevoflurane: 2, 2, 2-Trifluoro-1-[trifluoromethyl] ethyl fluoromethyl ether.

Sevoflurane also contains a number of fluorine atoms in its structure. It possesses sweet smelling, non-inflammable properties. It is used for induction and maintenance of general anaesthesia in surgical operation.

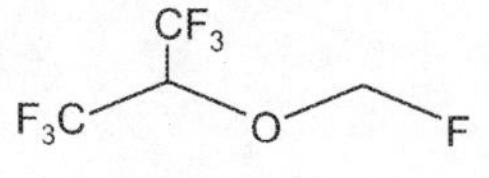

Fig. 3.23

(iii) Minaxolone: 1-(11-dimethylamino)-2-ethoxy-hexadecahydro-3-hydroxy-10, 13-dimethyl-1H-cyclopenta[a]phenenthrene-17-yl)ethane

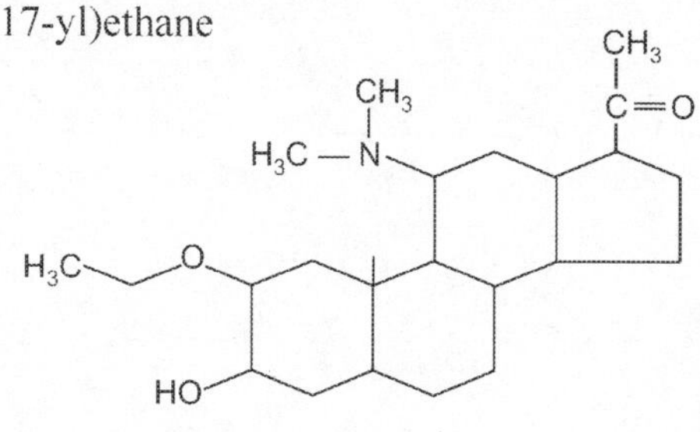

Fig. 3.24

Minaxolene is new water soluble steroidal anaesthetic. It is a safe and effective intravenous anaesthetic agent.

EXERCISE

1. Write a note on intravenous anaesthetic and ultra short acting barbiturates.
2. Write in detail on different stages of anaesthesia.
3. Write the synthesis of
   (a) Ketamine
   (b) Thiopental sodium.
4. Classify general anaesthesia with examples.
5. Write a note on narcotic anaesthetic agents.

4

Local Anaesthetics

4.1 INTRODUCTION

The term anaesthesia is loss of sensation with or without loss of consciousness. Local anaesthetics are those chemical agents which are applied topically or administered parenterally to a localized area and produce a state of local anaesthesia by reversible blocking the nerve potential that transmits the feeling of pain.

Local anaesthetics are used for soreness, itching, insect bites, wounds, cut, rashes, allergic condition and fungal infection and in minor surgical operation. Method of administration of local anesthetics is governed by the following properties like toxicity, stability, duration of action, water solubility, membrane permeability and point of application. On the other hand, local anaesthetics mode of action depends on their lipid solubility, PKa, vasodilation and protein binding characteristics. The drug exerts a systemic effect due to transport into blood from the site of administration. This effect depends on its concentration in the blood. Number of unwanted side effects may occur like sedation, restlessness, nausia and anxiety. Sometimes convulsions, respiratory and cardial depressions may occur due to high toxicity.

4.2 METHODS OF ADMINISTRATION

Table 4.1: Method of administration of local anaesthetics

S. No.	Types	Site of action	Drugs used (conc.)
1.	Topical anaesthesia	Mucous membrane of nose, throat, tracheobronchial, oesophagus, urinary tract	Phenlephrine (0.005%) Lignocaine (2-10%) Cocaine (1–4%) Tetracaine (2%)
2.	Infiltration anaesthesia	Directly injected into tissue (skin, intra-abdominal organs)	Epinephrine (5 μg/ml) Lignocaine (0.5–1.0%) Procaine (0.5–1.0%) Bupivacaine (0.125–0.25%)

3.	Nerve block anaesthesia	Peripheral nerves	Lignocaine (1–1.5%) Mepivacaine (1–2%
4.	Field block anaesthesia	Subcutaneous area	Lignocaine (0.5–1.0%) Procaine (0.5–1.0%)
5.	Intravenous anaesthesia	Nerve trunks ending of forearms, hand	Lignocaine (0.5%) Procaine (0.5%)
6.	Spinal anaesthesia	Cerebrospinal fluid in lumber space	Lignocaine (1–1.5%) Tetracaine Bupivacaine
7.	Epidural anaesthesia	Epidural space	Bupivacaine (0.5–0.75%) Lignocaine (2%) Chloroprocaine (2–3%)

4.3 CLASSIFICATION OF LOCAL ANAESTHETICS

I. Ester Linkage Derivatives

1. Benzoic Acid Derivatives: Basic structure

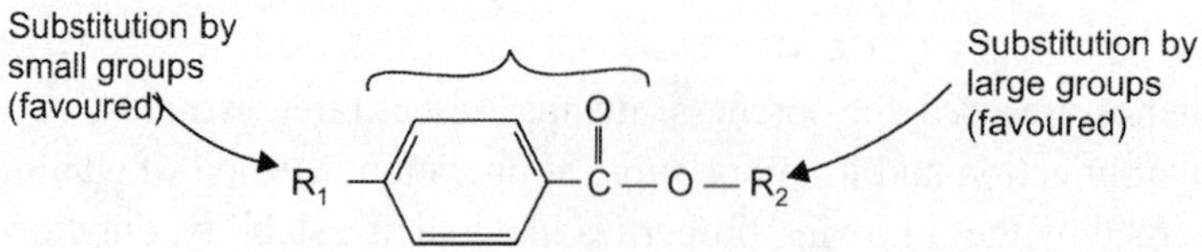

Fig. 4.1

	Derivatives	R_1	R_2
(i)	Cocaine	H	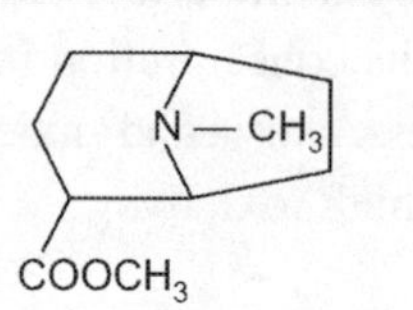
(ii)	Isobucaine	H	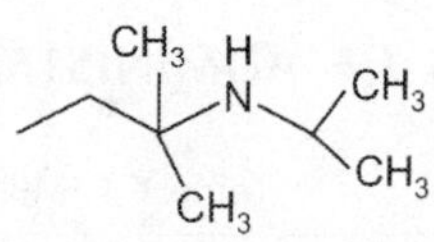
(iii)	Hexylcaine	H	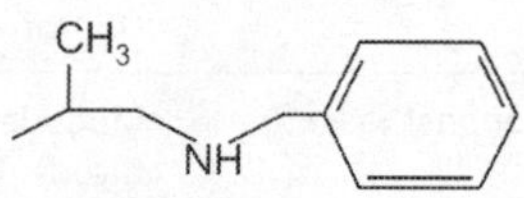
(iv)	Meprylacaine	H	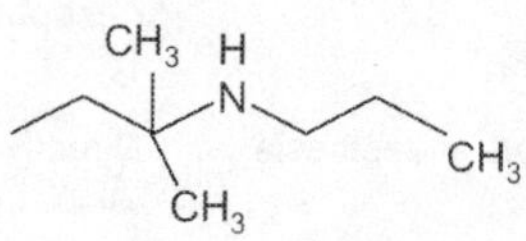
(v)	Piperocaine	H	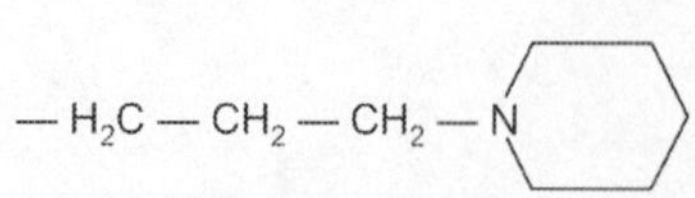

2. P-amino-benzoic Acid Derivatives

Basic Structure

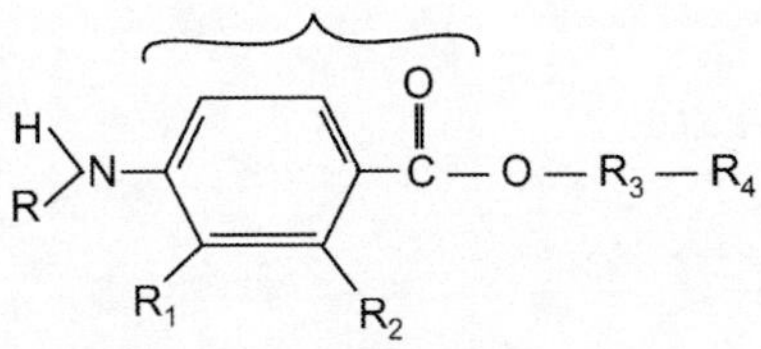

Fig. 4.2

	Derivatives	R	R_1	R_2	R_3	R_4
(i)	Benzocaine	H	H	H	$-C_2H_5$	—
(ii)	Procaine	H	H	H	$-C_2H_4-$	$-N(C_2H_5)_2$
(iii)	Chloroprocaine	H	H	Cl	$-C_2H_4-$	$-N(C_2H_5)_2$
(iv)	Tetracaine	Butyl	H	H	$-C_2H_4-$	$-N(C_2H_5)_2$
(v)	Butacaine	H	H	H	$-C_2H_4-$	$-N(C_4H_9)_2$
(vi)	Benoxinate	H	H	Butoxy	$-C_2H_4-$	$-N(C_2H_5)_2$
(vii)	Propoxycaine	H	H	Propyloxy	$-C_2H_4-$	$-N(C_2H_5)_2$

II. Amide Linkage Derivatives

Basic Structure

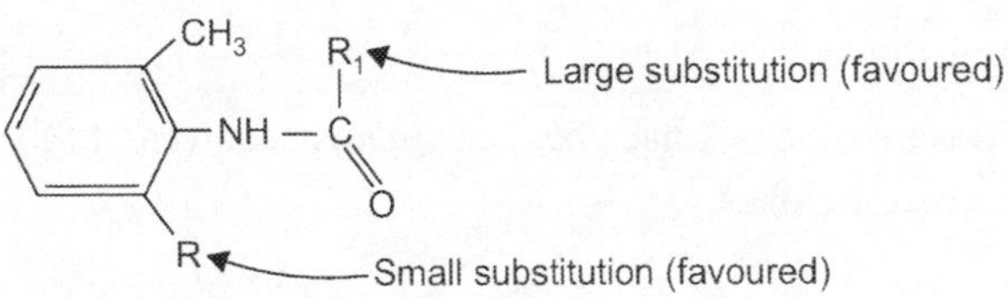

Fig. 4.3

	Derivaties	R	R_1
(i)	Lignocaine	CH_3	$-CH_2-N(C_2H_5)_2$
(ii)	Mepivacaine	CH_3	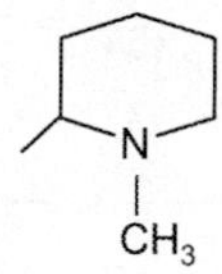
(iii)	Bupivacaine	CH_3	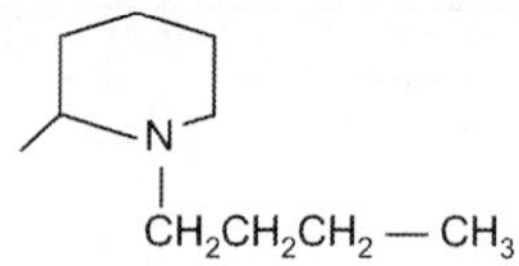
(iv)	Prilocaine	H	$-CH(C_2H_5)-NH-CH_2-CH_2-CH_3$
(v)	Etidocaine	CH_3	$-CH(C_2H_5)-N(C_2H_5)(C_3H_7)$

4.4 MECHANISM OF ACTION OF LOCAL ANAESTHETICS

Local anaesthetics supposed to act by decreasing the excitability of nerve cells which is associated with the movement of sodium ions across the nerve membrane. Local anaesthetics act by interfering with the movement of sodium ions by blocking the voltage sensitive sodium ion channels. Blocking of Na^+ channels is believed to achieved either by physical blocking by drug molecules like cork in a bottle or by destroying the channels enough like sodium ions channels that lead to preventing the conduction and release of neurotransmission at presynaptic site.

4.5 STRUCTURAL ACTIVITY RELATIONSHIP (SAR) OF ESTER LINKAGE DERIVATIVES

Structural activity relationship of local anaesthetics explains according to the classification, and the basic structure as represented below:

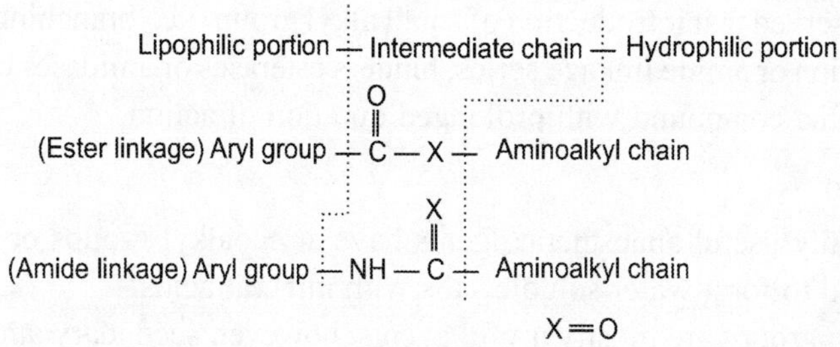

Fig. 4.4

General Points

(1) Isosteric replacement of alkoxy oxygen of–COOR group by –NH[–CONHR] or CH_2– [$COCH_2R$] makes the compound less hydrolysed by enzymatic hydrolysis.

(2) For amide linkage: Additional steric hindrance occurs at amide linkage which makes –NH-CO– group that is less susceptible from hydrolysis.

(3) Clinically useful local anaesthetics possess tertiary amines with PKa value 7.0–9.0. These compounds show their anaesthetic properties by virtue of binding of onium ions to a selective site with sodi ion channels. Therefore, any structural modification that alters the lipid solubility, PKa value and metabolic inactivation.

$$pH = Pka - Log[B\overset{+}{H}]\,[B]$$

(arrow to $B\overset{+}{H}$: Onium ion)

Fig. 4.5

Lipophilic Portion

(i) Lipophilic portion is supposed to be essential for anaesthetic activity. Aromatic group directly attached to carbonyl function or dimethyl substituted phenyl group attached to carbonyl portion through –NH– series makes the compound highly lipophilic and appears to play an important role in binding with receptor site.

(ii) In case of amino ester series, electron donating groups at position 2 and 4 of aryl ring leads to increased potency. Such group as an amino group, e.g., procaine and chloroprocaine. An alkyl amino, e.g., tetracaine, an alkoxy (paparacaine and propoxycaine). These groups contribute e-density to aromatic ring by both resonance and inductive effects respectively.

(iii) Introduction of electron withdrawing groups such as $–NO_2$, –CO–, Nitrite (C ≡ N) in the aromatic ring leads the compounds with reduced activity.

(iv) Aryl aliphatic radicals that contains a methylene group between aromatic moiety and carbonyl function leads to greatly decrease activity in procaine.

Intermediate Chain

(1) In the intermediate chain X may be carbon, oxygen, nitrogen or sulphur.

(2) A short alkyl chain of alkylene of one to three carbon atoms in length to aromatic ring via several possible organic functional groups, influence the duration of action and relative toxicity.

(3) It has been observed that introduction of small alkyl group (i.e., branching) especially around the ester function or amide linkage series, hinders esterases or amidases catalysed hydrolysis which makes the compound with prolonged duration of action.

Hydrophilic Portion

(1) Mostly clinically useful anaesthetic agents have aminoalkyl groups or tertiary alkylamine which are used to form water soluble salts with mineral acids.
(2) Tertiary amino groups are mostly useful agents, however, secondary amines appear to have longer activity but exhibit irritating properties. Tertiary amino groups may be diethyl amino, piperidine or pyrrolidine groups.
(3) Introduction of more hydrophilic morpholino groups usually are respoponsible to diminish the anaesthetic activity.

4.6 LOCAL ANAESTHETIC AGENTS

(1) Cocaine: (-)3-(Benzoyloxy)-8-methyl-8-azabi-cyclooctane-2-carboxylic acid methyl ester. Cocaine is an alkaloid obtained from the extraction of the leaves of *Erythroxylon coca*. Cocaine is obtained from the acid hydrolysis of esteralkaloid to (–)ecgonine then after purification followed by esterification with methanol.

Synthesis:

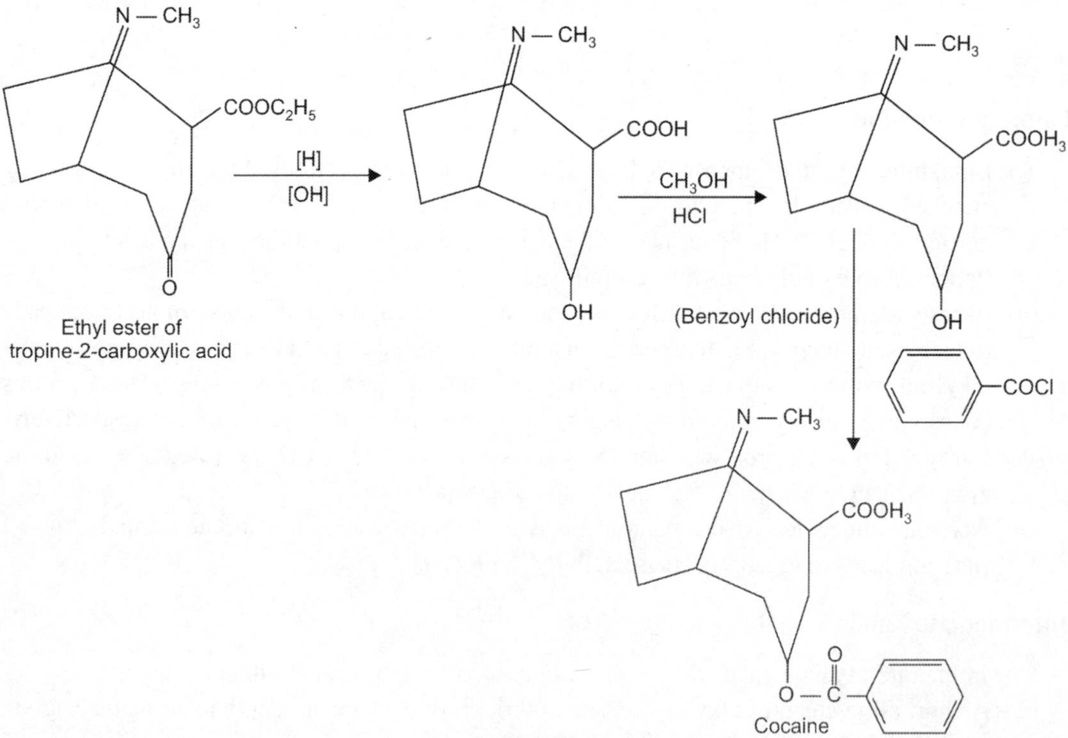

Fig. 4.6

(2) Hexylcaine: 1-(Cyclohexylamine)-2-propanol benzoate

Synthesis:

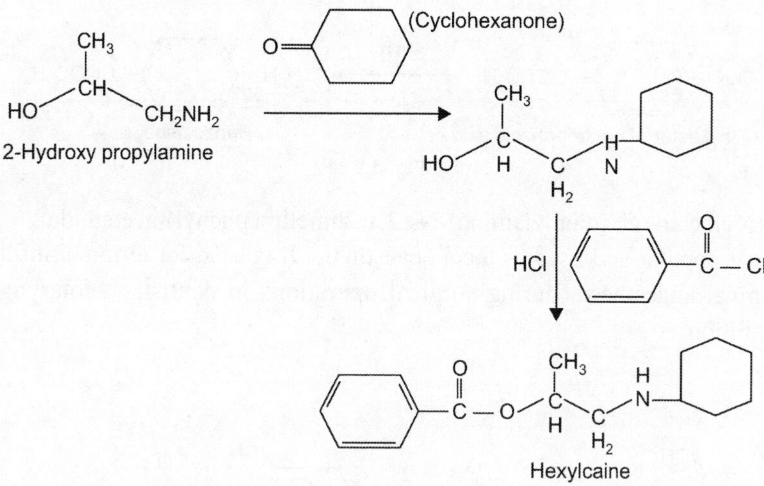

Fig. 4.7

(3) Procaine (novocaine)®: 2-(Diethylamino) ethyl-p-amino benzoate. Procaine is a short acting local anaesthetic agent. It is clinically useful for reducing the pains of various types and is widely used as infiltration, block, epidural, and spinal cord anaesthesia.

Synthesis:

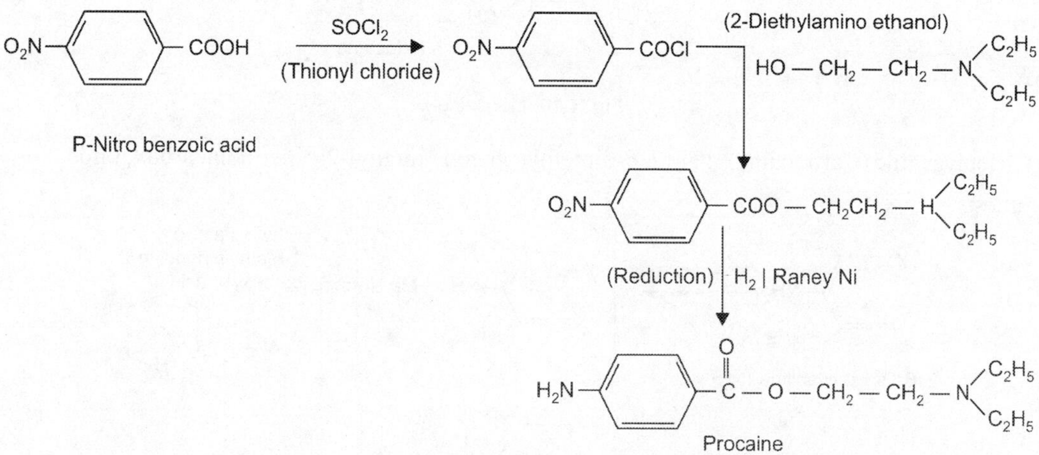

Fig. 4.8

(4) Benzocaine (Americaine)®: Ethyl-p-amino benzoate

Benzocaine is suggested in topical anaesthesia on the skin and mucous membranes in the form of aerosols or as creams for reducing pain caused by iching, cuts and insect bites.

Synthesis:

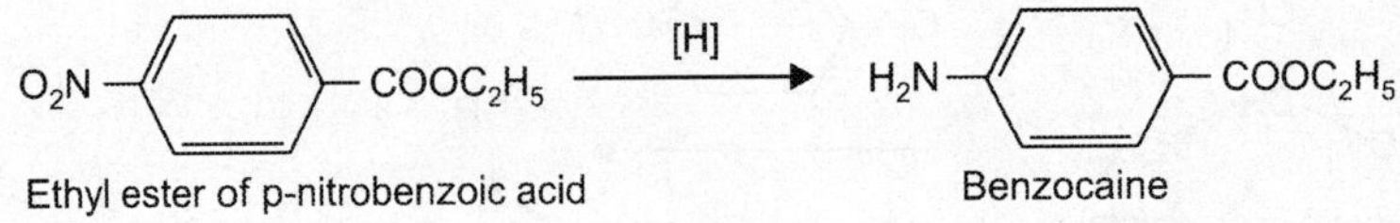

Fig. 4.9

(5) Lidocaine (Xylocaine)®: (diethylamino)-N-(2, 6-dimethyl phenyl)-acetamide

Lidocaine is most popular and used as local anaesthetic. It is used for terminal infiltration, block, epidural and spinal anaesthesia during surgical operations in dentistry, otolaryngology and in gynecology condition.

Synthesis:

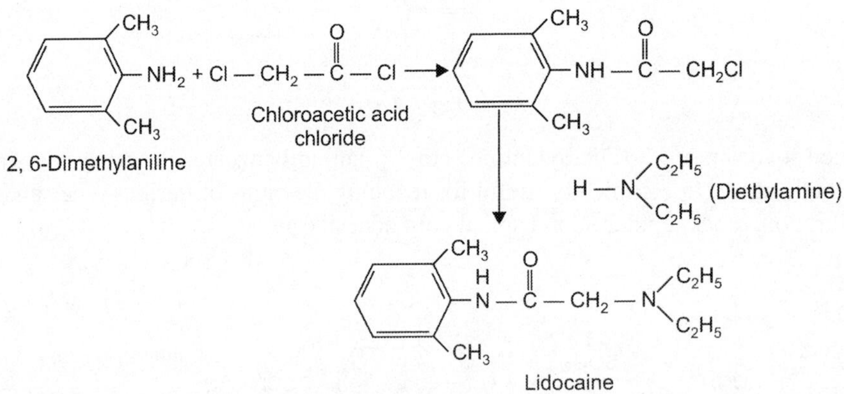

Fig. 4.10: Lidocaine.

(6) Mepivacaine (Carbonaine)®: N-(2, 6-dimethy(phenyl)-methyl-2-piperindincarboxamide.

Synthesis:

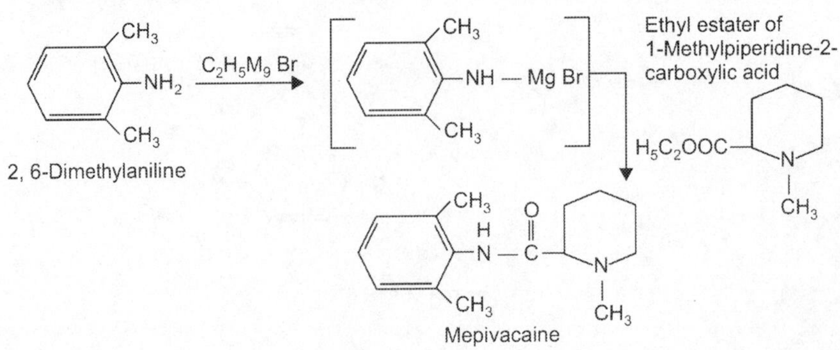

Fig. 4.11

(7) Prilocaine: N-(2-methylphenyl)-2-(propylamino) propanamide

Synthesis:

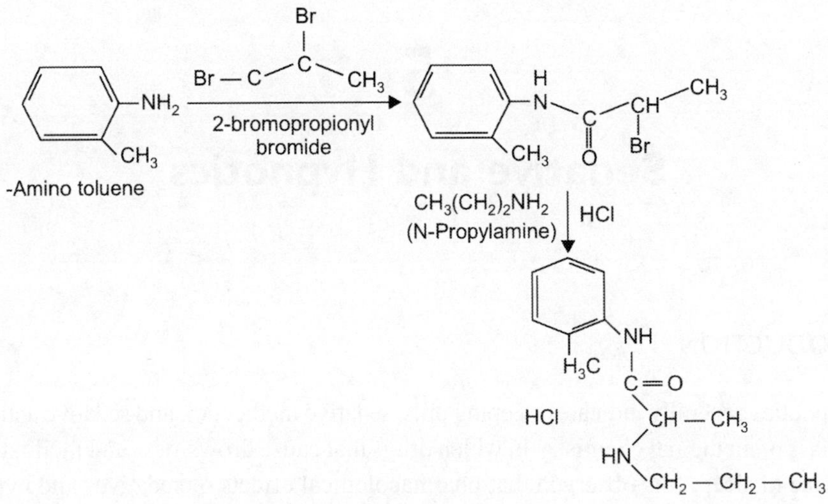

Fig. 4.12

4.7 NEWER AGENTS

(1) Ropivacaine: (S)-N-(2, 6-Dimethyl phenyl)-1-propyl-piperidine-2-carboxamide
Ropivacaine is used as local anaesthetic including infiltration, nerve block, epidural and intrathecal anaesthesia.

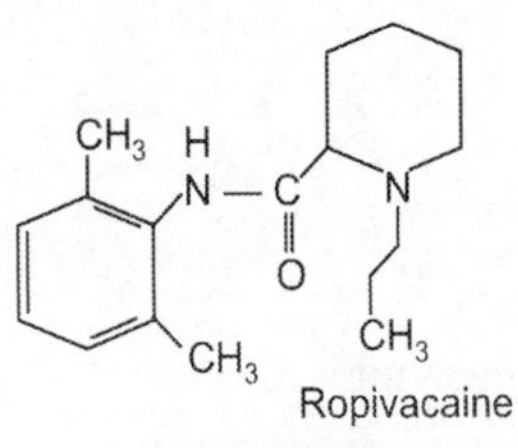

Fig. 4.13

EXERCISE

1. Write SAR of ester and amide derivatives of local anaesthetics.
2. Write the synthesis and mode of action of lignocaine and benzocaine.
3. Give chemical classification of local anaesthetics with examples.
4. Discuss details of various categories of anaesthetics.
5. Discuss details of ester linkage derivatives on local anaesthetics.

5

Sedative and Hypnotics

5.1 INTRODUCTION

The term hypnotics generally indicates sleeping pills, sedative medication and sedative and hypnotics. This is the class of medicinal chemistry in which drugs that cause drowsiness and facilitate initiation of maintenance of sleep. It is observed that pharmacological effects of sedatives and hypnotics are closely related. Small doses cause sedation while large doses cause hypnosis and further larger doses bring about surgical anaesthesia.

Sedative and hypnotics are operative in the short-term 'situational stress' to facilitate sleep. Drugs used as hypnotics are often sedative and depending on the dose called axiolytics. There are many autonomic, physiological and biochemical changes directly associated to wakefulness, non-rapid eye movement (NREM) as well as rapid eye movement sleep (REM). Apart from that of classical drugs, newer drugs available in the market and newer information available on the role of sleep factors such as haemostatic regulation of sleep, cardiac rhythm, chronotherapy and role of immunology, etc.

5.2 IDEAL HYPNOTICS DRUGS

Characteristics of ideal drugs are as following:

(i) They induce sleep promptly after administration.
(ii) They should maintain sleep for an adequate period without undesired awakening.
(iii) Hypnotic drugs should promote a sleep state identical to a non-drug induced or natural sleep.
(iv) Drugs should not produce undesired daytime sedation or drowsiness, etc.
(v) They should not lead to dependency.

5.3 CLASSIFICATION OF SEDATIVE AND HYPNOTICS

5.3.1 Barbiturates

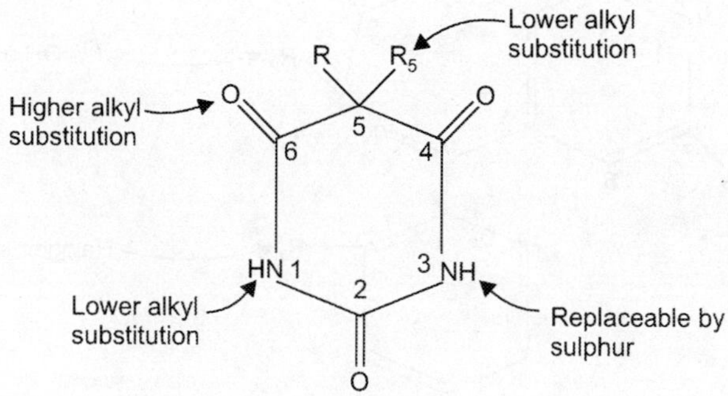

Fig. 5.1: General structure of barbiturates.

Substitution

	Derivatives	R_5	R	Other modification	Duration of action (h)
(i)	Amobarbital	C_2H_5	$(CH_3)_2$-CHC_2H_5		2-8
(ii)	Barbital	C_2H_5	C_2H_5		4-12
(iii)	Hexobarbital	CH_3	$(CH_2)_4CH = Ca$	1-CH_3	1-4
(iv)	Mephobarbital	C_2H_5	C_6H_5	1-CH_3	1-4
(v)	Methohexital	CH_2=$CHCH_2$	C_2H_5C=$CCH(CH_3)$	1-CH_3	10-30 min
(vi)	Pentobarbital	C_2H_5	$CH_3(CH_2)_2CH(CH_3)$		2-4
(vii)	Phenobarbital	C_2H_5	C_6H_5		4-12
(viii)	Secobarbital	CH_2=$CHCH_2$	$CH_3(CH_2)_2CH(CH_3)$		1-4
(ix)	Thiopental	C_2H_5	$CH_3(CH_2)_2CH(CH_2)$	2-5	4-12
(x)	Allobarbital	CH_2=$CH{\cdot}CH_2$	CH_2=$CH{\cdot}CH_2$		2-8
(xi)	Butabarbital	C_2H_5	$CH_3{\cdot}CH_2CH(CH_3)$		2-4
(xii)	Cyclobarbital	C_2H_5	$(CH_2)_4$-CH=C-		2-8
(xiii)	Heptabarbital	C_2H_5	$(CH_2)_5CH$=C-		2-4

Note: Long Action barbiturates (6-8 h)
Intermediate Acting (4-6 h)
Short acting Barbiturates (2-4 h)
Ultra Chort Acting (2) (10-30 min)

5.3.2 Benzodiazepines

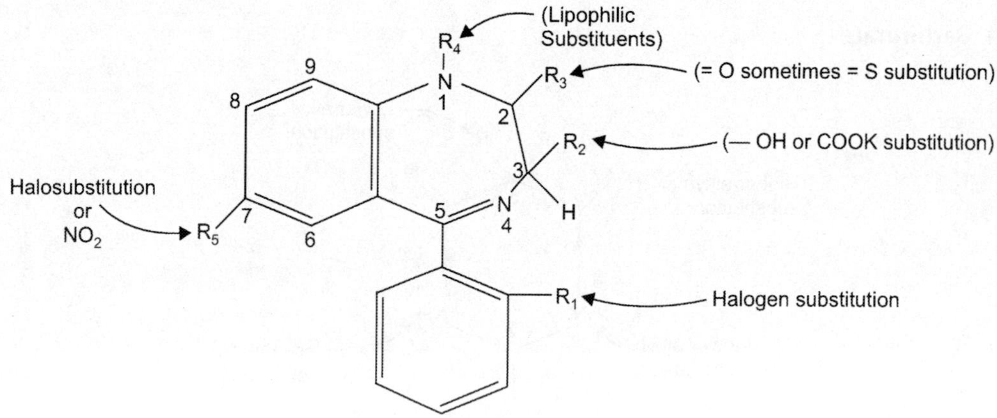

Fig. 5.2: General structure of Benzodiazepine.

	Derivatives	R_1	R_2	R_3	R_4	R_5
(i)	Cinolaze pam	F	OH	= O	$(CH_2)_2CN$	Cl
(ii)	Deloraze pam	Cl	H	= O	H	Cl
(iii)	Fluraze pam	F	H	= O	$(CH_2)_2N(C_2H_5)_2$	Cl
(iv)	Lorazepam	Cl	OH	= O	H	Cl
(v)	Lormetazepam	Cl	OH	= O	CH_3	Cl
(vi)	Nitrazepam	H	H	= O	H	NO_2
(vii)	Nordazepam	H	H	= O	H	Cl
(viii)	Temazepam	H	OH	= O	CH_3	Cl
(ix)	Diazepam	H	H	= O	CH_3	Cl

5.3.3 Halogenated Sedative and Hypnotics

(i) Chloralbetain: Chloral hydrate compound with betaine

$$Cl_3CCH(OH)_2 \cdot (CH_3)_3\overset{+}{N}CH_2CO_2^-$$

Fig. 5.3

(ii) Chloralodol: 2-methyl-2-(2, 2, 2,-trichloro-1-hydroxyethoxy)-2-pentanol

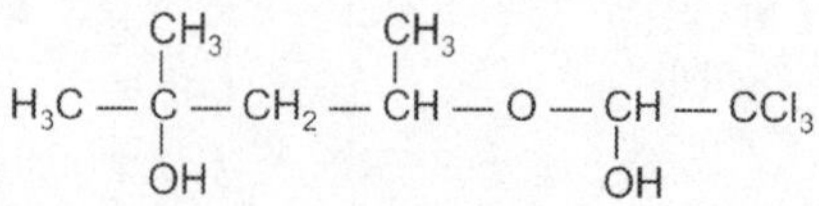

Fig. 5.4

(iii) Carbamol: 2-chloro-2-ethyl-butyryl urea

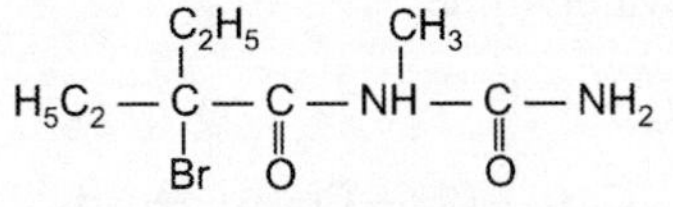

Fig. 5.5

(iv) Petrichloral: Pentacrythritol Hemiacetal with chloral

$$C(CH_2OCH(OH)CCl_3)_4$$

Fig. 5.6

5.3.4 Heterocyclic Sedative and Hypnotics

(i) Glutethimide: 3-ethyl-3-phenyl-2,6-piperidinedinone

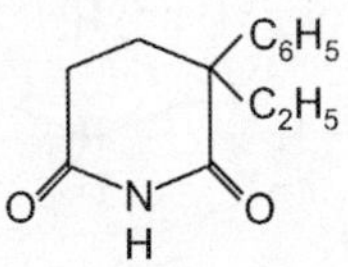

Fig. 5.7

(ii) Clomethazole: 5-(2-chloroethyl)-4-methylthiazole

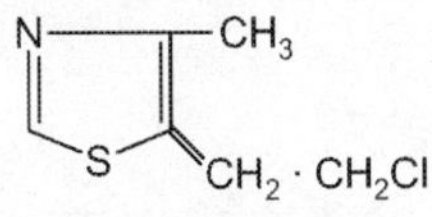

Fig. 5.8

(iii) Zopiclone: 4-methyl-1-piperazine carboxylic acid ester of 6-(5-chloro-2-pyridyl)-6, 7-dihydro-7-hydroxy-5-H-pyrrolo-(3, 4, b) pyrazine-5-one

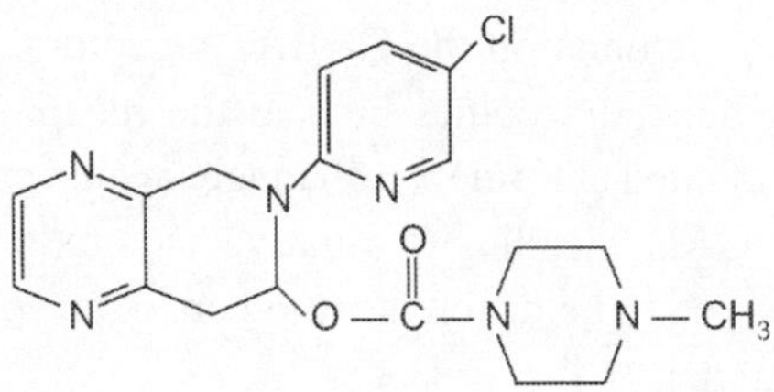

Fig. 5.9

5.3.5 Miscellaneous Sedative and Hypnotics

(i) Valnoctamide: 2-ethyl-3-methyl valeramide

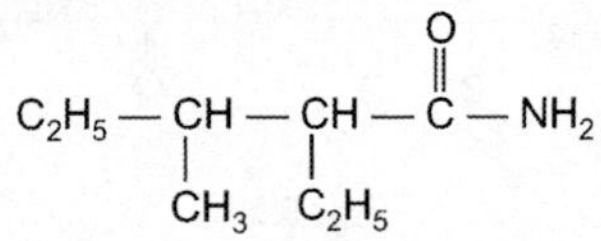

Fig. 5.10

(ii) Propofol: 2,6-Bis(1-methylethyl) phenol

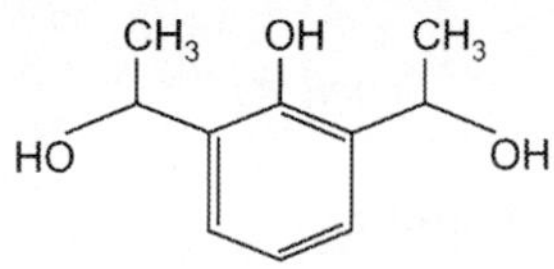

Fig. 5.11

(iii) Melatonine: N-[2-(5-methoxy-1H-Indol-3yl)ethyl] metamide.

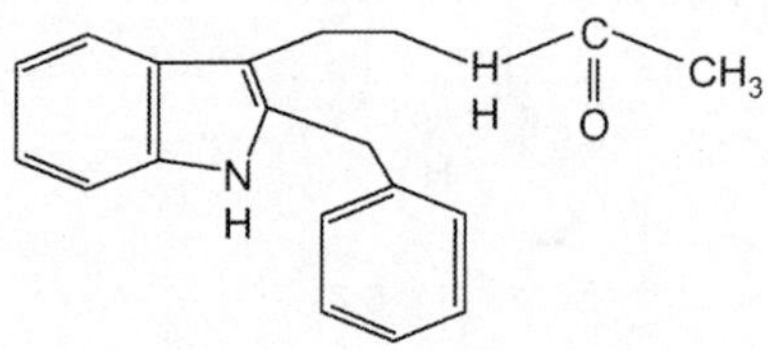

Fig. 5.12

5.4 MECHANISM OF ACTION

Both barbiturates and benzodiazepines bind to allosteric site of $GABA_A$ receptors followed by activating the receptors and alter the transmission of ions by opening the receptors and increase the affinity of GABA for its receptors by altering the structure. Receptors activation increases the membrane permeability to chloride ions, after binding of GABA to its receptors, thereby produces hyper polarization in the cell body. This, in turn, inhibits neuronal activity. However, benzodiazepines bind at the interface of α and γ subunits on the $GABA_A$ receptors. α subunits contain a histidin amino acid residue. However, benzodiazepines bind to the receptor followed by increasing the frequency of opening of the associated chloride ion channels, which causes hyperpolarization which potentiates inhibitory action of GABA leading to sedative and anxiolytic effect. On the other hand barbiturates binding site is believed to be different from benzodiazepines, that occurs at picrotoxin binding site on chloride ion channels.

5.5 BARBITURATES

Chemically barbiturates are the derivatives of barbituric acid and are synthesized from the condensation of malonic acid derivatives and urea derivatives. Barbiturates exert their pharmacological action on CNS, expressed in patient behaviour to the obvious effects like sedation, sleep or general anaesthesia depending on the administered dosage. Barbiturates are further classified on the physical presence into the cell membrane and exhibit drug-receptor interaction. Barbiturates are used for a period of time for the treatment of insomnia, controlling several convulsant condition and for the treatment of various forms of epilepsy. They are also used for pre- and post-operational sedation, for relieving patient from anxiety, nervousness and tension. Barbiturates are considered lethal at about 10 times the hypnotic dose that is believed to be 6–10 gm of phenobarbital or 2–3 gm of secobarbital, pentobarbital or amobarbital.

5.5.1 General Synthetic Approach of Sedative and Hypnotics

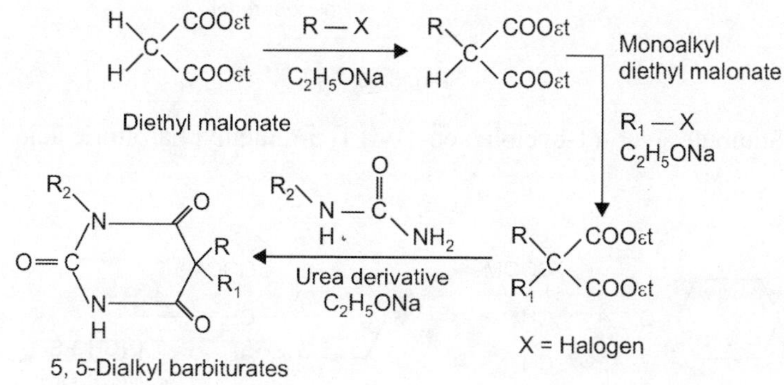

Note: This synthetic approach is meant for all barbiturates except hexobarbital and phenobarbital.

Fig. 5.13

1. Phenobarbital (Luminal)®: 5-ethyl-5-phenyl barbituric acid or 5-ethyl-5-phenylhexahydropyrimidine 2, 4, 6-trione phenobarbital is suggested in the treatment of epilepsy, cholera, spastic paralysis, etc.

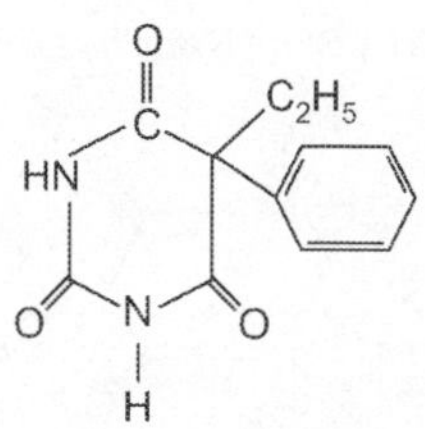

Fig. 5.14

Synthesis:

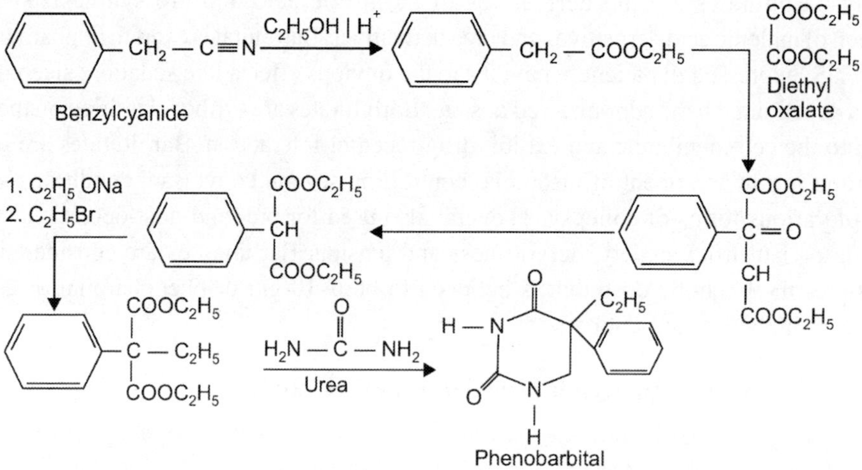

Fig. 5.15

2. Hexobarbital (Sombulex)®: 5-(1-cyclohexen-1-yl) 1, 5-dimethyl barbituric acid

Synthesis:

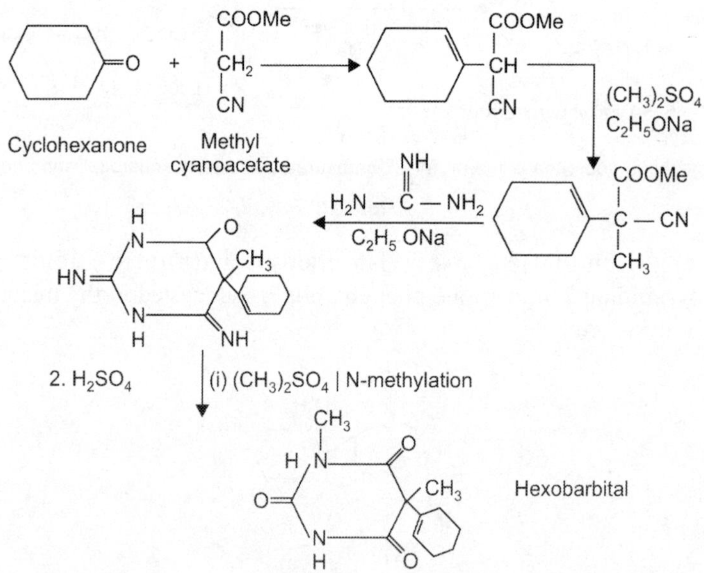

Fig. 5.16

Hexabarbital is a short acting barbiturate which undergoes within 15 minutes and hypnotic action remains for 1 to 2 hrs. It is used in the treatment of insomnia and pre-operative medication.

3. Mephobarbital (Mebaral)®: 5-ethyl-1-methyl-5-phenylbarbituric acid. Mephobarbital is used as a sedative for relieving anxiety and tension. It is also used in the treatment of major or minor epilepsy.

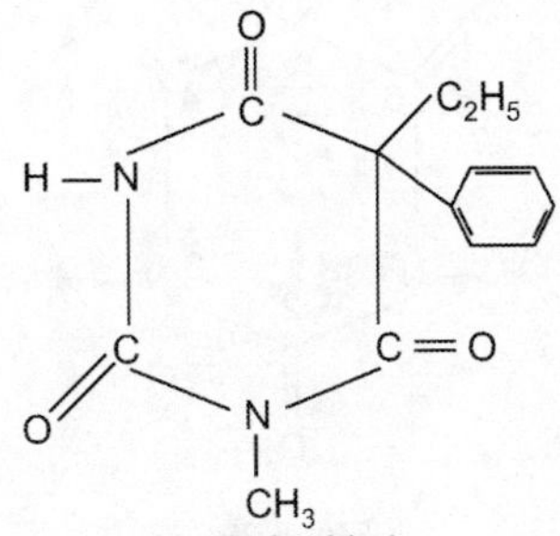

Mephobarbital

Fig. 5.17

4. Amobarbital (Amytal)®: 5-ethyl-5-isoamyl barbituric acid. Amobarbital is suggested as a sporific drug in various forms of insomnia and used as a sedative and an anticonvulsant drug.

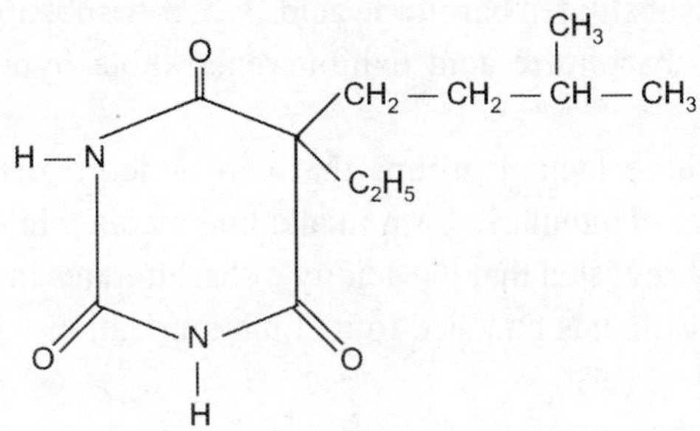

Fig. 5.18

5. Pentobarbital (Nembutal)®: 5-Ethyl-5-(2-amil)barbituric acid. It is a isomer of amobarbital. It is clinically useful as a relaxant and used for short-term insomnia.

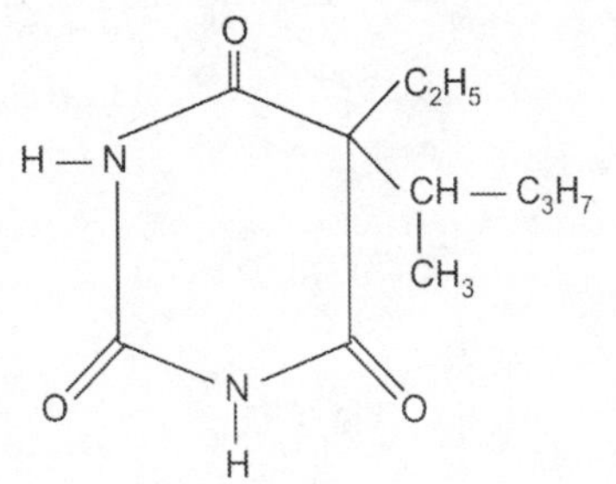

Fig. 5.19

5.6 SAR OF (BARBITURATES) SEDATIVE AND HYPNOTICS

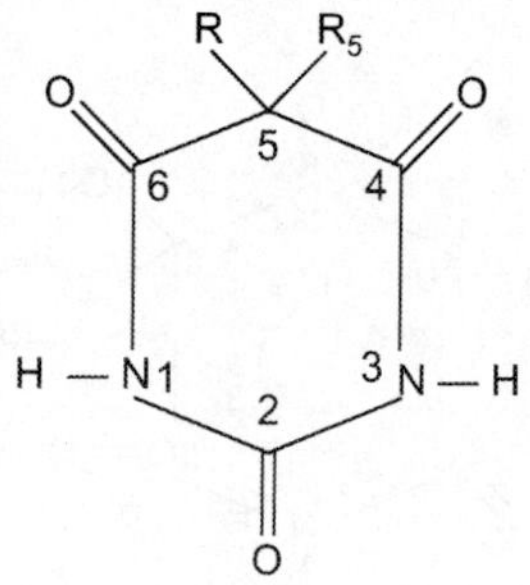

Fig. 5.20

1. PKa Aspects of Barbiturates/Structural Features
   (i) After several studies of synthetic approaches of barbiturates in 1951, scientist Sandberg postulated the fundamental point for exhibiting better hypnotic activity. Barbituric acid must be a weak acid and must have a lipid/water partition coefficient to a certain limit. This is only the reason that 5, 5-disubstituted barbituric acid, 1, 5, 5-trisubstituted thiobarbituric acid and 1, 5, 5-trisubstituted barbituric acid exhibit remarkable hypnotics, anticonvulsant and anaesthetic activities.
   (ii) There are possible three lactam groups that can undergo pH dependent lactim-lactam tautomerization—di-oxotautomaric form in alkaline media while trioxo tautomaric form in acid media. The study revealed that the acidity of barbiturates in acqueous solution depends on the number of substituents attached to it. Tautomerization is shown below.

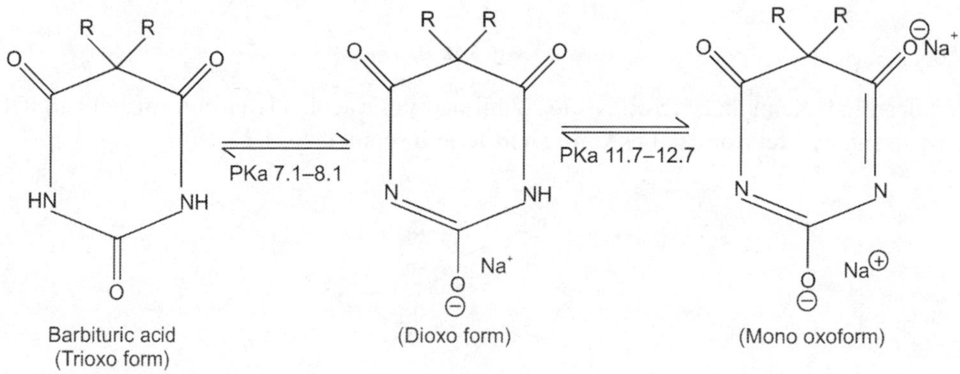

Fig. 5.21

2. 5, 5, Disubstitution
   (i) Introduction of homologue series of carbon atom at fifth position of carbon resulting, increases the lipophilic character of substituted barbituric acid.

(ii) Replacement of branching groups, unsaturation, alicyclic or aromatic substitution for alkyl substituents, as well as introducion of halogen derivatives, increase the lipid solubility of barbituric acid, subsequently easy penetration into blood brain barrier.

(iii) Introduction of polar group into alkyl substituents decrease the lipid solubility below the desirable expect.

3. Substitution on Nitrogen Group

(i) Introduction of imide hydrogen by alkyl groups leads to increase in lipid solubility, quick onset of action and shortening of duration of action.

(ii) Increase the size of alkyl substituent at nitrogen, further increase (C_1–C_3) is responsible for decrease hydrophilic character, further increase alkyl substituents, imparts convulsant character to barbiturates.

(iii) Alkyl substituents at both N^1 and N^3 render the drug non-acidic which make the inactive compound.

4. Modification of Oxygen Atom

(i) Replacement of oxygen at carbon 2 position by sulphur atom leads to increase lipid soluble character and increase onset of action, e.g., thiopental is used as intravenous anaesthetics.

5.7 BENZODIAZEPINES

All benzodiazepines show sedative effect on low doses. Even in higher doses they induce sleep and further coma. The sedative effect caused by benzodiazepines impaires motor skills, memory and judgement. Benzodiazepines have advantages over barbiturates. They exert multiple pharmacological effects on CNS including sedation, hypnosis, decrease anxiety, muscle relaxation and anticonvulsant activity. Benzodiazepines enhance GABAergic transmission in all CNS structures. The major advantages of benzodiazepines that they show less tolerance and dependency than other classical sedative and hypnotics (barbiturates) and are safer at larger dose.

(i) Diazepam (Valium)®: 7-chloro-1, 3-dihydro-1-methyl-5-phenyl-1, 4-benzodiazepine-2-one. Diazepam is used for the several anxiety states and is useful in the treatment of convulsants, as premedication in anaesthesiology.

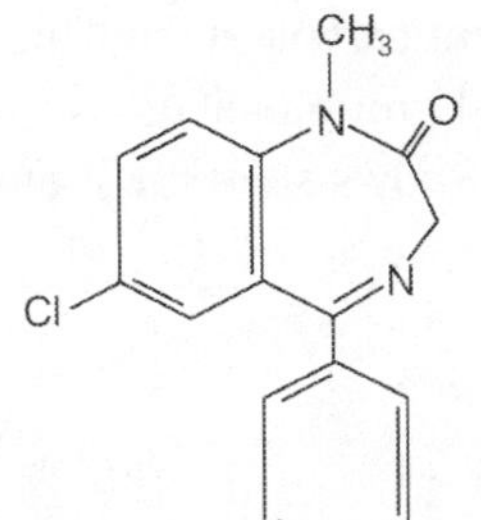

Fig. 5.22

Synthesis:

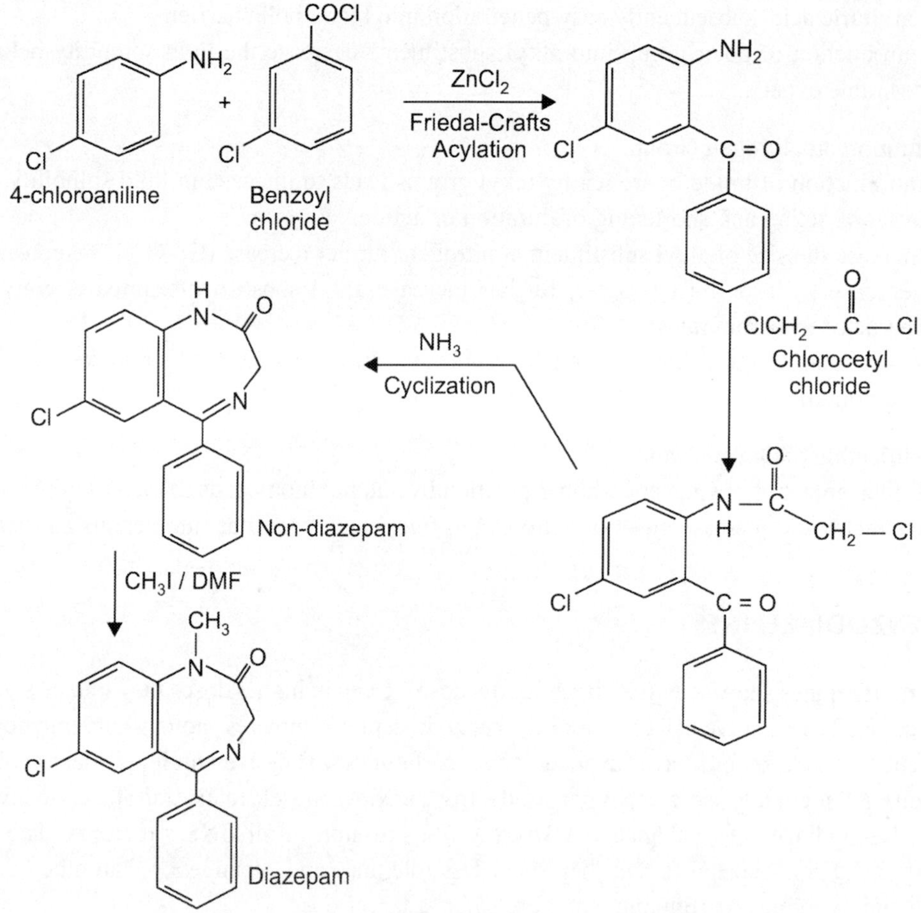

Fig. 5.23

2. Triazolobenzodiazepines

Triazolobenzodiazepines is the class of benzodiazepines. This category of drugs contains triazole ring in association of N^1 and its adjacent carbon atom. This class of sedative and hypnotics show remarkable activity. Presence of methyl group undergoes oxidative metabolism to methyl alcohol followed by conjugation. Therefore, it shows short duration of action:

Synthesis:

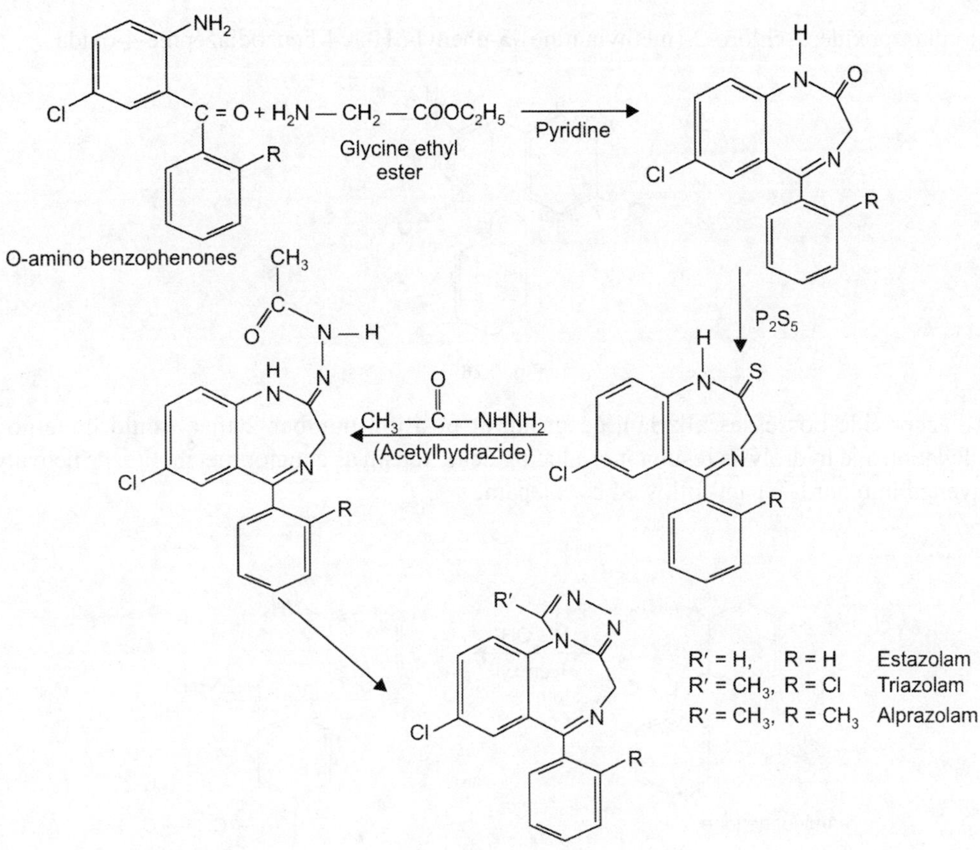

Fig. 5.24

Alprazoam (Xanax)®: 8-chloro-1-methyl-6-phenyl-4H,S. triazolo [4, 3-a] [1, 4] Benzodiazepine

Triazelam: 8-chloro-6-(2-chlorophenyl)-1-methyl-4 H,S-[4, 3-a] [1, 4] Benzodiazepine

3. Oxazepam (Serax)®: 7-chloro-1, 3-dihydro-3-hydroxy-5-phenyl-2H-1, 4-Benzodiazepine-2-one

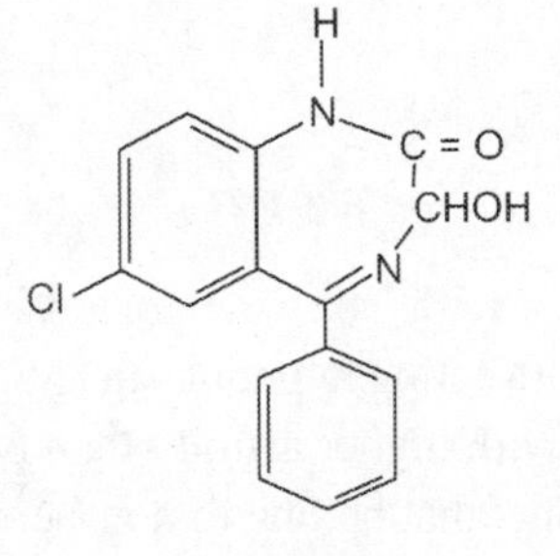

Fig. 5.25

Oxazepam is used as sedative and hypnotics. Its duration of action is short.

4. Chlordiazepoxide: 7-chloro-2-(methylamino)-5-phenyl-3H-1, 4-benzodiazepine-4-oxide

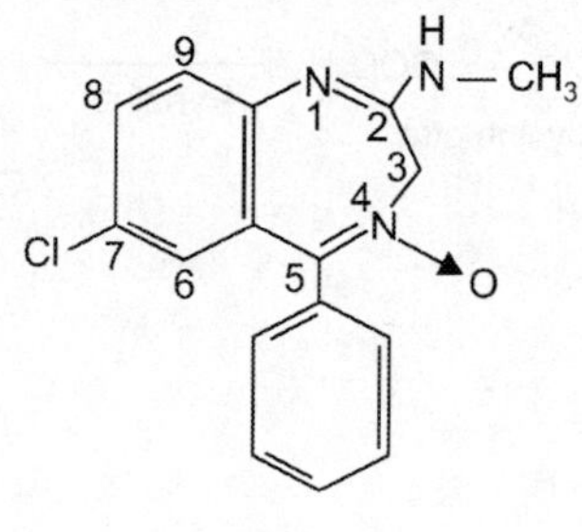

Fig. 5.26

Chlordiazepoxide possesses alkylamine group at position number 2 that would undergo N-demethylation and hydrolysis resultant producing demoxepam as a major metabolite. Principally, it is converted into nordazepam followed by oxepam.

Synthesis:

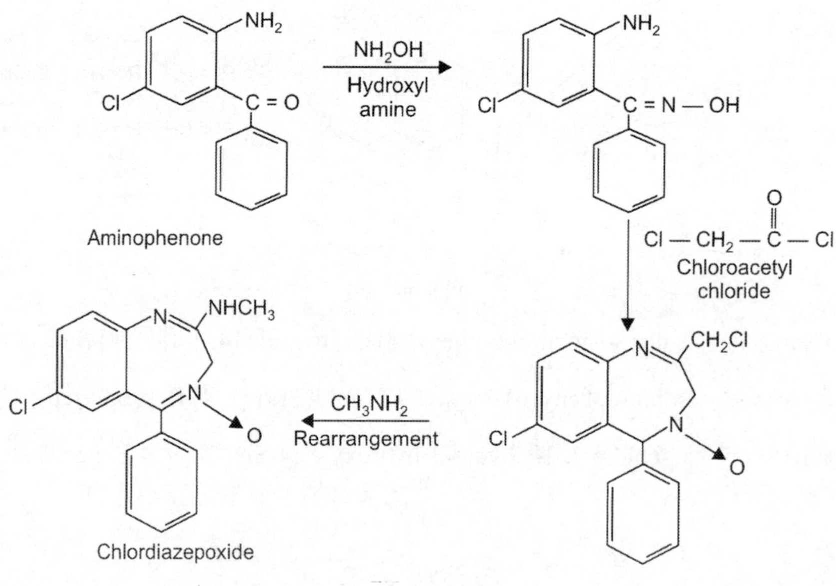

Fig. 5.27

1. Non-Benzodiazepines Derivatives

Non-benzodiazepines are used as short-acting hypnotics in U.S. and Europe. It is believed that these derivatives act on $GABA_A$ receptor with higher affinity than benzodiazepines. There are two drug examples, i.e., zolpidem and zopiclone, that belong to non-benzodiazepines class.

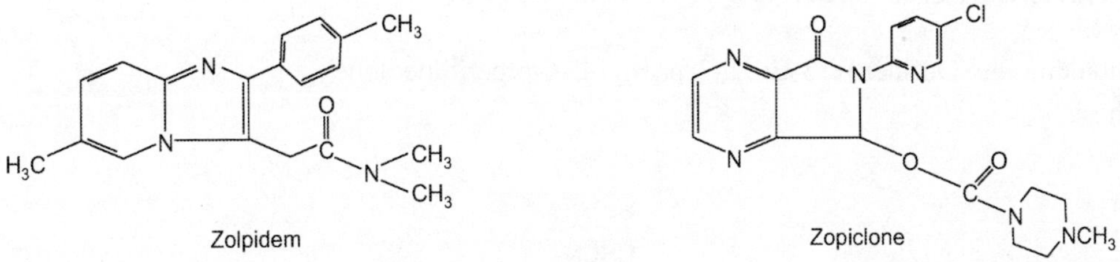

Fig. 5.28

5.8 SAR OF BENZODIAZEPINES

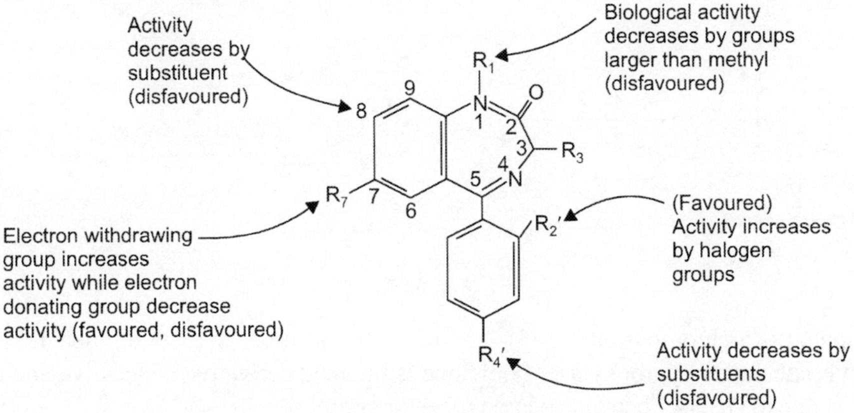

Fig. 5.29: Basic pharmacophore of benzodiazepines.

(i) As the Figure 5.29 indicates N-substituents at position 1 should be small. Larger group than methyl may decrease the activity.

(ii) Additional rings (triazole or imidazole) derivatives joining the diazepine nucleus at position 1 and 2 are supposed to be active.

(iii) Substitution of alkyl at position 3 of azepine ring generally decreases activity while introducing polar groups (–OH, COOH). It may retain the activity but may have short half life.

(iv) Saturation of 4, 5 position or shifting of double band to 3, 4 position, generally decreases the activity.

(v) Oxo group at position 2 is essential for activity.

(vi) Phenyl group or pyridyl group (bromazepam) at position 5 is essential for activity substituted with electron attracting group at position 2' of phenyl ring or 2'-6'-leads to increase in activity. On the other hand 3'-4'-substituents lead to decrease activity.

(vii) Introduction of electron withdrawing groups (Cl, Br, F, NO_2) at position 7, is the requirement for hypnotic activity while electron donating groups lead disfavoured for activity.

(viii) Positions 6, 8 and 9 should retain for unsubstitution for activity.

5.9 HETEROCYCLIC SEDATIVE AND HYPNOTICS

(1) Glutethimide (Doriden)®: 3-Ethyl-3-phenyl-2, 6-piperidinedione
Synthesis

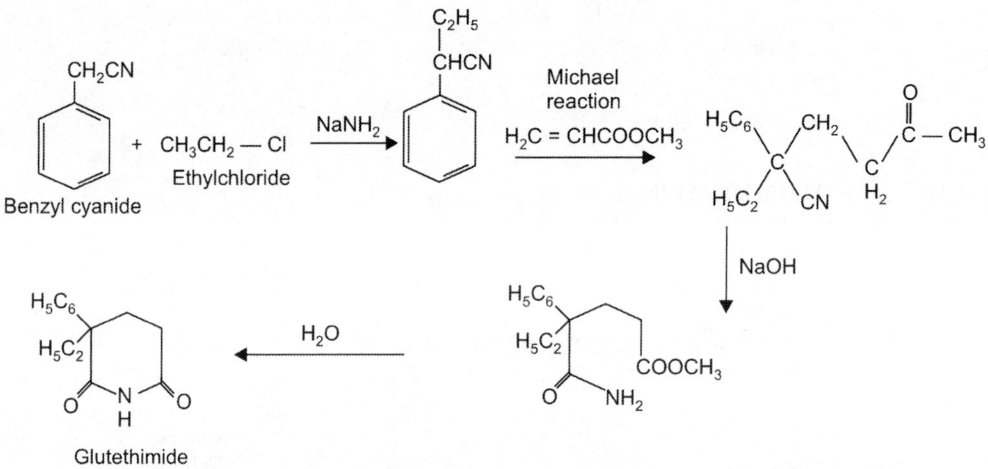

Fig. 5.30

1. Zopiclone: [8-(5-chloropyridine-2-yl)-7-oxo-2, 5, 8 triazabicyclo [4, 3, 0] nona-1, 3, 5-trien-9-yl] 4-methyl piperazine-1-carboxylase. Zopiclone is the new derivative of sedative and hypnotics. Its activity is due to $GABA_A$ benzodiazepine receptor complex.

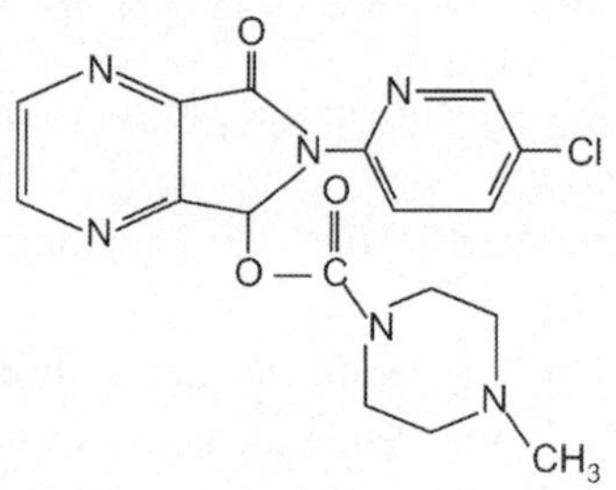

Fig. 5.31

5.10 HALOGENATED SEDATIVE AND HYPNOTICS

(1) Chloral Hydrate: (Sonos)®: 2,2,2 trichloro-1,1-ethandiol. Chloral hydrate is used as a sedative and hypnotics. The metabolic formation of chloral hydrate is trichloro ethanol which is synthesized after its reduction into the tissue. It is used in insomnia and as an altermative to benzodiazepine.

Synthesis:

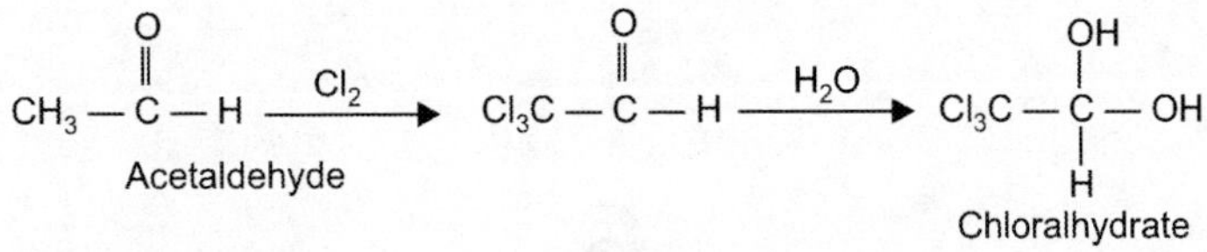

Fig. 5.32

5.11 RECENT DEVELOPMENT OF SEDATIVE AND HYPNOTICS

1. Melatonergic Agents

Melatonergic agents are new arrivals in the field of sedative and hypnotics. Melatonis is an endogenous hormone and biosynthesized in pineal glands. It is believed that it regulates the body circadian rhythm (sleep wake cycle). It acts on GABA receptor in forebrain membrane.

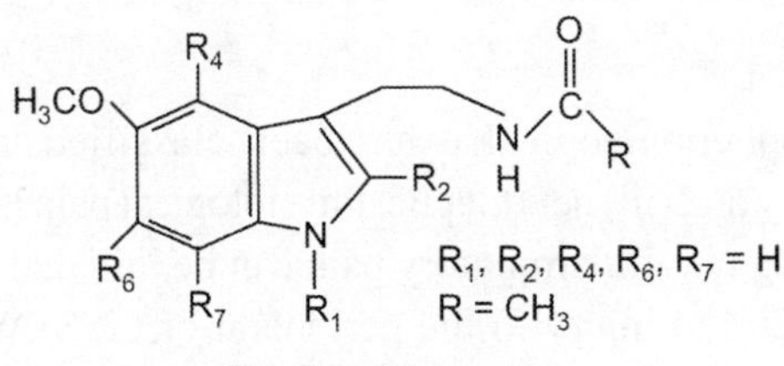

Fig. 5.33: Melatonine.

Indole containing and non-indole containing analogues of melatonine have been established, that are act on MT_2 (melatonine receptor) selective ligands and exhibit sedative and hypnotic activity.

EXERCISE

1. What are sedative and hypnotics. Write SARoF Barbiturates.
2. Write the synthesis of the following drugs.
   (i) Phenobarbitone
   (ii) Diazepam
3. Write SAR of benzodiazepine with examples.
4. Write the synthesis of chloridiazepoxide and mechanism of action and uses.
5. Write a note on ultrashort acting barbiturates.
6. Write the general synthesis of barbiturates.
7. Differentiate between sedative and hypnotics.
8. Explain the structural features of barbiturates.

6

Opioid Analgesics

6.1 INTRODUCTION

Pain is the most important phenomenon that has been classified into the following types (i) physiological, (ii) inflammatory and (iii) neuroleptic. Physiological pain is most common, for example, touching a hot object or getting cut. Inflammatory pain can be initiated by infection or tissue injury and neuroleptic pain occurs due to injury to the peripheral or CNS. Within these classes of pain, there are different levels of pain. These include acute, chronic, cancer, orthopathy, visceral, neuroleptic and diabetic pain. Thus, three major classes of drugs used to manage pains are opioids, NSAIDS (nonsteroidal anti-inflammatory drugs) and analgesic adjuvant (acetomenophen).

The term opioid was introduced to describe all the compounds both nonpeptide and peptides (Leucin Enkephalin and Methionine Enkephalin) with opiates. These peptides are structurally distinct from alkaloid opiates. Opioid analgesics are supposed to bind with their receptors at molecular level. Major opioid receptors are delta (δ) followed shortly mu (μ) and kappy (κ). Americal Pain Society advocates the uses of opioids analgesics in the treatment of chronic pain and cancer pain. Opioid agonists also have antitussive action attributed to the depression of cough reflex.

6.2 OPIOID RECEPTORS

Effect of narcotic analgesics is usually explained in terms of their interaction with specific opiate receptors (located in medial thalamus of CNS), are initiated the processes of deep chronic and burning pain.

The receptors are following:

(1) Mu (μ): μ receptors are responsible for supraspinal analgesia, euphoria, respiratory depression and physical dependence effect. They are subclassified in μ_1 and μ_2 receptors.

(2) Kappa (κ): These receptors are responsible for mediate spinal analgesia, sedative action and mitotic effects. They are further classified into κ_1, κ_2 and κ_3 receptors.

(3) Sigma (σ): These receptors are known as opiate receptors. These receptors are responsible for hallucinogic and cardiac stimulants properties.

(4) Delta (δ): Delta receptors are responsible for analgesia, euphoria and physical dependence. Delta receptors are further classified into δ_1, δ_2 and δ_3 receptors.

6.3 OPIOID PEPTIDES

A number of peptides located in the brain have tremendous pharmacological action similar to those of morphine. Although these peptides are not structurally similar to morphine but similar in mode of action of morphine. There are some neuronally located proteins to which opiates binds to initiate the pharmacological action (analgesic effect) are called opioid receptors. Some pentapeptides (isolated from mammalial tissue) exhibit opioid like action. There are met-enkephlin (Tyr-Gly-Gly-Phe-Met) and Lev-Enkephalin (Tyr-Gly-Gly-Phe-Lev) respectively. However, other endogenous peptides (beta-endorphine and dynorphin) were isolated from mammalian brain tissues. These are endogenous peptides. While exogenous alkaloids are opiods, alkaloids and their synthetic derivatives. Other μ-receptor agonist are hexapeptides and beta-casomorphrin (Tyr-Pro-Pro-Gly-Pro-Lle) was isolated from cow's milk.

6.4 CLASSIFICATION

1. Morphine and Its Derivatives

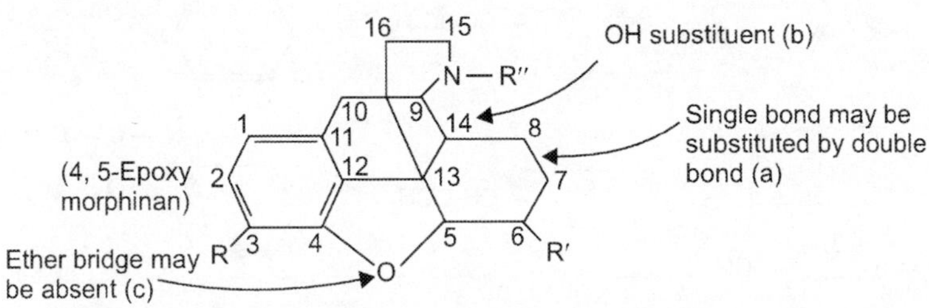

Fig. 6.1: Basic pharmacophore of morphines.

Derivatives	R	R'	R"	Other possible substitution
Morphine	—OH	—OH	$-CH_3$	—
Heroine	$-OCOCH_3$	$-OCOCH_3$	$-CH_3$	—
Hydromorphone	—OH	=O	$-CH_3$	(a)
Oxymorphone	—OH	=O	$-CH_3$	(a) (b)
Levorphanol	—OH	—H	$-CH_3$	(a) (c)
Codeine	$-OCH_3$	—OH	$-CH_3$	—
Hydrocodone	$-OCH_3$	=O	$-CH_3$	(a)
Oxycodone	$-OCH_3$	=O	$-CH_3$	(a) (b)
Buterphanol	— OH	— H	$-CH_2-$◇	(b) (c)
Nalbuphine	—OH	—OH	$-CH_2-$◇	(a) (c)

2. Piperidine analogues

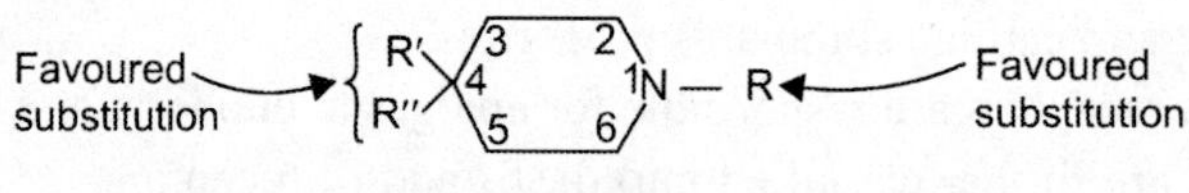

Fig. 6.2

Derivative	R	R'	R''	Other possible substituents
Meperidine	$—CH_3$	$—C_6H_5$	$—COOCH_3$	—
Bemidone	$—CH_3$	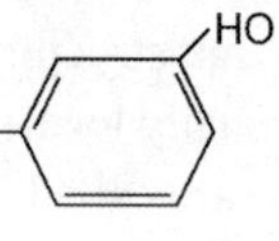	$—COOC_2H_5$	—
Ketobemidone	$—CH_3$	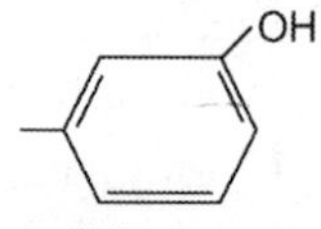	$—COC_2H_5$	—
Properidone	$—CH_3$	$—C_6H_5$	$—COOCH(CH_3)_2$	—
Alphaprodine	$—CH_3$	$—C_6H_5$	$—OCOC_2H_5$	$3—CH_3$
Fentanyl	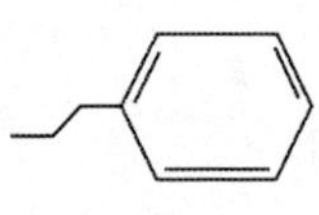	H	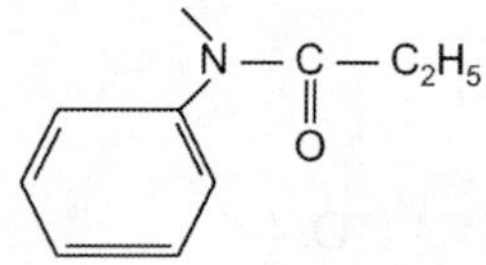	—
Lofentanil	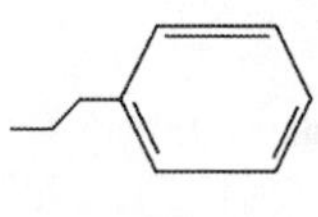	$—COOC_2H_5$	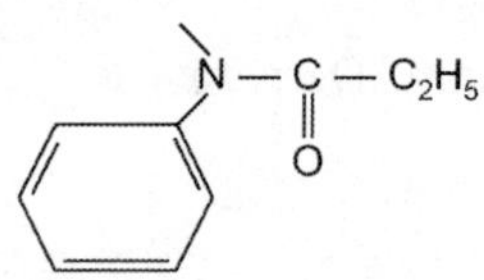	$3—CH_3$
Sufentanil (General anaesthetic)	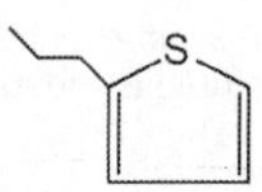	$—CH_2OCH_3$	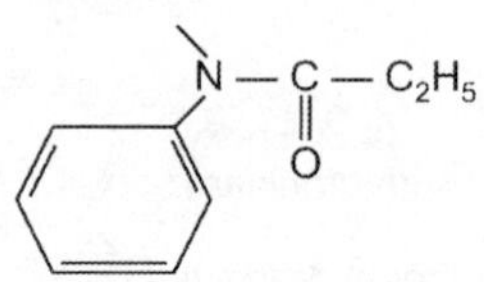	
Diphenoxylate (Antidiarrhoeal)	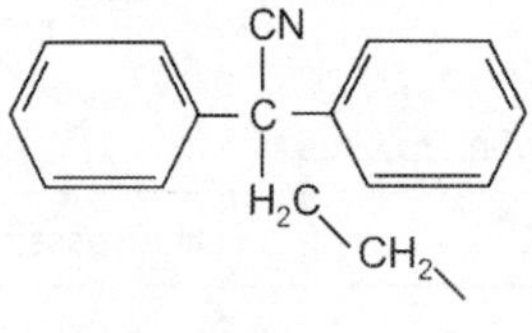	$—C_6H_5$	$—COOC_2H_5$	—
Loparamide (Antidiarrhoeal)	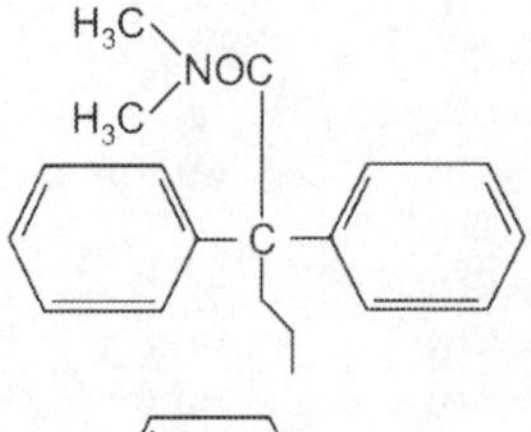	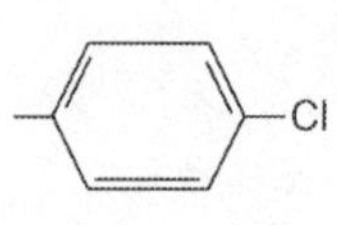	—OH	—
Anileridine	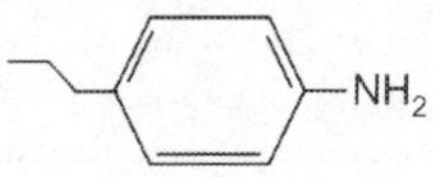	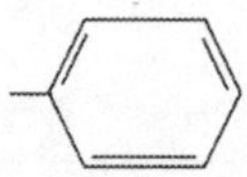	$—COOC_2H_5$	—

3. Diphenyl Haptanones Analogues

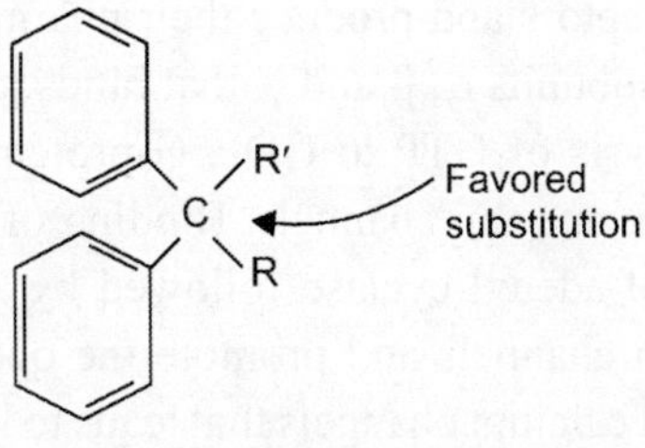

Fig. 6.3: Basic structure of diphenyl haptanones.

Derivative	R	R′
Methadone	$—COC_2H_5$	$—CH_2CH—N(CH_3)_2$ with CH_3 on CH
Isomethadone	$—COC_2H_5$	$—CH—CH_2—N—(CH_3)_2$ with CH_3 on CH
Normethadone	$—COC_2H_5$	$—CH_2—CH_2—N(CH_3)_2$
Dipanone	$—COC_2H_5$	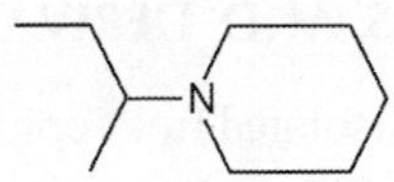
Phenadoxone	$—COC_2H_5$	
Dextromoramide	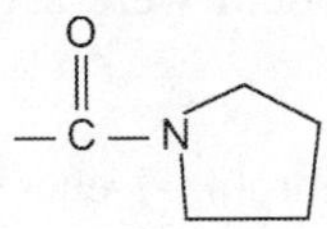	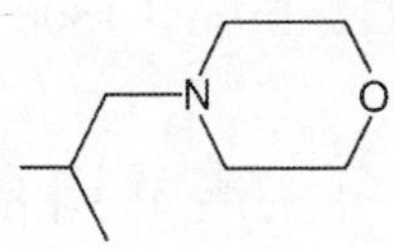

4. Benzazocin Derivates

Derivatives	R	R′
Pentazocin	$—CH_2—CH=C(CH_3)_2$	H
Phenazocin	$—CH_2—CH_2—C_6H_5$	H
Cyclazocin	$—CH_2—$	H
Ketazocin	$—CH_2—$	= O

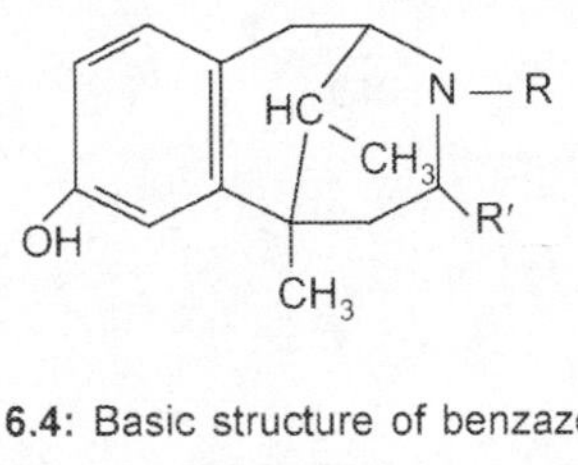

Fig. 6.4: Basic structure of benzazocin.

6.5 MECHANISM OF ACTION (SIGNAL TRANSDUCTION)

Opioid are G-protein coupled receptors and produce their pharmacological action. G-proteins are heterotrimers, which contain the subunits α, β and γ, that binds to Guanine Nucleotides to their α-subunit and promotes the hydrolysis of GTP to GDP. G-protein coupled interaction with opiate receptors including adenyl cyclase and ion channels. Binding of opioid receptors agonist to these receptors attribute to inhibition of adenyl cyclase followed by reduction in cAMP content. This binding initiates the effect on ion channels and promote the opening of potassium channels and thereby inhibition of voltage gated calcium channels that leads to hyper polarization and consequent reduction in excitement of neural activity.

6.6 CENTRAL SIDE EFFECTS OF OPIOID ANALGESICS

1. Respiratory depression
2. Tolerance, dependence and addiction liabilities.
3. Sedation and cognitive impairment.
4. Nausea, vomiting and constipation
5. Constipation, itching, contraindication and interaction with cytochrom P450 enzymes.

6.7 MORPHINES AND DERIVATIVES

Morphine was first isolated from opium in 1805 by Serturner. He suggested name Morphius (Greek, God of Sleep and Dreams). Further 3-methyl ether codein was isolated from poppy (*papaver somniferum*) in 1832. In mid-1800 morphine and codein were actively used in the medicinal preparation.

1. Morphine (Durapinue)®: 4, 5-Epoxy-17-methyl morphin-7-ene-3, 6-diol, morphine salt wth hydrochloric acid prepared by neutralization process with hydrochloric acid followed by

Synthesis:

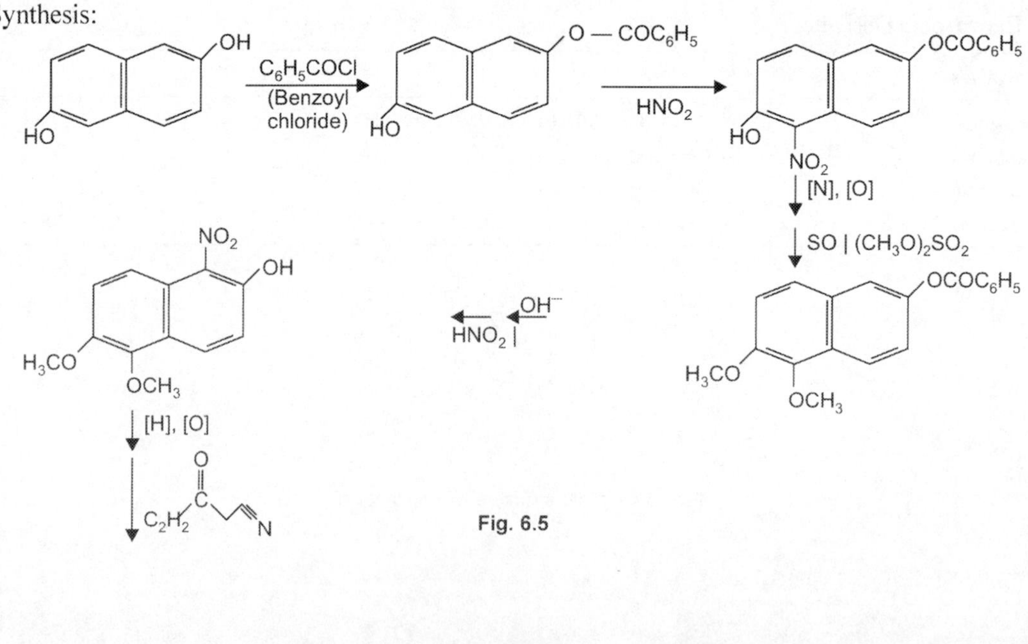

Fig. 6.5

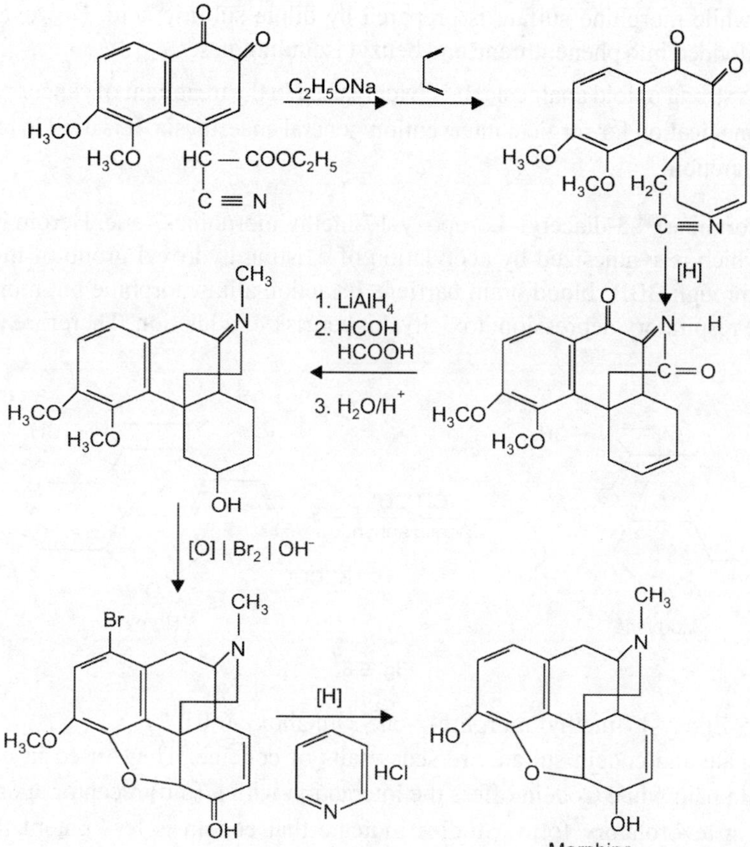

Fig. 6.6

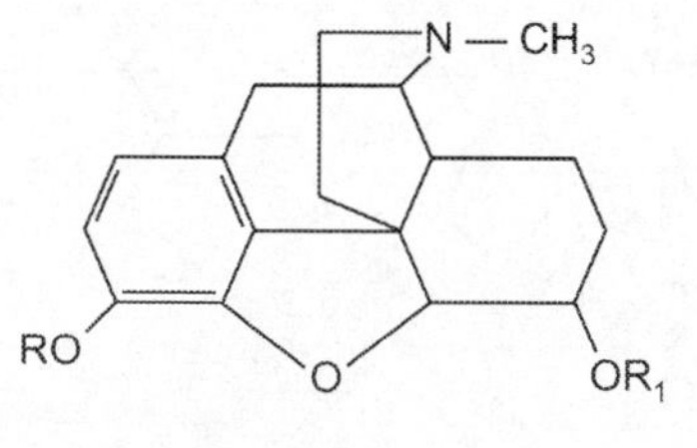

Phenenthrene nuleus
morphin R = R_1 = H

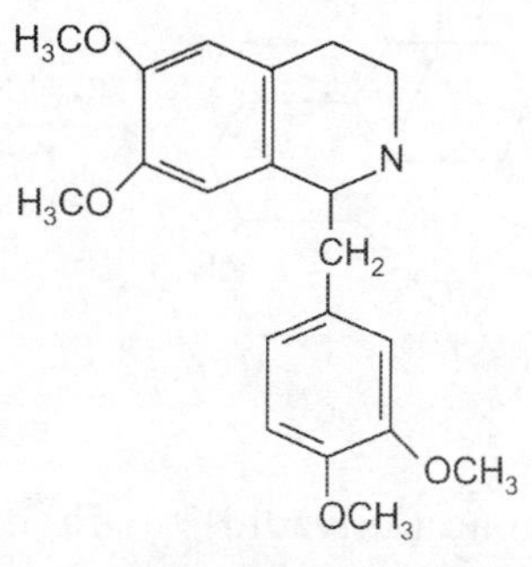

Benzyl iso quinoline nucleus
(papavarine)

Fig. 6.7

crystallization while morphine sulfate is prepared by dilute sulfuric acid. However, morphine is chemically subdivided into phenenthrene and benzyl isoquinolines.

Morphine is strong opioid analgesic. It is widely used in the treatment of cancer pain, in surgery. Preanaesthetic medication for surgical intervention/general anaesthesia. It is used in relieving pain in mycocardial infarction.

2. Heroin (Dimorphine)®: 3-diacetyl-4.5-epoxy-17-methylmorphine-7-ene. Heroin is another form of morphine, which is synthesized by acetylation of existing hydroxyl group of morphine. It has ability to pass through BBB (blood brain barrier). Its action alike morphine but moreover narcotic side effects like respiratory depression, toxicity, higher risk of addiction. Therefore it is prohibited.

Synthesis:

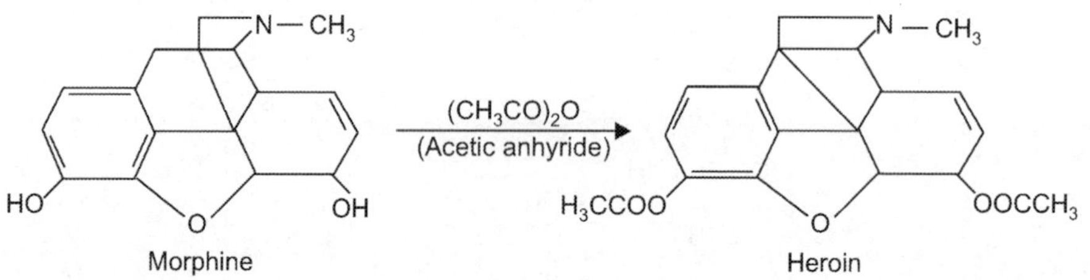

Fig. 6.8

3. Codeine: 4, 5-Epoxy-17-methyl morphin-7-ene-3-methoxy-6-01
Codeine phosphate and codein sulfate are sedas salts of codeine. They used in codein sulphate moderate to mild pain while Codein offers the interaction with μ (mu) receptor agonist. Generally, codein occurs in levorotatory form. Studies indicate that codein is less potent than morphine. Codein is used in the cough preparation as an antitussive agent.

Synthesis: (Semisynthetic method)

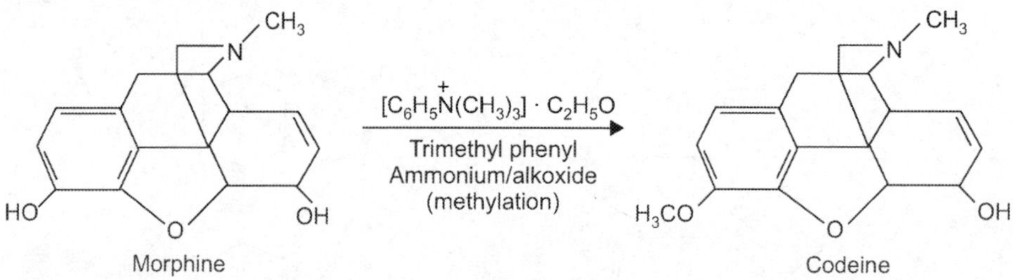

Fig. 6.9

4. Hydromorphone (Dilvadid)®: Dihydromorphine (4, 5-epoxy-3-hydroxy-N-methyl-6-oxymorphinan). Hydromorphone is synthetic derivative of morphine. It is synthesized by catalytic hydrogenation and dehydrogenation of morphine. Due to higher solubility than morphine, it is available in injected fluid. It is used in the same ways as morphine.

Synthesis

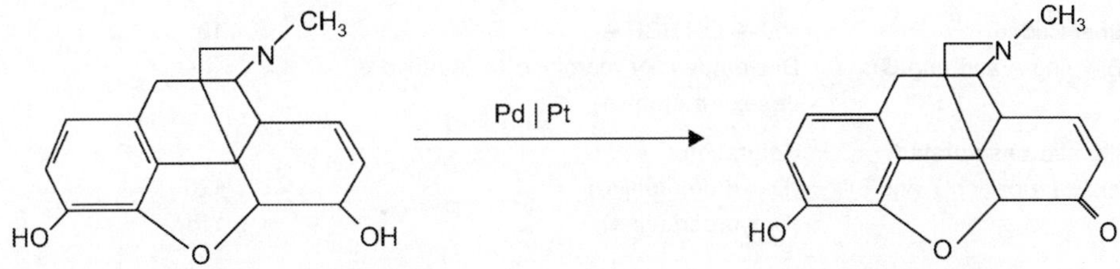

Morphine Hydromorphone

Fig. 6.10

5. Oxymorphones (Numorphan)®: (–)14-hydroxydihydromorphinone hydrochloride. Oxymorphone is used as a salt with hydrochloric acid. It is as effective as morphine is in 1/8 to 1/10 of the dosage. It has good duration of action but higher risk of addiction. It is widely used for postoperative Pain.

6. Tramadol (Vltran)®: (±)-cis-2-[dimethylaminomethyl]-1-(m-methoxy-phenyl) eyclohexanol hydrochloride. Tramodol is used as a salt with hydrochloric acid. Tramodol undergoes O-demethylated metabolites which is 6 times more potent than morphine. It is used in the treatment of moderate to savere pains.

6.8 SAR OF MORPHINES DERIVATIVES

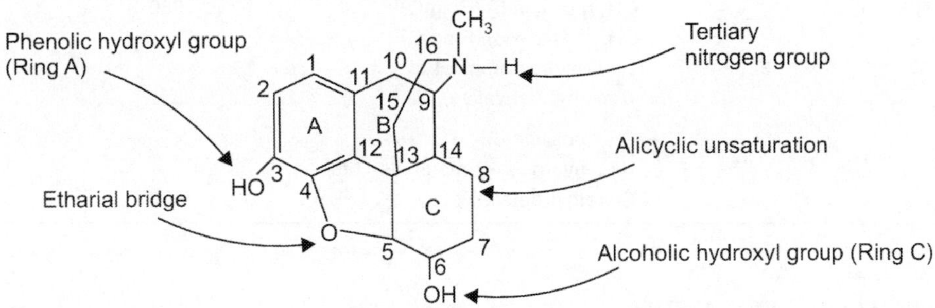

Fig. 6.11: Morphine nucleus.

S. No.	Attached peripheral groups of morphine	Possible modifications	Analgesic activity
1.	Ring-A Phenolic hydroxyl group	— OCH_3 (codeine)	15
		— OC_2H_5 (Ethylmorphine)	10
		— OCH_2 — CH_2 — N O (Pholocodine)	1
2.	Ring-C Alcoholic hydroxyl group	— OCH_3 (Heterocodein)	500
		= O (morphinone)	420
		= O (Dihydromorphine to dihydromorphnone)	600

		— H (dihydromorphine to dihydrodesoxy morphine-D)	1000
3.	Ether Bridge (B/w ring A and ring C)	= C — OH HCH — Dihydrodesoxy morphine to tetrahydro desoxy morphine)	13
4.	Alicyclic unsaturated linkage (position 7 and 8)	Saturation	
		(Dihydromorphine)	120
		(Dihydrocodeine)	115
5.	Tertiary nitrogen	$>N — H$	5
		$>N — CH_2 — CH_2 — C_6H_5$	1400
		$>N^{\oplus}(CH_3)_2\ Cl^-$	1 (strong curare action)
		$>N$ — (Propyl, Isobutyl, Allyl, Methalyl)	Morphine antagonist
6.	Centralised substitution	— NH_2 (at position-2) (Ring-A)	Decrease activity
		— Halogen (at position-1) (Ring-A)	50
		— OH (position-14) (Ring-A)	250–530
		— CH_3 (position-6) (Ring-C)	280
		CH_3 (position-6) (Ring-C) (in Dihydromorphine) to 6-methyl derivative)	33
		— CH_3 (position-6) (in dihydrodeoxy morphine-D to 6-methyl derivative)	490

6.9 PIPERIDINE DERIVATIVES

Piperidine derivaties regarded as A, D ring analogues of morphine. It has been proved that piperidine analogue heperidine is μ receptor agonist with ¼ potency of morphine.

1. Meperidine (Pethidine)® (Demerol)®: Ethyl-1-methyl-4-phenyl-isonipecoate/1-methyl-4-phenyl-peperidine carboxylate.

Meperidins or pethidine is a salt of hydrochloric acid. It was considered a spasmolytic agent but studies found them to have far greater analgesic activity. Meperidine also exhibits anticholinergic activity due to structural similar to morphine. Therefore, it causes histamine release and spasm of smooth muscles. It lacks antitussive properties. It has $1/8^{th}$ activity of morphine. It is widely used in premedical and stabilizing anaesthesia.

Synthesis:

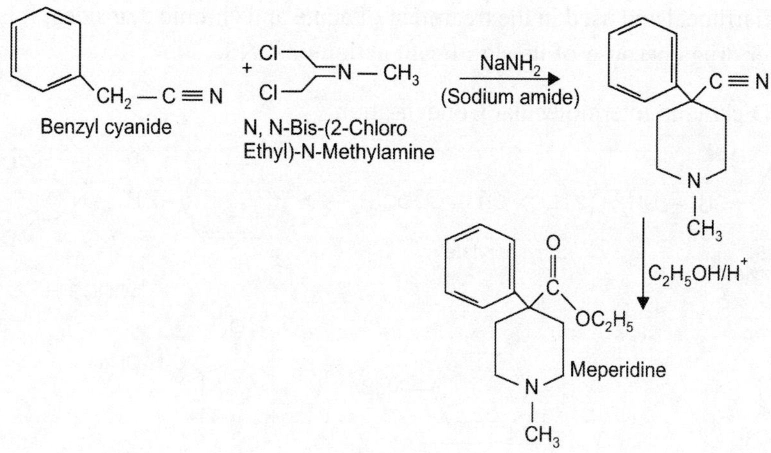

Fig. 6.12

2. Ketobomidone: 1-[4-(3-hydroxyphenyl)-1-methyl-4-piperidyl]-propane-1-one
Ketobomidone is used as potent analgesic.

Synthesis:

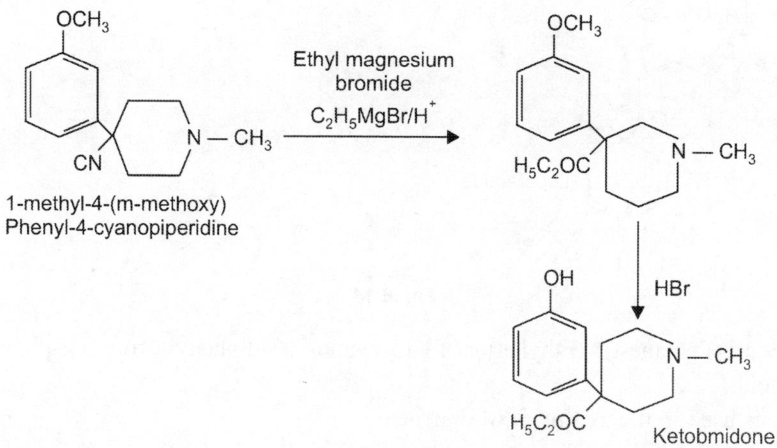

Fig. 6.13

3. Anileridine (Leritine)®: 1-(p-aminophenethyl)-4-phenyliso nipecate.
Anileridine is also used as salt of hydrochloric acid. It is more active than meperidine. It has advantage over meperidine due to its less dependency. Hence, it is used as suitable substitutes for meperidine.

4. Loperamide (Imodium)®: 1-(4-chlorophenyl)-4-hydroxy-N, N-dimethyl-α, α-diphenyl-1-piperidine butyramide.

Loparamide is considered a hybride of methadone nucleus as well as meperidine nucleus. Loparamide acts as an antidiarrhoeal and used in the treatment of acute and chronic diarrhoea. It is also listed in over the counter drugs because of its significant action on CNS.

Synthesis: By Dickmann Intermolecular Condensation

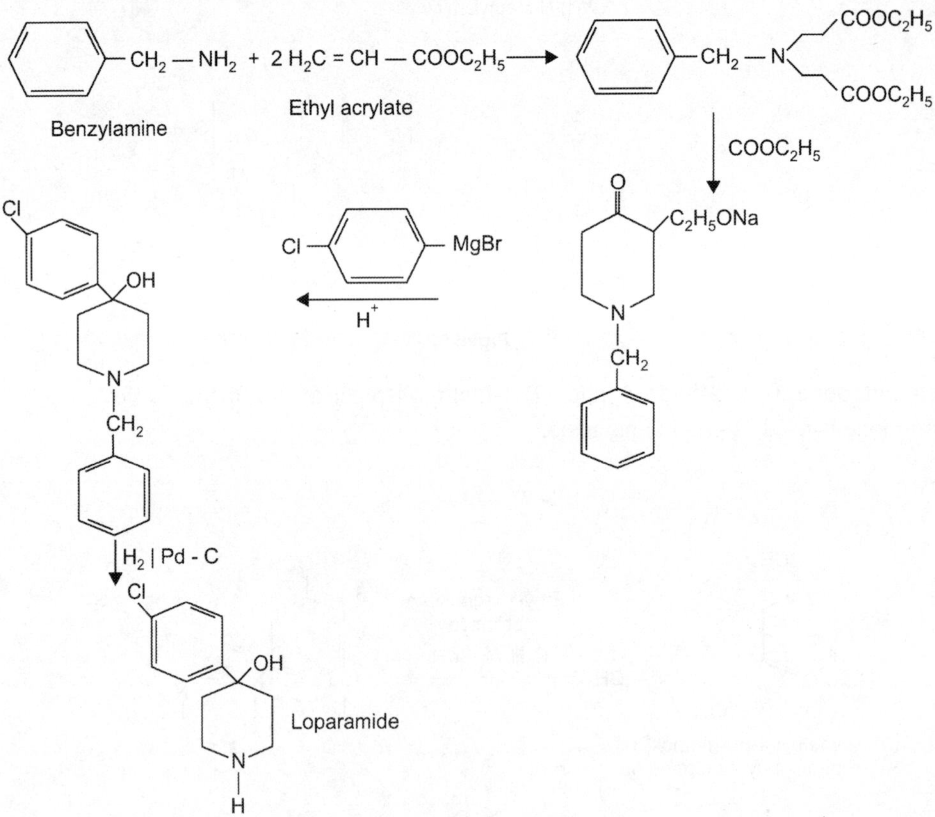

Fig. 6.14

(5) Diphenoxylate (Fentanest)®: Ethyl ester of 1-(3-cyano-3, 3-diphenyl-propyl)-4-phenyl piperidine-4-carboxylic acid.

Diphenoxylate is used in the treatment of diarrhea.

Synthesis:

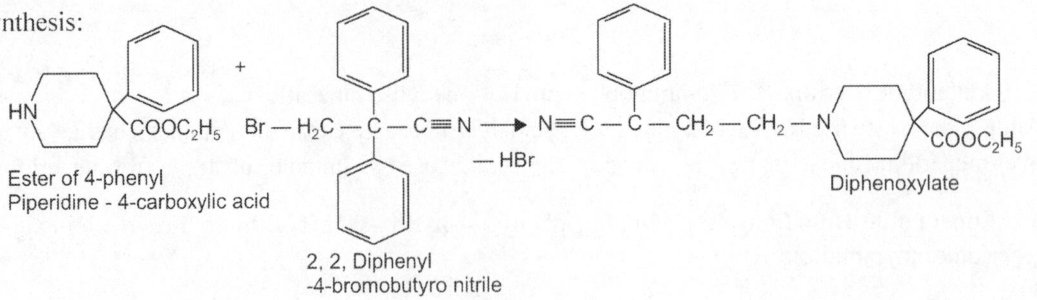

Fig. 6.15

6. Lofentanyl: Methyl-(3S-4R)-1-(2-cyclohexylethyl)-4-(cyclohexyl-propanoylamino)-3-methyl-piperidine-4-carboxylate.

Lofentanyl is 8,400 times more potent than meperidine and used as a potent analgesic.

7. Alfentanyl (Alfent®): N-[1-[2-(4-ethyl-4,5-dihydro-5-oxo-1H-tetrazol-1-yl)ethyl]-4-(methoxymethyl)-4-piperidinyl]-n-phenyl-propionamide.

Alfentanyl is a potent analgesic and used as a primary anaesthetic agent along with barbiturates.

Synthesis:

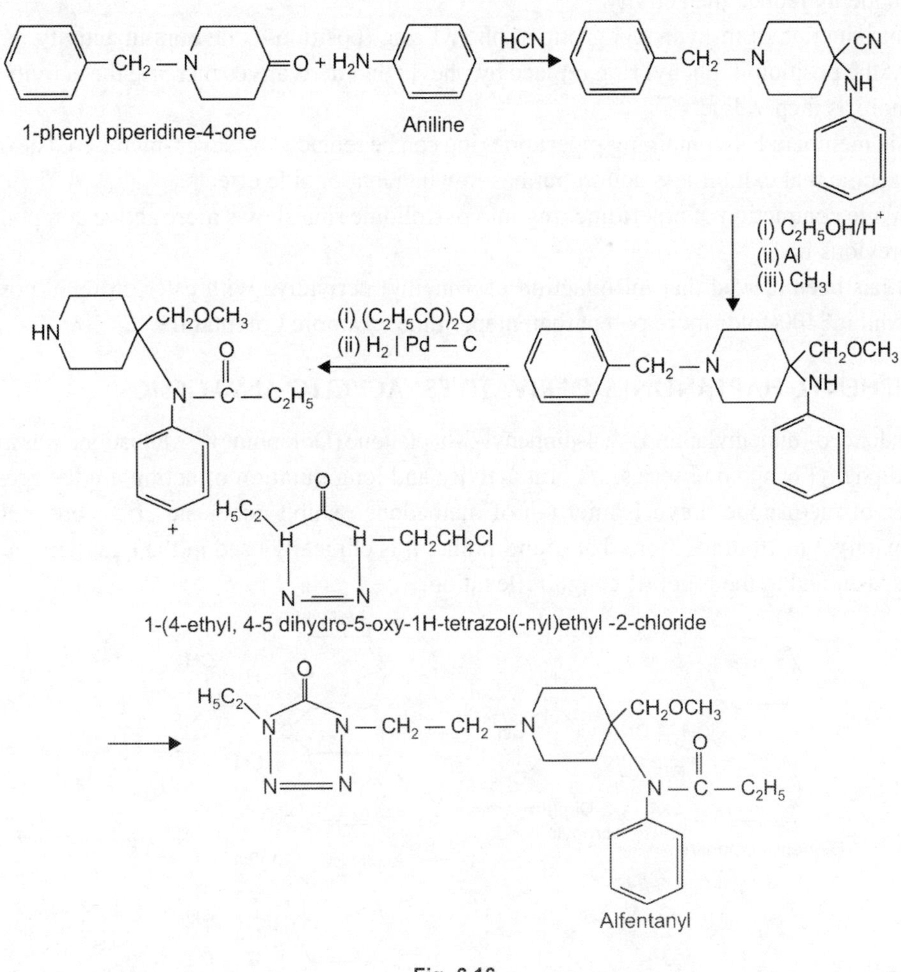

Fig. 6.16

6.10 SAR OF PIPERIDINES DERIVATIVES

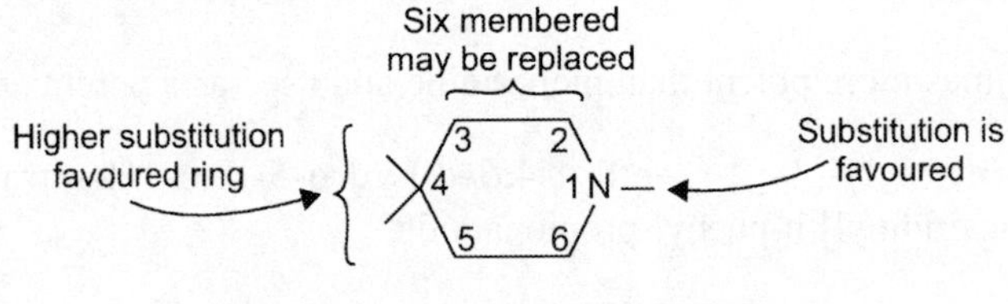

Fig. 6.17: Basic nucleus of piperidines.

1. It has been observed that replacement of 9-phenyl group of meperidine by any other groups, suddenly reduce the activity.
2. Introduction of m-hydroxyl group at phenyl ring (position-4) favours in activity.
3. At the position-4, phenyl ring replace by pheyl ethyl derivatives increase the activity 3 times more as meperidine.
4. Six membered N-containing piperidine ring can be replaced by seven-membered hexahydro-azepine that exhibit less action but has low incident of side effects.
5. While contraction of piperidine ring into pyrrolidine ring shows more active compound than previous one.
6. It has been proved that introduction of 3-methyl derivative with ester group at position 4, exhibit 8400-folds more potent than meperidine, example Lofentanil.

6.11 DIPHENYL HAPTANONES DERIVATIVES (ACYCLIC ANALGESICS)

1. Methadone: 6-(dimethylamino)-4, 4-diphenyl-3-haptanone (Dolophine)®: Methadone was a pioneer drug of diphenyl haptanone series. Its oral activity and long duration of action are the prestigious properties of methadone. Levo isomer (–) of methadone exhibit analgesic effect but methadone shows toxicity 3 to 10 times more that of morphine. It is clinically used in the treatment of urinary bladder spasm and in the relief of cough reflexation.

Synthesis:

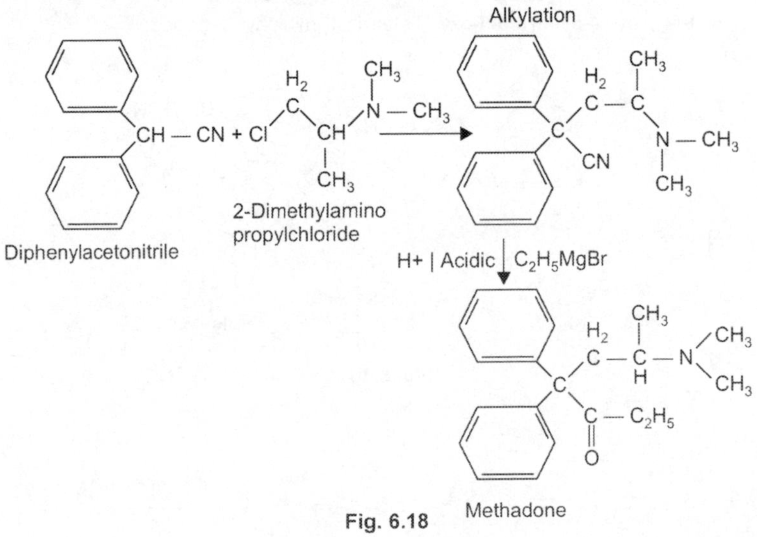

Fig. 6.18

2. Dextromoramide: (3 R)-3-methyl-4-morpholin-4-yl 1-2, 2-diphenyl-1-pyrrolidin-yl-butan-1-one

Synthesis:

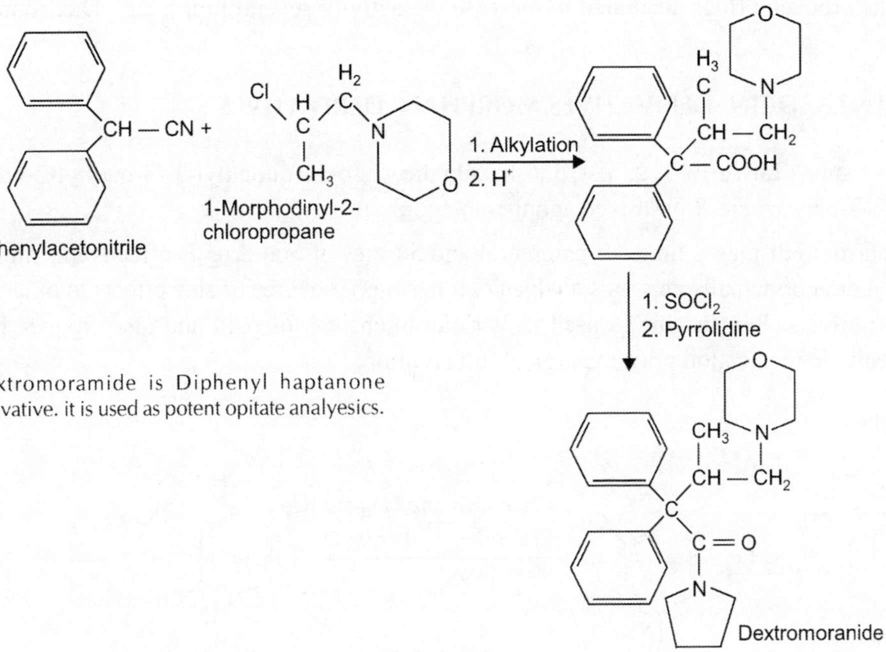

Dextromoramide is Diphenyl haptanone derivative. it is used as potent opitate analyesics.

Fig. 6.19

6.12 SAR OF DIPHENYL HAPTANONE DERIVATIVES

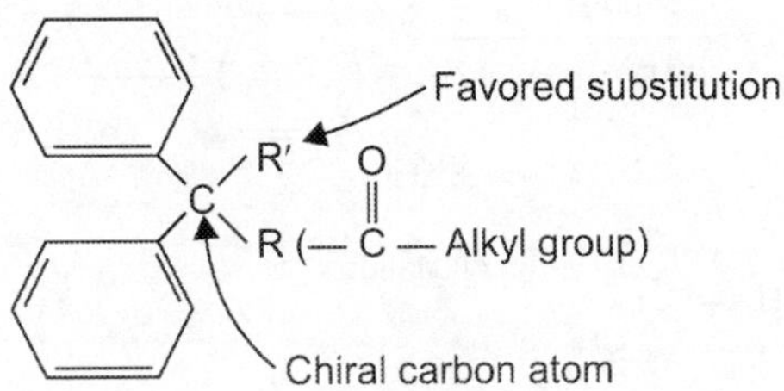

Fig. 6.20

1. It has been proved that levo isomer of methadone and isomethadone exhibit greater activity than recemic mixtures.
2. Both phenyl groups are essential for activity. Removal of any one results in decreased activity.
3. Terminal dimethyl amino group is also responsible for analgesic activity. Replacement of this group by peperidine heterocyclic ring leads to decrease in activity.

4. Replacement of propionyl group by other group like hydrogen, hydroxyl or acetoxyl group leads to decrease in activity. Whereas introduction of pyrrolidinoyl and terminal morpholino heterocyclic rings attributed to increase the activity several times, e.g., Dextromoramide.

6.13 BENZAZOCIN DERIVATIVES/MORPHAN DERIVATIVES

1. Pentazocine (Talwin)®: 1, 2, 3, 4, 5, 6-hexahydro-cis-6-11-dimethyl-3-(3-methyl-2-butenyl)-2,6 methano-3-benzazocin-8-ol (mixed agonist-antagonist)

Pentazocin display 3 times of parenteral and 5 times of oral drug is effective as morphine. Its levo isomer is principally active as analgesic. It has higher degree of side effects to other morphine like side-effects. Pentazocine is used in various intensities of pain and also frequently used as preanaesthetic medication prior to surgical intervention.

Synthesis:

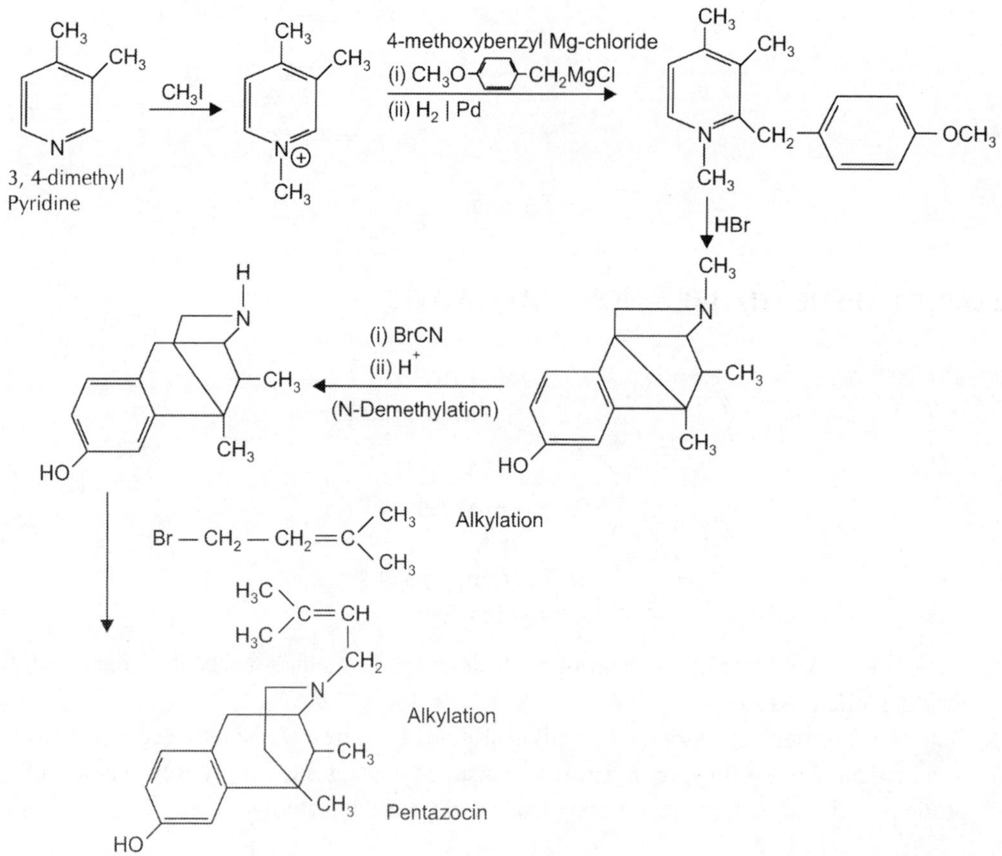

Fig. 6.21

2. Cyclazocine (-3-(cyclopropyl-methyl)-1, 2, 3, 4, 5, 6- hexahydro-6, 11-dimethyl-2, 6-methano-3-benzazocin-8-ol (mixed action). Cyclazocine is an analogue of benomorphan and its clinical action is 40 times more potent than morphine as an analgesic effect and 100 times more potent than nalorphine as antagonist. However, it attributes less addictive effect than morphine.

6.14 NARCOTICS ANTAGONISTS/OPIOID ANTAGONISTS

Opioid antagonists are clinically useful to reverse the effects of opiates in overdose or postoperative sedation. On the other hand, opioid antagonists differ from mixed agonist-antagonists agents in that they do not show agonist action.

1. Naloxone (Narken)®: (–) 17-(allyl)-4, 5-epoxy -3, 14-dihydroxy morphinane-6-one

Naloxone is considered 7-fold more active than nalorphine hydrochloride in antagonistic effect of opioid analgesics. It is free from withdrawal effect after long-term administration. It is having ability to block opioid receptors as well as eliminate the central and peripheral action of opiates. Naloxone is used in the treatment of overdose of narcotic analgesics.

Synthesis:

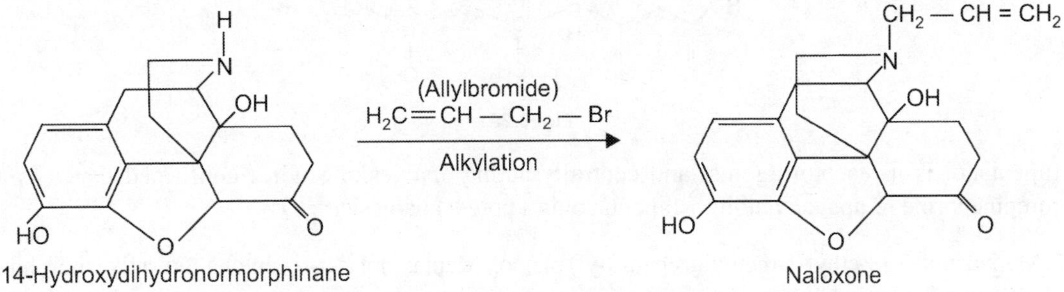

Fig. 6.22

2. Naltrexone (Revia)® (Trexan)®: 17-(cyclopropylmethyl)-4, 5-α epoxy-3, 14-dihydroxymorphinan-6-one.

Naltrexone is another example for treating opiate addiction. Its action towards long lasting effect due to its metabolite 6-β-naltrexol that shows strong antagonist action.

Synthesis:

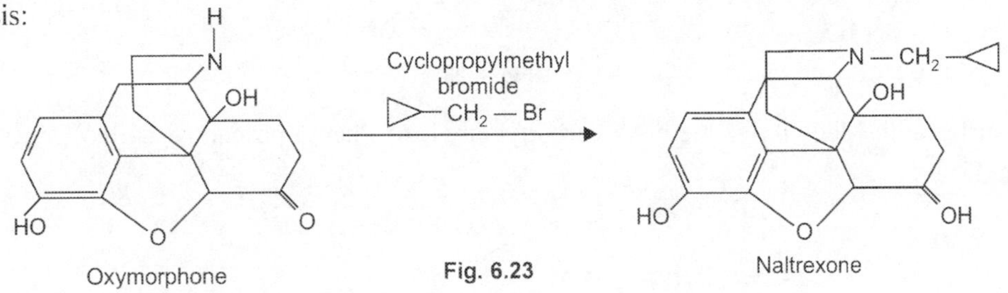

Fig. 6.23

3. Nalorphine: N-Allylmorphine

Nalorphine exhibits direct antagonist effect against opioids analgesics such as morphine, meperidine methadone and levorphanol. Nalorphine is a strong analgesic but it has higher degree of undesirable psychotic effects although nalorphine has antagonist effect against barbiturates or general anaesthetic depression.

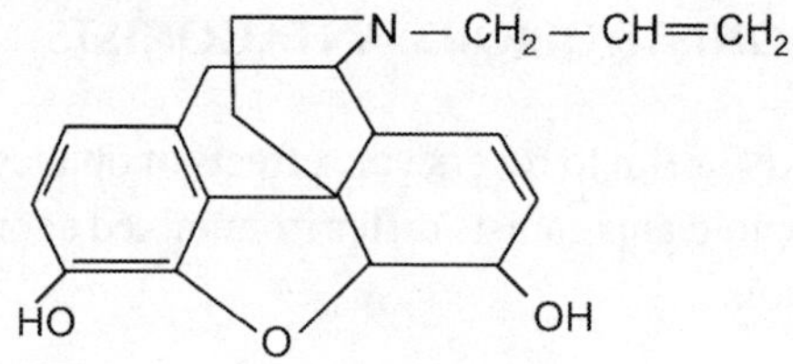

Fig. 6.24. Nalorphine

6.15 NEWER DRUGS

1. Tapentadol: 3-[(1R, 2R)-3-(dimethylamino)-1-ethyl-2-methyl propyl] phenol

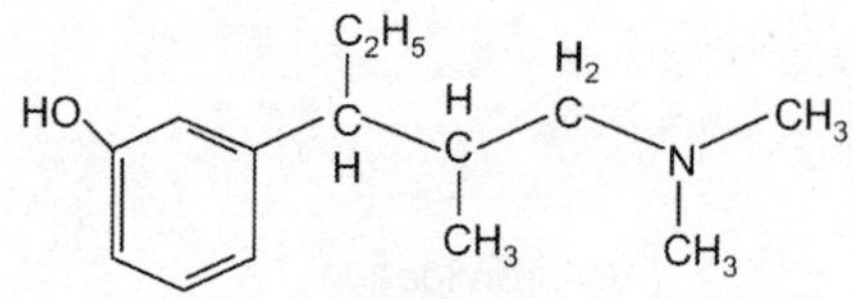

Fig. 6.25

Tapentadol is μ-receptor agonist and centrally acting analgesic. Studies confirmed that it is a norepinephrine re-uptake inhibitor. Tapentazol is a potent analgesic.

2. Meptazinol: 3-(3-ethyl-1-methylazepan-3-yl) phenol. Meptazinol is μ-receptor agonist like morphin. Meptazinol is suggested to have mixed agonist/antagonist action with lower degree of dependency and abuse. Meptazinol is a potent analgesic and used in the treatment of severe pain.

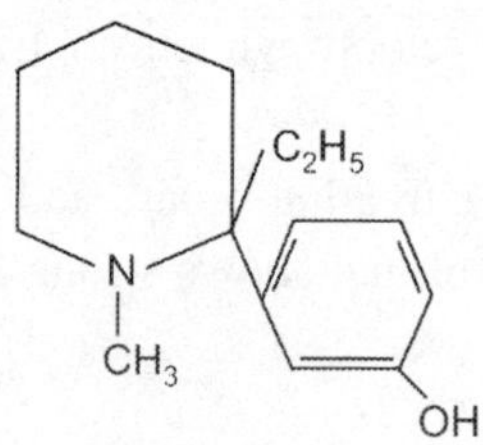

Fig. 6.26

3. Remifentanil: Methyl-1-(2-methoxy carbonyl ethyl)-4-(Phenyl-propanoyl-amino)-piperidine-4-carboxylate.

Remifentanil is a synthetically derived ultra-short acting opioid analgesic. Remifentanil contributes synergistic effect with other general anaesthesia. Thus, it is frequently used during surgery for relieving pain.

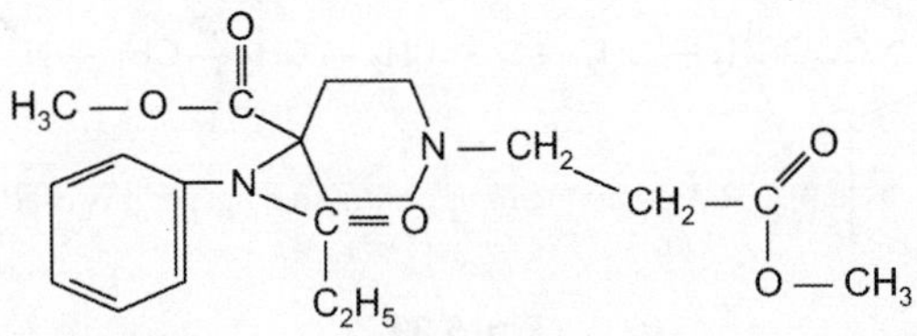

Fig. 6.27

6.16 ANTITUSSIVE AGENTS

Antitussive agents are those which are employed in the symptomatic control of cough by the mechanism of depressing cough-centre located in the medulla of brain. Antitussive agents are also known as anodynes, cough suppressants and centrally acting antitussives.

> Note: Etiology of cough: Cough is regarded as a protective and physiological reflux that play an important role in the health and disease states. There are many eitiological reasons that make unusual symptons and longer duration of cough persistence.

A number of narcotic analgesics have ability to depress the cough centre by symptomatic control. The most important and widely used ones are morphine, codein, hydromorphane, methadone and levorphanol.

1. Noscapine (Tusscapine): (–) narcotine

Noscapine is an alkaloid and isolated by extraction. Noscapine is widely used as antitussive agent. It is employed in the treatment of and control of cough due to pulmonary disorders, asthma, etc. It shows weaker bronchodilatory effect and stimulates respiration.

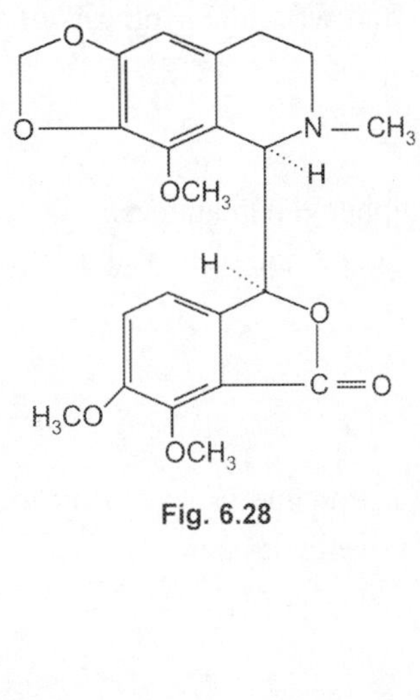

Fig. 6.28

2. Carbetapentane (Toclase)®: 2-[2-(diethylamino) ethoxy] ethyl-1-phenylcyclopentane.

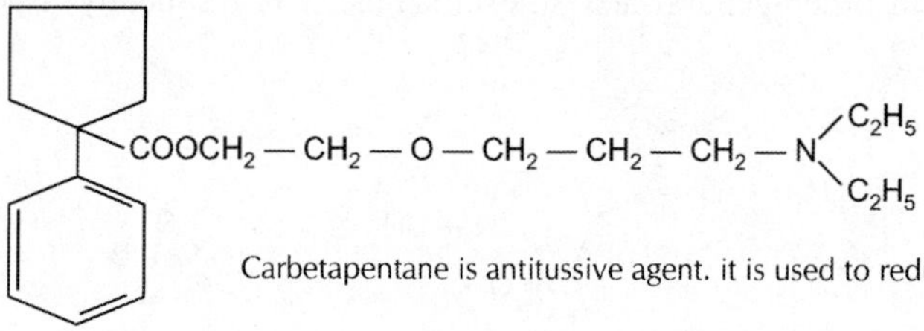

Carbetapentane is antitussive agent. it is used to reduce bronchial secretion.

Fig. 6.29

Carbetapentane is a centrally acting antitussive agent. It is frequently used as cough suppressant and helps reduce bronchial secretion.

3. Levopropoxyphene (Novrad)®: (–)-α-[2-(dimethyl-amino)-1-methyl ethyl]-α-phenylphenethyl. Propionate. Levopropoxyphene also acts as an antitussive and cough suppressant agent.

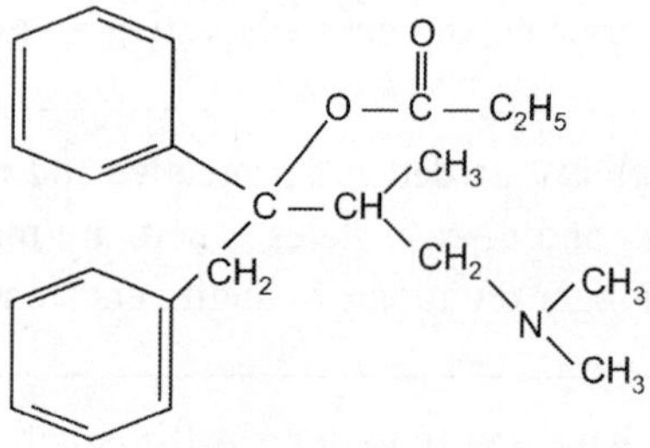

Fig. 6.30

EXERCISE

1. Explain narcotic analgesics.
2. Write a note on chemical structure, name and uses of morphine.
3. Classify opioid analgesics.
4. Discuss morphine analogues and write the synthesis of any two.
5. Give SAR of the following:
   (1) Morphine and its derivatives
   (2) Piperidine analogues
   (3) Acyclic analgesics or Diphenyl haptanones.
6. Write the synthesis, mode of action and uses of the following drugs:
   (i) Morphine
   (ii) Pethidine
   (iii) Loparamide
   (iv) Methadone
7. Write the synthesis, mode of action and uses of pentazocine.
8. What are antitussive agents? Give their names.

7

Anticonvulsant Agents

7.1 INTRODUCTION

The term seizure comes from the Latin word 'sacire' means 'to take possession of'. Seizure is clinically a paroxymal event caused by abnormal, excessive, hypersynchronous discharges from central nervous system neurons. This is the abnormal function of CNS, depends on the distribution of discharges that cause various manifestations, ranging from dramatic convulsant activity to experimental one. Epilepsy is a clinical phenomenon and described as a condition in which as subject undergoes recurrent seizures due to chronic and underlying process. On the other hand, the term convulsant is described as the experimentally induced seizures in the laboratory animals.

According to the studies recurrent incident of epilepsy is 0.3 to 0.5% in different populations around the world while prevalence of epilepsy has been estimated at 5-10 persons per 1000 subject of populations. All types of seizures are clinically characterized by their pattern reading of EEG (electroencephalogram).

7.2 CLASSIFICATION OF EPILEPSY

Table 7.1: Classification of epilepsy/seizure

S. No.	Type	Characteristics/features
1.	**Partial Seizures**	**Discharge spread through the cortex**
A.	Simple partial seizures	Motor, sensory, automatic, psychic symptoms Affected area: facial expression, abnormal movement of face and hand (left). Movements are typically clonic.
B.	Complex partial seizures	Visual and verbal commands impairment involuntary and automatic behaviours arrested (automatism) chewing, lip smacking, swallowing movement and hands and projection of emotion and running.
C.	Partial seizures associated with secondary generalization	Discharge covers cerebral hemisphere generate generalized seizures it includes focal symptoms.
2.	**Generalized seizure**	**Arise from both cerebral hemisphere**
A.	Absence seizures (Petit mal seizures)	Failure of consciousness without loss of postural control. It includes rapid blinking of eyelids, chewing movements,

S. No.	Type	Characteristic features
		clonic movements of hands. Mainly occurs in early adolescence (4-8 years)
B.	Atypical absence seizures	Longer duration of loss of consciousness caused by multifocal structural abnormalities of brain, mental retardation.
C.	Generalized, Tonic-clonic Seizures (Grand mal)	It results from metabolic malfunctions, contraction of muscles throughout the body and caryxlictalong impairment of respiration, oropharynx secretion, cyanosis development.
D.	Atonic seizures	It includes sudden loss of postural muscles tone, impairment of consciousness.
E.	Myoclonic seizures	It is characterized by sudden and brief muscles contraction that may include a part or entire body, sudden jurking movement observed, metabolic disorders, degenerative CNS disease.
3.	Unclassified Seizures	Occurs in infants and neonates.

7.3 SEIZURE INITIATION AND PROPAGATION

Partial seizure is initiated from the cortex region of CNC and is discrete to the neighbouring regions. This is called seizure initiation and propagation stages. There are basic reasons of seizure initiation. It includes:

(i) Uncontrolled action potential due to long lasting depolarization, causes the influx of extracellular Ca^{++} followed by opening of voltge dependent Na^+ channel by influx of Na^+ ions, further repetitive action potential generation. Hyperpolarization exhibit by γ-amino-butyric acid (GABA) receptors activation or K^+ ions channels activation. This prevents the spread of uncontrolled or bursting action potential. Repetitive neuronal discharge leads to increased extracellular K^+ which blunts hyperpolarization and makes the other neurons depolarization.

(ii) Hypersynchronization makes synchronized burst from the sufficient number of neurons which lead spiky waves pattern in EEG. Role of NMDA receptors for the antiepileptic drugs is shown in the schematic representation.

7.4 RECEPTORS INVOLVED IN THE MECHANISM

1. $GABA_A$ Receptors: They are linked to chloride ion channels. Modulated by benzodiazepines and barbiturates and second generation antiepileptic drugs, it includes NMDA receptors.
2. $GABA_B$ Receptors: Beclofen is the agonist to $GABA_B$ receptors.
3. T-Type Ca^{++} Channels
   (i) Voltage dependent calcium channels divides into L, P-types and Q-types (high voltage activated)
   (ii) R-type: Intermediate voltage activated Ca^{++} channels
   (iii) T-type: Low voltage activated channels. These channels are further classified by the different subunits α, β and γ composition.

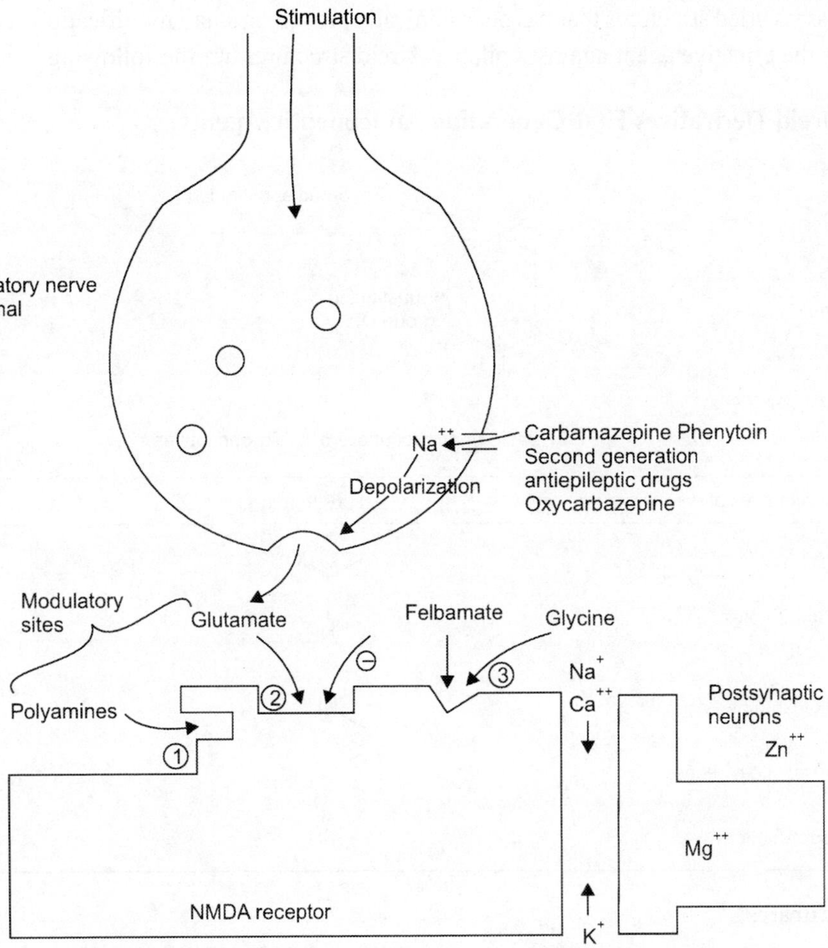

Fig. 7.1: Schematic diagram of action of drugs on glutamate mediated neurotransmission.

Note:

1. Indicate Antagonist Action
2. Glycine show absolute requirement for receptor, channel complex to enter the open state
3. NMDA Receptor (N-methyl-D-aspartate Receptor)
4. Signs 1, 2, 3 are modulatory sites

4. Glutamate Receptors: Glutamate receptors include G-protein couped excitatory amino acid receptors (m Glu Rs). Glutamic acid is the main (amino acid) excitatory neurotransmitter in CNS.

7.5. CLASSIFICATION OF ANTIEPILEPTIC AGNENTS

Potassium bromide was the first effective remedy that was introduced in 1857. Hauptmann discovered phenobarbiton in 1912, initially it was used as sedative agents in epilepsy. So many antiepileptic drugs possess uried structures that has been clinically proved on small modification in ureid structure to makes the effective agent against epilepsy. Ureid structures are the following.

7.5.1 Ureid Derivatives/First Generation Antiepileptic Agents

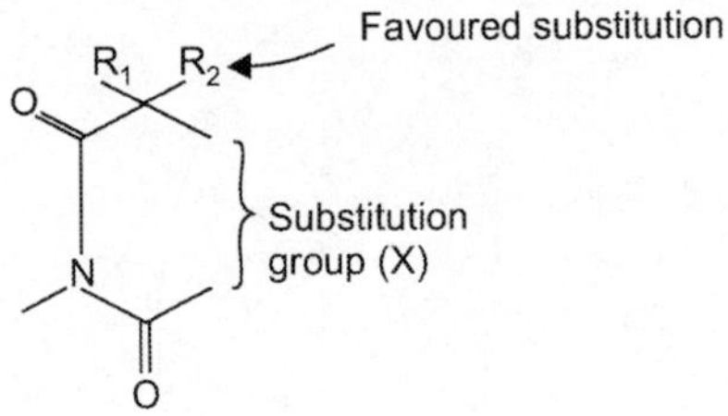

Fig. 7.2: Basic pharmacophore of ureid derivatives.

S.N.	Major class of derivatives	Substitution group (X)
1.	Barbiturates	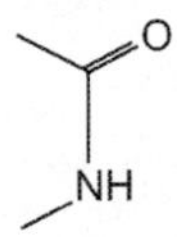
2.	Hydantoins	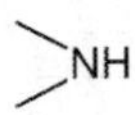
3.	Oxazolidinediones	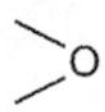
4.	Succinimides	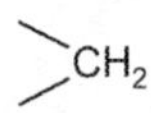

1. Barbiturates

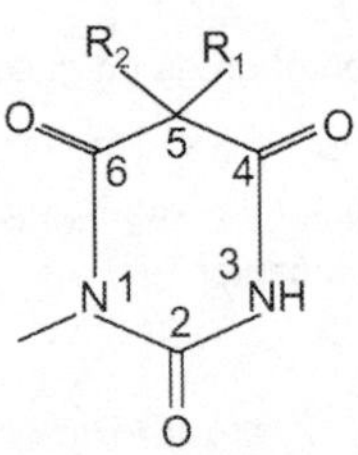

Fig. 7.3: Barbiturates.

	Derivatives	R_1	R_2	Other possible modification
(i)	Phenobarbital	C_2H_5	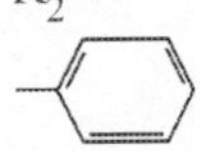	
(ii)	Mephobarbital	C_2H_5	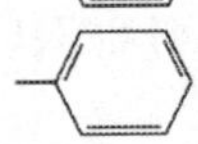	$1 - CH_3$

(Discussed in Chapter-5, Sedative and Hypnotics)

2. Hydantoins

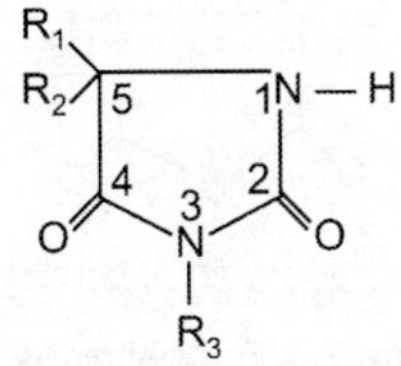

Fig. 7.4: Hydantoins.

	Derivatives	R_1	R_2	R_3
(i)	Phenytoin	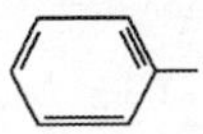	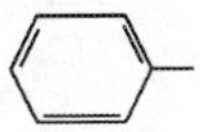	H
(ii)	Mephetoin	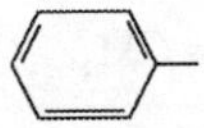	C_2H_5	CH_3
(iii)	Ethotoin	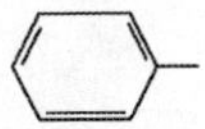	H	C_2H_5
(iv)	Phosphenytoin	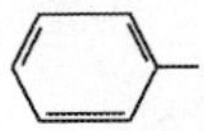	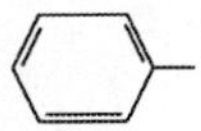	$CH_2 — OPO_3^{++}$
(v)	Phenyl ethyl hydantoin	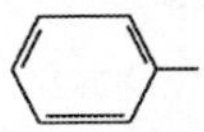	C_2H_5	H

3. Oxazolidinedinones

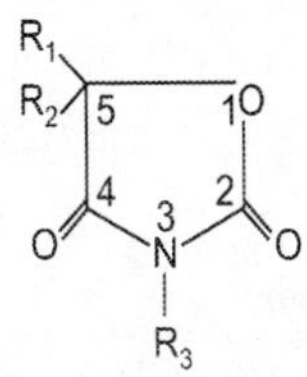

Fig. 7.5: Oxazobidinedinones.

	Derivatives	R_1	R_2	R_3
(i)	Trimethdine	CH_3	CH_3	CH_3
(ii)	Parameth dione	CH_3	C_2H_5	CH_3
(iii)	Aloxidone	CH_3	$CH_2=CHCH_2$	CH_3

4. Succinimides

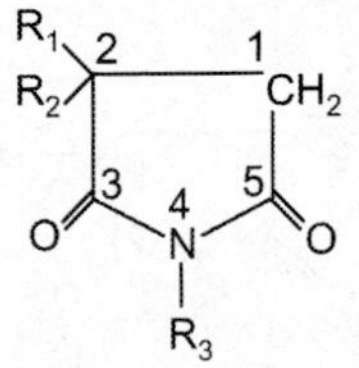

Fig. 7.6: Succinimides.

	Derivatives	R_1	R_2	R_3
(i)	Phensuximide	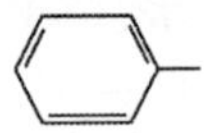	H	CH_3
(ii)	Methsuximide	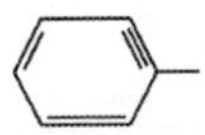	CH_3	CH_3
(iii)	Ethosuximide	C_2H_5	CH_3	H

7.5.2. Benzodiazepines

Derivatives (See in chapter 5)

(i) Clonazepam
(ii) Diazepam
(iii) Chlorazepate

7.5.3. Urea and MonoAcyl Urea-Derivatives/Iminostilbenes

Derivatives

(i) Carbamazepine
(ii) Oxycarbazepine

7.5.4. Miscellaneous Agents

(i) Primidone
(ii) Valproic Acid/Aliphatic acids
(iii) Gabapentin (Second Generation)
(iv) Tiagabine (Second Generation)
(v) Felbamate (Second generation)
(vi) Lamotrigine (Second generation)
(vii) Zonisamide and topiramate (Second generation)

7.6 HYDANTOINS AGENTS

Hydantoins are structurally similar to barbiturates and are assumpted by lacking an 6-oxogroup in structure. Therefore, they are considered cyclic monoacyl urea derivatives rather than cyclic diacyl urea derivatives. They are weaker acids than barbiturates (due to loss of carbonyl group). Hydantoins are mainly used in the treatment of tonic-clonic seizures.

General Synthetic Approach of Hydantoins

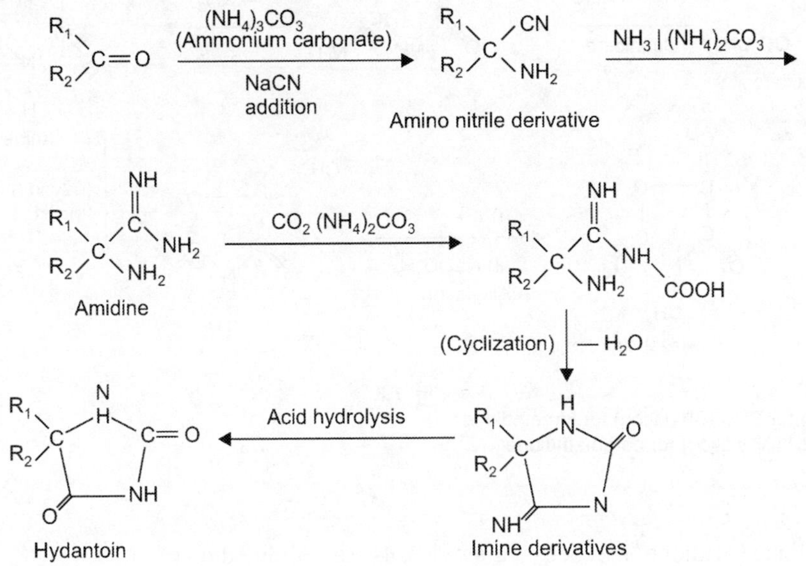

Fig. 7.7

Note: R_1, R'_2 are the substituents according to their derivatives.

1. Phenytoin (Dilantin ®): 5, 5-diphenyl hydantoin

Phenytoin is considered a first generation anticonvulsant drug. It blocks the Na^+ channels followed by decrease in glutamic acid release and give the anticonvulsant action. Phenytoin is used in the treatment of all types of seizures except absence sezires.

2. Mephetoin (Mesantoin®): 5-ethyl-3-methyl-5-phenyl hydantoin.

Mephetoin was earlier suggested as sedative and hynotics agents because of its toxicity. It was withdrawn due to haematological disorders as well as skin disorders.

3. Ethotoin (Peganone®): 3-ethyl-5-phenylhydantoin. Its dealkylated metabolite is considered an active metabolite. Ethotoin is mainly used in the treatment of generalized seizures.

7.7 OXAZOLIDINEDIONES AGENTS

The trophism oxazolidinediones come at the platform by the replacement of –N-H group at the position-1 of the hydantoin moiety with oxygen atom that makes oxazolidine-2, 4-dione system. The derivatives of this trophism are mainly used in the treatment of generalized tonic-clonic seizures.

General synthetic approach of oxazolidinediones

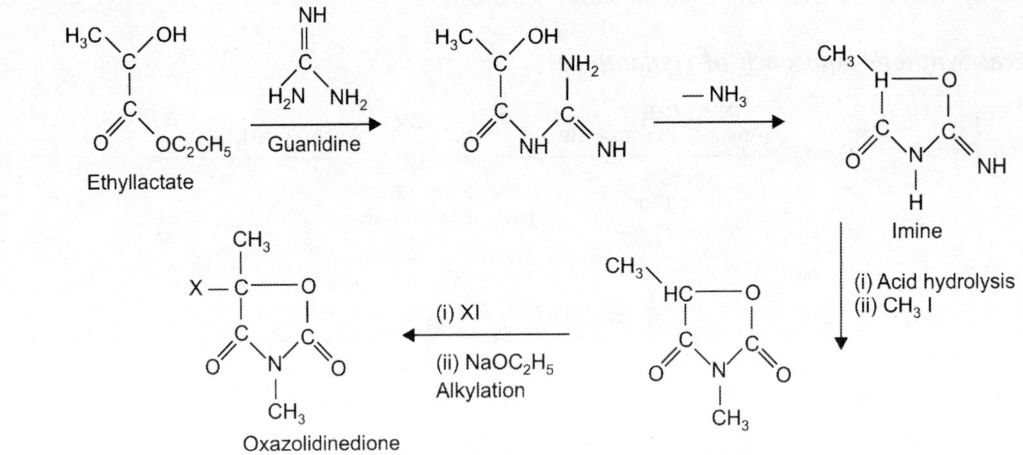

Fig. 7.8

X = Substitution for CH_3I (Alkylation) for trimethdione derivatives C_2H_5I (Alkylation) for paramethdione

1. Trimethdione (Tridione®): 3, 5, 5-trimethyl-2, 4-oxa-3olidinedione.

Trimethdione was the first drugs that was used in the treatment of absence seizure. Dimethdione is the N-demethylated form of trimethdione and is active metabolite due to its dermatological and haematological toxicity, it is used only in the refractory cases of absence seizures.

2. Paramethdione (Paradione®): 5-ethyl-3, 5-dimethyloxazolidine-2, 4-dione.

Paramethdione is just different from trimethdione by absence of methyl group to ethyl group. Paramethdione is used in minor form of epilepsy condition.

7.8 SUCCINIMIDES AGENTS

Succinimides trophism represents the compounds that replace oxazolidinediones oxygen to $-CH_2$ group.

General synthetic approach of succinimides

1. Ethosuximide (Zarontin®): 2-ethyl-2-methyl-succinimide ethosuximide is T-type calcium channel blockers. Ethosuximide is used in the treatment of absence seizures.

2. Phensuximide (Milontin®): N-methyl-2-phenylsuccinimide. Phensuximide exhibit antiabsence activity by the attribution against absence seizure. Its N-demethylated form is active metabolite. It is also useful in the treatment of generalized tonic-clonic and partial seizures.

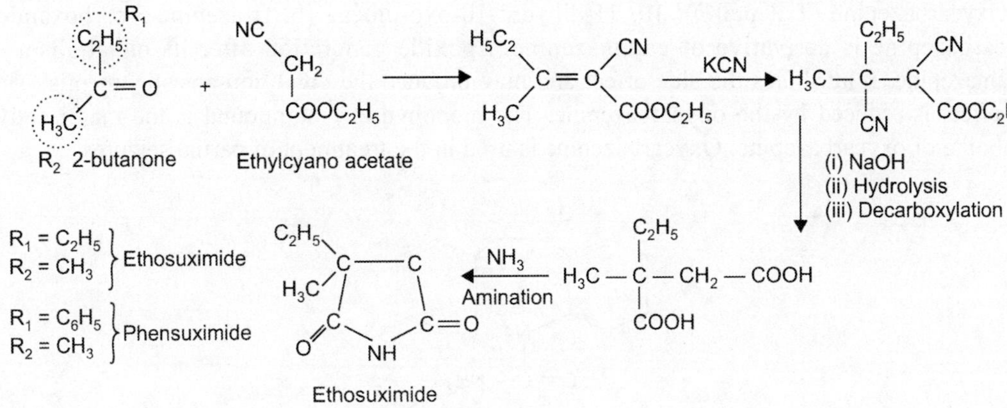

Fig. 7.9

7.9 IMINOSTILBENS AGENTS

1. Carbamazepine (Tegretol®): 5H-dibenz[b,f] azepine-5-carboxamide.

Carbamazepine is sodium ion channel blockers. Epoxide form of carbamazepine generates after its metabolism (z) cis-stilben double bond. Carbamazepine is clinically useful in the treatment of generalized tonic-clonic and partial seizures.

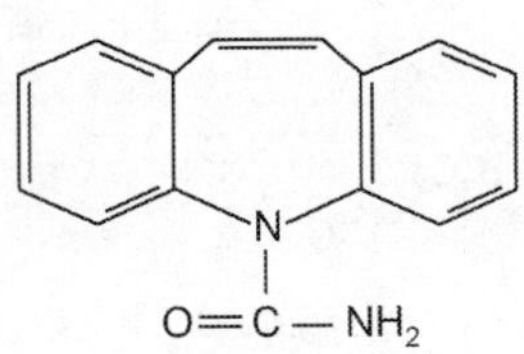

Fig. 7.10

Synthesis:

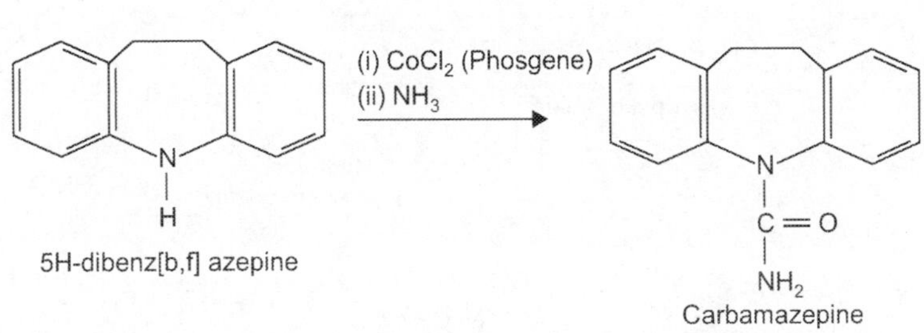

Fig. 7.11

2. Oxycarbazepine (Trileptal®): 10, 11-dihydro-10-oxo-dibenz [b, f] azepine-5-carboxamide Oxycarbazepine is derivative of caramazepine. Epoxide generation after its metabolism of carbamazepine. It attributes the side effect and may produce the condition aplastic anemia. This side effect is reduced by the oxycarbazepine. The monohydroxy compound is the major active metabolite of oxycarbazepine. Oxycarbazepine is used in the treatment of partial seizures.

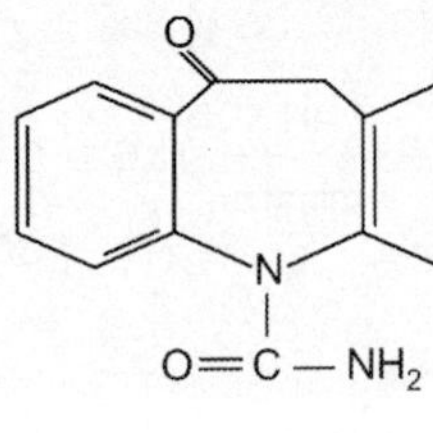

Fig. 7.12

7.10 ALIPHATIC ACID AGENT

1. Valproic Acid (Depakene®): 2-prophylpentanoic acid. Valproic acid represents the anticonvulsant action by blocking Na^{++} channels and increases the level of GABA in the brain. Valproic acid is employed in the treatment of myoclonic and generalized tonic-clonic seizures.

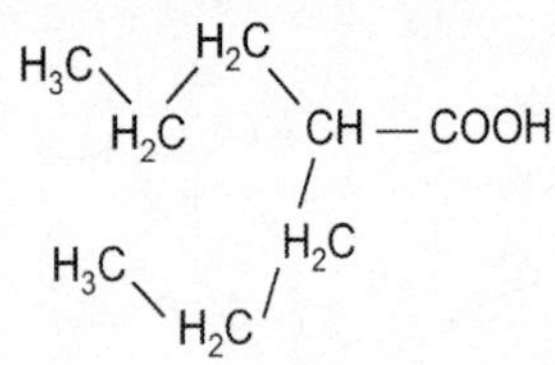

Fig. 7.13

Synthesis:

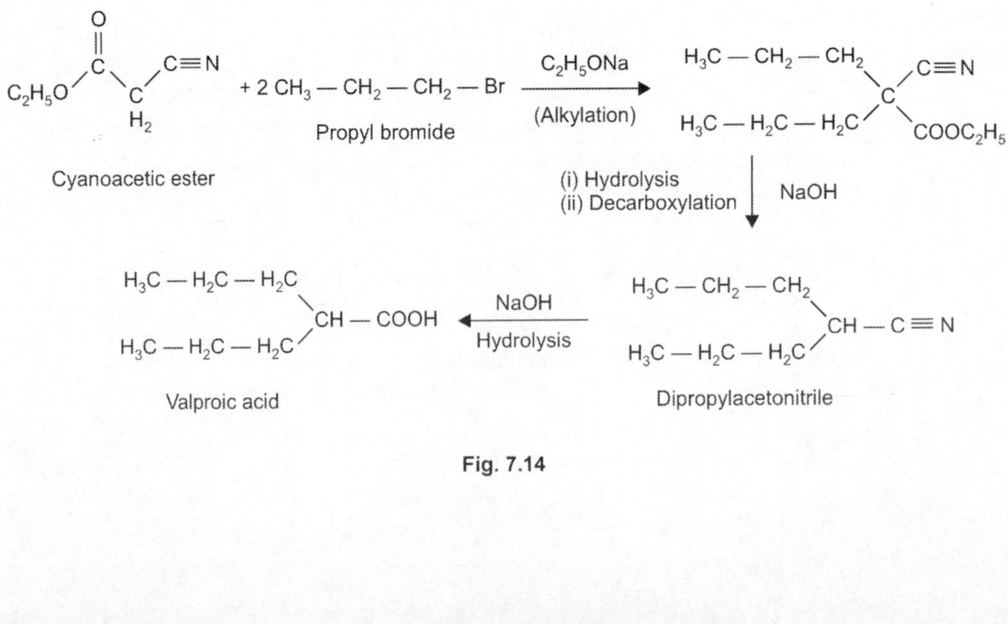

Fig. 7.14

7.11 SECOND GENERATION ANTIEPILEPTIC DRUGS

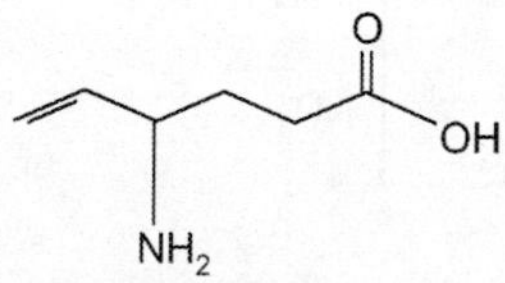

Vigabatrin
(Hexanoic acid derivative)

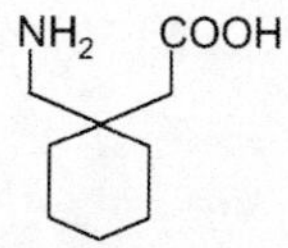

Gabapentin
(Cyclohexane derivative)

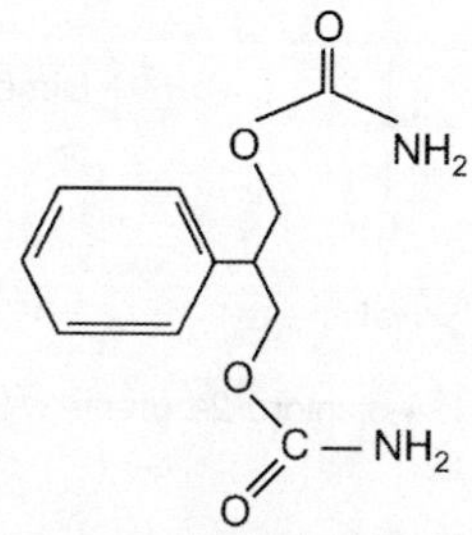

Felbamate
(Carbamate) derivative

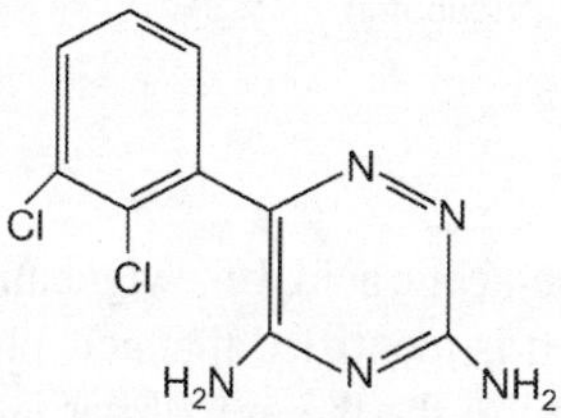

Lamotrigine
(Triazine Derivative)

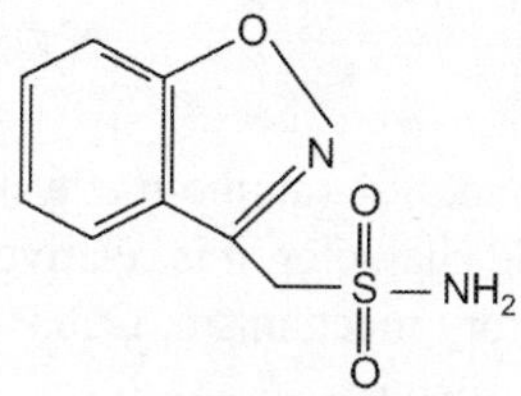

Zonisamide
(Sulfonamide derivative)

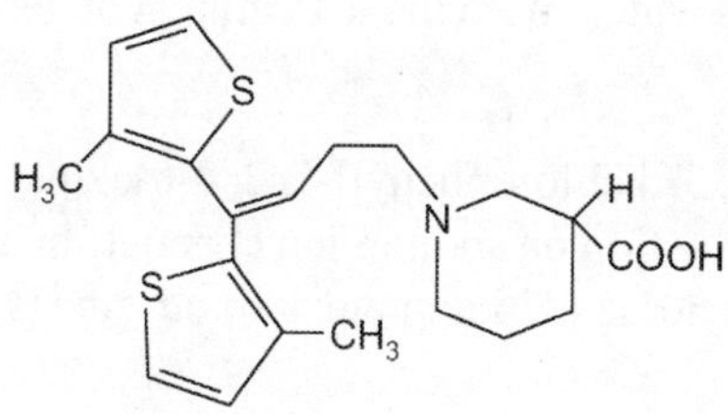

Tiagabine
(Nipecotic acid derivative)

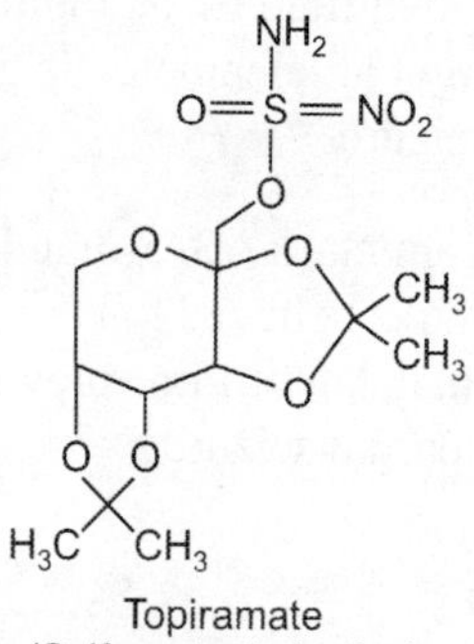

Topiramate
(Sulfamate derivative)

Fig. 7.15: Second generation anticonvulsants.

Second generation anticonvulsant are GABA analogues.

1. Vigabatrin (Sabril®): 4-amino-5-hexanoic acid. Vigabatrin is 4-amino-5-hexanoic acid. It is a recemeate of R and S form of vigabatrin. Its isomer is an active component. Vigabatrin inhibits GABA-transaminose enzyme therefore increases the level of GABA in the brain, hence it acts as anticonvulsant agent.

Synthesis:

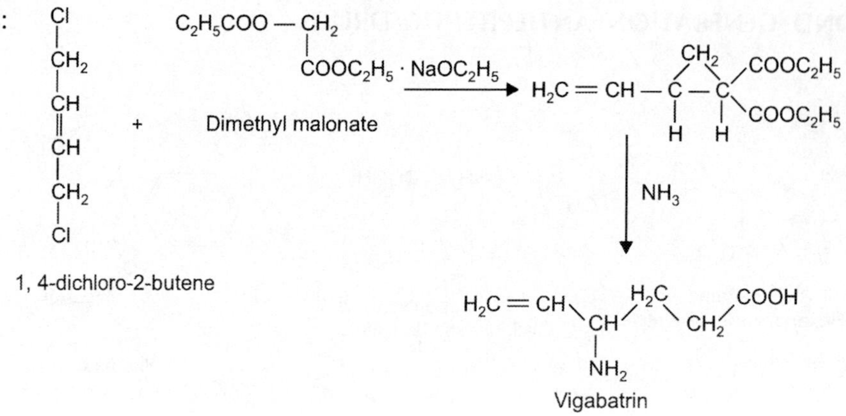

Fig. 7.16

2. Gabapentin (Neurontin®): 1-(amino-methyl)-cyclohexane-acetic acid. Like vigabatrin gabapentin increase the hydrophobic character, it is relative to GABA. It is identified that acts on the Ca^{++} ion channels by their inhibitory mechanism. Gabapentine is used in the therapy of partial seizures and generalized tonic-clonic seizures.

3. Felbamate (Felbatol®): 2-phenyl-1, 3-propandiol-dicarbamate. Felbamate is carbamate derivative. Its action has been identified by its inhibitory action on voltage dependent Na^{++} channels. NMDA receptor is also a target of felbamate. Felbamate is widely used in the treatment of generalized tonic-clonic and partial seizures.

4. Lamotrigine (Lamictal®): 3, 5-diamino-6-[2, 3-dichlorophenyl]-1, 2, 4-triazine. Lamotrigine is triazine analogue and is believed to have primary action on sodium ion channels by their inhibitory action. Apart from that, lamotriqine prevents the release of glutamate. Lamotrigine is clinically used in the treatment of partial seizures.

Synthesis:

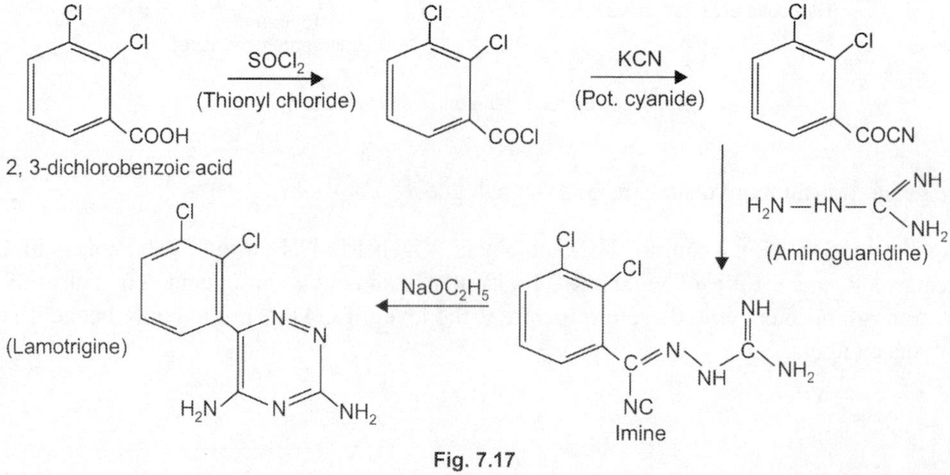

Fig. 7.17

5. Tiagabine (Gabitril®): (R)-(–)1-[4, 4-bis-(3-methyl-thienyl-3-butenyl]-3 piperidine carboxylic acid.

Tiagabine is a nipecotic acid derivative. It is GABA-re-uptake inhibitor and increase the required level of GABA in the brain. Tiagabine is used against partial seizures.

6. Sulfonamide and sulfamate derivatives: Zonisamide (Zonegran®): 1, 2-benzisoxazole-3-methanesulfonamide.

Topiramate (Topamax®): 2, 3, 4, 5-bis-O-(1-methylethyl-idene)-β-D-fructopyranose sulfamate. Both are sodium ion channel blockers. Zonisamide is calcium-T-channels blocker while Topiramate inhibits the action of glutamate. Kainic acid thus increases the level of GABA. Both drugs are used against partial seizures.

7.12 OTHER MISCELLANEOUS ANTIEPILEPTIC AGENTS

1. Primidone (Mysoline®): 5-ethyldihydro-5-phenyl-4, -6-(1H, 5H)-Pyrimidinedione

Primidone is 2-deoxyanalogue of Phenobarbital. Primidone shows a potent anticonvulsant agents in conjunction with other antiepileptic drugs, it exhibits tremendous action against seizures. It is clinically used in the treatment of psychomotor condition, grandmal and focal epileptic stages.

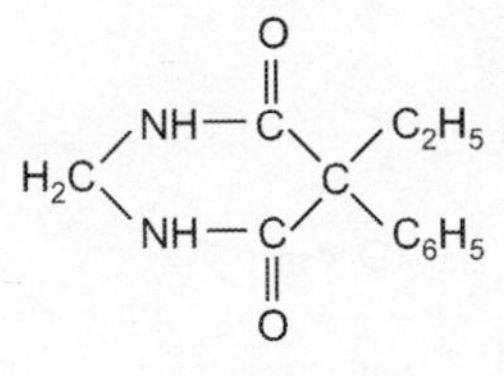

Fig. 7.18

Synthesis:

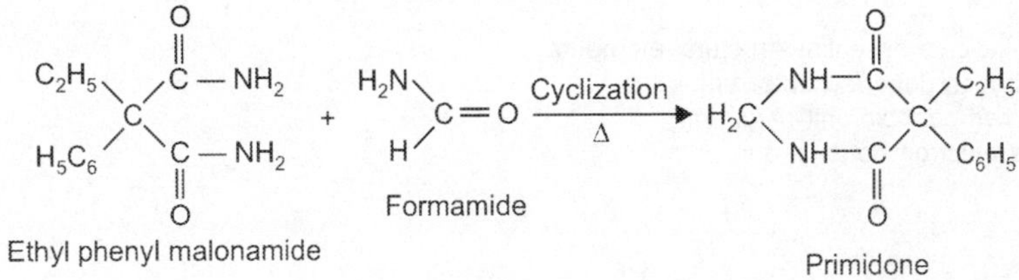

Fig. 7.19

7.13 MODELS REQUIREMENTS FOR ANTIEPILEPTIC AGENTS/SAR

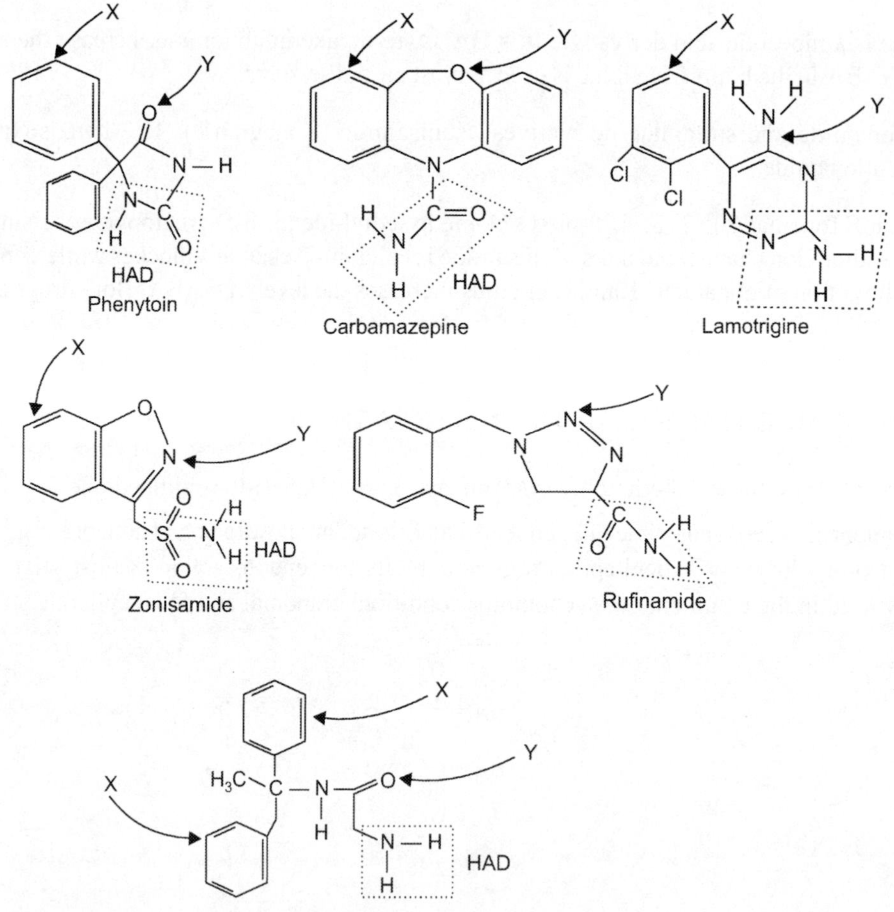

Fig. 7.20: Remacemide.

Note

1. Dotted lines indicate essential structural elements
2. HAD → Hydrogen donor/acceptor unit
3. X represent hydrophobic unit (Aryl ring)
4. Y represents electron donor group

Salient Features of Models

1. All the pharmacophores are based on the structure-activity data of MES-receptors.
2. All the anticonvulsant classes of pharmacophores are concerned with distinguished anti-MES activity in the testing animals studies. All the classes include benzodiazepines, barbiturates, triazolines, semicarbazones and enaminones respectively.
3. All distinguished classes of anticonvulsants show similar anticonvulsant profiles.

4. The Unverferth model postulates the favourable representation towards anticonvulsant activity and this model also favours suitable possible candidate among the derivatives.
5. The required features include the anticonvulsant properties.
   (i) Pharmacophores must have at least one aryl ring I hydrophobic unit (X)
   (ii) One electron donor (Y) serve as important role.
   (iii) A second donor atom in close proximity to the N-H group required for generation of hydrogen bond acceptor/donor (HAD) unit.

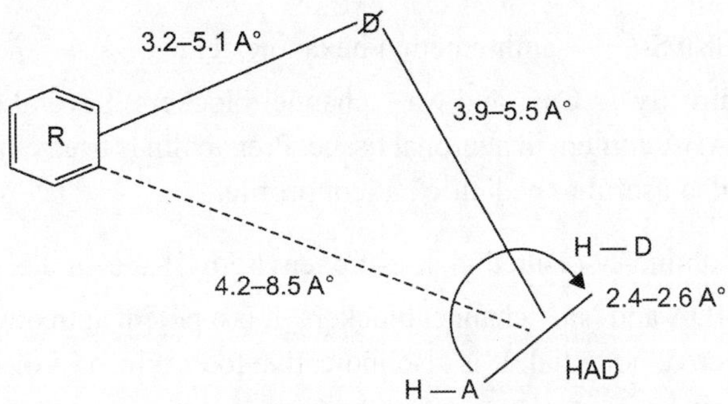

Fig. 7.21: Unverferth model.

Salient Features of Unverferth Model

(i) This model is suggested for pharmacophore model for anticonvulsants in respect of voltage dependent sodium-ion channels on the basis of molecular dynamic simulation of classes of anticonvulsant agents.

Upcoming model of semicarbazones

The extensive series of semicarbazones postulate the lead compound among the (Aryloxy) Aryl semicarbazones found by the several studies, is the potent Na^{++} channels blockers and in future it would be developed for the treatment of neuroleptic pain. The model is shown in figure.

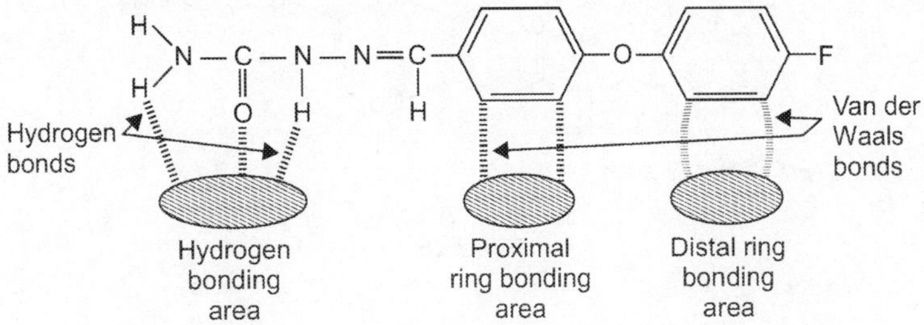

Fig. 7.22: 4-(4'-Fluorophenoxy) Benzaldehyde semicarbazone.

7.14 NEWER DRUGS DEVELOPMENT

(1) Harkose ride: (R)-2-Acetamido-N-benzyl-3-methoxy propionamide.

Harkoseride is an amino acid derivative. Studies postulated its excellent anticonvulsant activity.

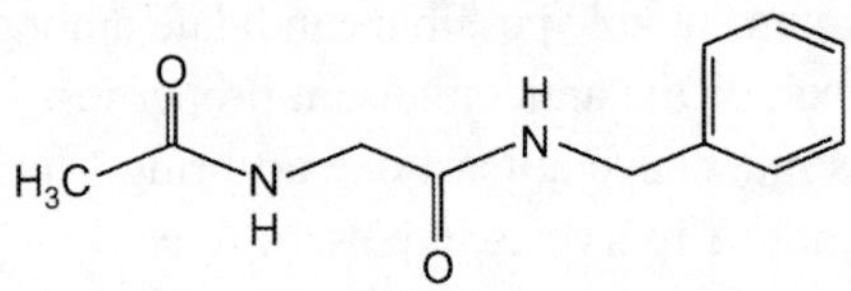

Fig. 7.23

(2) Pregabelin (C1-1008): S-(+)-3-aminomethyl hexanoic acid.

Pregabelin acts indirectly as Ca^{++} and Na^{++} channel blockers. Several studies confirm that it increases the level of GABA content in neuronal tissue. Pregabalin is used clinically in the treatment of partial seizures and also useful to pediatric patient profile.

(3) Remacemide: (±)-2-amino-N-(1-methyl-1, 2-diphenyl-ethyl) acetamide.

Remacemide is NMDA and Na^{++} channel blockers. It is a potent anticonvulsant agent due to its ability of its neuroprotective potential. It is also indicative to Parkinson's disease.

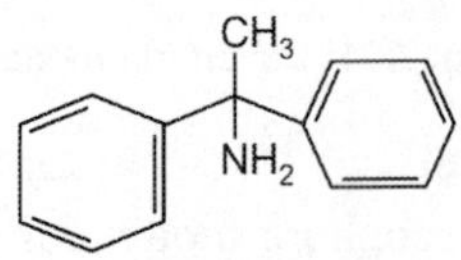

Fig. 7.24

4. Retigabine (D-23129): ethyl-N-(2-amino-4-(4-fluorobenzalamino)-phenyl carbamate. Retigabine shows the anticonvulsant action by increasing K^+ conductance in neuronal cells. It is an active antiepileptic agent.

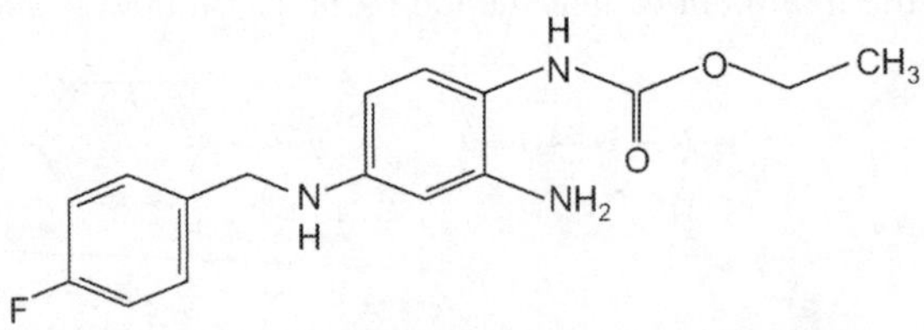

Fig. 7.25

EXERCISE

1. Classify anticonvulsant agents with their various structures.
2. Differentiate between antiepileptic and anticonvulsant drugs.
3. Explain hydantoins as anticonvulsant agents.

4. Write the synthetic approaches, mode of action and uses of the following.
   (A) Hydantoins
   (B) Trimethdione
   (C) Ethosuximide
   (D) Valproic acid
5. What are iminostilbent? Write the synthesis of carbamazepine.
6. Give a note on second generation antiepileptic drugs.
7. Give a note on model requirements for antiepileptic drugs. (SAR of antiepileptic drugs).

8

Antiparkinsonism Agents

8.1 INTRODUCTION

Parkinsonism or Parkinson's disease (PD) is the most common form of progressive neurodegenerative disorders. About 1 million individuals in United States (over 55 years) are affected by the PD. However, it has been studied that its peak age of its onset 60s of affected person's life. The clinical features of Parkinson's disease include bradykinesia (slowness of cognitive activities), impairment of movement, rest tremor (85% of patients), muscular rigidity, shuffling gait and flexed posture, etc. Parkinson's disease is postulated by a variety of non-motor symptoms like autonomic, sensory, sleep, cognitive and psychiatric disturbances. All types of Parkinson's diseases result from a dramatic reduction of dopaminergic stimulation within the basal ganglia of brain. PD affects about 1% of the general population.

Parkinson's disorders generally characterized by an earlier age of onset (before age 45 years) and a longer course than cases of "sporadic PD. This type of Parkinson's disorder is generally represented by genetic form LLRK-2 in the same age range. There are also a number of factors which cause PD like family history, gender, head injury, exposure of pesticides, consumption of well water, rural living factors. Other factors include coffee drinking, smoking, NSAIDS, estrogen replacement therapy.

8.2 ETIOLOGY OF PARKINSON

Exact etiology of PD is still not known. The most likely cause of the aforementioned motor stimulation problems may be a lack of dopamine neurotransmitter (inhibitory action on spinal cord). On the other hand, cholinergic neurons act in regulating the extrapyramidal system. Various alkaloid drugs like belladonna, which have characteristics of cholinergic action (reduce the sensitivity to Ach) have been used in the treatment of Parkinson's disorders.

8.3 INITIATION OF THERAPY

In the present time, definition of Parkinson's disease have been changed. Parkinsons disorder is the consequence of an imbalance between dopaminergic and cholinergic system. Further treatment of PD may include either blocking excessive stimulation of cholinergic system or normalize the functional activity of popaminergic system. Figure 8.1 represents the agonist therapy by dopamine or COMT or MAO inhibitors that are useful in the treatment of PD.

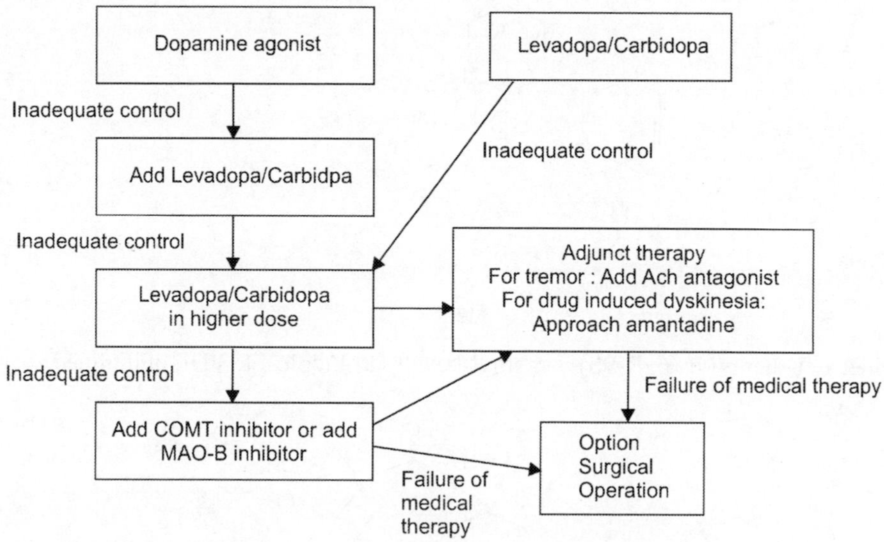

Fig. 8.1: Schematic diagram of initiation of therapy.

8.4 CLASSIFICATION OF ANTIPARKINSONISM AGENTS

8.4.1 Dopaninergic Treatment

8.4.1.1 Dopamine-D_2 Receptor Agonists

1. Pergolide Mesylate (Permax®): Ergoline-8-(methylthio) methyl)-6-Propyl-8-monomethane sulfonate.

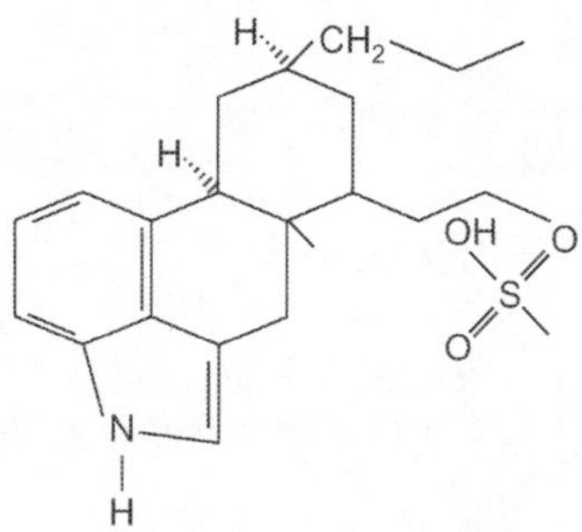

Fig. 8.2

2. Cabergoline (Dostinex®): 1-[16-allylergolin-8β-yl)-carbonyl]-1-[3-(dimethylamino)propyl]-3-ethylurea

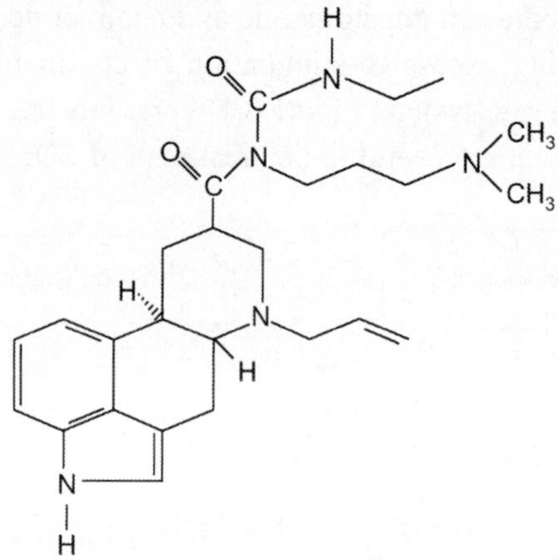

Fig. 8.3

3. Lisuride: 1, 1-diethyl-3-(6aR-9S)-7-methyl-hexahydroindolo-[4, 3 f9] quinoline-9-yl)-urea.

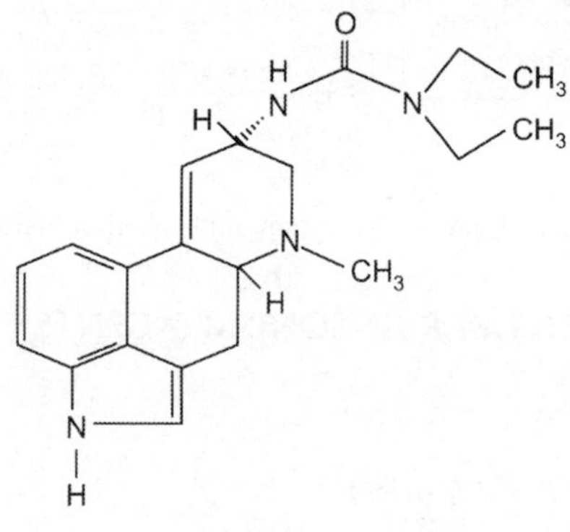

Fig. 8.4

4. Apomorphine

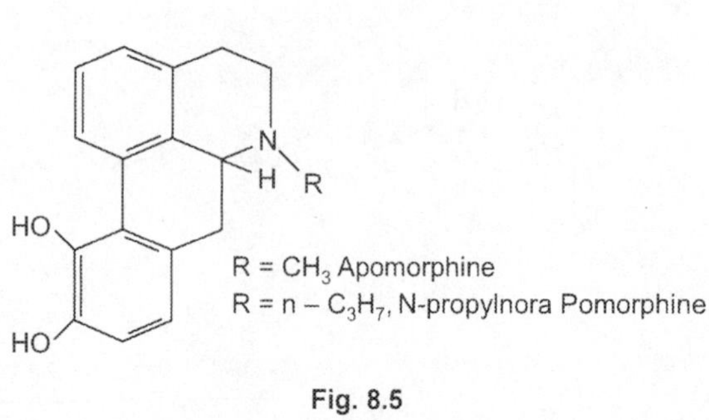

Fig. 8.5

8.4.1.2 Di-Receptor Agonists:

1. Dihydrexidine: 5, 6, 6a, 7, 8, 12b-hexahydro-benzo-phenanthridine 10, 11-diol

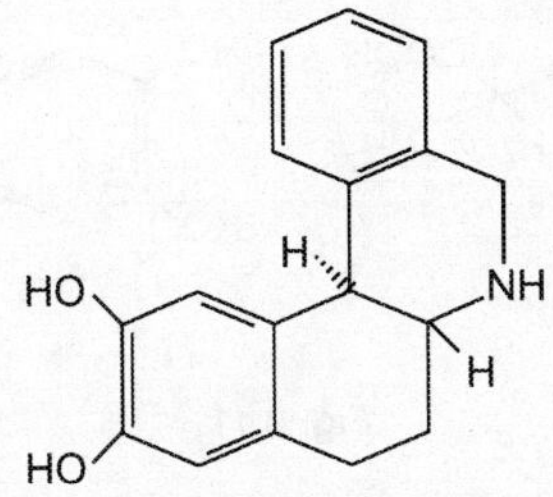

Fig. 8.6

2. Dinapsoline: 8, 9-dihydroxy-2, 3, 7, 11b-tetrahydro-1H naphisoquinoline

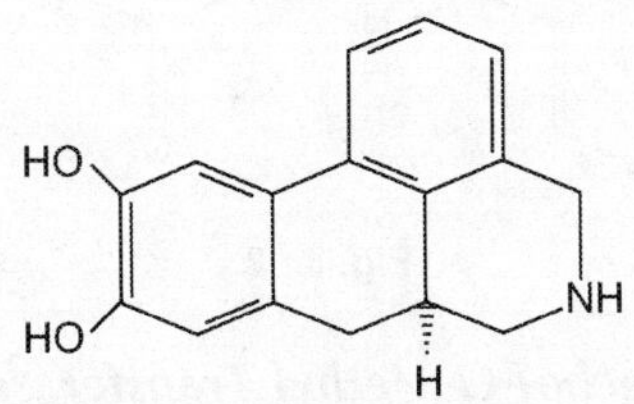

Fig. 8.7

8.4.1.3 Mao inhibitors

1. Phenelgine (Nardil®): (2-phenylethyl) hydrazine

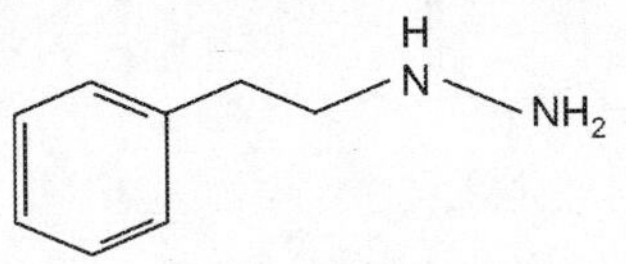

Fig. 8.8

2. Selegiline (Eldepryl®) –N-methyl-N-(2-propionyl)-2-methyl-Phenylethylamine

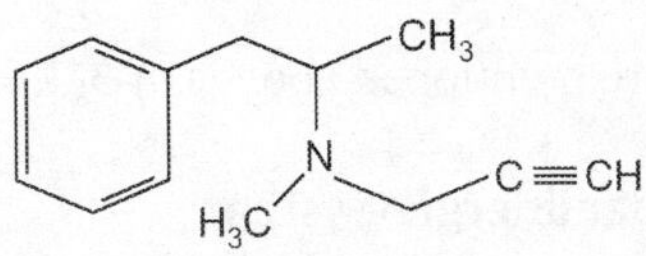

Fig. 8.9

3. Tranylcypromine (Parnate®): (±)-trans-2-phenylcyclopropane-1-amine.

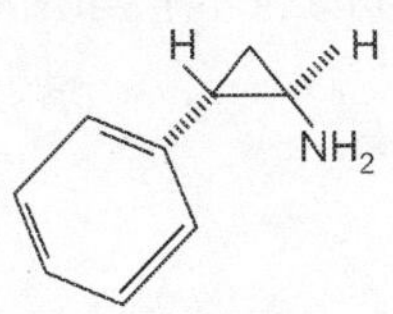

Fig. 8.10

8.4.1.4 Decarboxylase Inhibitors (Peripheral)

(1) L(–) carbidopa (Lodosyn®): (–)-3-(3, 4-dihydroxy phenyl)-L-alanine)

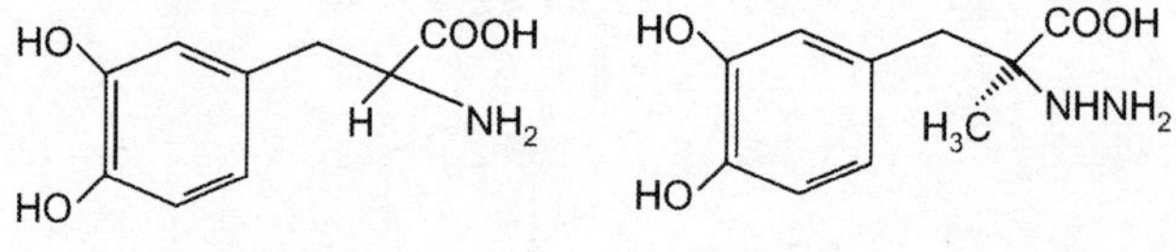

Levodopa

Fig. 8.11

(2) Benserazide (Serazide®): 2-amino-3-hydroxy-N (trihydroxybenzyl)-propane hydrazide.

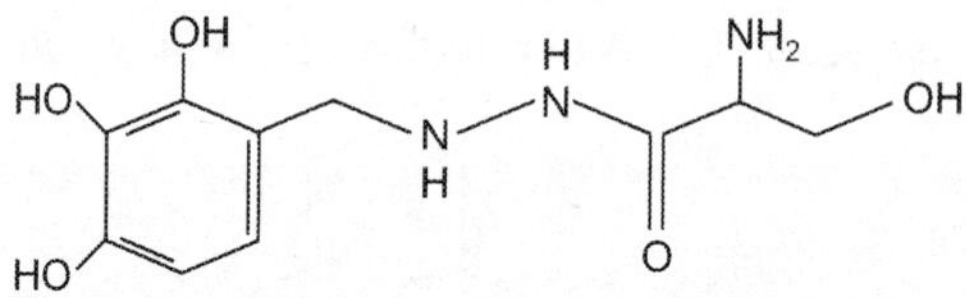

Fig. 8.12

8.4.1.5 COMT Inhibitors: (Catechol-O-Methyl Transferase Inhibitor)

1. Entacapone: 2-cyano-3-(3, 4-dihydroxy-5-nitrophenyl)-N, N-diethyl-propenamide

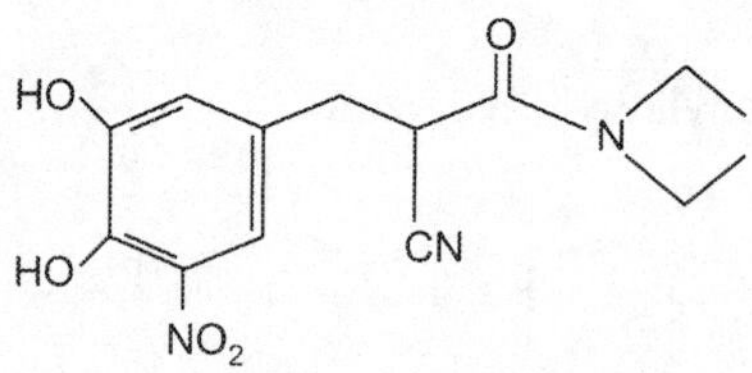

Fig. 8.13

8.4.1.6 Dopamine Agonist

1. Levodopa (Sinemet®): Structure mentioned above (–)-3(3, 4-dihydroxy phenyl)-L-alanine.

8.4.2. Agents Action on Nondopaminergic System

8.4.2.1 Anticholinergic Agent

1. Benztropine (Cogentin®): 3-(diphenylmethoxy)-8-methyl-8-azabicyclo [3, 2, 1] octane.

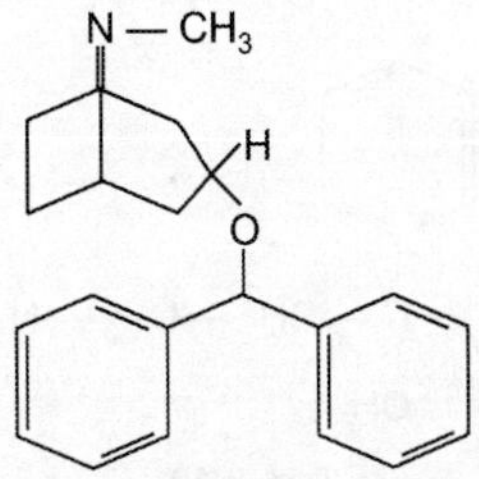

Fig. 8.14

2. Biperiden (Akineton®): 1-(6-bicyclo[2, 2, 1]hept-2-enyl)-1-phenyl-3-piperidine-1-yl-propane-1-ol.

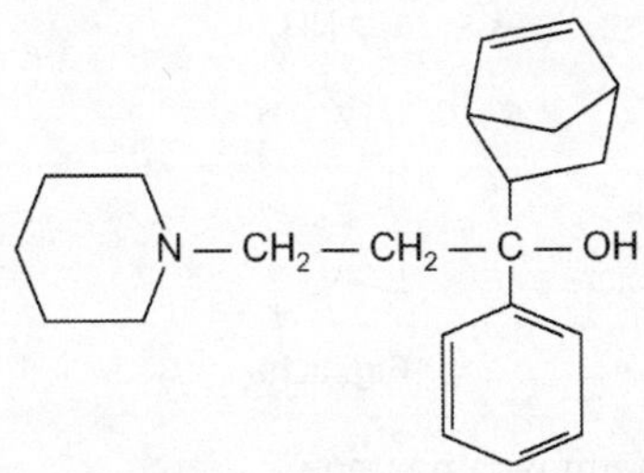

Fig. 8.15

3. Diphenhydramin (Benadryl®): 3-(diphenylmethoxy)-N, N-dimethyl ethanamine

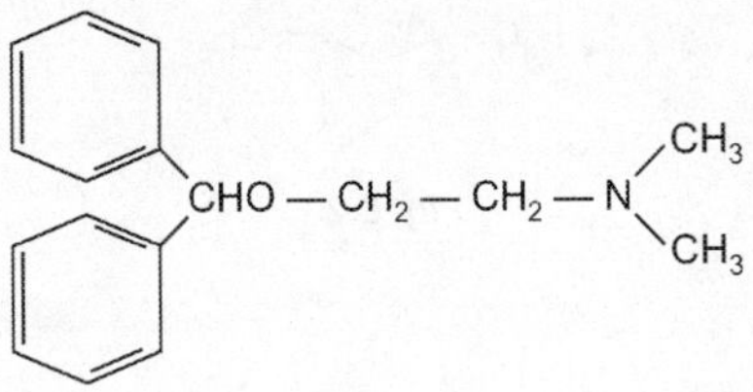

Fig. 8.16

4. Procyclidine (Kemdrin): 1-cyclohexyl 1-phenyl-3-pyrrolidin-1-yl-propan-1-ol.

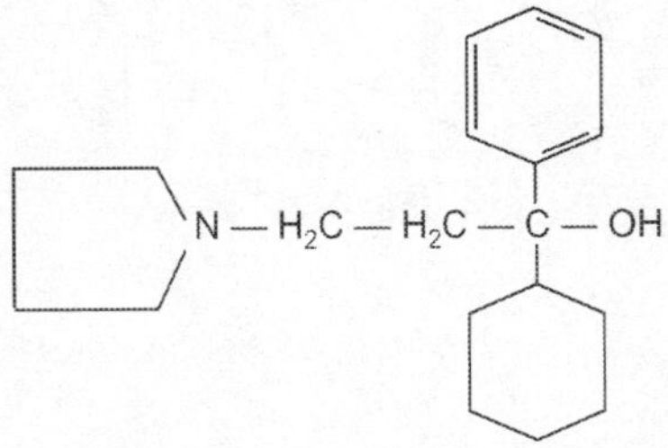

Fig. 8.17

5. Trihexyphenidyl (Parkopan®): 1-cyclohexyl-1-phenyl-3-piperidinopropan-1-ol.

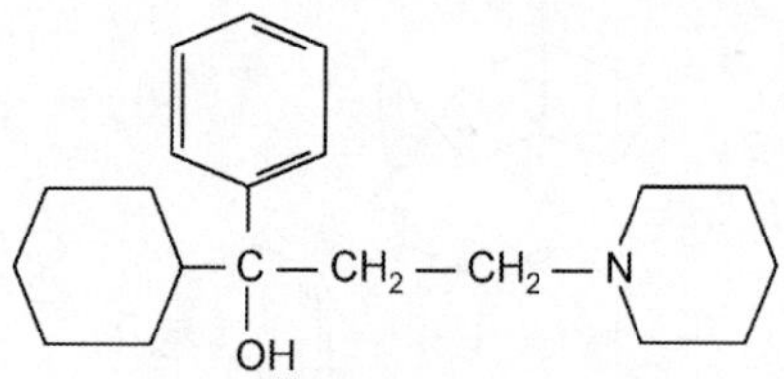

Fig. 8.18

8.4.2.2 *Glutamate Antagonist*

1. Amantadine (Symmetrel®): Adamantan-1-amine

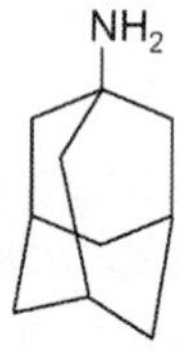

Fig. 8.19

2. Memantine: 1-amino-3, 5-dimethyladamantane

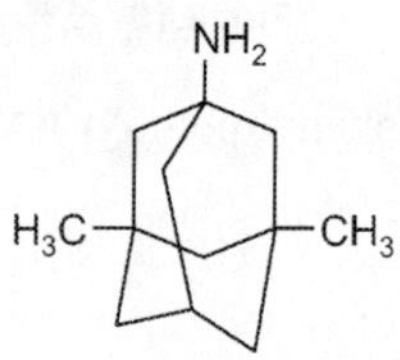

Fig. 8.20

8.4.2.3 *Serotonin Agonists*

1. Sarizotan: 1-[3, 4-dihydro-2H-chromen-2yl]-N-[5-(4-fluorophenyl) pyridin-3-yl]methyl) methanamine

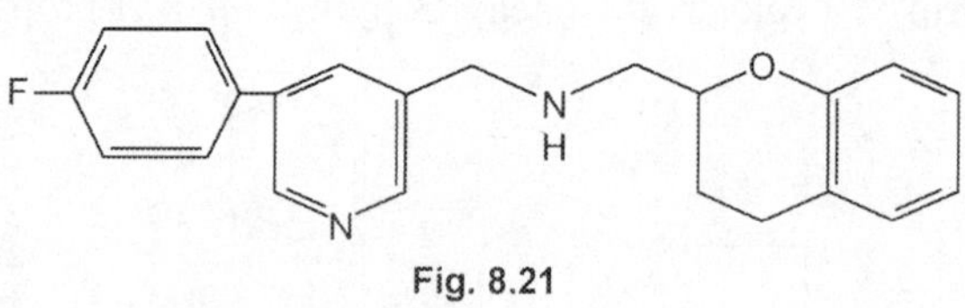

Fig. 8.21

8.5 DOPAMINERGIC AGENTS

In the treatment of PD, the compensation of the dopamine deficiency in neostriatum, usually by direct stimulation of dopamine receptors. Another option is to enhancing the synthesis of dopamine by decrease its catabolism. Thus, dopaminergic agents are given to the patient during treatment. It has been proved that parkinsonism produced by MPTP that has ability to prevent MAO-B inhibitors action therefore selective MAO-B inhibitors or antioxidants are the right choice to slow progressive of idiopathic PD.

1. Pergolide (Permax®): Pergolide is ergoline derivative.

Synthesis:

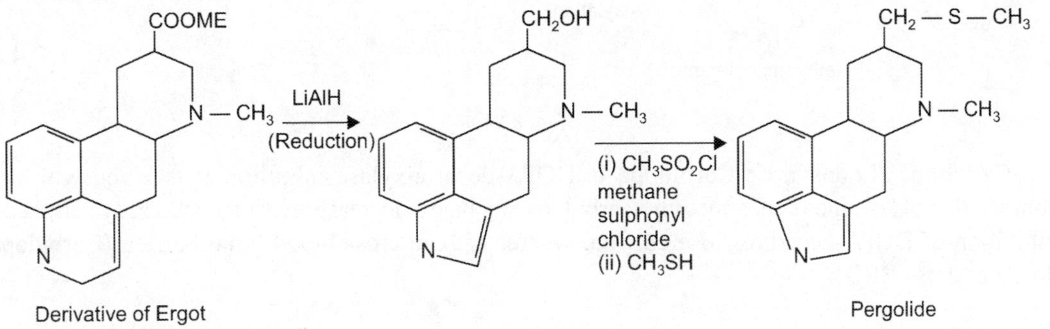

Fig. 8.22

Pergolide is selective D_2 receptors agonist and used in the treatment of parkinsonism.

2. Levodopa (Sinemet®): Levodopa is a levorotatory isomer of dioxyphenylalanine and mainly used as a precursor of dopamine. Therapy behind the assumption that Levodopa does not cross the blood brain barrier. While it turns into dopamine and it has ability to normalize the dopamine level in the brain.

Synthesis:

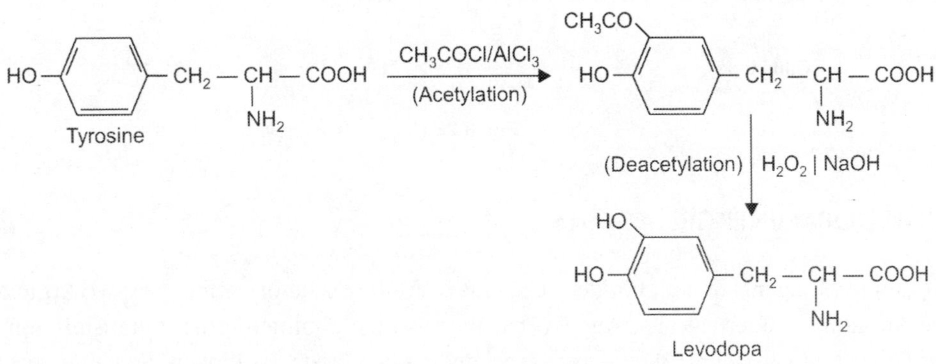

Fig. 8.23

3. Selegiline (Eldepryl®): Selegiline is selective MAO-B inhibitor which suppress dopamine inactivation process and increase the level of dopamine in brain as well as Selegiline is used in the combination with Levodopa in the treatment of Parkinson disorder. It also oxidizes MPTP to neurotoxic by-product.

Synthesis:

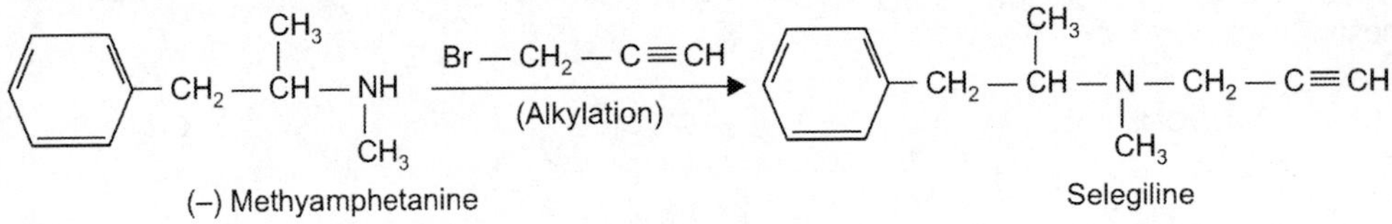

Fig. 8.24

4. Carbidopa (Lodosyn®): Carbidopa is DOPA-decarboxylase inhibitor. It is given with the combination of levodopa concomitantly. Carbidopa falicitate to increase level of levodopa by peripheral inhibition of DOPA-decarboxylose enzyme whereby it can cross blood brain barrier. (Carbidopa does not cross BBB).

Synthesis:

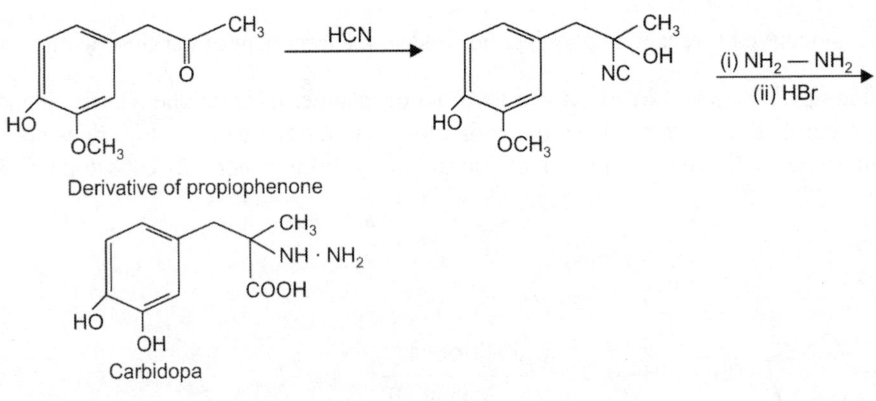

Fig. 8.25

8.6 NON-DOPAMINERGIC AGENTS

Non-dopaminergic agents or anticholinergic agents exert the excitatory effect on GABArgic output from the striatum in interneurons. Agents that increase the cholinergic neurotransmission (e.g., physostigmine and carbachol) would be associated to potentiate Parkinson character in human while centrally active antimuscarinic agents (e.g., belladonna alkaloids) are concerns to deactivate cholinergic transmission and manage the deficiency of dopamine release transmission. In the concern of antimusarinic activity, antihistaminic agents also exert central muscarinic antagonist action but the adverse effects also put by central antimuscarinic agents like peripheral blocked of parasympathetic functions (dry mouth, impaired visual accommodation, constipation, urinary retention, tachycardia). nonpaminergic drugs sometimes given with L-dopa to prevent tremor.

1. Benzotropin (Congentin®): Benzotropin is synthetic anticholinergic drug. Benzotropin is used in the treatment of Parkinson's disease.

Synthesis:

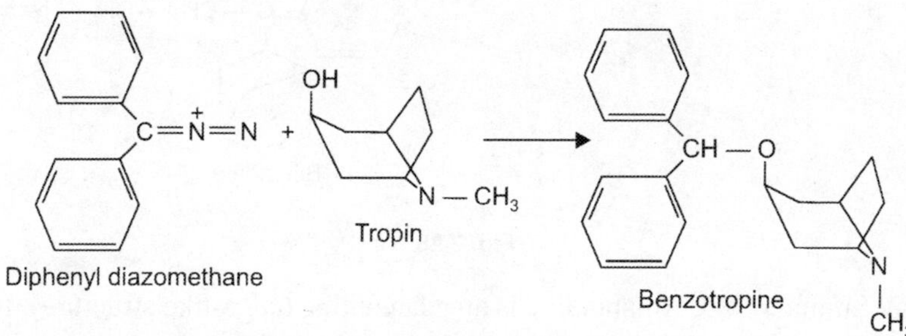

Fig. 8.26

2. Procyclidine (Kemdrin®): Procyclidine is also synthetic anticholinergic drug. It is to be suggested in the treatment of Parkinson's disorder.

Synthesis:

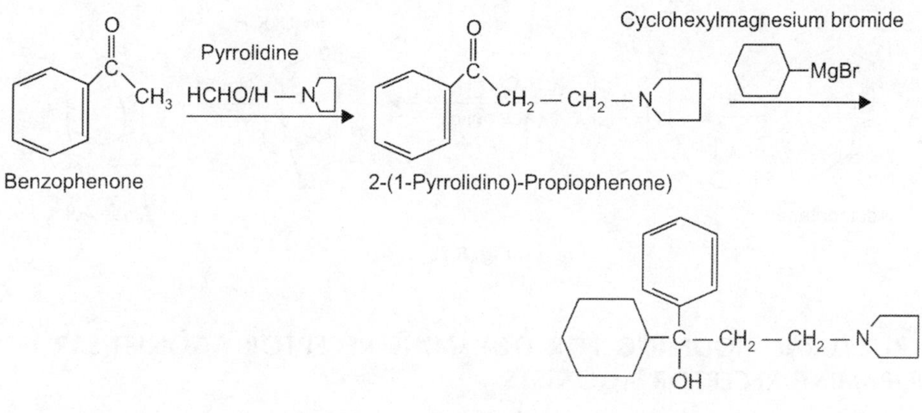

Fig. 8.27

3. Biperiden (Akineton®): Biperiden also possesses same properties as the aforementioned agents.

Synthesis:

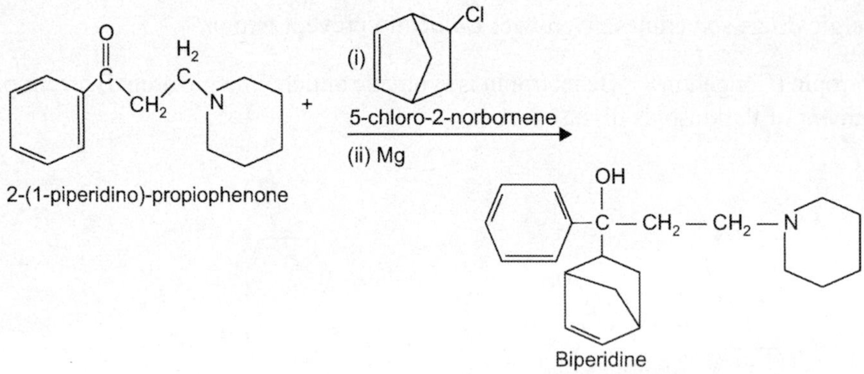

Fig. 8.28

4. Amantadine (Simmetrel®): Amantadine is an adamantine (cage-like structure). It was initially developed as an antiviral agent. Amantadine has ability to release dopamine and Nor-adrenaline from the interneuronal storage place and block-re-uptake of dopamine, therefore the appropriate availability of dopamine as neuro-transmitter. Amantadine also enhances the effect of L-Dopa, on the other site, the derivative of *Memantine* that has been used as spasmolytic agent in the treatment of both Parkinson's and dementia disorders.

Synthesis: Ritter Reaction

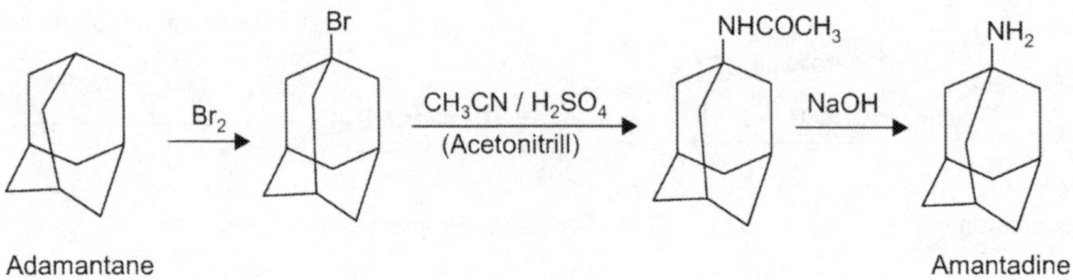

Fig. 8.29

8.7 STRUCTURAL MODELING FOR DOPAMINE RECEPTOR AGONIST/SAR OF DOPAMINE RECEPTOR AGONISTS

8.7.1. Conformational Aspects

1. Development of selective dopamine receptor proximity are the major concerns according to QSAR studies of lead molecule.
2. It should be clear that agonist action exerted by the compounds at dopamine D_1 type and especially D_2-type receptors properly.

3. The side chain of Dopamine possess unlimited flexibility and unrestricted rotation in respect of β-carbon phenyl bond. Therefore, it must be clear that the drugs concerning conformational requirements for activation of Dopamine receptors with the help of endogenous ligands, would possess prime statement.
3. However, catechol ring containing compounds and amino-ethyl side chain exhibit rigid conformation but this point is required for essential pharmacological activity.
4. Studies reveals that rigid dopaminomimetic agents strongly suggest for suitable conformation of D_2-like Dopamine Agonist models in its extended trans-α-conformation which is likely to be an important detriment of agonist activation of D_1-type receptors.

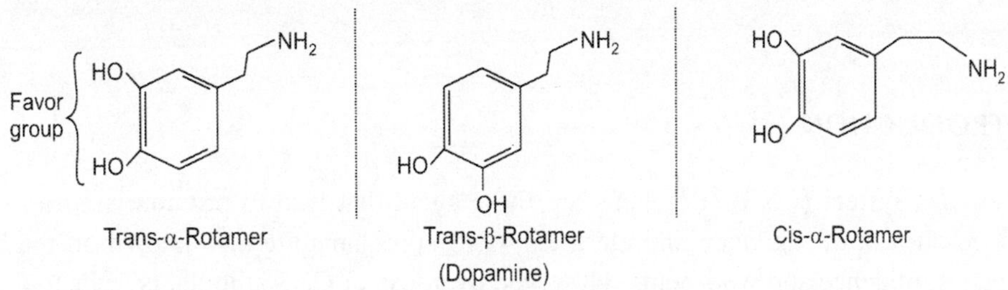

Fig. 8.30

8.7.2 PKa Aspects

1. It should be essential, one protonated nitrogen (at physiological pH) and at least one electronegative centre would be able to participate in hydrogen bonding (such groups are catechol critically positioned hydroxyl groups) for the interaction of agonists to dopamine receptors.
2. PKa of apomophine is about 9 (protonated at physiological pH). It would assist to cross the blood brain barrier whereas Dopamine (PKa = 10.6), failure to cross blood brain barrier.

EXERCISE

1. What do you mean by Parkinsonism. Explain the initiation therapy of PD.
2. Classify Antiparkinson's agents with various structures.
3. Give the mechanism of action of dopaminergic agents.
4. Write the synthesis of the following drugs:
   (i) Pergolide
   (ii) Seligeline
   (iii) Carbidopa
   (iv) Amantadine
5. Why anticholinergic agents are suggested for the treatment of Parkinson's disease. Explain.
6. Write any two synthesis of synthetic anticholinergic drugs. Give their MOA and uses also.
7. Explain structural model requirements or SAR dopamine receptors agonists.

9

Central Nervous System Stimulants

9.1 INTRODUCTION

Central nervous system (CNS) stimulants are those agent that lead to promote arousal, improve performance on task of vigilance and alertness. CNS stimulants are also known on the basis of sense of self-confidence and well-being. However, overdose of CNS stimulants leads to feeling of elation or euphoria. Naturally occurring chemical agents also have stimulating properties. They are known as psychostimulants, natural products are ephedra sinica (Ephedra) and khat, (–)cocain, caffeine, all act as CNS stimulants on the specific dose, overdose may cause harmful effects. These drugs also cause dependency because they make user's feel 'good' sometimes they are taken for extended period of time to maintain the elevated mood.

It has been proved clinically that most CNS stimulant agents also have appetite-suppressant effects, therefore they are used to control obesity. Studies have proved that amphetamine exhibits favourable results rather than placebo in promoting weight loss effect.

CNS stimulants encompass a number of physiological mechanisms of action and distinguished biologically active substances.

Xanthin derivatives show CNS stimulant effects and take part to relax smooth muscles. Strychnine also shows CNS stimulant effects by antagonist action of glycine receptors. Benzodiazepines concern with decreasing the inhibitory action of GABA receptors on inhibitory chloride ion channels.

9.2 CLINICAL USES OF CNS STIMULANTS

1. To improve the level of activity, alleviate fatigue, increase alertness, elevated mood.
2. CNS stimulants are used in the treatment of attention deficit hyperactivity disorders (ADHD).
3. In the treatment of depression in terminal illness such as amphetamine and methylphenidate.
4. In the treatment of obesity (anorectics) in the treatment of AOP (apnea of prematurity) of infants, to stimulate respiration in infants, drugs like methylxanthines are extensively used.

9.3 CLASSIFICATION

CNS stimulants are generally classified on the basis of their mode of actions.

(1) Psychomotor stimulants: examples are caffeine (xanthine derivatives), amphetamine, methyl phenidate, pemoline, etc.

(2) Respiratory stimulants or analaptics: Doxapram, almitrine are the examples of analeptics

(3) Appetite suppressants or anorectics: examples are phentermine, diethylpropion, etc.

The chemical classification of CNS stimulants is the following.

Classification

1. Phenyl isopropylamine/amphetamine related agents/sympathomimetics agents

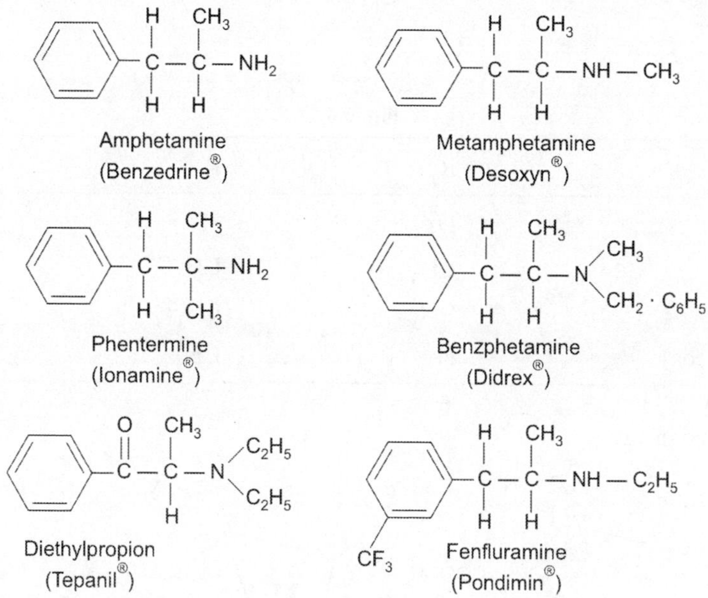

Fig. 9.1

2. Morpholine Derivative (Phendimetrazine Tartrate) (Plegine®)

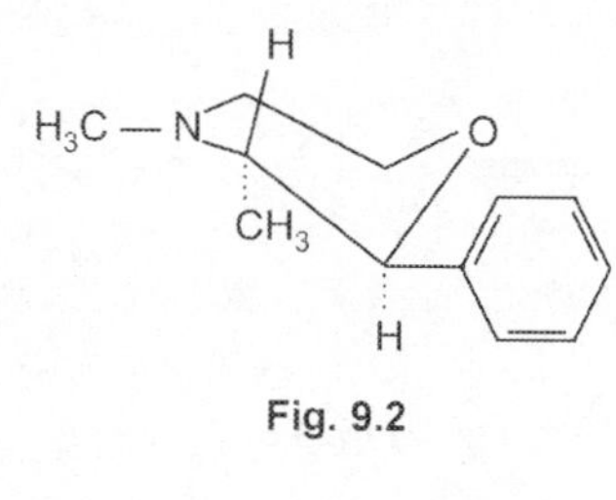

Fig. 9.2

3. Pyrrolidine Derivative (Methyl Phenidate) (Ritalin®)

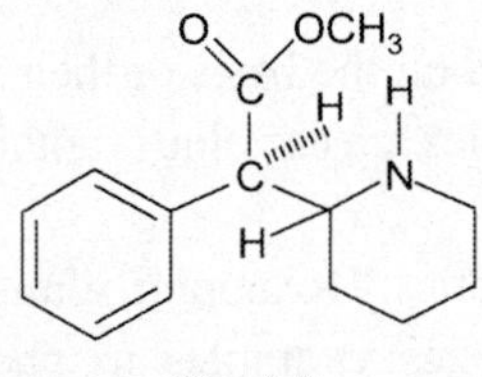

Fig. 9.3

4. Methylxanthines/xanthine alkaloids

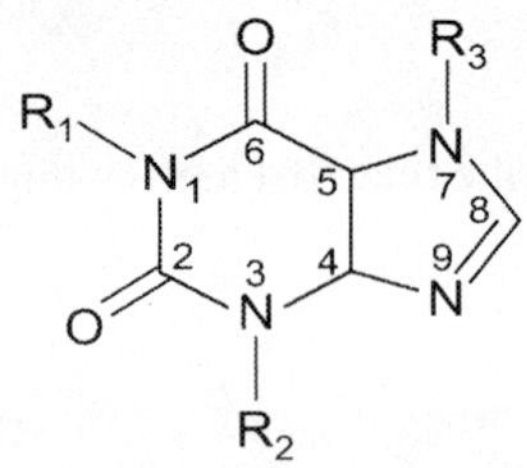

Fig. 9.4

Derivatives	R_1	R_2	R_3
Xanthine	H	H	H
Caffeine	CH_3	CH_3	CH_3
Theophylline (tea)	CH_3	CH_3	H
Theobramine (cocoa)	H	CH_3	CH_3

5. Cocaine related drugs

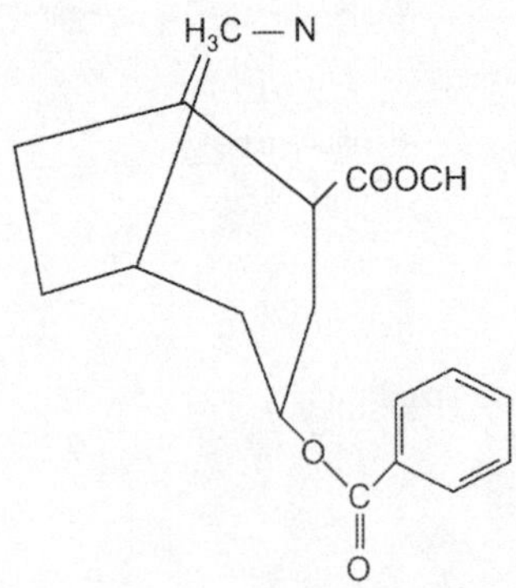

Cocaine

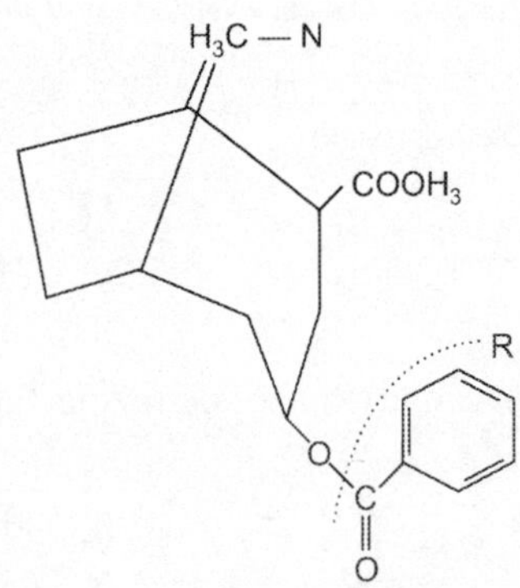

Benzoylecgonine

Ecgonine

R = H

Fig. 9.5

6. Miscellaneous CNS stimulants

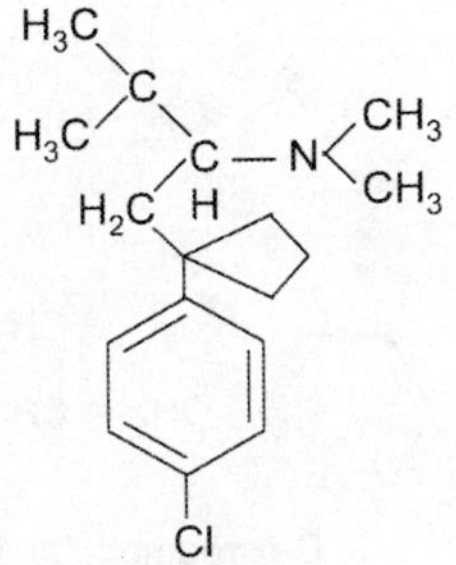

Sibutramine
(Meridia®)

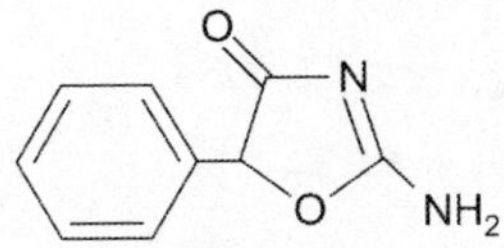

Pemoline
(Cylert®)

Fig. 9.6

9.4 MECHANISM OF ACTION

1. According to the various studies of mechanism of action of psychostimulants in animals uniformly focus the role of dopamine pathway to betterment of locomotor activity and reinforce the properties.
2. Clinically tested events concluded that several stimulants using *in vivo* brain imaging either with SPECT (single photon emission tomography) or PET (positron emission tomography) technique renders the evidence for elevation of extracellular dopamine in response to central nervous system stimulants therapy.
3. Serotonin reuptake inhibitors at presynaptic cleft or 5-HT_{1B} receptor potentiation plays important role to stimulate psychomotor properties, for example, cocain.
4. Increase the level of norepinephrine by the potent NE-reuptake inhibitors are another factors of psychostimulation e.g., amphetamine, cocain and xanthine derivatives.

9.5 CNS STIMULANTS AGENTS

9.5.1 Amphetamine Related Agents

Amphetamines are considerred powerful synthetic psychostimulants associated with high potential of addiction. They help to improve vigilance, ability to concentrate, elevation of mood and betterment of motor activity. Apart from that, they may cause euphoria at various stages, increase BP and mydriasis. Prolonged use may cause irritability, insomnia and hyperhidrosis.

1. Dextroamphetamine: D-2-amino-1-phenyl propane dextroamphetamine is a potent stimulant of nervous system. It is responsible for release of dopamine and NE at presynaptic nerve ending. It is used in the treatment of narcolepsy, encephalitis, drowsiness, asthenia, etc.

Synthesis: By Leucart Reaction

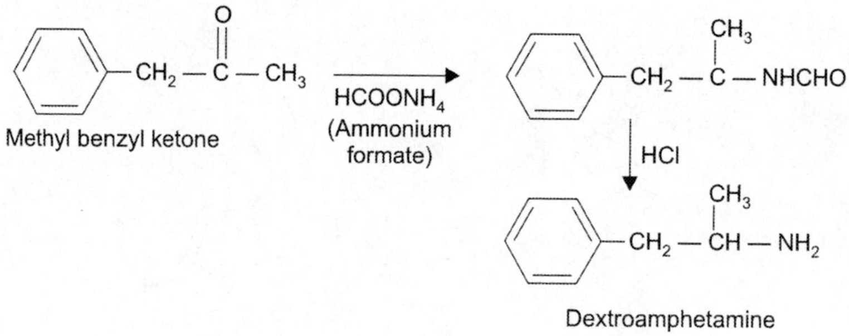

Fig. 9.7

2. Methamphetamine: (+)-N-α-dimethyl phenylethyl-amine Methamphetamine possesses all features as in dextroamphetamine.

Synthesis:

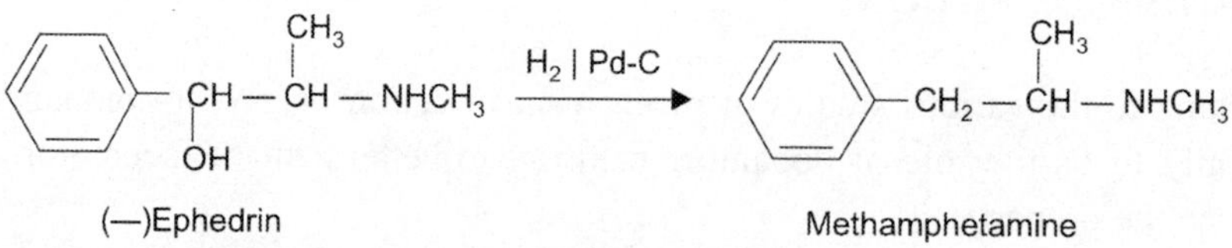

Fig. 9.8

3. Phentermine: α, α-Dimethyl Phenylethylamine, in the contrast of amphetamine, phentermine has an additional methyl group at α-position of amino group. Phentermine is the agent of anorectic category. Its action leads to activation of satiation centre of hypothalamus and resulting reduction in appetite, therefore phentermine is used in the treatment of obesity.

Synthesis:

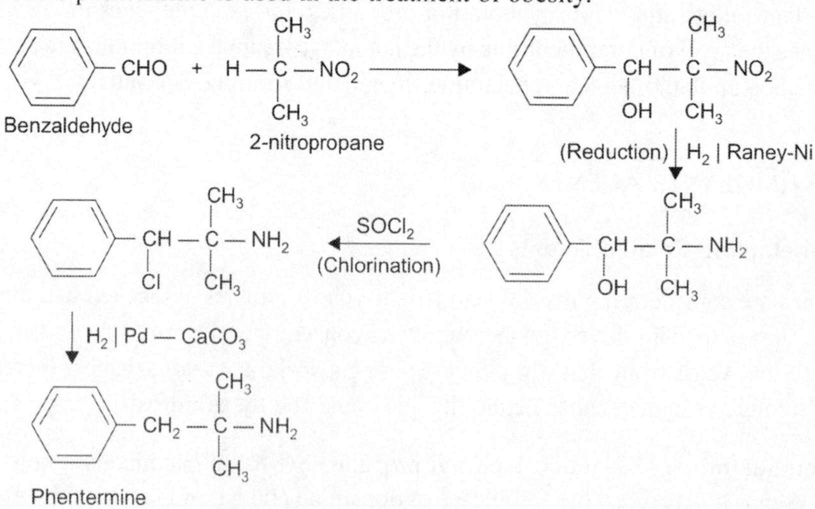

Fig. 9.9

Synthesis:

4. Diethylpropione: 1-Phenyl-2-diethylaminopropane Diethylpropione is the drug which comes in the anorectic category. It is used in the control of obesity by limiting calory intake.

Synthesis:

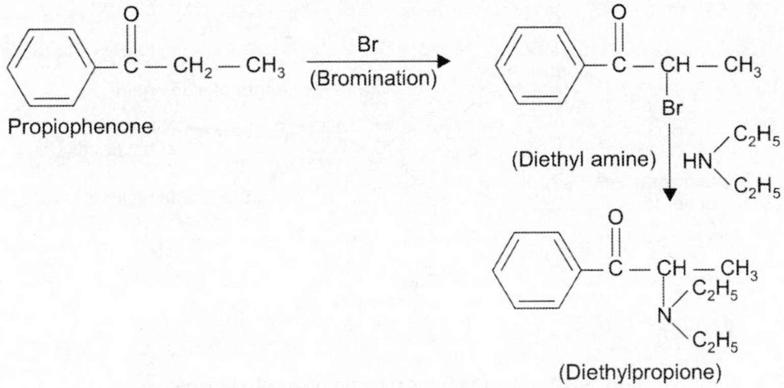

Fig. 9.10

5. Methylphenidate: Methyl ester of α-phenyl-2-piperidilactic acid

Methylphenidate is a pyrrolidine derivative. Its properties are similar to amphetamine. It is a potent CNS stimulant. Its action lies in mental activity rather than motor activity. It has advantage over others than in therapeutic dose. It does not increase the blood pressure. Methylphenidate is used in the treatment of moderate depression and apthetic conditions.

Synthesis:

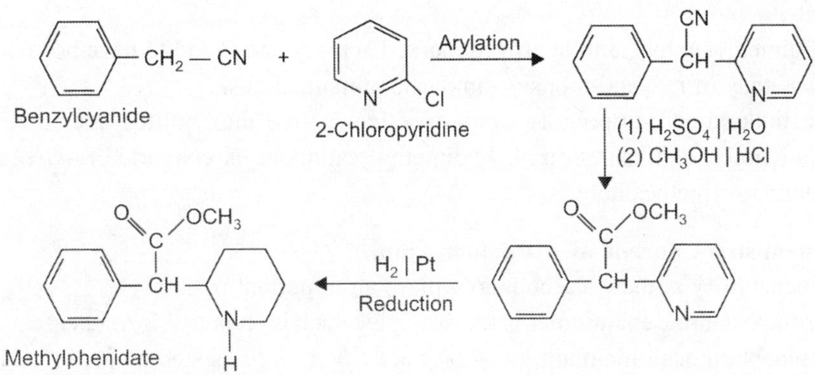

Fig. 9.11

9.5.2 SAR or Structural Requirements of Amphetamines

1. The basic skeleton of amphetamine compete the structural requirements of dopamine (and norepinephrine).
2. Various locations of substitution for a general phenethylamine structure have been studied for a potent CNS stimulant activities.
3. Essential requirements of amphetamines review by Biel and BoPP.

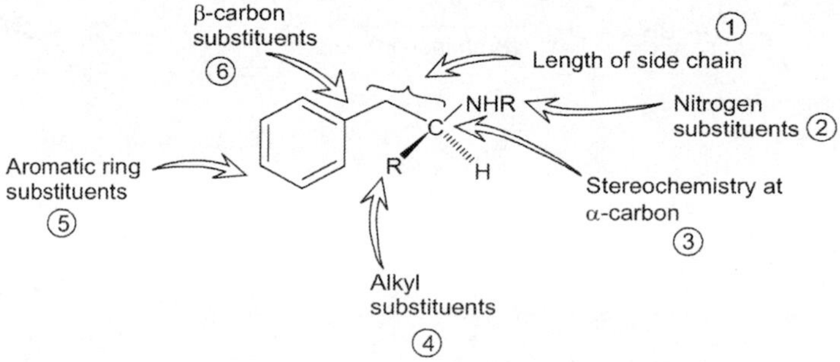

Fig. 9.12: Favoured substitution of amphetamines.

1. Length of Side Chain

(i) To exhibit optimum pharmacophoric template, the length of two carbon (α, β to nitrogen atom) atoms are essential.

(ii) This is essential core that bear by dopamine, norepinephrine and serotonine.

2. Nitrogen Substituents

(i) Primary amine (amphetamine) and N-methylamine (methamphetamine) are the most potent agents.

(ii) Introduction of larger alkyl groups or N, N-dialkylation substitution dramatically reduce the activity.

(iii) N, N-dimethyl amphetamine proved marked activity and found to have behavioural effects *in-vivo* study of rats and monkey similar to amphitamines.

(iv) N, N-dialkylated compounds show excellent active metabolites due to bearing of β-ketofunction. For illustration N, N-dimethyl cathinone is converted *in-vivo* to N-methyl compound of methcathenone.

3. Stereochemistry Concept of α-Carbon Atom

(i) Stereochemistry at the α-carbon atom plays an important role.

(ii) S-(+) amphetamine enantiomer is the example which is responsible to release dopamine and nor-epinephrine and maintain locomotor activity as well as stereotypic effects rather than R(–) isomer.

(iii) Study suggested that (+) enantiomer shows excellent activity than (–) isomer.

4. Presence of α-alkyl substituents

(i) Lower alkyl group shows greater activity than higher alkyl group at α-alkyl substitution.

(ii) Lacking of side chain α-methyl group is inactive *in vivo* study because of its rapid inactivation by mono-amine oxidase enzyme, therefore addition of α-methyl group abolish the inactivation process and increase the oral bioavailability of drugs.

5. Other Side Chain Substituents

(i) Limited substitution of side chain at β position would be tolerable.

(ii) β-hydroxy substitution on methamphetamine give ephedrine which is a CNS stimulant but degree of adverse effects are less than the previous one.

(iii) Introduction of β-hydroxy group (polar) at β-position results in reducing the hydrophobicity therefore decreases in CNS penetration.

(iv) On the other hand, invitation of methyl group, e.g., N-methyl ephedrine increase lipid solubility, thus ephedrine is greater CNS stimulant rather than phenyl propanolamine.

6. Aromatic Ring Substitution

(i) Any changes in aromatic ring results in decrease in the activity.

(ii) Study reveals that no ring modification retains the amphetamine-like psychostimulant activity.

(iii) However, introduction of halogens at para position leads to compounds that have marked serotonin releasing potency for illustration fluoroamphetamine.

9.5.3 Xanthine Alkaloids

Xanthine alkaloids are naturally occurring alkaloids. Principal compound of xanthine alkaloid is caffeine. Caffeine is 1, 3, 7-trimethyl xanthine. Other similar alkaloids are theophylline and theobromine. Caffeine is most widely accepted as CNS stimulant. It is an alkaloid obtained from the tea leaves (*Thia sinensis*), from coffee beans (*Coffea arabica*), from cocoa beans (*Theobroma cocoa*). Xanthin alkaloids clinically accepted for treating pulmonary disease.

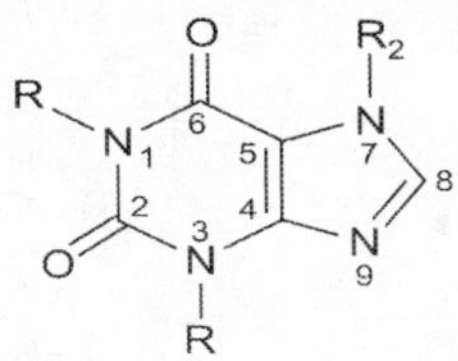

	R	R_1	R_2
Calleine	CH_3	CH_3	CH_3
Theophylline	CH_3	CH_3	H
Theobromine	H	CH_3	CH_3

Fig. 9.13: Xanthine ring.

Alkyl substitution on position 1, 3, 7 are suitable for xanthine derivatives.

Caffeine: 1, 3, 7-trimethyl xanthine

Theophylline (Elixophylline®): 1, 3-dimethyl xanthine

Theobromine (Theosalvosa®): 3, 7-dimethyl xanthine

A cup of coffee may contain 50 – 150 mg of caffeine and used as CNS stimulants (weak in small dose). Caffeine is used for relieving headaches associated with blood flow disorders of brain. It also leads stimulatory effect on respiratory and vasomotor centres. In large dose caffeine directly acts on myocardium therefore leads to tachycardia and arrhythmia.

Theophylline is the principal alkaloid of tea while theobromine is an example of alkaloid of cocoa.

Theophylline is used in the treatment of asthma and used as bronchodilator.

Mechanism of action: CNS stimulating effects of methyl xanthine is attributed to their phosphodiesterase-inhibiting ability on therapeutic dose. Evidence reveals that antagonist action on adenosine at A_1 and A_{2A} receptors of the CNS-stimulants agents. They increase the level of cyclic adenosine monophosphate (cAMP) therefore initiate the second messenger and several biological active substances.

Synthesis of Caffeine

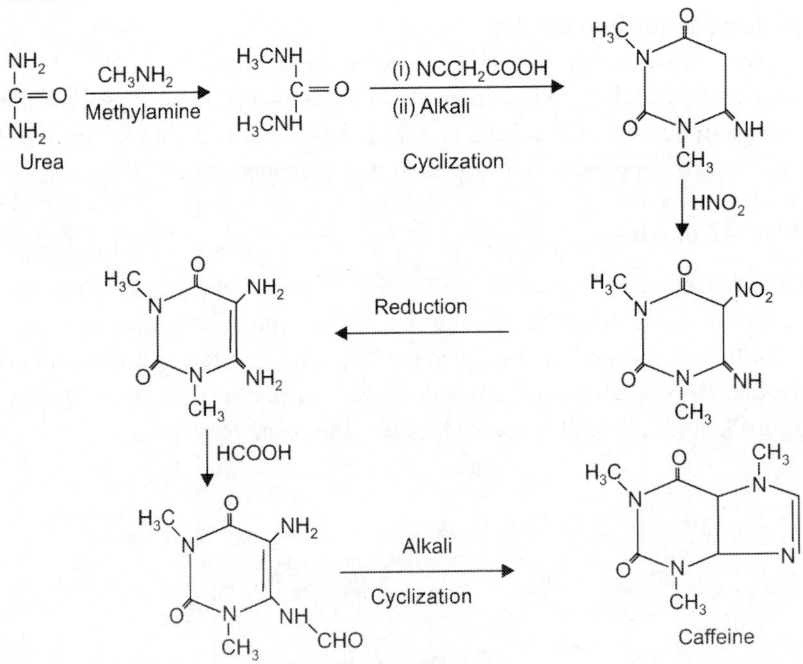

Fig. 9.14

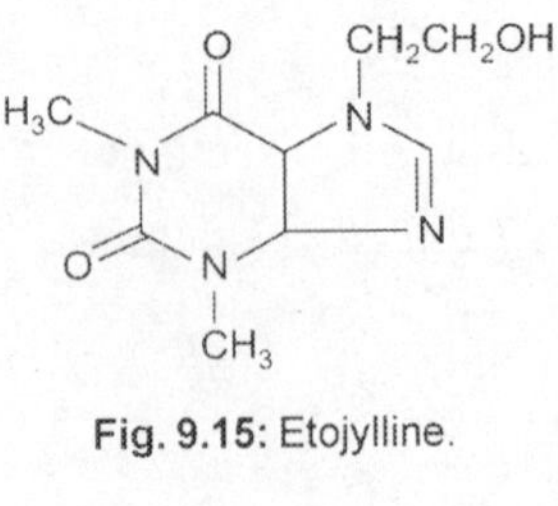

Fig. 9.15: Etojylline.

Other Xanthine Derivatives

(i) Aminophylline (phyllocontin®): 3, 7-dihydro-1, 3-dimethyl xanthine
Aminophylline is synthesized from theophylline and ethylenediamine treatment in ethanol. It is used in the treatment of congestive heart failure, in asthma and cardiac paraxymol dyspnea.

(ii) Etofylline (bio-phyllin®): 7-(2-hydroxyethyl)-theophylline. Etofylline is considered a better tolerable drug than its parent drug theophylline.

9.5.4 Cocaine Related CNS Stimulating Agents

Cocaine is chemically 2R-Carbomethoxy-3.S benzyloxy-1 R-tropane. It is naturally occurring alkaloid obtained from the plant of erythroxyloncoca species, (described in local anaesthetics) and moiety detailed in opioid analgesics). Cocaine possesses CNS stimulants activity in addition to vasoconstriction action and local anaesthetic activity. Cocaine containing preparations are available in the market, vincoca and soft drinks Coca-Cola. Cocaine is therapeutically useful agents as 5-HT_3 and serotonin antagonistic. Cocaine is marked to block re-uptake action of norepineprine, serotonine and dopamine therefore increase the level of neurotransmitters in synaptic cleft, thus exhibit CNS stimulant activity.

9.5.5 Miscellaneous CNS Stimulants Agents

(i) Sibutramine (Medridia®): Sibutramine is norepinephrine and 5-HT reuptake inhibitor. It is reported as antidepressant and anorexiant agent. Sibutramine is widely used in the control of obesity.

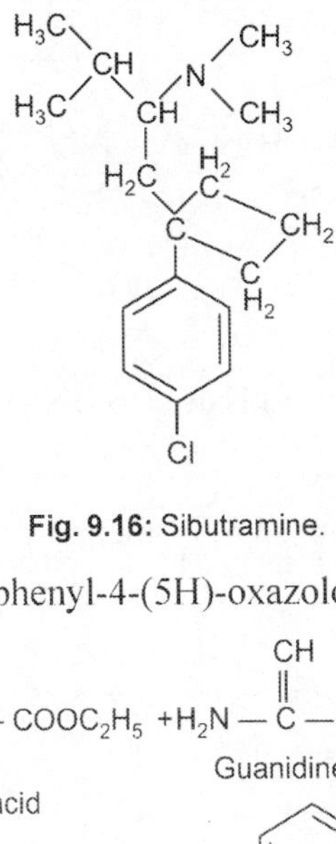

Fig. 9.16: Sibutramine.

(ii) Penoline (Cylert®): 2-amino-5-phenyl-4-(5H)-oxazolone

Synthesis:

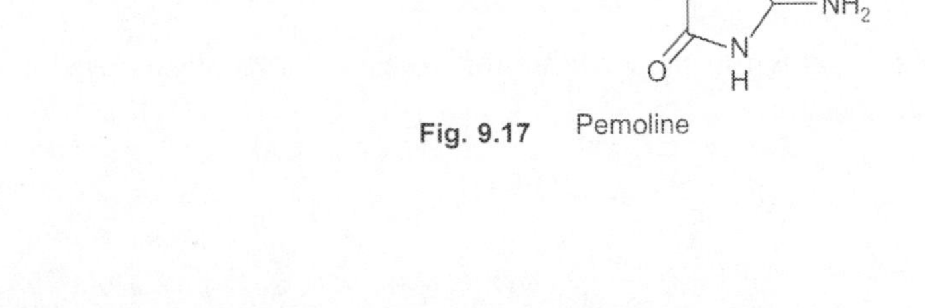

Fig. 9.17 Pemoline

Pemoline is nor-epinephrine and dopamine re-uptake inhibitors therefore it increases the rate of synthesis of dopamine. Pemoline is used as potent CNS-stimulating agents.

Synthesis:

9.5.6 Newer Agents

1. **Modafinil (Provigil®):** D-modofinil is derivative of non-amphetamine moiety. It is approved for the use in Nacrolapsy. It is more effective than amphetamine with lower incidence of side effects.

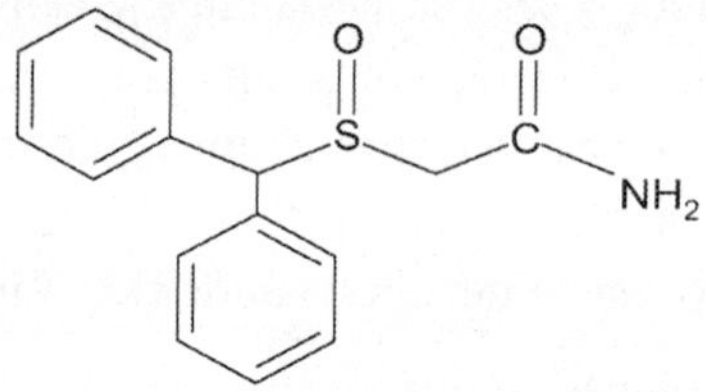

Fig. 9.18

2. **Pramipexole:** N-Propyl-tetrahydro-1, 3-benzothiazole-2,6-diamine

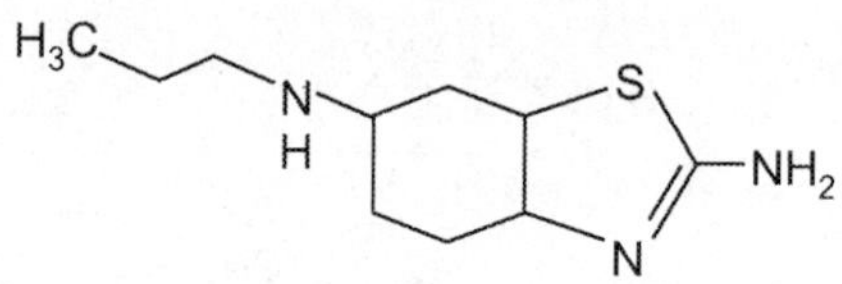

Fig. 9.19

Pramipexole is dopamine receptor agonist. It is used as CNS stimulant as well as in the treatment of Parkinsonism disease.

1. **Tolcapone:** (3, 4-dihydroxy-5-nitrophenyl) (4-methyl-phenyl) methanone.

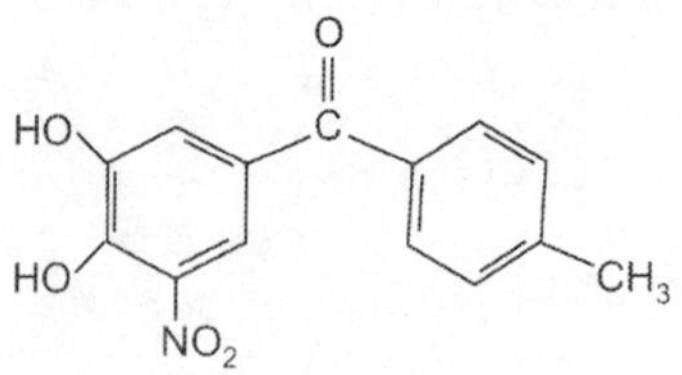

Fig. 9.20

Tolopone is catechol-o-methyltransferase inhibitor. Tolcapone is basically used in the treatment of Parkinson's disease.

9.6 METABOLIC PATTERNS OF SOME CNS STIMULANTS AGENTS

(i)

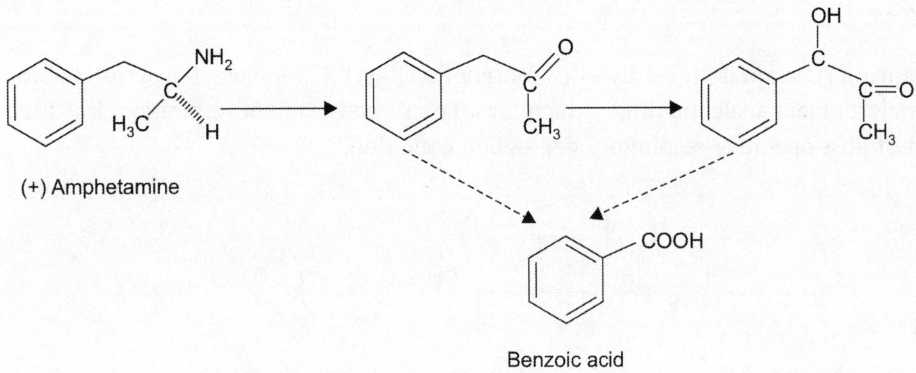

(ii)

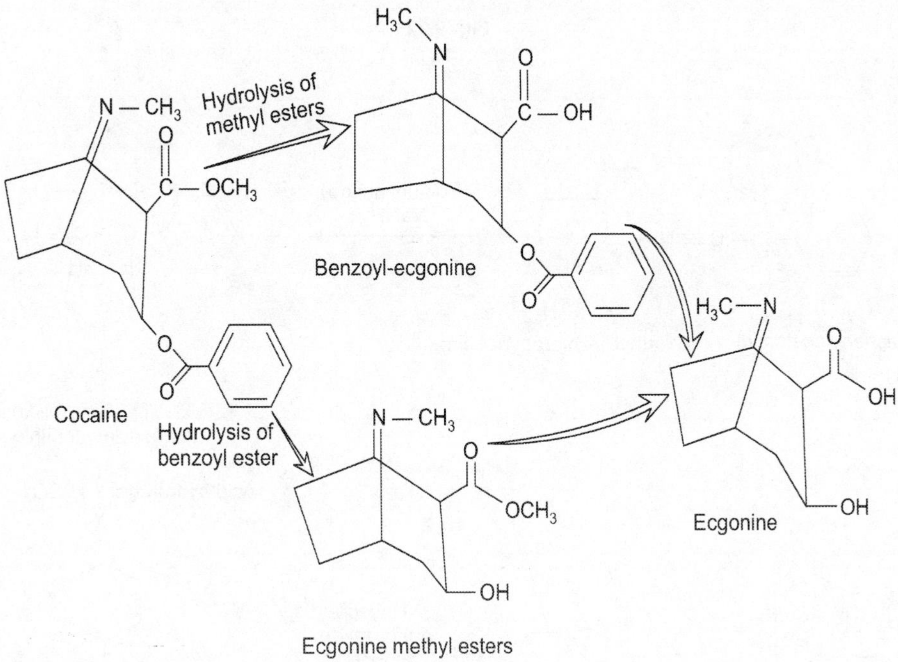

Fig. 9.21

9.7 ANALEPTICS OR RESPIRATORY STIMULANTS

Analeptic is the category of the drugs that possesses the stimulatory effects on the respiratory and vasomotor centres situated in the medulla of brain. Analeptics are concerning primarily used as antagonist in the overdose cause sedative and hypnotics. While at the small dose of therapeutic action, they can stimulate other portion of centre nervous system. Further, in minor overdose they lead a number of undesirable side effects like stimulation of cardiovascular system, convulsions, vomiting, etc.

1. Doxapram (Doxapril®): 1-ethyl-4-(2-morpholinoethyl)-3, 3-diphenyl-2-pyrrolidinone. Doxapram is a potent analeptic drug. It increases the rate and depth of respiration. It is used in the treatment of post-operative respiratory depression condition.

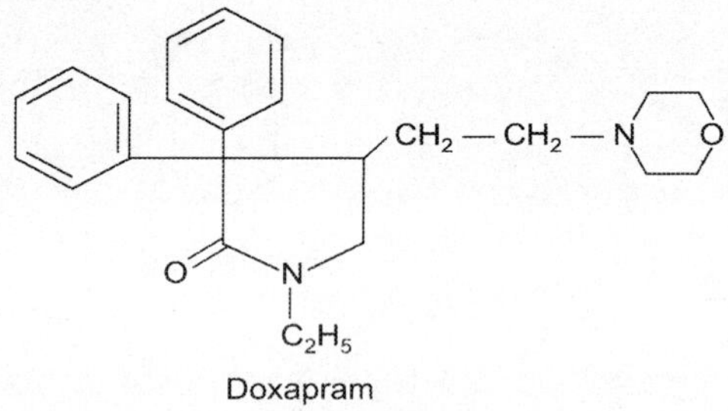

Fig. 9.22

Synthesis:

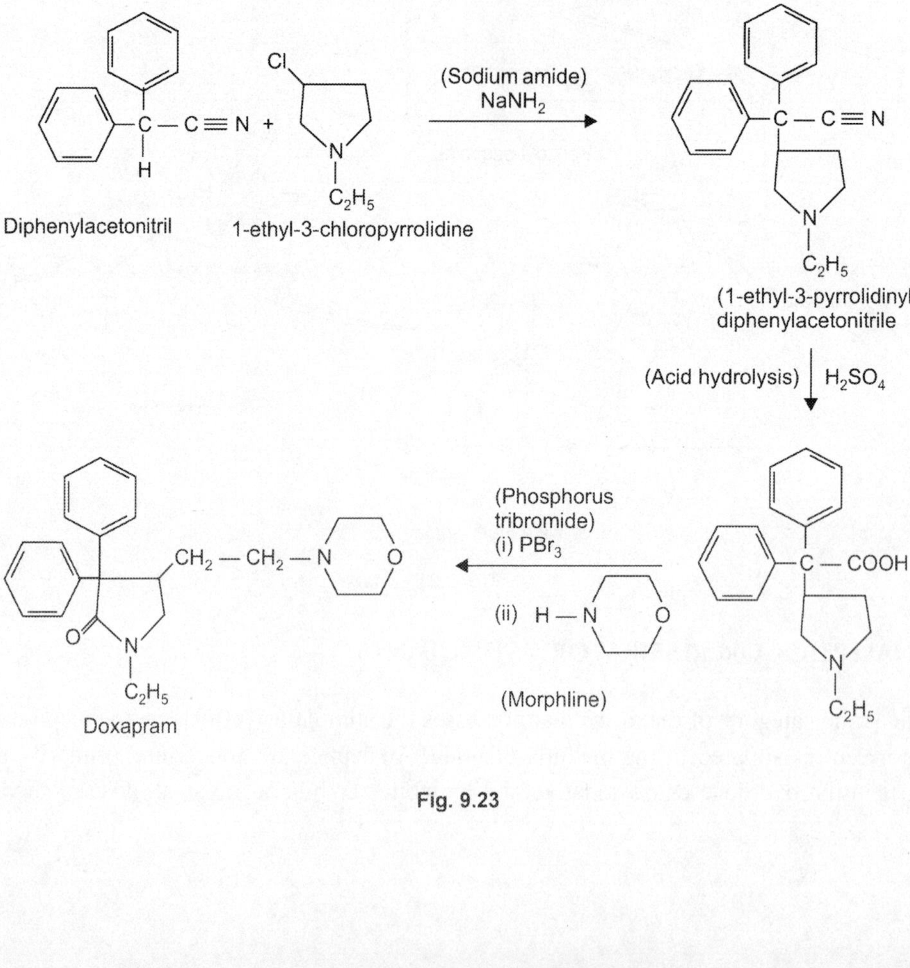

Fig. 9.23

2. Almitrine (Duxil®): -2, 4-bis (allylamino)-6-[4-[bis-p(fluorophenyl) methyl]-1-piperazinyl-s-triazine.

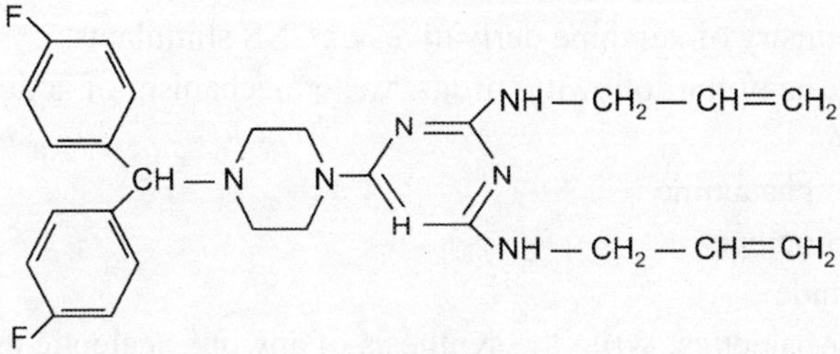

Fig. 9.24

Almitrine is a potent analeptic drug. It has similar properties like doxapram. It increases the rate and depth of respiration. It increases the blood circulation in alveoli for better pulmonary ventilation.

3. Nikethamide (Corazone®): N, N-diethyl Pyridine-3-carboxamide.
Nikethamide is nicotinic acid derivative. It is considered a weak analeptic, used as respiratory stimulants.

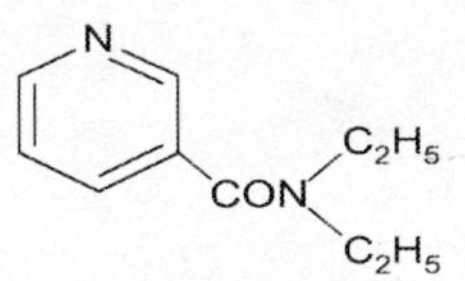

Fig. 9.25

Nikethamide is synthesized from nicotinic acid and Benzene-sulfonic acid.

Synthesis:

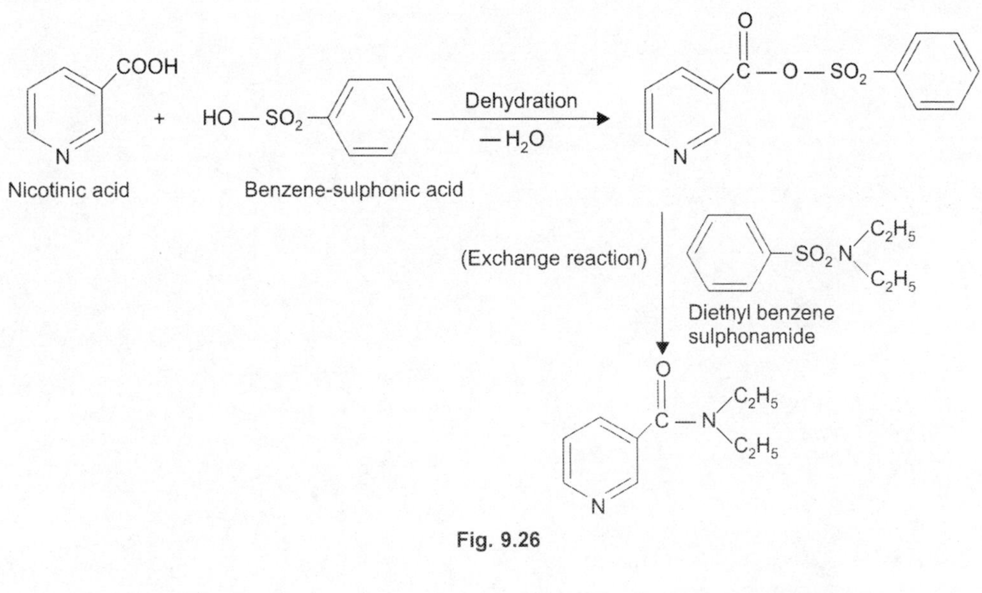

Fig. 9.26

EXERCISE

1. Classify CNS-stimulating agents.
2. Explain the chemistry of xanthine derivatives of CNS stimulants.
3. Write the synthesis of the following drugs, write mechanism of action also.
   (i) Caffeine
   (ii) Dextroamphatamine
   (iii) Diethyl propione
   (iv) Nikethamide
4. Give a note on Analeptics. Write the synthesis of any one analeptic drug.
5. What are the structural requirements of amphetamines? Explain.
6. Discuss the drugs which come in the anorectics or appetite suppressants.
7. Discuss the cocaine related agents as CNS stimulants.
8. Give a note on sibutramine as CNS stimulants.

10

Antipsychotic Drugs

10.1 INTRODUCTION

Schizophrenia is characterized by a severe, long life, idiopathic psychiatric disorder with multigenic component. It is constructed by severe thoughts and disorders called psychoses. Psychosis is characterized by illogical, delusional or paranoid thoughts. It is clinically seen that its occurrence lead onset its early adulthood stage with remissions to throughout life. There are no pathogenomic syndrome features. The syndrome is commonly seen at late adolescence stage, further social withdrawal, perception disturbance to recurrent delusions and hallucinations. Patient may get covered with positive symptoms as well as negative symptoms but negative symptoms predominate in 1/3 population of schizophrenic patients, are associated poor long-term outcomes and poor response to drug treatment.

Schizophrenia is further classified into the subcategories as given below.

Table 10.1: Schizophrenia subtypes

	Type	Salient Features
1.	Catatonic type	Change in motor activity, negativism, echolalia and/or echopraxia
2.	Paranoid type	Delusion, negative thoughts
3.	Disorganized type	Disorganized speech, behaviour superficial affects
4.	Residual type	Absence of delusion, hallucination, motor disturbance

10.2 CAUSES OF SCHIZOPHRENIA

Genetic causation is an evidently proof that is responsible for schizophrenia, genes includes, neurogulin-1, dysbindin, proline dehydrogenase, D-amino oxidase activator, COMT (Catechol-o-methyl transferase), nevregulin-1, dysbindin and D-amino-oxidase activator involves in glutamatergic function further stimulate N-methyl-D-Aspartate (NMDA), in the concerning of treatment for D_2-receptors and possess much higher affinity for the D_4 receptors than D_2-receptors.

10.2.1 Adverse Effects

1. EPS (extrapyramidal symptoms): include Parkinson-like symptoms
2. Tradive pyskinesia
3. Neuroleptic malignant syndrome
4. Seizures
5. Sedation
6. Cognitive impairment
7. Anticholigergic effects
8. Cardiovascular effects
9. Sexual side effects and hyperprolactemia.
10. Weight gain
11. Haematological side effects
12. Hepatic effect
13. Dermatological side effects.

Possible mechanism involved in mediated glutamate responce would be therapeutic target for the treatment. While COMT involves to deactivate the excess action of dopamine from the synapse.

10.3 ANTIPSYCHOTIC AGENTS

Antipsychotic agents are those which are used to treat acute and for maintenance of schizophrenia. Antipsychotic agents are effective in the treatment of hallucinations, delusions and thought disorders.

10.3.1 Mechanism of Action

Antipsychotic drugs are prominent to bind dopamine D_2/D_3 receptors. Newer atypical antipsychotic agents also have ability to bind D_2 receptor and block them. All neuroleptics bind to dopaminergic sites, however newer typical antipsychotic drugs may involve NMDA receptor blocked, α_1 and α_2 non-adrenergic activity.

Conventional neuroleptics like chlorpromazine and thioridazine are anticholinergic agents and show sedative properties while higher potency of antipsychotics, some drugs haloperidol, perphenazine are more likely to induce extrapyrimidal side effects. (EPS: extrapyrimidal symptoms), while Benzodiazipines has greater potency towards $5HT_2$ receptors.

10.4 CLASSIFICATION

1. Phenothiazines (Tricyclic Neuroleptics)

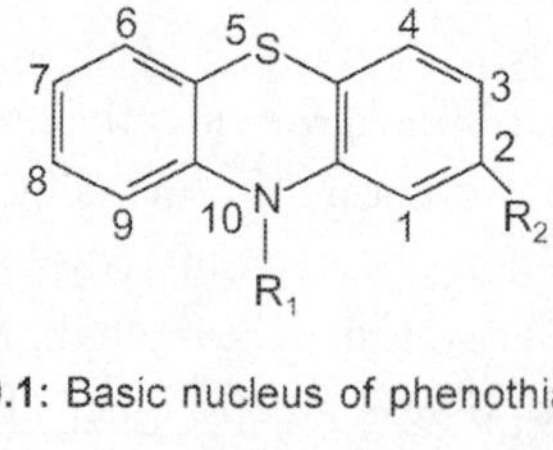

Fig. 10.1: Basic nucleus of phenothiazine.

S. No.	Derivatives	Substitution R_1	R_2
1.	**Propyl Dialkyl Amino Substitution**	(10-(3-Dimethyl aminopropyl) phenothiazine	
A.	Promazine (Sparine®)	$—(CH_2)_3$-N-$(CH_3)_2$	H
B.	Chlorpromazine (Thorazine®)	" "	Cl
C.	Triflupromazine (Vesptin®)	" "	CF_3
2.	**Piperidyl ring substitution**		
A.	Thioridazine (Mellaril®)	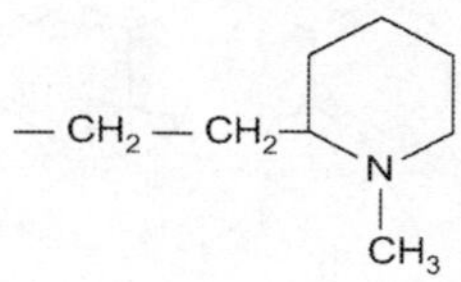	$—SCH_3$
B.	Mesoridazine (Serentil®)	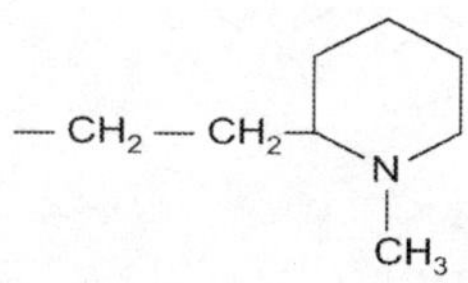	O ↑ SCH_3
3.	**Piperazine ring substitution**		
A.	Prochlorperazine (Compazine®)	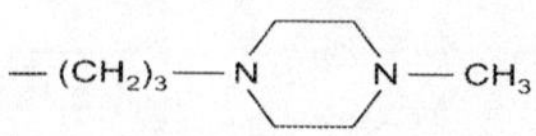	Cl
B.	Trifluoperazine (Stelazine®)	"	CF_3
C.	Perphenazine (Trilaphone®)	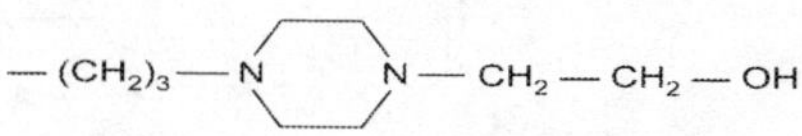	Cl
D.	Flujenazine fenazine (Permatil®)	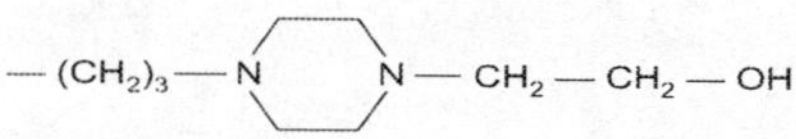	CF_3

Fig. 10.2

2. Pehnothiazine ring modification system

A. Thiothixene (Navane®)

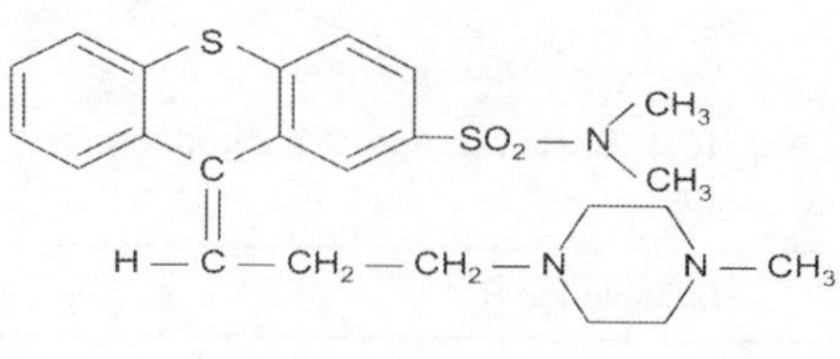

B. Chlorprothixine (Clothixene®)

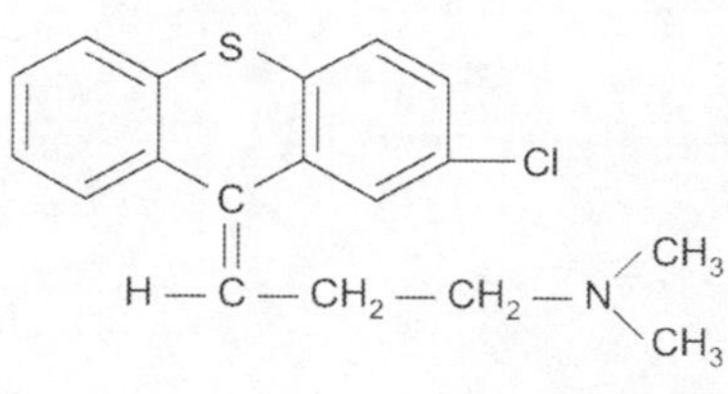

C. Chlopenthixole X = Cl
Flupentixol X = CF_3

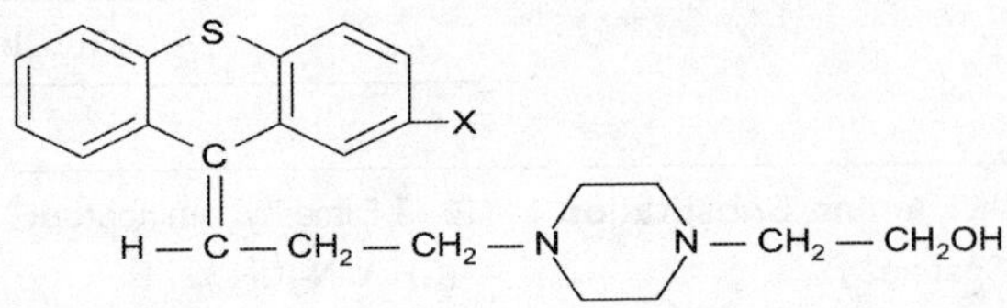

D. Dibenzoxazine (Loxapine®)

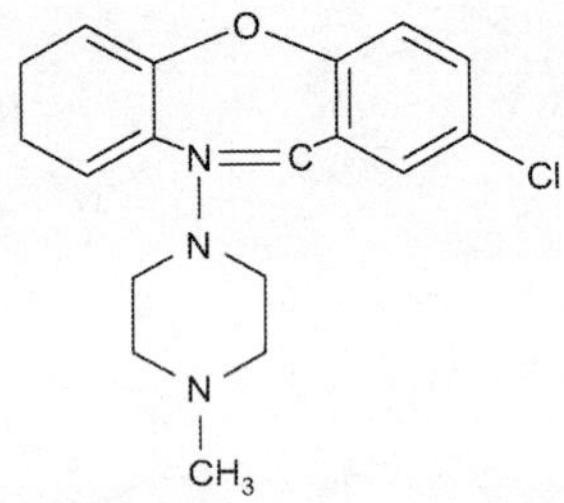

E. Dibenzodiazepine Clozapine (Clozaril®)

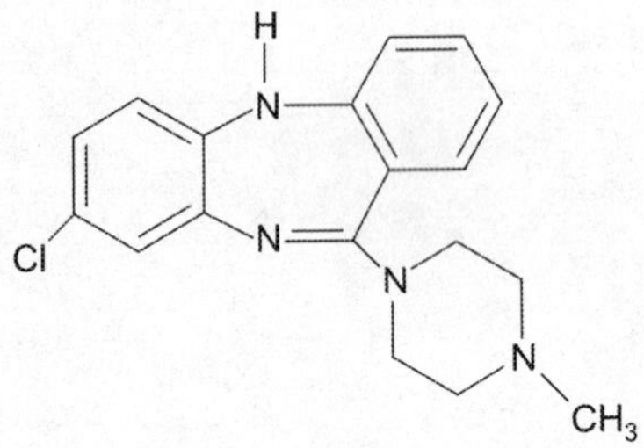

Fig. 10.3

(3) Butyrophenones Derivates

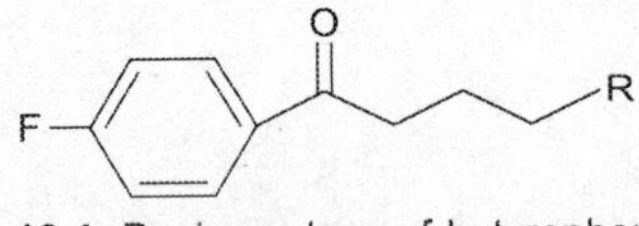

Fig. 10.4: Basic nucleus of butyrophenone.

	Derivatives	Substitution R
(i)	Haloperidol (Haldol®)	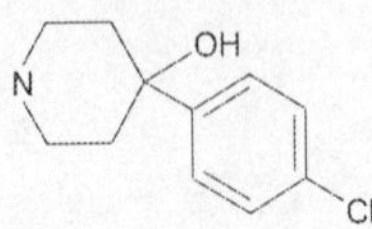

(ii) Methylperidol

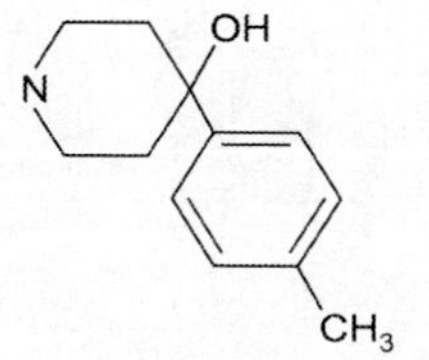

(iii) Trifluperidol

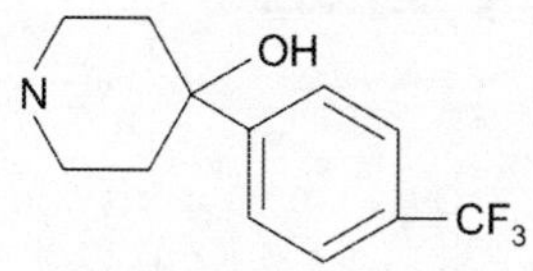

(iv) Peridol

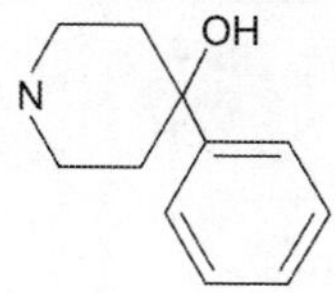

(v) Haleperidide

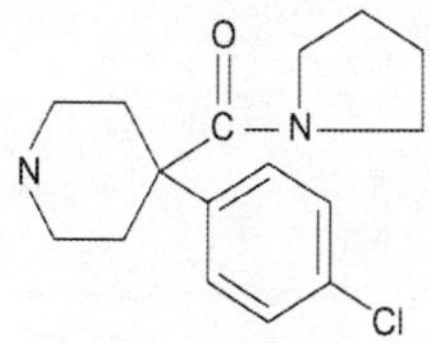

(vi) Spiperone

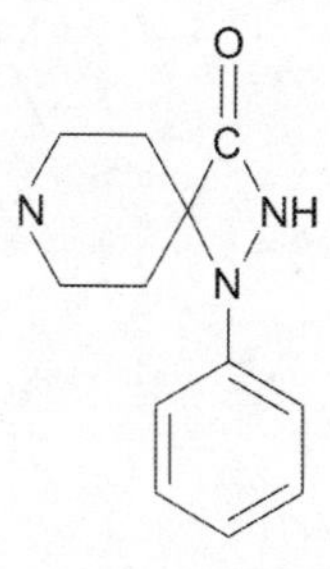

4. Benzamides

(i) Metoclopramide

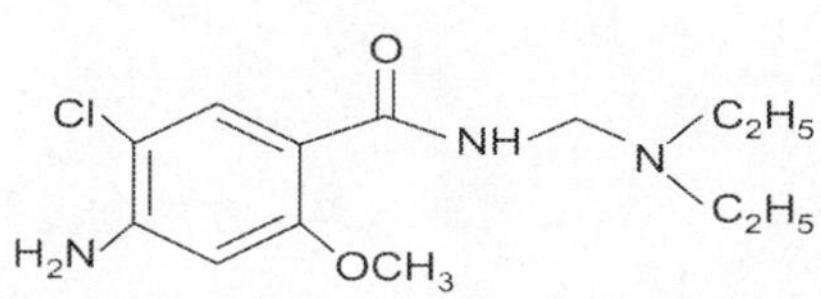

(ii) Sulpiride

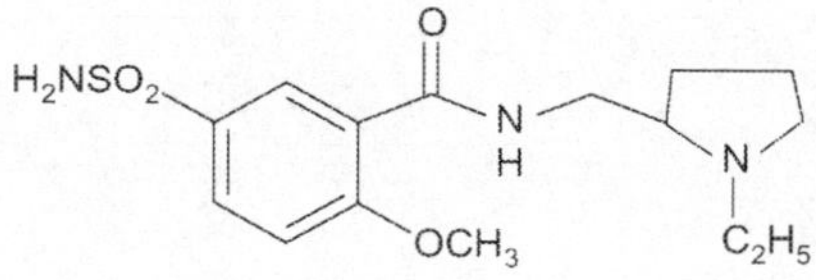

(iii) Ramoxipride

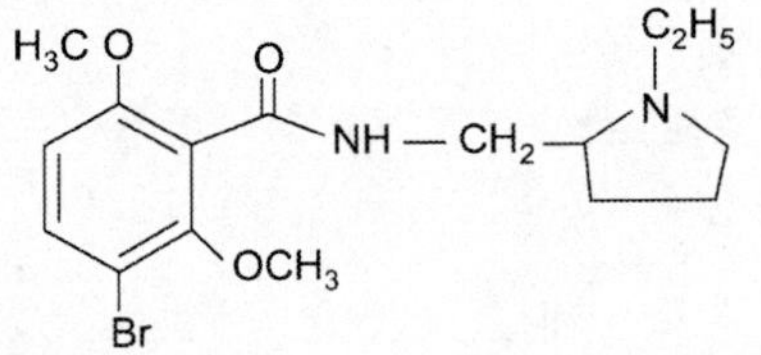

Fig. 10.5

5. Dihydroindolone Derivatives

(i) Molindone (moban®)

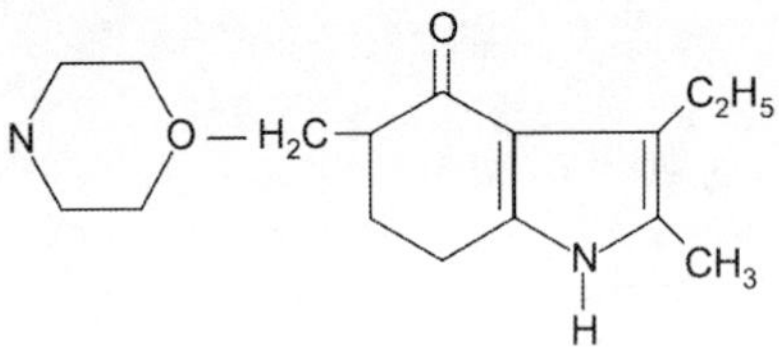

Fig. 10.6

6. Diphenyl Butyl Piperidine Derivatives:

(1) Pimozide (Orar®)

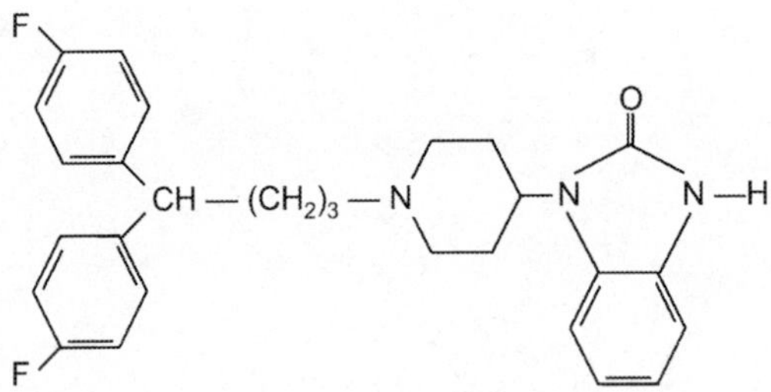

(2) Fluspirilene (Imap®)

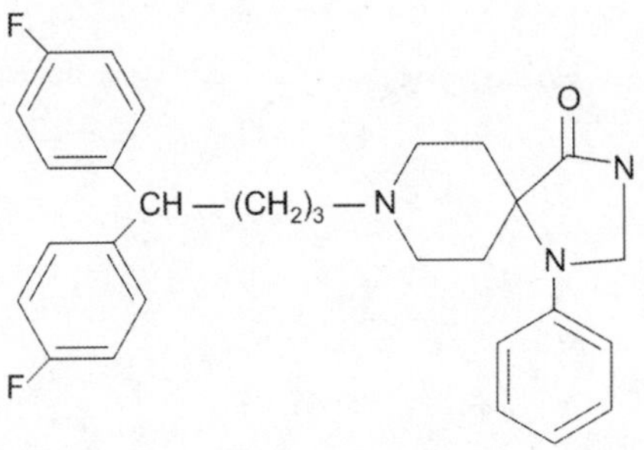

(3) Penfluridol (semap®)

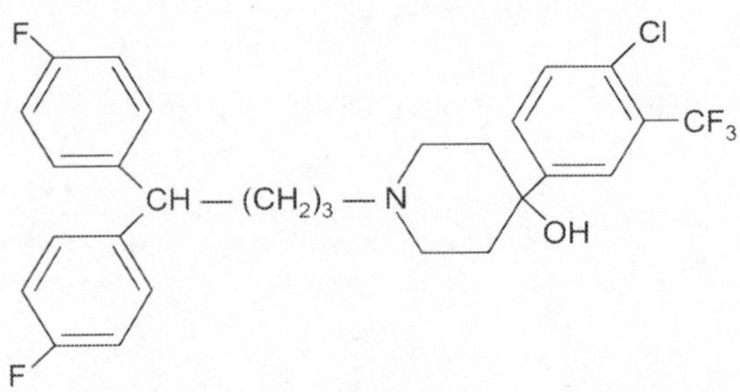

Fig. 10.7

(7) Miscellaneous Class

(i) Resperidone (Resperdal®)

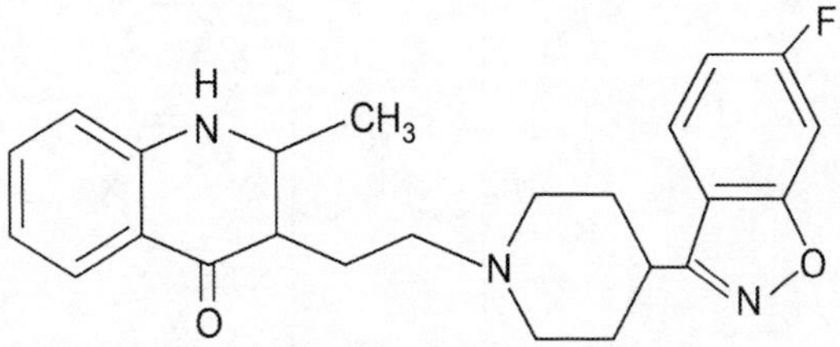

(ii) Butaclamol

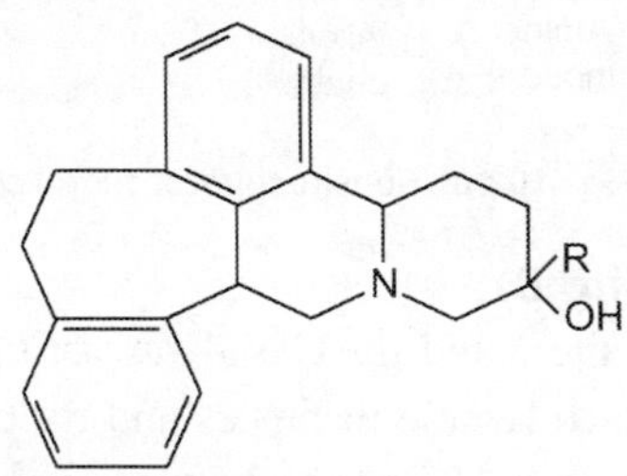

R = t-butyl Dexclamol R = isopropyl

Fig. 10.8

10.5 TRICYCLIC NEUROLEPTICS/PHENOTHIAZINES

1. Tricyclic Neuroleptics/Phenothiazines

Phenothiazines are the class of neuroleptics. These are non-selective, competitive D_1 and D_2 receptors antagonists which falls to block the action of dopamine in addition to their action exerts by blocking α-adrenoceptors, serotonine, cholinergic, nicotinic and muscarinic receptors. Phenothiazines act on CNS hence causing moderate sedative and antiemetic effects. They also affect thermoregulatory system, skeletal muscles and endocrine system as well as action of analgesia.

10.5.1 Phenothiazines Structure and Structure Activity Relationship

1. Tricyclic neuroleptics may be observed into three substructures one is amine functional group site (site A), second is intervening alkyl chain (site B) and third is diarylheterocyclic ring (site C). All these sites affect the pharmacological action further.

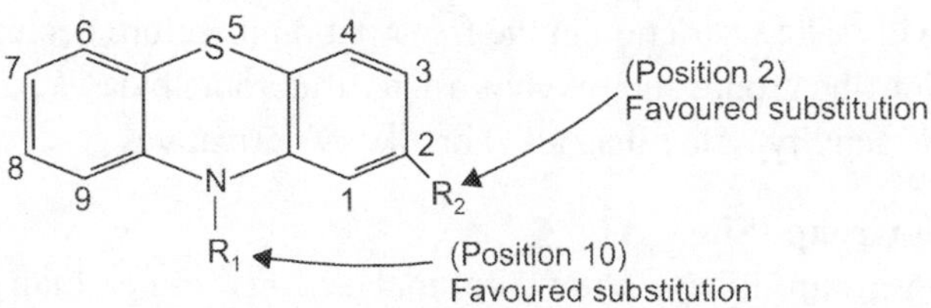

Fig. 10.9: Basic structure.

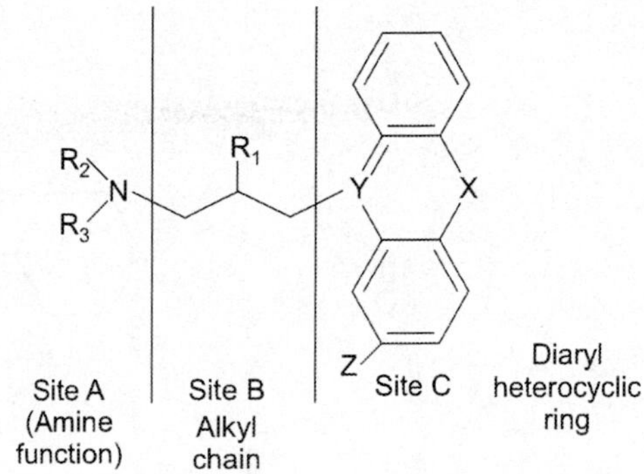

Fig. 10.10: Substructure of tricyclic neuroleptics.

2. **Alkyl Chain Effects (Site B)**
   (i) The distance between site A and site C is always critical for pharmacological activity. Length of carbon atom is 3 for tolerable instances and for optimal activity.
   (ii) While shortening the carbon chain (two carbon atoms) leads to amplifying the anticholinergic and antihistaminic activity, e.g., diethazine for Parkinson's disease. Promethazine acts as antihistamine

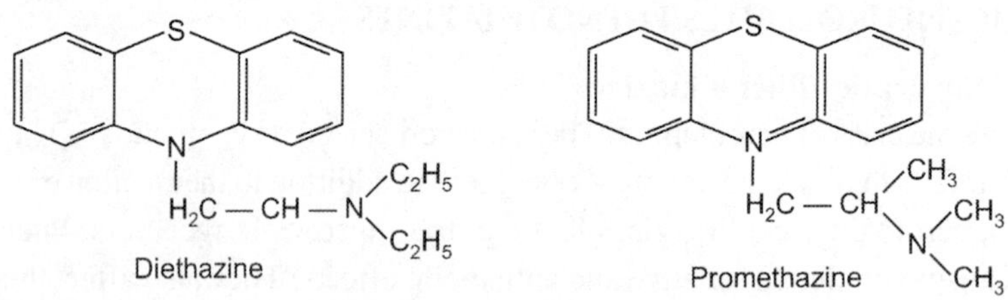

Fig. 10.11

3. **Effects on Aromatic substitution**
   (i) Introduction of aromatic substitution makes the effective tricyclic antipsychotic compounds and influences the activity.
   (ii) Substitution at position No. 2 of aromatic ring provides the compounds with enhanced potency. Small alkyl substitution at position 2 would be tolerable while larger substitution (R = phenyl) will make restriction in the free rotation therefore decreases the activity.
   (ii) Electron withdrawing groups such as chloro and trifluoromethyl (CF_3) show greater response to antipsychotic activity rather than alkyl or alkoxy derivatives.

4. **Nature of Amino Group (Site – A)**
   (i) Tertiary amine group always shows optimal activity rather than mono or desmethyl chlorpromazine respectively.
   (ii) In many cases larger N-alkyl group leads to decrease in the neuroleptic activity.

(iii) Groups of diethyl, pyrrolidinyl, morpholine and thiomorpholine of chlorpromazine show lack of activity.
(iv) In contrast to other groups like N-methylpiperazine leads to enhance potency as well.

5. **Alteration within Tricyclic Topology**
   (i) Deviation of sulphur group with other element produce the corresponding phenoxazine and phenoselenazine derivatives.
   (ii) Phenoxazine exhibits as less active compounds than phenothiazines.
   (iii) Replacement of nitrogen atom within the phenothiazine ring system with methane carbon made another series of neuroleptics.
   (iv) Introduction of double bond within propylamino chain provides the compounds geometric isomers.

10.5.2 Specific Derivatives

1. Promazine (Sparine®): 10-(3-dimethylamine propyl) phenothiazine.
Promazine is a potent antipsychotic agent. Its one milligram of dose is sufficient to counteract psychoses condition. Promazine is used in the treatment of psychomotor excitement in schizophrenics and manic depressive conditions.

Synthesis:

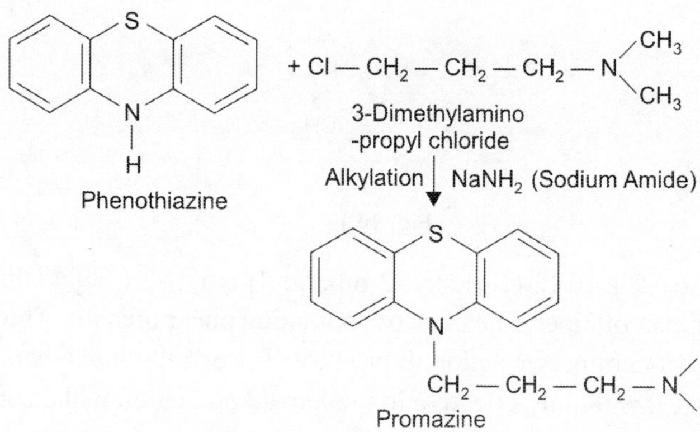

Fig. 10.12

2. Chlorpromazine (Megaphen®): 2-chloro-10-(3-dime thylaminopropyl)-phenothiazine
Chlorpromazine is active in the salt with HCl. It was the first phenothiazine compound introduced in the therapeutic manner. Chlorpromazine is used in various psychomotor conditions like schizophrenia, chronic paranoid condition, alcohol psychosis, neurosis, etc.

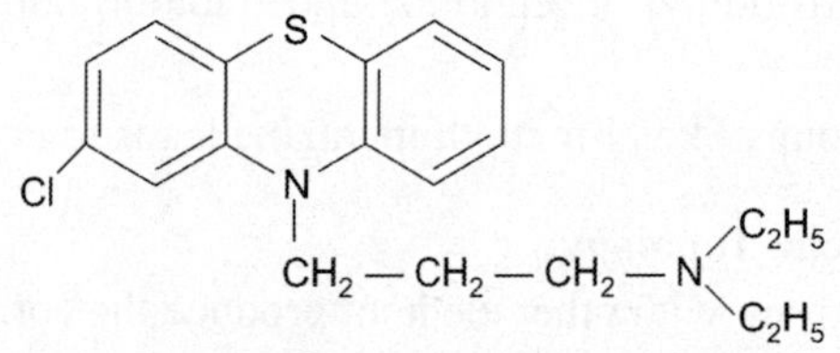

Fig. 10.13: Chlorpromazine.

3. Triflupromazine (Vesprin®): 2-fluoromethyl-10-(3-dimethyl amino propyl)-Phenothiazine. Triflupromazine is an active antipsychotic drug. Its few milligram of dose is sufficient for antipsychotic action. It is used in the treatment of various psychomotor excitements in patients with psyzophrenia for paranoid and manic-depressive conditions.

Synthesis:

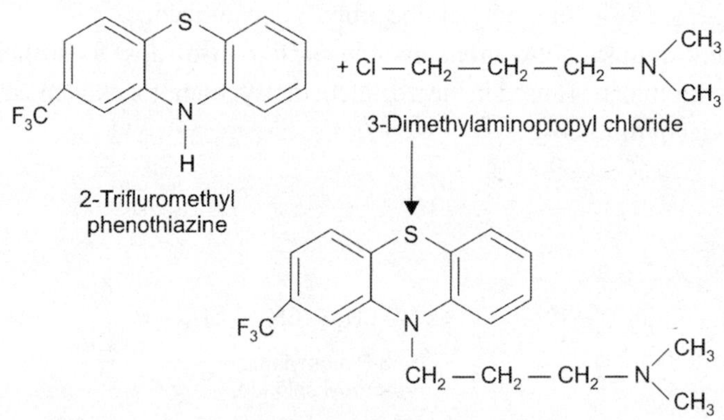

Fig. 10.14

4. Thioridazine (Conapax®): 10-[2-(1-methyl-2-piperidyl)-ethyl]-2-(methyl thio)-phenothiazine. Thioridazine is the congener of piperidine nucleus containing phenothiazine. Thioridazine also have the character of higher anticholinergic action. It produces lower EPS than others. It shows sedative and hypotensive activity. It is mainly effective in the mental and emotional disorders accompanied by fear, stress and excitement conditions.

Synthesis:

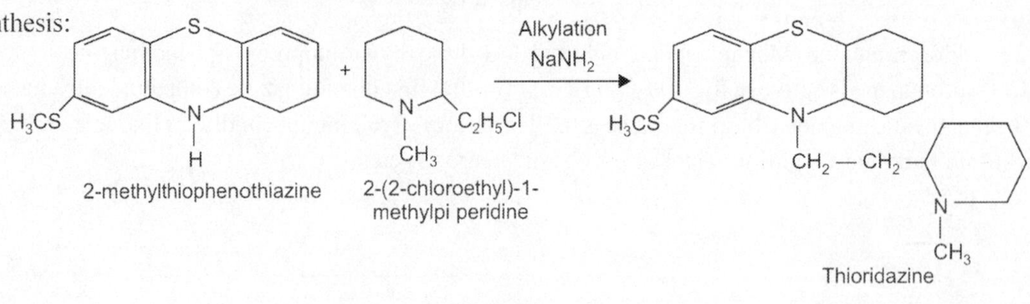

Fig. 10.15

5. Mesoridazine (Serentil®): 10-[2-(1-methyl-2-piperidyl)-ethyl]-2-(methyl sulfinyl) phenothiazine.

Mesoridazine also contains piperidine nucleus in phenothiazine. Mesoridazine is used in the treatment of schizophrenia-psychoneurotic problems, in severe and chronic alcoholism.

Synthesis:

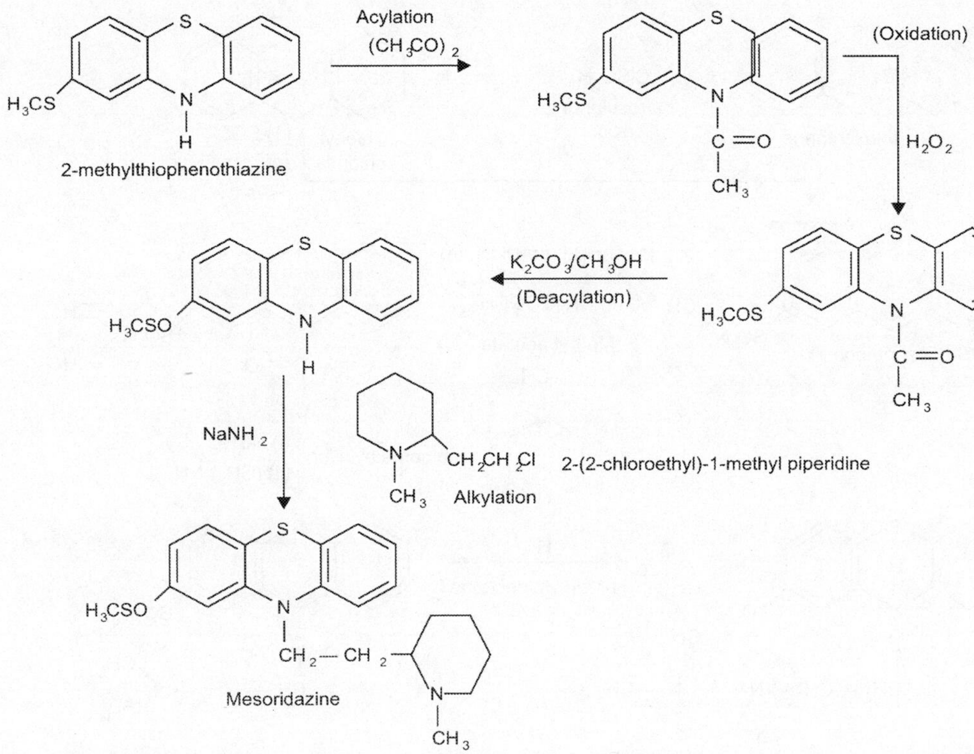

Fig. 10.16

10.6 THIOXANTHENE DERIVATIVES

Thioxanthenes are structurally different from phenothiazene in respect of the nitrogen atom of the centrally ring situated of the tricyclic system replaced by carbon which is joined to the side chain with double bond. Pharmacological concerns are similar to that of phenothiazines.

1. **Thiothixene (Navane®):** N, N-dimethyl-9-[3-(4-methyl-1-piperazinyl)-propylidine]-thioxanthane-2-sulfonamide. Thiothixene displays a fundamental chemical and pharmacological

similarities to the piperazine derivatives of phenothiazines. Due to presence of carbon atom and side chain double bond, the possible Z and E isomers are produced. Thiothixene is used in the treatment of psychotic disorders.

Synthesis:

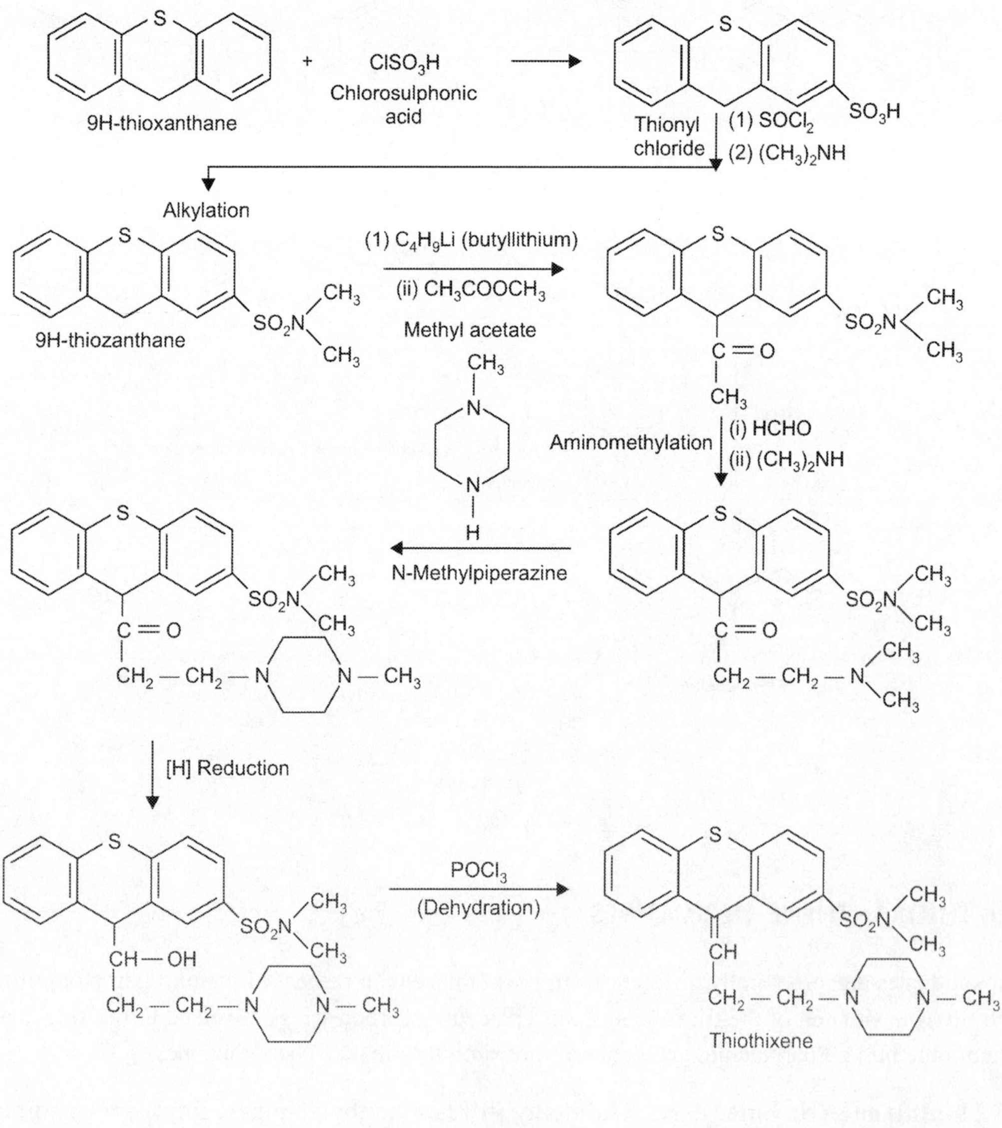

Fig. 10.17

2. Chlorproxithene (Clothixene®): 2-chloro-9 (1-dimethyl-amino)-3-propyliden]-thioxanthene. Chlorproxithene is a potent antipsychotic agent and possesses sedative like action. It is supposed to be an antiemetic agent. Chloroproxithene is used in the treatment of distinguished psychoses conditions, psyzophrenia, reactive and neurotic depression, in the condition of excitement concerning with fear and stress.

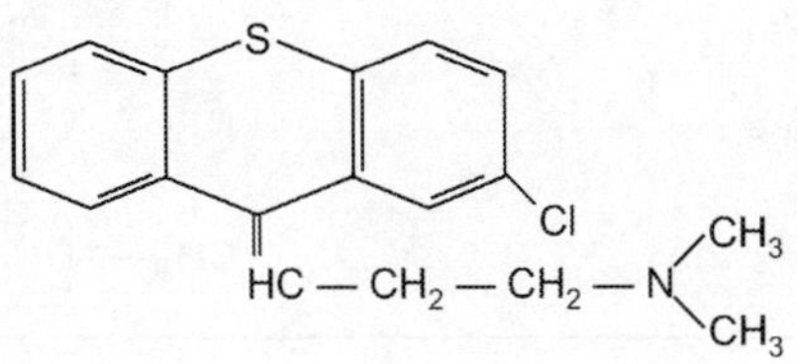

Chlorproxithene

Synthesis:

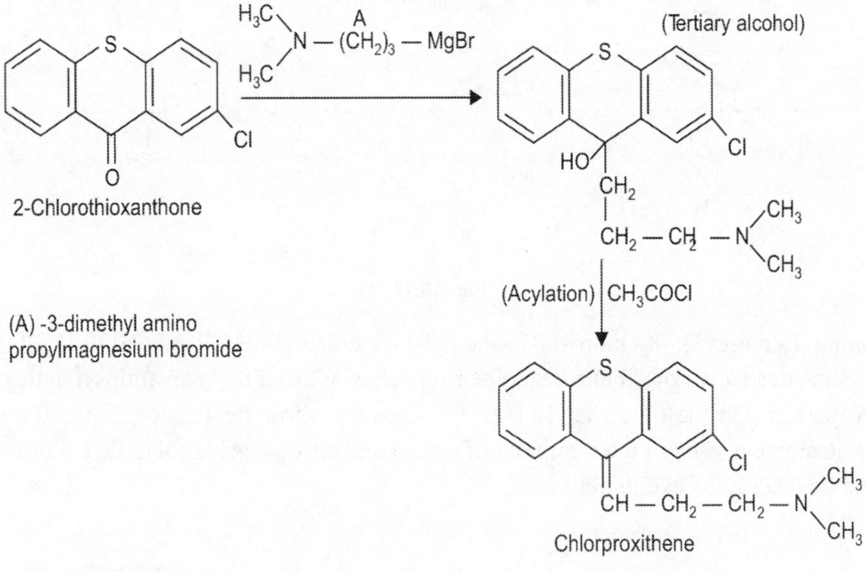

Fig. 10.18

10.7 DIBENZOXAZEPINE AND DIBENZDIAZEPINE DERIVATIVES

Dibenzoxazepine and dibenzdiazepine derivatives are structurally different from other derivatives. However, their mode of action, uses and side effects are analogue to phenothiazine derivatives.

1. **Loxapine (Loxapac®):** -2-chloro-11-[4-methyl-1-piperazinyl]-dibenz-[b, f] [1, 4] exazepine. Loxapine is the derivative of dibenzoxazepine series. It is a potent psychotic agent rather than chlorpromazine. It shows less sedation than chlorpromazine. Loxapine is suggested in the treatment of psychotic conditions particularly in the case of chronic and severe schizophrenia.

Synthesis:

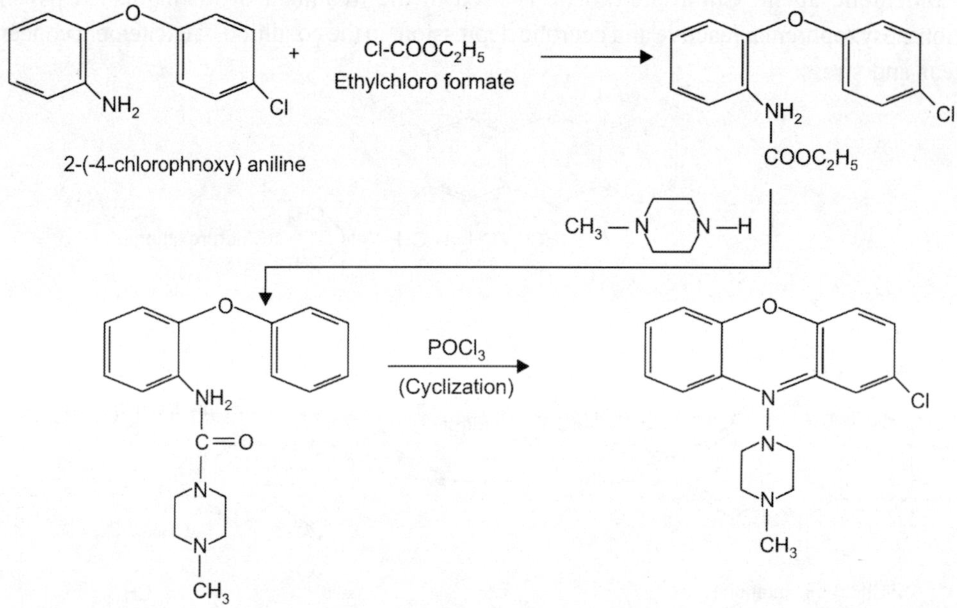

Fig. 10.19

2. Clozapine (Leponex®): -8-chloro-11-[4-methyl-1-piperazinyl-5H-dibenzo] [b, e][1, 4]-diazepine. Clozapine is the derivative of dibenzdiazepine ring series. Clozapine is an antipsychotic drug with sedative properties. On the other hand clozapine does not show depression and extrapyraimidal disorders. Clozapine is used in the treatment of severe and chronic schizophrenic conditions, in the condition of psychomotor excitement also.

Synthesis:

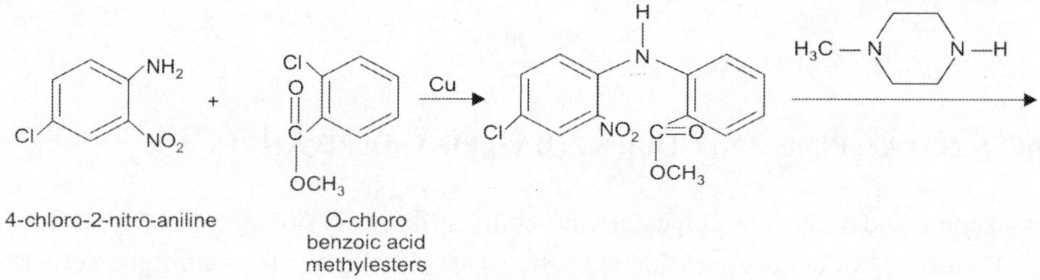

Fig. 10.20 (Contd.)

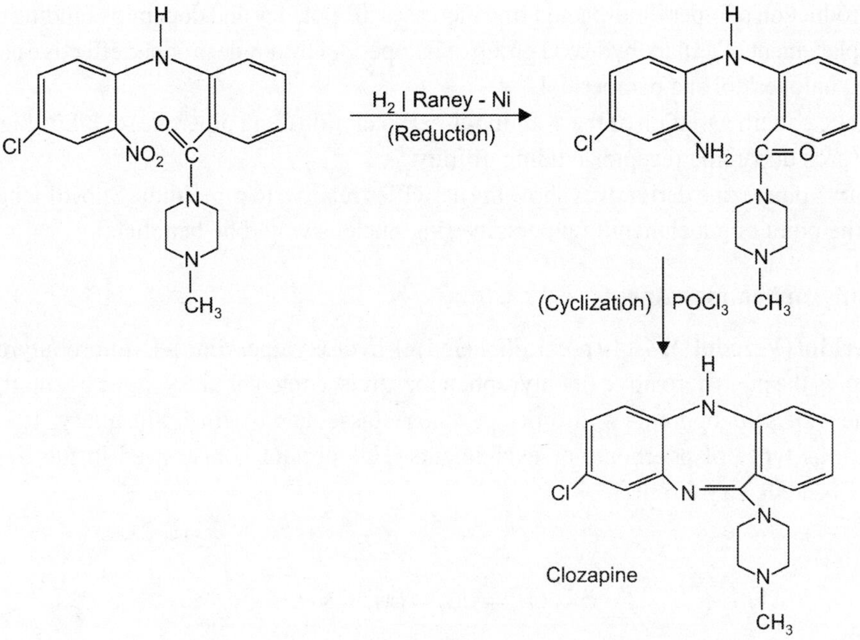

Fig. 10.20

10.8 BUTYROPHENONES

Butyrophenones were inspired by the modified meperidine analogues. Butyrophenones possess a tertiary amine at fourth carbon of the butyral chain. Butyrophenones act as antipsychotic agents as well as anaesthetics. Butyrophenones' mechanism of action are similar to those of phenothiazines and thioxanthenes, they are dopamine (D_2) receptor antagonists.

10.8.1 SAR of Butyrophenones

1. These compounds possess a tertiary amine analog at fourth carbon of butyral chain that is responsible to show analgesic activity by the addition of a substituent at fourth position of aryl ring.

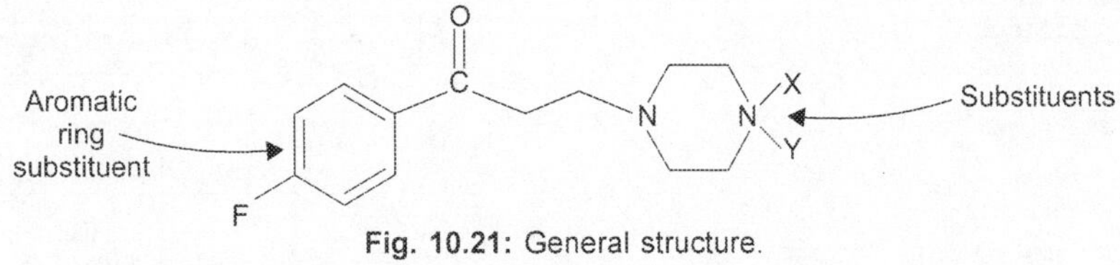

Fig. 10.21: General structure.

2. All the compounds of this series are highly neuroleptic relative to chlorpromazine.
3. First compound, haloperiodol show less toxic and widely accepted drug.

4. Introduction of piperidino-phenyl ring increases the potency and dopamine binding properties.
5. Replacement of tertiary hydroxyl group of haloperidol by amide produce effective compounds, e.g., haloperidol and paraperidol.
6. Cyclic pattern variation of the 4-amino-4-carboxamido (e.g., piperone) exhibits high potency towards dopamine receptor binding affinity.
7. N-aryl-piperazine derivatives show the activities relative to piperidine. Substituents adjacent to the point of attachment to piperazine ring nucleus would be beneficial.

10.8.2 Butyrophenones Agents

1. Haloperidol (Vezadol®): 4-[4-(p-chlorophenyl)-4-hydroxypiperidino]-4'-fluorobutyrophenone. Haloperidol is the main derivative of butyrophenones. It is a potent antipsychotic agent. It is used in the treatment of schizophrenia and other psychoses associated with brain injury. It effectively retards various types of psychomotor excitements. Haloperidol is also used in the treatment of Gilles dela Tourette's syndrome.

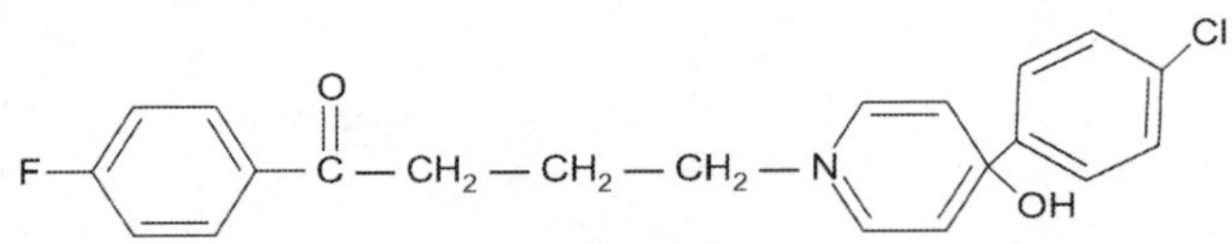

Fig. 10.22

Synthesis:

Step – I: Synthesis of 4'-chloro-4-fluorobutyrophenone

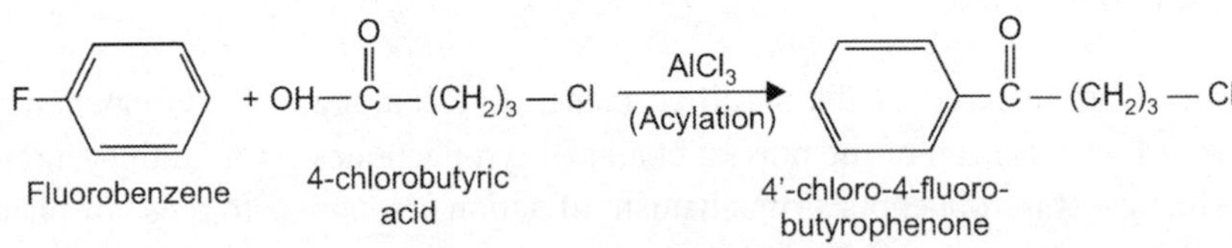

Step – II: Synthesis of 4-(4-chlorophenyl)-4-hydroxypiperidine (Prime reaction)

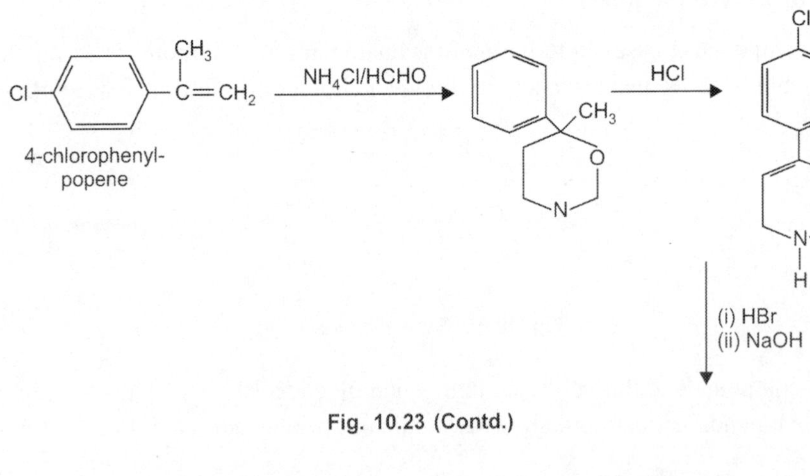

Fig. 10.23 (Contd.)

Step – III:

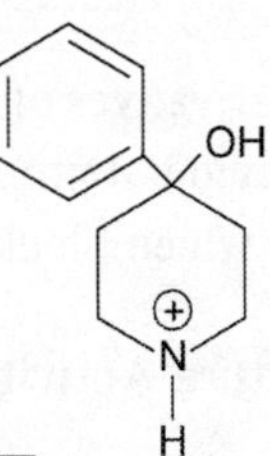

Step – I + step – II

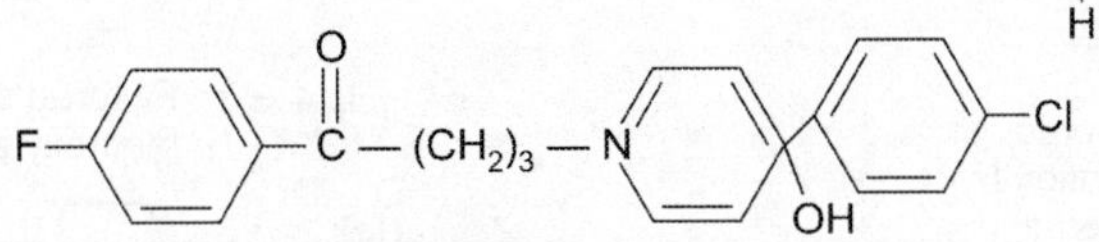

Haloperidol

Fig. 10.23

2. Droperidol (Talamonal®): 1-[1-3-(p-fluorobenzoyl)-propyl]-1, 2, 3, 6, 4-pyridyl]-2-benzyimidazolinone.

Droperidol is another example of butyrophenones. It possesses benzimidazole moiety attached to piperidine nucleus. Droperidol is a potent neuroleptic agent. It possesses sedation and antishock like action. Its action is towards synergistic to drugs for nacrosis. Droperidol is used in the treatment of various psychomotor conditions and hallucinations.

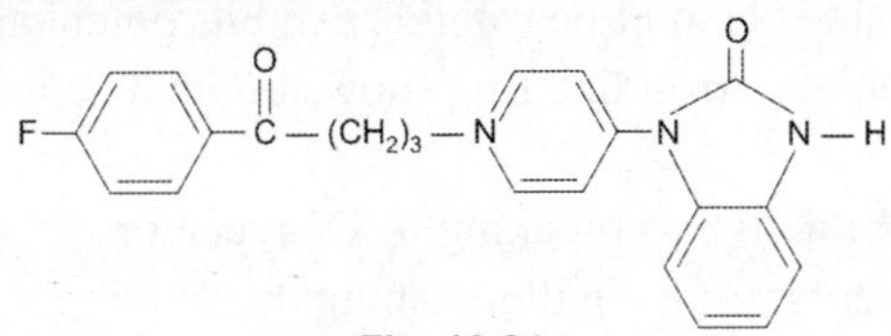

Fig. 10.24

Synthesis:

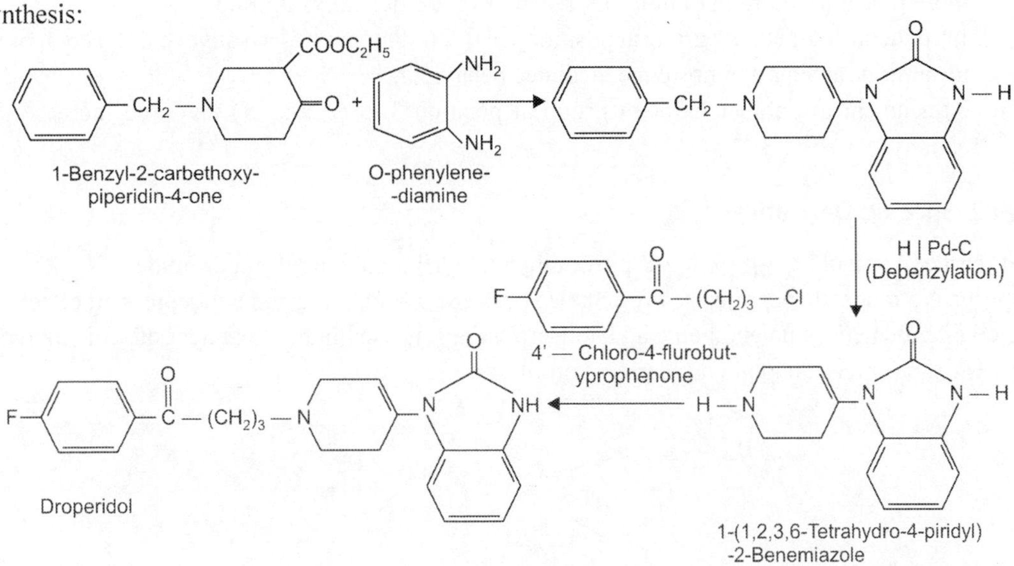

Fig. 10.25

10.9 BENZAMIDES DERIVATIVES

Benzamides derivatives of antipsychotic class are concerned with antiemetic agent, metaclopramide exhibited behavioural properties consistent with dopamine receptor antagonist. Metaclopramide is the main drug which blocks Dopamine (2-methoxy benzamide) receptor in striatum.

10.9.1 Structure Activity Relationship of Benzamides

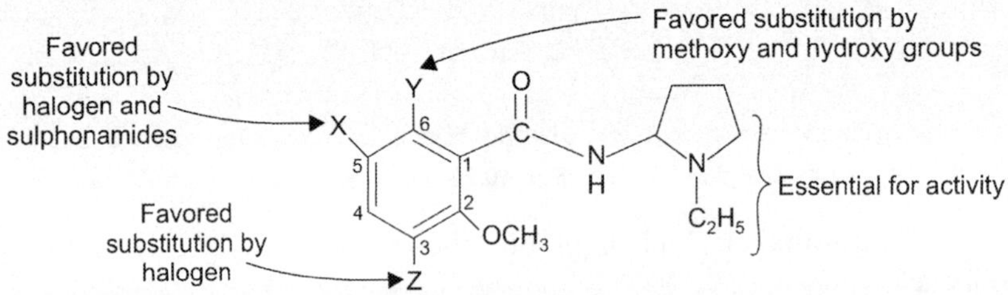

Fig. 10.26: Basic pharmacophore.

1. 2-methoxy benzamide is core nuleus of benzamides that has ability to block dopamine receptors.
2. 2-pyrrolidinyl methyl side chain of benzamide exhibits potent antipsychotics.
3. Small alkyl substitution at pyrrolidine ring shows optimal activity, e.g., alkyl, cycloalkyl or benzyl groups.
4. S-enantiomers are more active to blocking of D_2-receptors.
5. Removal of N-substituents results in diminishing the activity.
6. Replacement of halogen at position X(5) by sulfonamide group that makes the compound more potent to (more lipophilic) D_2 receptor blockade, e.g., sulpiride.
7. Introduction of methoxy group at position y(6) (2, 6-dimethoxy derivative) exhibit equipotent to sulpiride against apomorphine mediated behaviour.
8. Attachment of another halogen group at position 3, 5 (Z and X) favoured increase the activity.

10.9.2 Specific Derivatives

1. Sulpiride (Arminol®): N-[(1-ethyl-2-pirrolidinylmethyl)]-5-sulfamoyl-o-anizamide

Sulpiride is a moderate neuroleptic drug. It also possesses stimulating and antidepressant effects. It is used as antiemetic, cataleptogenic and antiserotonin. It is used in the treatment of schizophrenia depression, migrains, abdominal ulcerative problems.

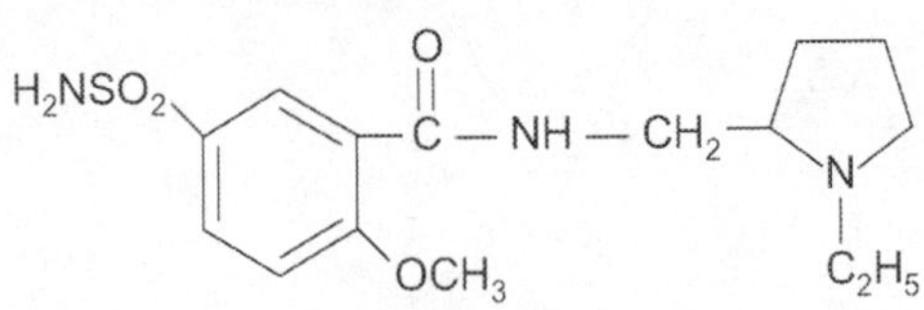

Fig. 10.27

Synthesis:

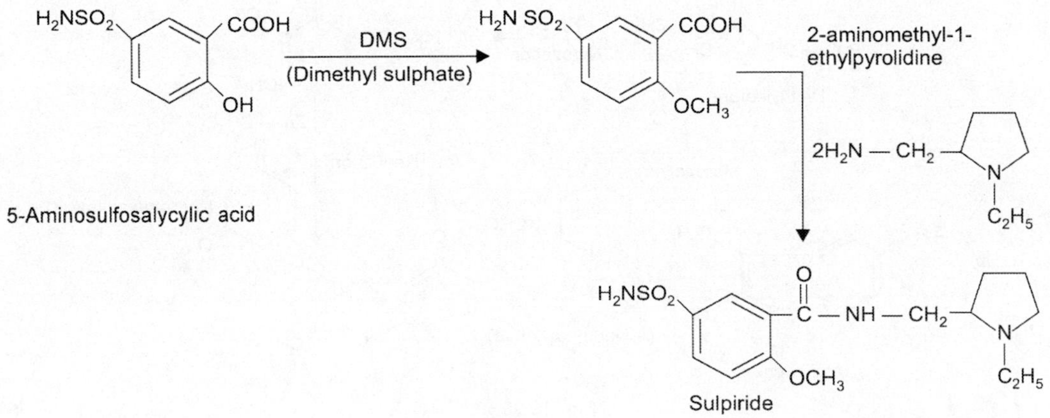

Fig. 10.28

2. Remoxipride (Roxiam®): Remoxipride is D_2 receptor antagonist. Remoxipride is used in the treatment of various symptoms of schizophrenia and mental disorders.

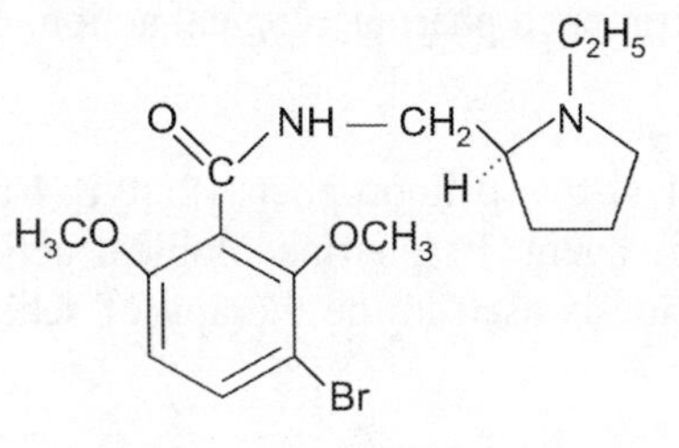

Fig. 10.29

10.10 DIHYDROINDOLONE DERIVATIVE

Dihydroindolone derivatives are not structurally similar to other class but their mechanism of action, uses and side effects are very close to phenothiazine moeity respectively.

(1) Molindone (moban®): 3-ethyl-6, 7-dihydro-2-methyl-5-(morpholinomethyl)-indol-4(5H)-one. Molindone is a potent neuroleptic. Molindone is clinically approved to be more potent than chlorpromazine. It is used in the treatment of various psychomotor conditions, in chronic and severe schizophrenic conditions.

Synthesis:

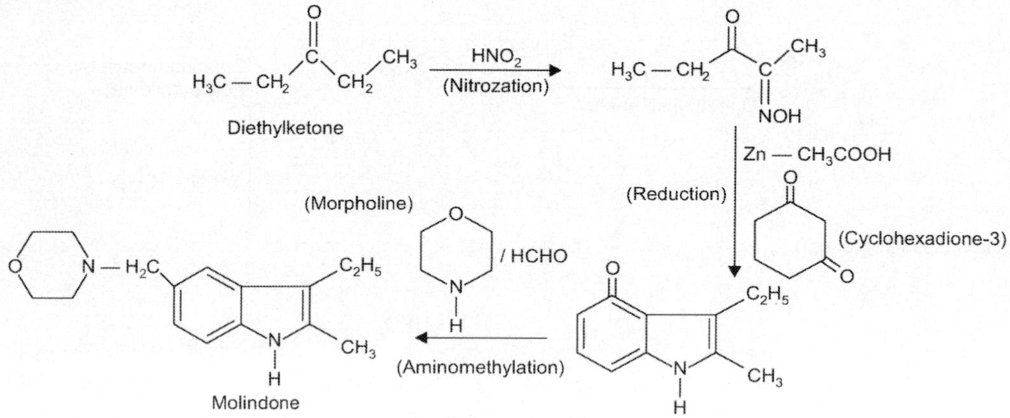

Fig. 10.30

10.11 DIPHENYLBUTYLPIPERIDINE DERIVATIVES

Diphenylbutylpiperidines are powerful neuroleptics. Their antipsychotic properties are very close to butyrophenones. They having prolonged pharmacological action. They are considered D_2-receptor antagonist.

(1) Pimozide (orap®)): 1-[1-[4, 4-bis-(p-fluorophenyl)butyl]-4-piperidyl]-2-benzimidazolinone. Pimozide is a potent neuroleptic agent. Its pharmacological action is similar to butyrophenone derivative (Haloperidol). Pimozide is used in the therapy of schizophrenia, paranoid conditions, mental and neurotic disorders.

Synthesis:

Step - I: Synthesis of 1, 1-bis-(4-fluoro-phenyl)butylchloride

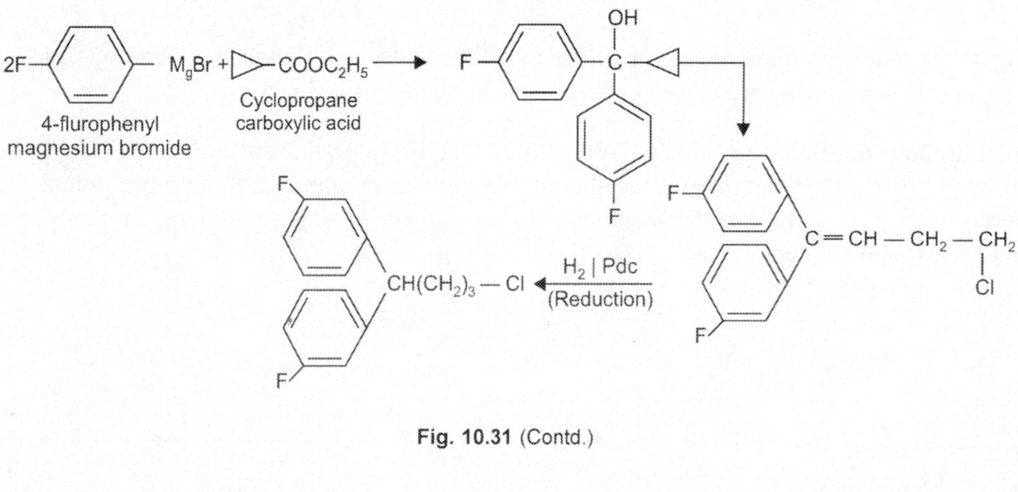

Fig. 10.31 (Contd.)

Step - II:

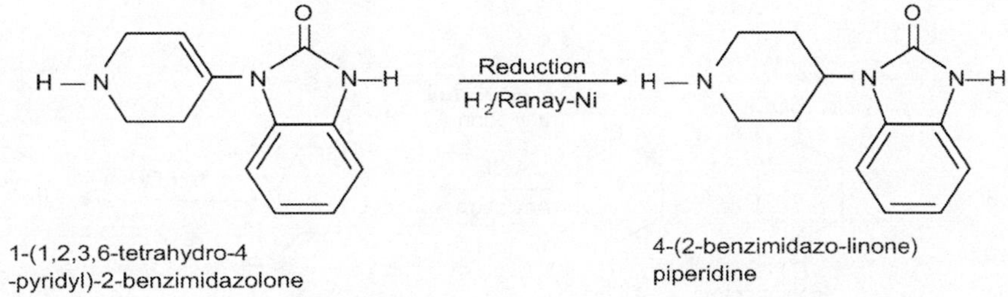

Step - III:

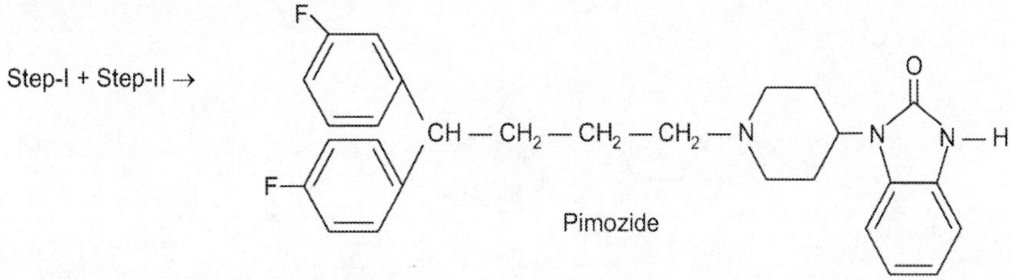

Fig. 10.31

(2) Fluspirilene (Redeptin®): 8-[4, 4-bis (p-fluoro-phenyl)butyl]-1-phenyl-1, 3, 8 triazaspiro [4, 5]-decan-4-one. Fluspirilene is a potent antipsychotic drug. It is used in the patients of mental illness and psychotic condition. Fluspirilene is used frequently in ambulatory practice due to absence of hypno-sedative effects.

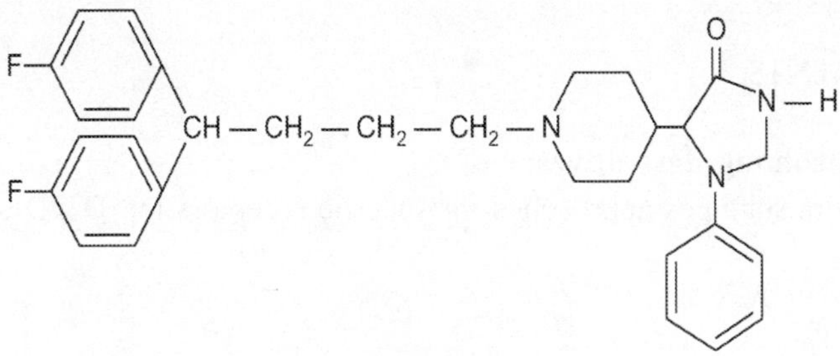

Fig. 10.32

10.12 MISCELLANEOUS AGENTS

1. Risperidone (Risperdal®): 4-[2-[4-(6-Fluorobenzo-[d]-isoxazol-3-yl)-1-piperidyl] ethyl]-3-methyl-2, 6-diazabicyclo[4, 4, 0] deca-1, 3-dien-5-one.

Risperidone is an atypical antipsychotic (to treat schizophrenia) which is a new generation antipsychotics. Risperidone is first line drug to treat schizophrenia. It is reported in hybrid molecular structure of butyrophenone and trazodone-like antidepressant. Risperidone acts as $5HT_2$ as well as D_2 receptor antagonist.

Synthesis:

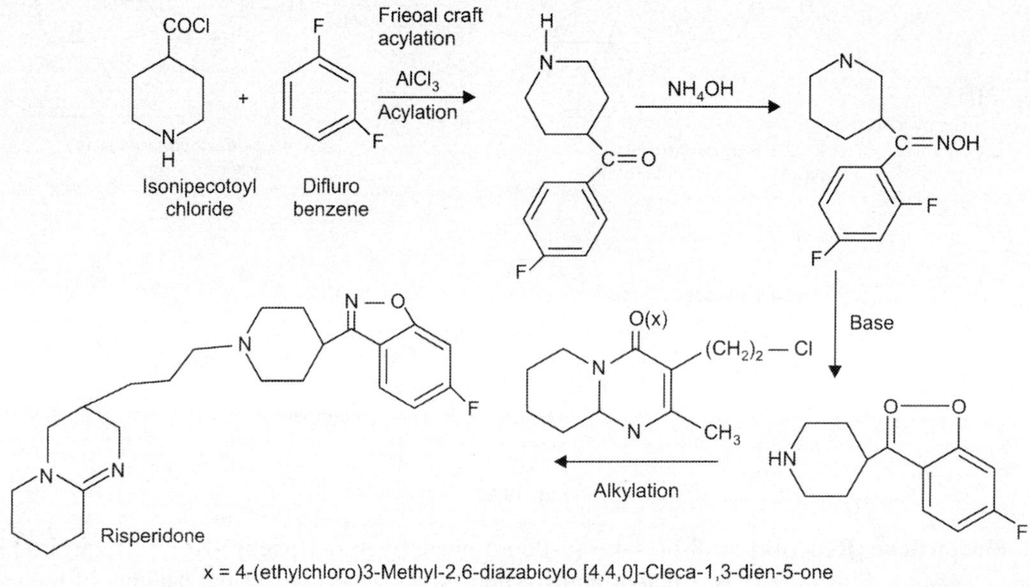

Fig. 10.33

10.13 NEWER AGENTS

1. Strategies for upcoming derivatives:

A. Dopamine receptors subtypes approaches for specific receptors D_4, D_2, D_3, e.g., nemonapride.

Nemonapride

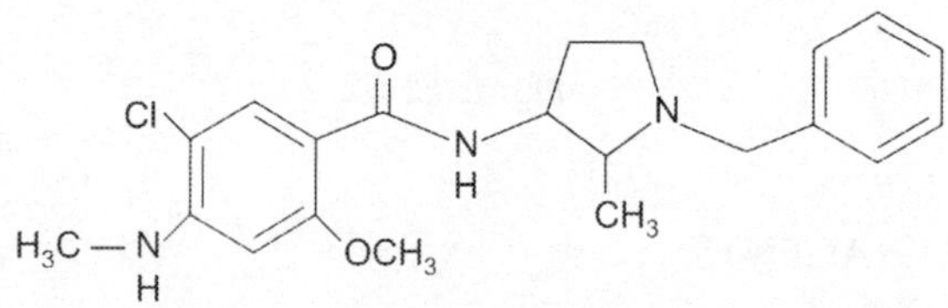

Fig. 10.34

B. Newer aryl piperazines/piperidines derivatives would be more effective for $5HT_2/D_2$ receptor antagonist, e.g., sertindole.

Sertinodole

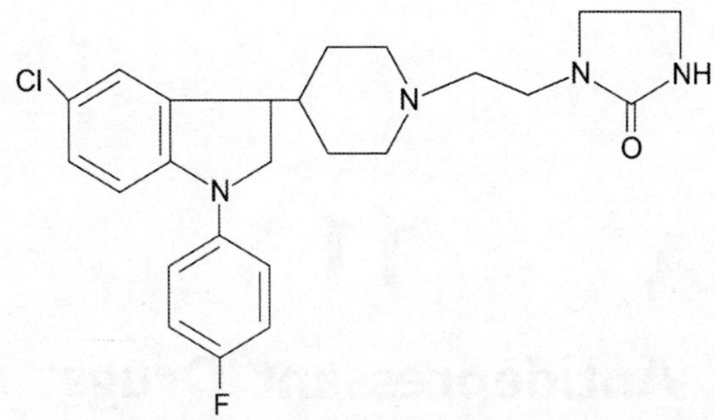

Fig. 10.35

C. 5HT2A receptor antagonist: e.g., Fananserin.

Fnanserin

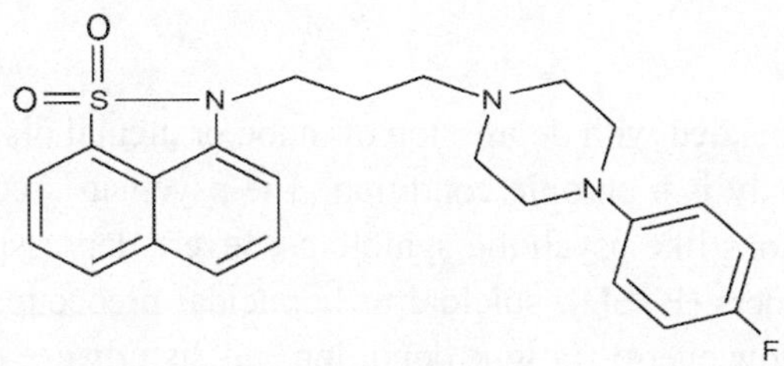

Fig. 10.36

D. Inotropic Glutamate receptors Antagonist: e.g., Milacemide (Glycine Prodrug)

Milacemide

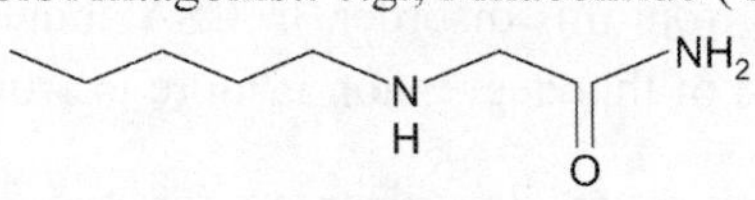

Fig. 10.37

EXERCISE

1. Classify antipsychotic agents. Write the structural features of tricyclic antipsychotic molecule.
2. Discuss phenothiazines, their mode of action and write any one synthesis of phenothiazines.
3. Write the synthesis of following drugs. Give their uses also.
    A. Promazine
    B. Thioridazine
    C. Mesoridazine
    D. Loxapine
    E. Haloperidol
4. Write structural activity relationship of phenothiazines.
5. Write a note on Butyrophenone derivatives. Give their structural requirements for favourable activity.
6. Discuss the molecular Benzamide as antipsychotics. Enumerate the examples too.
7. Write the synthesis of resperidone and MOA as well as uses.

11

Antidepressant Drugs

11.1 INTRODUCTION

The term depression is associated with depression of mood or mental disorders. It may than can be transient but more commonly is a chronic condition. The psychiatric condition may be reflected with many medical conditions like psychotic symptoms, severe depression, anxiety, symptoms of post-traumatic stress disorders (PTSD), suicidal or homicidal preoccupation. Depressed mood is typically accompanied by low energy, fatigue condition and disturbance of sleep. Factors affecting may be physical, social or personal functioning. Physical illness for illustration diabetes or arthritis also are responsible for depression. It is a important and common disorder, study says that 100 million people/year are affected from this disorder. In USA about 30% people are threatened by mental disorders. The prevalence of this depression is more in women than men.

Depression is also associated with psychiatric disorders for, e.g.

1. Anxiety disorders
2. Panic disorders
3. Gneralized anxiety disorders
4. Phobic disorders
5. Stress disorders
6. Obsessive-compulsive disorders
7. Mood disorders
8. Depressive disorders
9. Bipolar disorders
10. Somatoform disorders
11. Personality disorders

However, partial remission of symptoms are usually outcome of drug treatment. Recently, availability of antidepressant drugs (safe and effective) are the achievement of twentieth century at the platform of psychopharmacology. Most common drugs which are accepted worldwide, e.g., fluoxanthin (Prozac). However, serotonin selective reuptake inhibitors (SSRI) are accepted and approved well but due to adverse side effects in men and women, fluoxanthines are much better than SSRI and have pronounced clinical recognition and treatment of depression to a much greater number of people than before.

11.2 MANAGEMENT OF DEPRESSION

By SSRI, SNRIs and NSRIs

1. Mostly used SSRIs drugs
2. Dose maintenance: Continuous use of drugs
3. Patient counselling by pharmacist and physician.
4. Regular monitoring of patient to decrease suicidal tendency.
5. Adverse drug reaction aspects monitoring regularly.

Note: SSRIs: Selective serotonin re-uptake inhibitors.
SNRIs: Selective norepinephrine re-uptake inhibitors.
NSRIs: Nor-epinephrine/serotonin re-uptake inhibitors.

11.3 CLASSIFICATION

1. Tricyclic Antidepressants (TAs)

2. Serotonin-selective Reuptake Inhibitors (SSRIs)

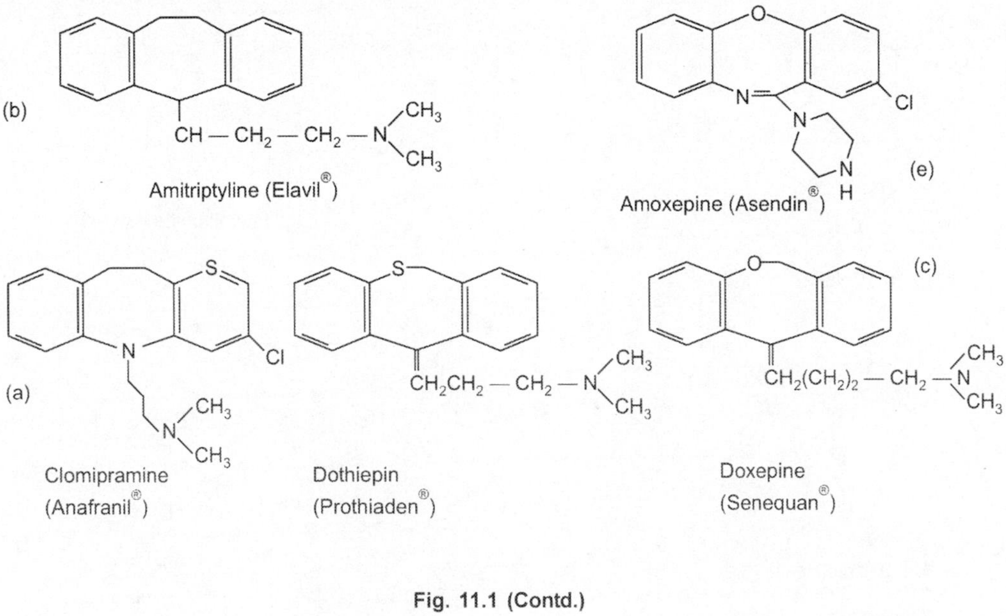

Fig. 11.1 (Contd.)

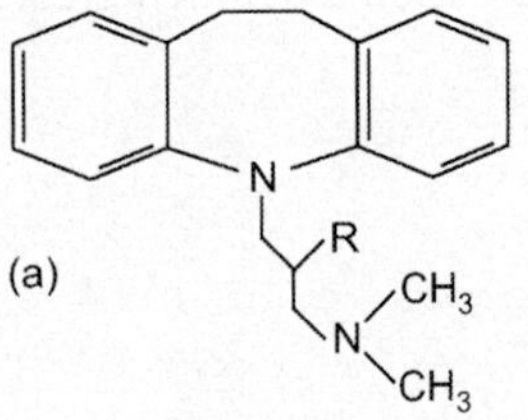

R = Imipramine (Anafranil®)
R = CH_3 Trimipramine (Surmontil®)

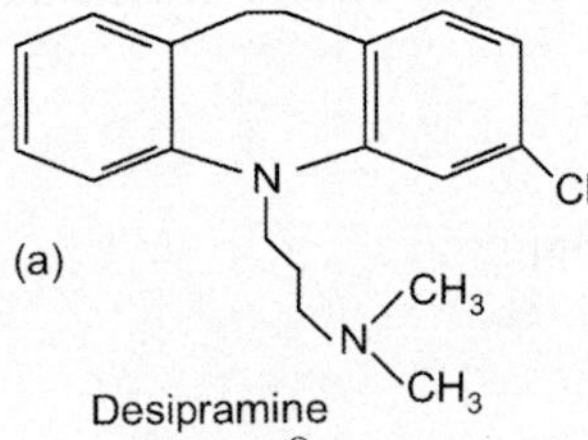

Desipramine
(Norpamin®)

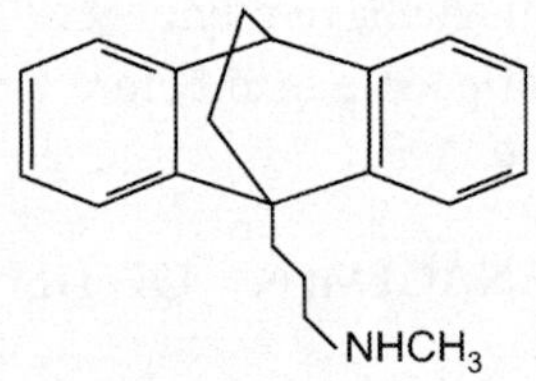

Maprotiline
(Ludiomil®)

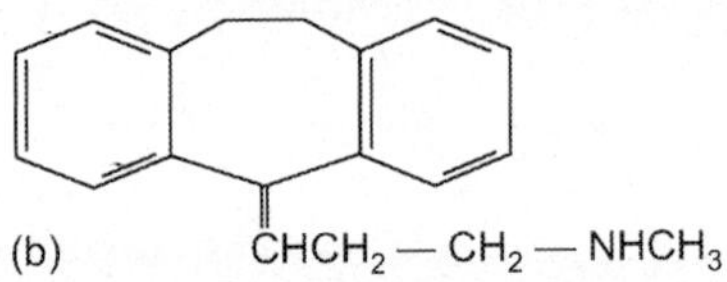

Nortriptyline (Aventyl®)

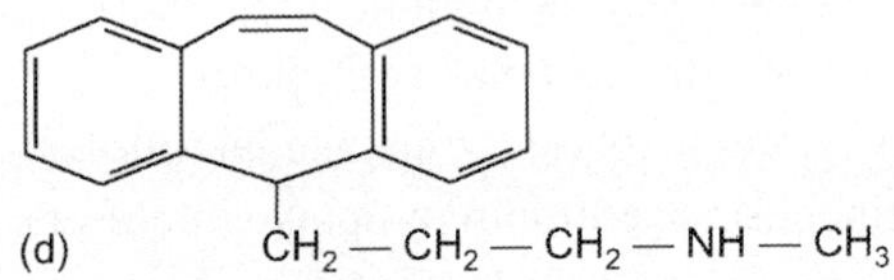

Protriptyline (Vivactil®)

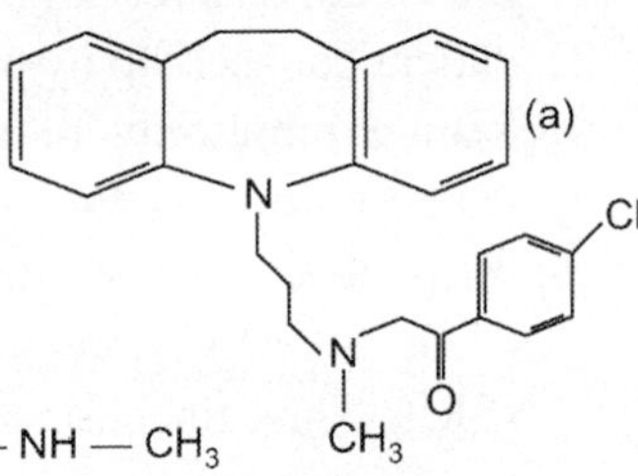

Lofepramine
(Gamanil®)

Note
(a) = dihydrodibenzazepine ring
(b) = dibenzocycloheptene ring
(c) = dibenzoxepine ring
(d) = dibenzocycloheptriene ring
(e) = dibenzazoxepine ring
(f) = tetracyclic ring

Fig. 11.1

3. Irreversible MAO (Monoamine Oxidase) Inhibitors

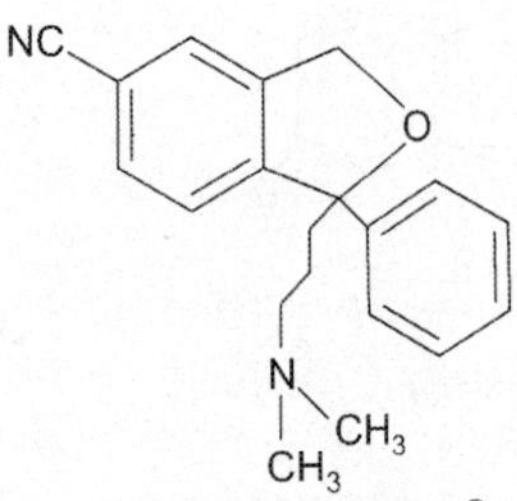

Citalapram (Celexa)®

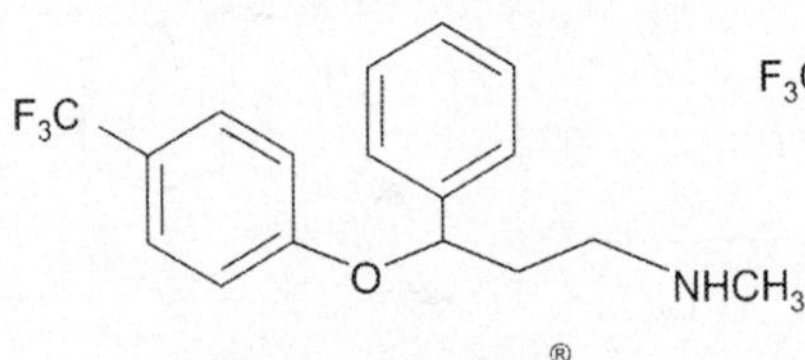

Fluoxetine (Prozac)®

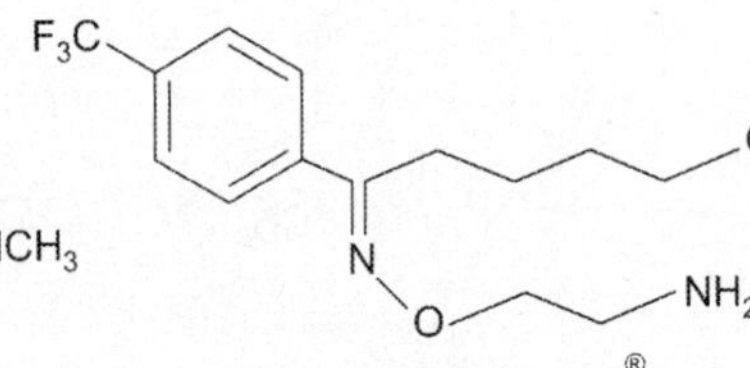

Fluvoxamine (Luvox)®

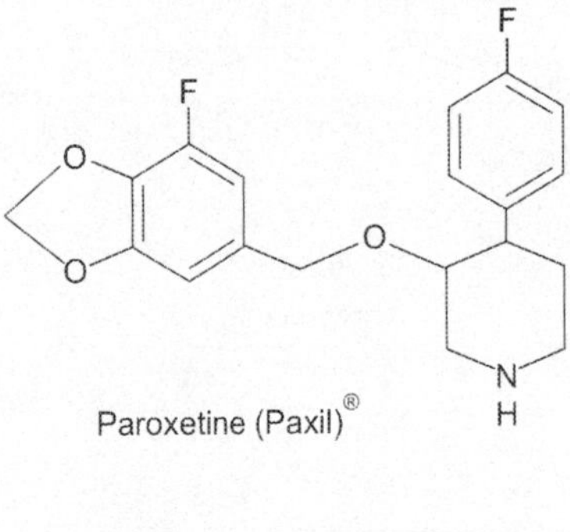

Paroxetine (Paxil)®

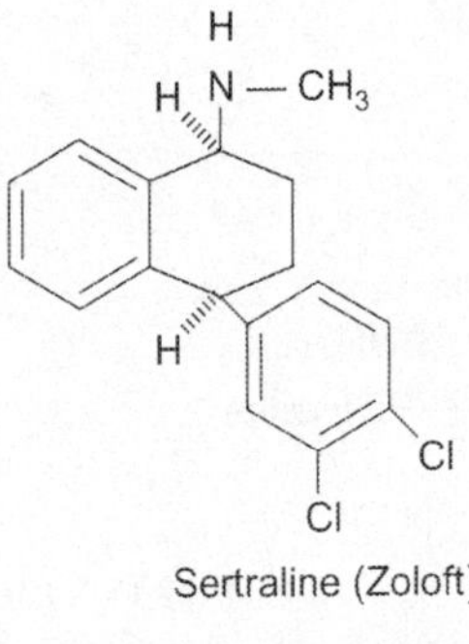

Sertraline (Zoloft)®

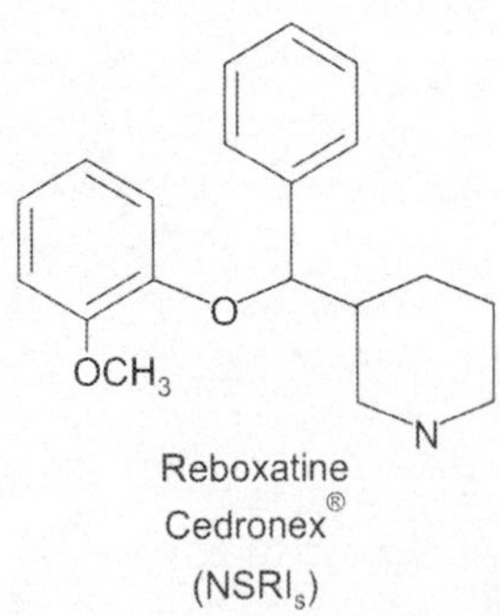

Reboxatine
Cedronex®
($NSRI_s$)

Fig. 11.2 (Contd.)

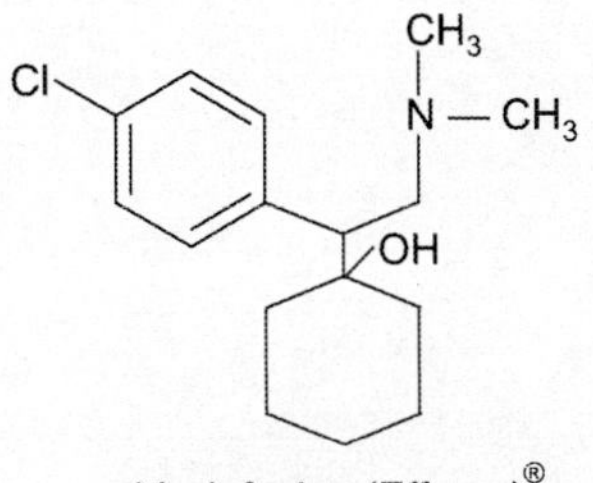

(NSRI$_s$) = Norepinephrine selective reuptake inhibitors

Venlafaxine (Effexor)®

Fig. 11.2

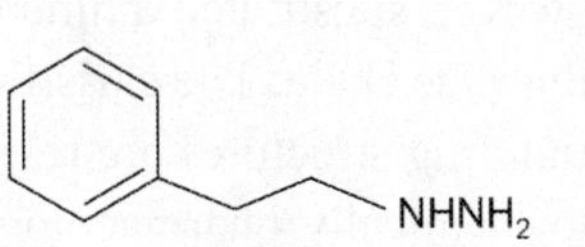

Phenelzine (Nardil)®

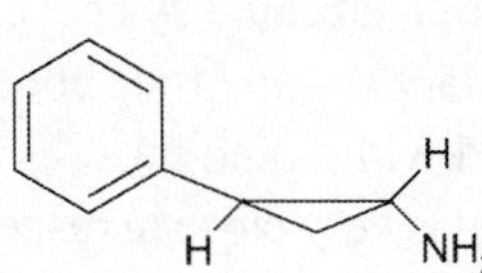

Tranylcypromine (Parnate)®

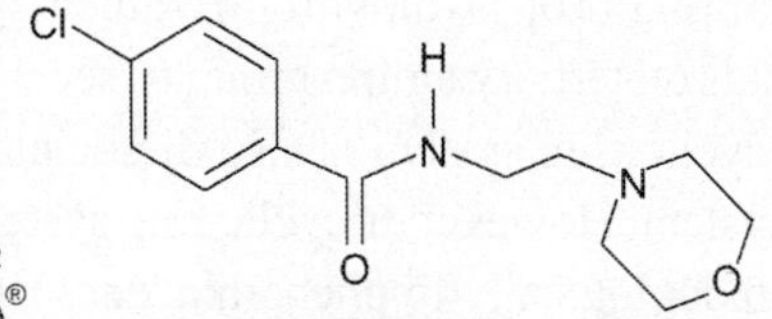

Moclobemide
(Manerix)®
(Reversible RIMAs)

Fig. 11.3

4. Newer Atypical Antidepressants

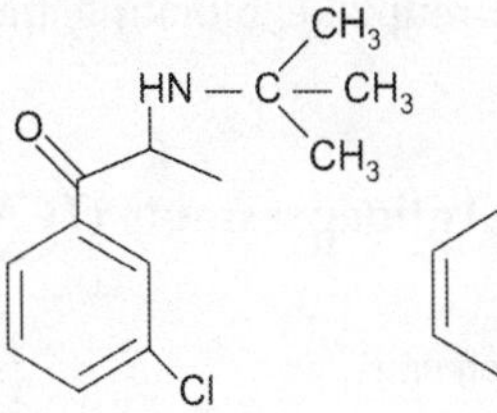

Bupropion
(wellbutrin®)

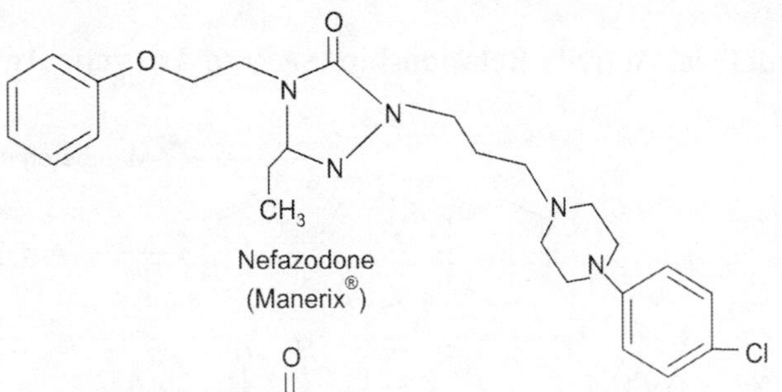

Nefazodone
(Manerix®)

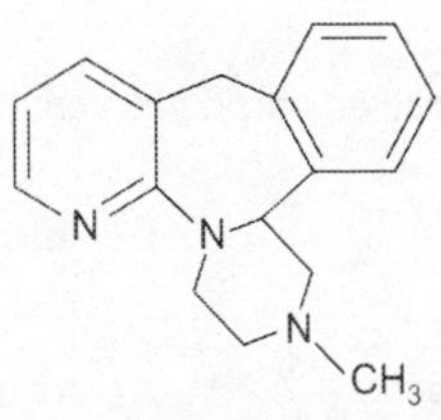

Mitrazepine (Remeron)®

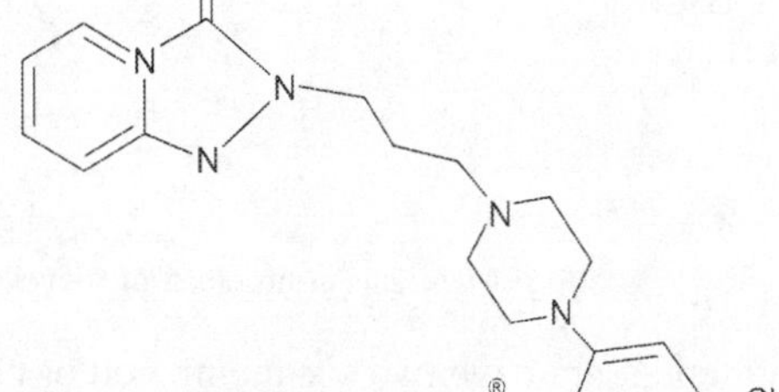

Trazodone (Desyrel)®

Note: Serotonergic agents = Trazodone, Nefazodone

Fig. 11.4

5. Mood Stabilizers:

   Lithium (Eskalith®)

   Valproic Acid (Depakote®)

   Carbamazepine (Tegretol®)

11.4 TRICYCLIC ANTIDEPRESSANTS

As the name indicated tricyclic antidepressants, the compound bearing two benzene ring system which is joined to a central 7-membered ring with substituted amines group is connected to the central ring (10, 11-dihydro-5H-dibenz [b, f] azepine system). However, substituted amino group depends on tertiary amino group or secondary amino group containing side chain. The classification of tricyclic ring system is also dissociated with it and most essential drug structures are related to this system. However, tricyclic ring system is very close in respect to chemically, pharmacologically and toxicologically to phenothiazenes ring system of antipsychotic drugs.

11.4.1 Mechanism of Action

Research depicted that mechanism of actions of tricyclic antidepressants are based on to inhibit reuptake of nor-epinephrine (Nor Adr) and serotonin by presynaptic nerve ending, thus, blocking of leading mechanism of their inactivation process and thereby increase the concentration of neurotransmitters at the nerve ending this indicates amines potentiating their effects. It is to be noted that presence of amine functional group potentiate the reuptake blocking mechanism of neurotransmitters.

11.4.2 Structural Activity Relationship (SAR) of Tricyclic Antidepressants (TCAs)

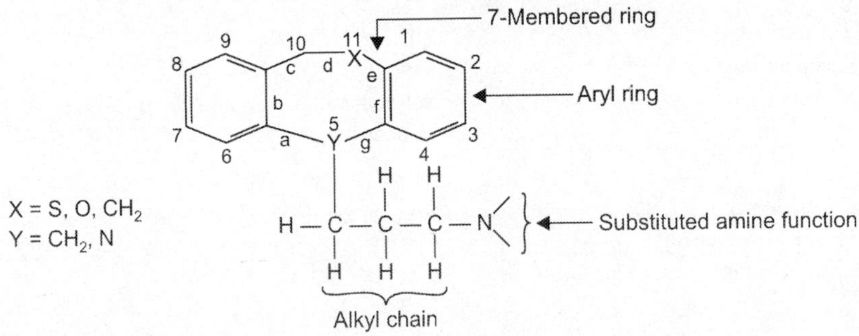

Fig. 11.5: Basic structure of tricyclic antidepressant.

1. Tricyclic antidepressant have two six-membered or benzenoid ring system with flanked seven-membered ring.

2. Aromatic ring system:
   (i) Introduction of electron withdrawing group to the aryl ring, enhance the activity
   (ii) Both aromatic rings are essential for activity.
   (iii) Oxidation of central ring to planner structure leads to inactive compound.

3. Replacement of ring nitrogen atom to sp^2 hybridized carbon atom to which the aminopropyl substituent is attached, results in decreased antipsychotic action and increase in antidepressant activity.

4. However, ring nitrogen atom or carbon atom (sp^2 hybridized) is not required to antidepressant effect, e.g., nortryptyline and protryptyline.

5. Terminal amine function:
   (i) The tricyclic ring system is generally attached to a basic amine function by three carbon chain.
   (ii) Shorten chain length (two carbon) preserve the antidepressant effect.
   (iii) Tertiary amines are more nonselective for inhibition of 5HT and norepinephrine reuptake mechanism while secondary amines are more selective for inhibition of NE reuptake. However, tertiary amines undergo metabolism by demethylation to secondary amines.

6. TCAs are subdivided into, dihydrodibenzazepine ring (e.g., imipramine), a dibenzacycloheptene ring (e.g., amitryptylene), a dibenzoxepine ring (e.g., doxepine), a dibenzocycloheptrieine ring (e.g., protryptylene) and a tetracyclic ring (e.g., maprotilene) derivatives.

11.4.3 Metabolic Pattern of Tricyclic Antidepressants

1. Most of the tricyclic antidepressants have longer half lives. Many of the compounds are metabolized in liver by demethylation to yield active desmethyl metabolites that would be responsible for further longer duration of action, e.g., imipramine is metabolized into desipramine that has longer elimination half life than parent compound.

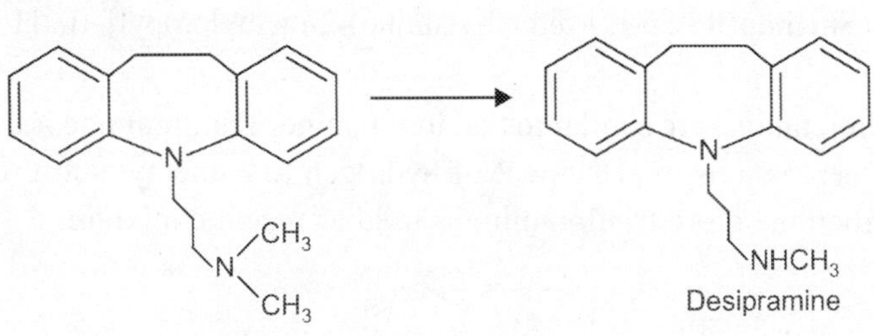

Fig. 11.6

2. Slow metabolizers are associated with hydroxylation by cytochrome P450 (CYP2D6)enzyme, that plays an important role in the aromatic hydroxylation of tricyclic antidepressants.

11.4.4 Specific Drugs

1. Imipramine (Tofranil®): 5-[3-(dimethylamino)Propyl]-10, 11-dihydro-5H-dibenz [b, f]-azepine. Imipramine is the lead representative drug belonging to TCAs category. Its structure is nearby to the antipsychotic agent phenothiazene in that way to replace the 10, 11 bridge with sulphur and compound is called promazine. Its mechanism of action represents blocking the reuptake of biogenic amines. It is a weak D_2 receptor blockers. Due to presence of dimethyl amine group. It marked the anticholinergic and sedative (H_1 receptor blocker) activities. It tends to block 5HT and NE uptake, hence increase the concentration of these neurotransmitters therefore may be called serotonin transport inhibitors (SERTI). Its demethylated compound is less anticholinergic, less sedative but has more SNERI action. Imipramine is oftenly used in the treatment of depression caused by motor clumsiness, associated with Parkinson's disease.

Note: SNERI (selective nor-epinephrine reuptake inhibitor)

Synthesis:

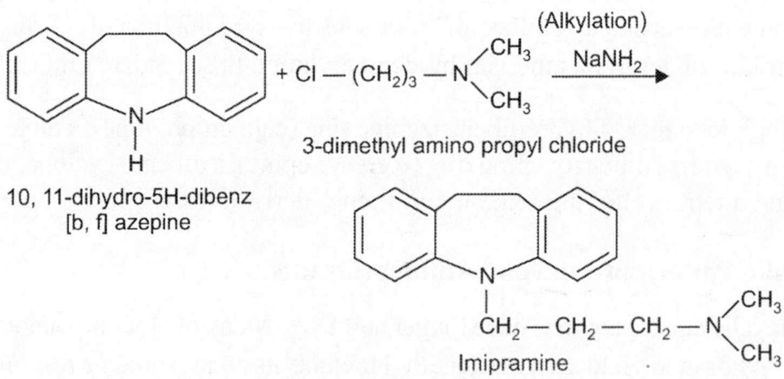

Fig. 11.7

2. Trimipramine (Surmontil®): 5-[3-(dimethylamino)-2-methylpropyl]-10, 11-dihydro-5H-dibenz [b, f] azepine.

Properties of trimipramine are similar to that imipramine. Trimipramine is used in the treatment of various types of depressions, replacement of hydrogen to α-methyl substituent of imipramine produce a chiral carbon therefore trimipramine is used as recemic mixture.

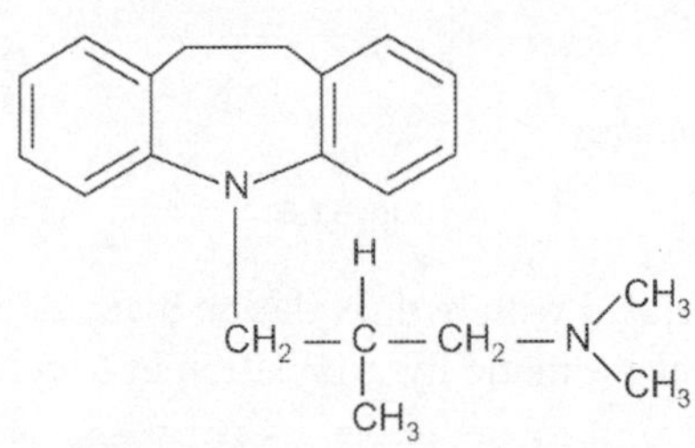

Fig. 11.8

3. Amitryptyline (Elavil®): 5-(3-dimethylamino propyliden)-10, 11-dihydrodibenzocycloheptene. Amitryptyline is known more often its anticholinergic and sedative properties of TCAs. The main difference between imiramine is that the nitrogen atom in the central ring is replaced by carbon atom that is bound to side chain by double bond. N-demethylation occurs by metabolism formed nor- triptyline that is less anticholinergic, less sedative and more effective than amitryptyline as well as SNERI.

Amitryptyline is used for anxious-depressive conditions, and various types of depressive situations.

Synthesis:

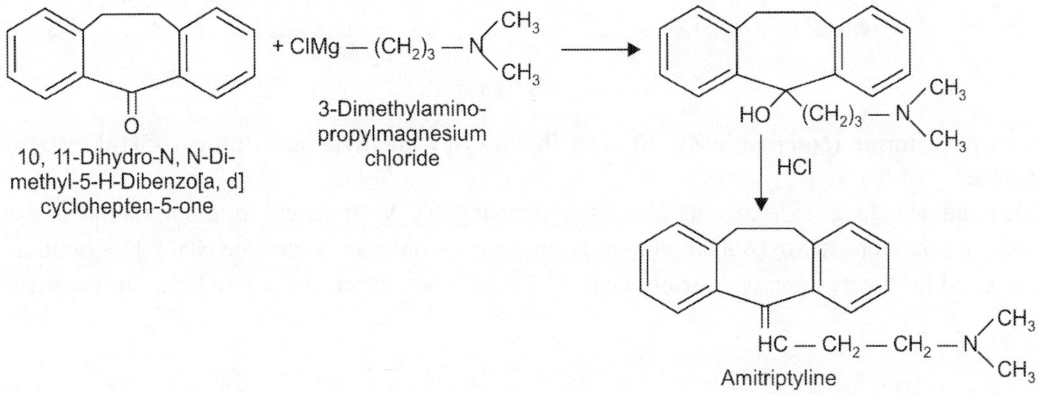

Fig. 11.9

4. Doxepine (Senequan®): (11-[16H]-(3-dimethyl amino-propyliden)-6, 11-dihydrobenz [b, e] oxepine).

Doxepine is an important drug of TCAs which is a oxa proximity of amitryptyline. Introduction of oxygen at the place of 11 position makes the compound more effective towards postsynaptic and presynaptic binding affinities. Oxygen atom influence oxidative metabolism. Its Z isomer is considered as more effective. It is available in recemic mixture. Doxepine retards the reuptake of 5HT and NE as well and also exhibits anticholinergic and sedative properties.

Doxepine is suggested for the treatment of anxious depressive and anxious conditions, neuroses, alcoholism, illness of CNS and psychoses conditions.

Synthesis:

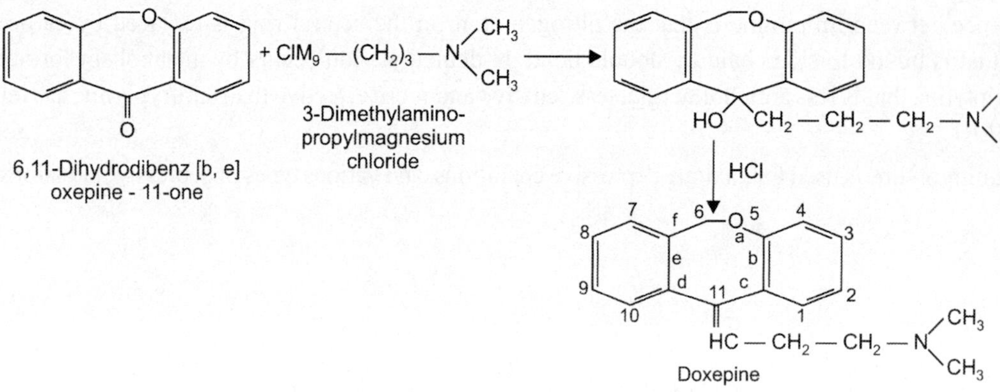

Fig. 11.10

5. Desipramine (Norpamin®): 10, 11-Dihydro-5-[3-(methylamino) propyl]-5H-dibenz[b, f] azepine

Desipramine belongs to secondary amines derivative of TCAs. It differs from imipramine in that it possesses only one methyl substituent on propylamine's nitrogen atom side chain. Desipramine is suggested for the treatment of various causes of depression, especially for endogenous depression.

Synthesis:

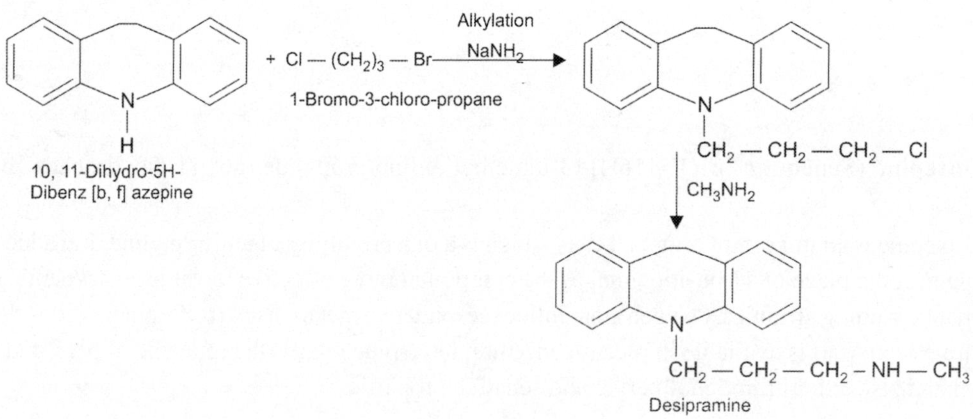

Fig. 11.11

6. Norptryptyline (Aventil®): 5-(3-methyl amino propyliden) 10, 11-dihydrodibenzocycloheptene. Nortryptyline is also a congener of secondary amines TCAs. Nortryptyline differs from Desipramine in that way nitrogen atom of central (seven member) ring is replaced by carbon atom (sp^2 hybridized) which is bound with double bond to side chain. Nortryptyline is selective Nor-epinephrine (NE) inhibitor. It has shorter duration of action and mainly used in the treatment of manic-depressive psychoses as well as all causes of endogenous depression.

Synthesis:

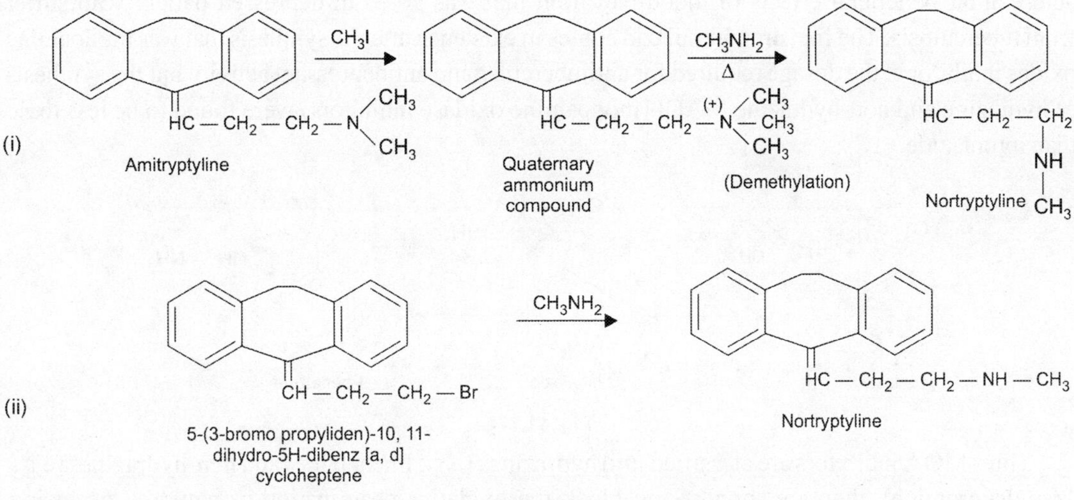

Fig. 11.12

7. Protryptyline (Vivactil®): N-methyl-5H-dibenz [a, d]-cyclohepten-5-propylamine. Protryptyline belongs to secondary amine containing TCAs. Protryptyline differs from previous drugs structures in that. It possesses double bond at the position C_{10}-C_{11} of central seven-member ring of tricyclic rings system, and excyclic double bond is excluded. Localized double bond is expected to inactivate the molecule. Its sedative property is low than others.

It is considered a powerful antidepressant. It is clinically used in the treatment of various depressive conditions.

Synthesis:

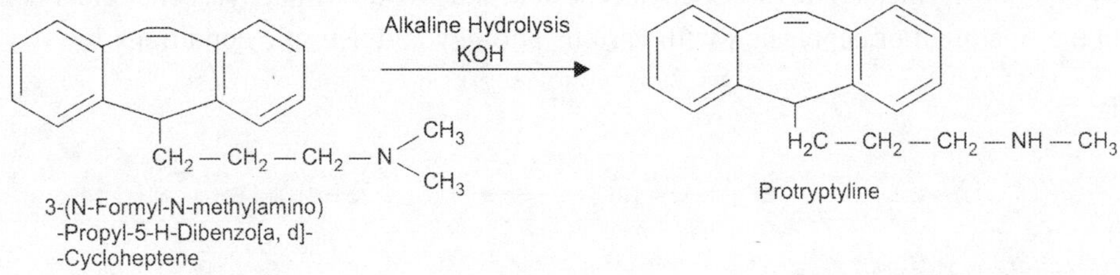

Fig. 11.13

11.5 MONOAMINE OXIDASE INHIBITORS (MOI)

Monoamine oxidase inhibitors became popular after discovery of isonicotinic acid derivative (isoniazide) that is antitubercular drug. Clinically tested of the hydrazine derivative of isonicotinic acid exhibit beneficial effects of mood elevation that was noted in depressed patient who suffer from tuberculosis. The first drug iproniazid comes in existence after its synthesis that was monoamine oxides inhibitor at the dosage required for antitubercular and antidepressant activity but the synthesis of hydrazine and non-hydrazine MAOI (monoamine oxidase inhibitors) were found to be less toxic than iproniazide.

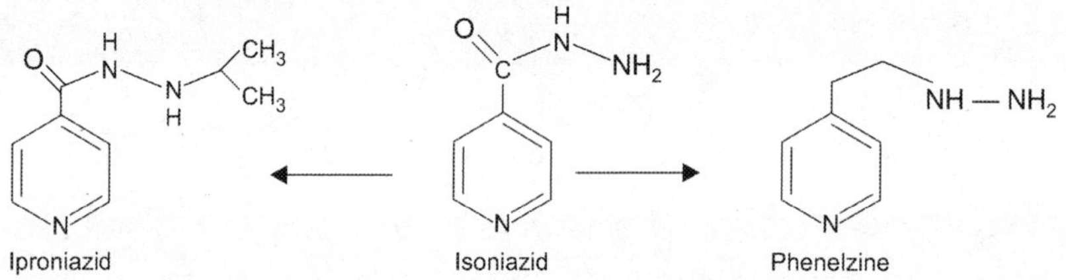

Fig. 11.14

Thus, MOA inhibitors are classified into hydrazines (e.g., Phenelzine) and non-hydrazines (e.g., Tranylcypromine), they are supposed to block the oxidative-deamination of naturally occurring monoamines, including epinephrine, NE, Dopamine and 5HT. Above-mentioned classified drugs are irreversible non-selective inhibitors of MAO by covalent bond they increase the concentration of free 5HT and NE in synaptic cleft in the CNS.

MAOIs are suggested in the treatment of severe endogenous, exogenous and reactive depressions.

11.6 SPECIFIC DRUGS

1. Phenelzine (Nardil®): 2-phenylethyl hydrazine. Phenyelzine is a potent antidepressant drug. Phenelzine action is prolonged, non-selective and is irreversibly. It inhibits the MAO enzyme by covalently bonding. Its metabolites are phenylacetic acid and p-hydroxy phenyl acetic acid. Phenelzine is used in the treatment of depression with various etiology and neurotic conditions.

Synthesis:

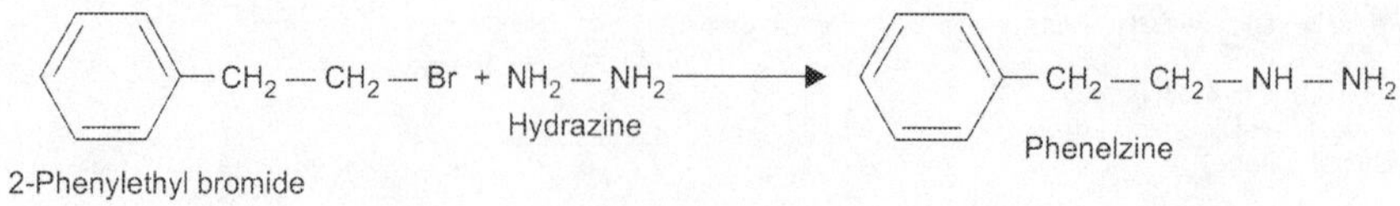

Fig. 11.15

2. Tranylcypromine (Manerix®): (±) trans-2-penylcyclo-prophylamine.
Tranylcypromine is another important example of MAO inhibitor antidepressant drug. It is rapidly absorbed and extensively degraded in liver by first pass metabolism. Tranylcypromine is used in the treatment of such depressions that are failure to other drugs.

Synthesis:

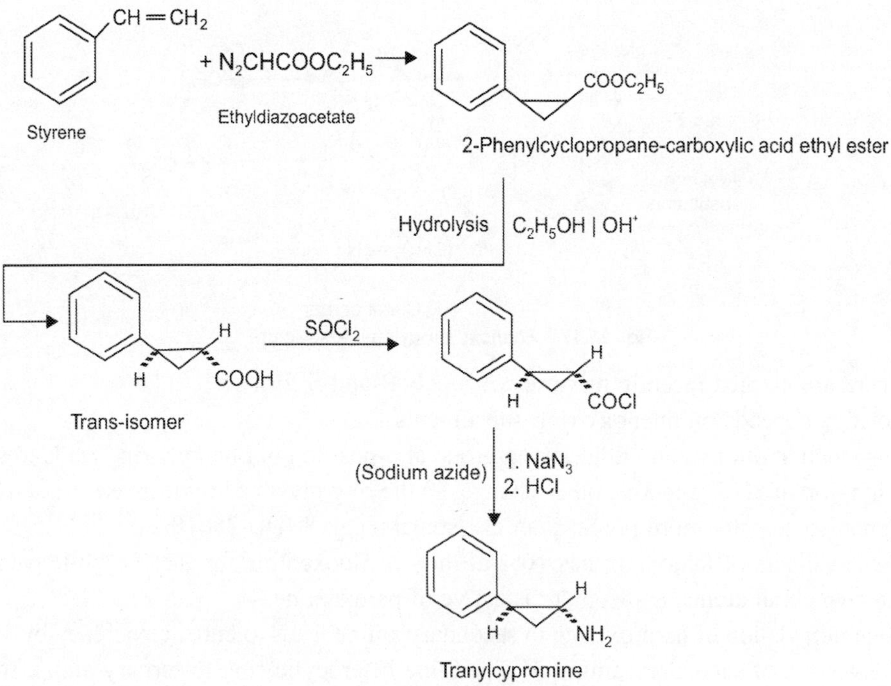

Fig. 11.16

11.7 SELECTIVE SEROTONINE REUPTAKE INHIBITORS (SSRI)

5HT is a major neurotransmitter in depressive illness and serotonergic pathway is very close to the mental disorders like depression. Therefore SSRIs drugs affecting the levels of 5HT and serotonergic pathways in neuronal spaces of synapse lead to effective therapy of depressive illness.

SSRIs are clinically proved in the treatment of which depression, obsessive-compulsive disorders and panic disorders as well. The most substantial advantage of SSRIs over TCAs are to have lower incidence of side effects and better tolerability. The meaningful differences between the individual drugs of SSRIs are largely associated to their pharmacokinetics, metabolism of active metabolites, inhibition of CYP isoforms and drug-drug interaction behaviour as well as half life of drugs.

11.7.1 Mechanism of Action

SSRIs epinephrine preferentially inhibit the reuptake transport of 5HT (SERI) with minimum or no affinity to nor-epinephrine and Dopamine transporters respectively. SSRIs have higher affinity for SERT therefore block 5HT from binding to SERT and being absorb into presynaptic cells, because excess of 5HT cause over-activation of post-synaptic receptors.

11.7.2 Structure Activity Relationship (SAR) of (SSRIs)

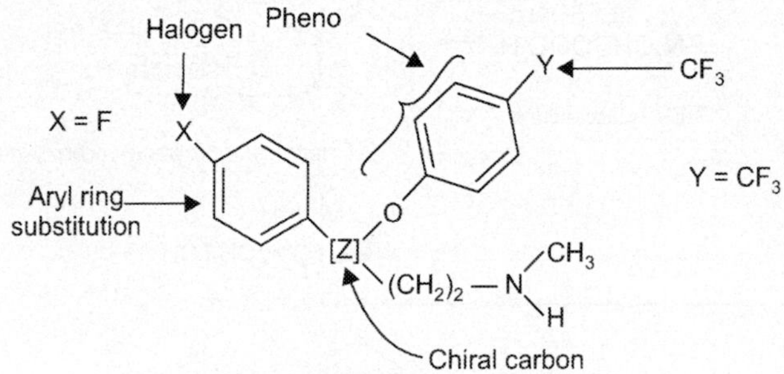

Fig. 11.17 Phenoxy propylamine structure.

1. There are marked racemic mixtures of SSRIs R and S forms.
2. Potency depends on phenoxy ring substituents.
3. Introduction of electron withdrawing group at p-position of phenoxy ring leads to selective inhibition of 5HT reuptake, other changes in the ring may lead to decrease the activity.
4. Trans-isomers are more potent than cis-isomers, e.g., MDL-28618A.
5. Trans-isomers of famoxatin has 10% affinity of fluoxentine for SERT. While paroxentine has two chiral atoms, (–) isomere is active of paroxentine.
6. N-demethylation of faxmoxetine to secondary amine leads to enhance potency by 10 times.
7. Conversion of secondary amine of piperidine heterocylic ring to tertiary amine with alky group decrease the activity for 100 times.
8. Introduction of 4-fluoro is essential for SERT affinity, replacement of 4-fluoro substituent with hydrogen or alkyl group leads to sudden decrease in activity.

11.7.3 Specific Drugs

1. Fluoxantine (Prozac®): 3-phenoxy-3-phenyl propyl-3-[p-(trifluoro methyl)-phenoxy]-N-methyl-3-amine is the derivative of phenyl propylamine. Fluoxantine is active drug of SSRIs. Its S isomer is more potent for SERT. Its N-demethyl derivative is more potent than parent compound and more selective for SERT. Fluoxantin inhibits the re-uptake of 5HT in the presynatic cleft and increase the abundance in an anxious state.

Synthesis:

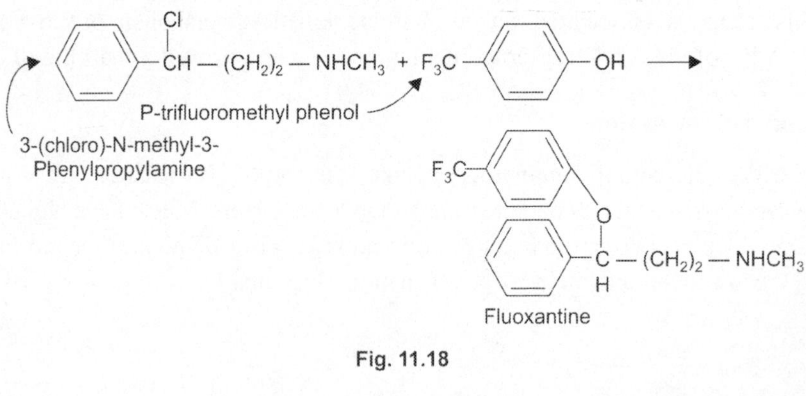

Fig. 11.18

2. Paroxetine (Paxil®): (3S, 4R)-3-[2H-(1, 3-benzodiazol-5-yl oxyl) methyl]-4-(4-fluorophenyl) piperidine

Paroxetine is a potent SSRIs antidepressant drug. It has β-Aryl amine like structure with extra aryl group. Paroxetine is highly selective towards SERT. Paroxetine is suggested in the treatment of various etiological depressive illnesses.

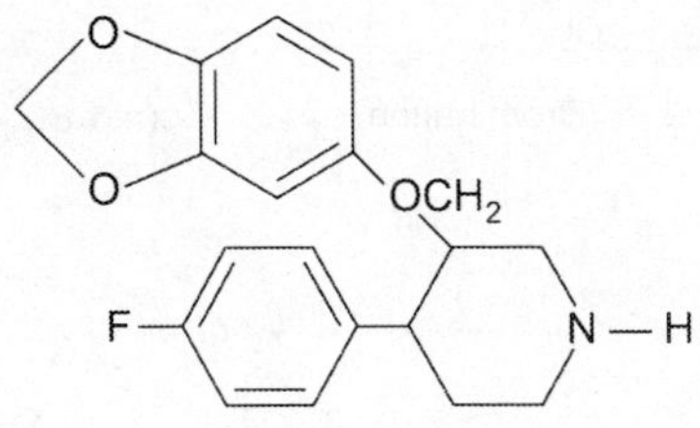

Fig. 11.19

3. Citalopram (celexa®): (RS)-1-[3-(dimethylaminopropyl)]-1-(4-fluorophenyl)-1, 3-dihydroisobenzofuran-5-carbonitrile, citalopram is available in racemic mixture. It is highly selective for SERT. Its N-mono demethylated derivative is less potent than parent compound. Its aryl moiety makes the compound more potent. It is used in the treatment of various types of depressions.

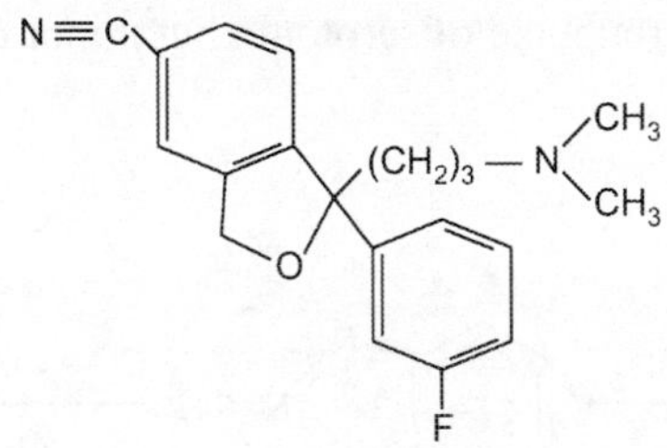

Fig. 11.20

11.8 NEWER ATYPICAL ANTIDEPRESSANTS

Newer atypical antidepressant agents are second generation antidepressant drugs. It is very likely that their mechanism of action is due to their ability to inhibit the re-uptake of norepinephrine or serotonine.

11.8.1 Specific Drugs

1. Bupropion (Wellbutrin®): 1-3-(chlorophenyl)-2-[(1, 1-dimethylethyl) amino]-1-propanone. Bupropion is structurally similar to amphetamine because of having α-amino ketone. Bupropion is a dopamine re-uptake inhibitor via dopamine transporter (DAT).

Synthesis:

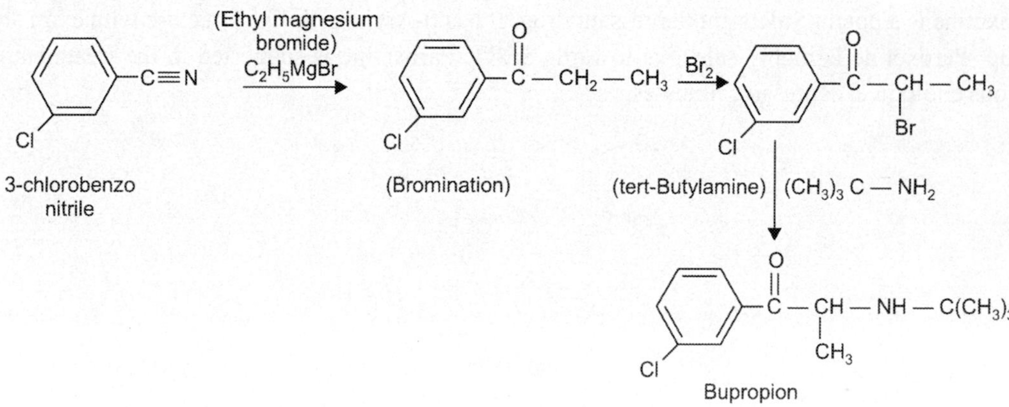

Fig. 11.21

2. Trazodone (Thrombran®): 2-[3-[4-(m-chlorophenyl)-1-piperazineyl]-propyl]-S-triazolo [4, 3-a]-piperazinyl-3-(2H)-one trazodone is a potent atypical antidepressant drug. It is believed that it has the ability to inhibit neuronal reuptake of serotonin but minor effect on NE and Dopamine.

Synthesis: (Semicarbazide)

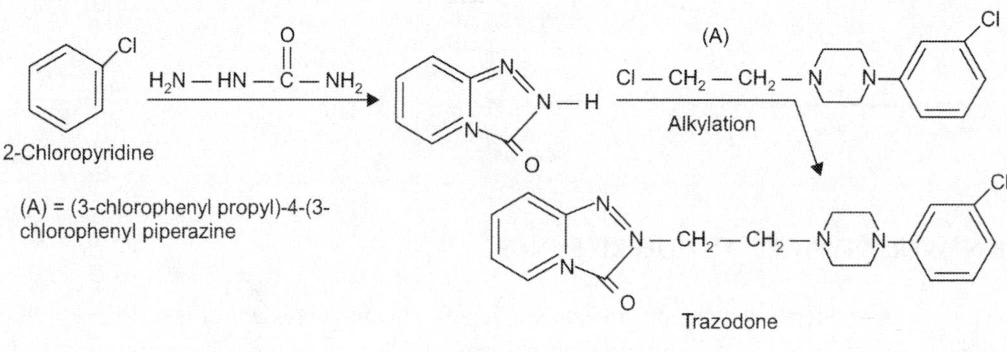

Fig. 11.22

11.9 NEWER AGENTS

1. Escitalopram: Escitalopram is (S) isomer of citalopram. It has higher affinity for the human serotonin transport (SERI). It is suggested for the treatment of several anxiety and depressive states, in panic disorder and obsessive compulsive disorder.

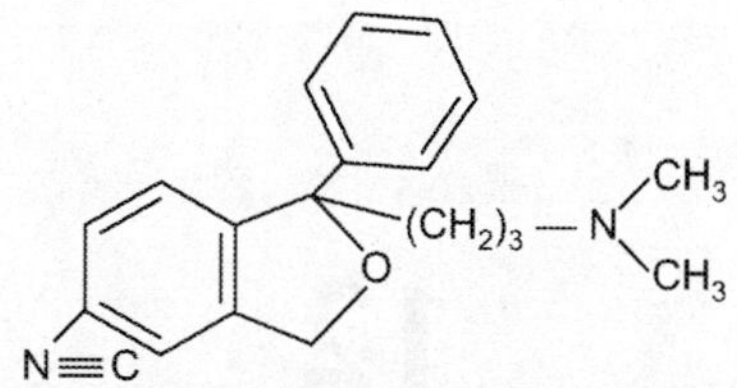

Fig. 11.23

2. Things to come: Drugs acting on Neuropeptide receptors
   (i) Substance P-NKI receptor antagonists: e.g., MK-0869.
   (ii) CRF antagonist (corticotrophin releasing factor antagonist) e.g., CP-614, 526, DMP-696.

EXERCISE

1. Classify antidepressant drugs. Give their mechanism of action.
2. What are trycyclic antidepressants? Discuss in detail.
3. Write SAR on tricyclic antidepressant with various illustrations.
4. Write the synthesis, MOA and uses of the following drugs:
   (i) Imipramine
   (ii) Amitriptyline
   (iii) Doxepine
   (iv) Protrypyline
5. Discuss in detail the monoamine oxidase inhibitors. Write any one synthesis of this class.
6. Discuss in details the SAR of SSRIs compounds.
7. Write a note on a typical antidepressant. Give the chemistry of bupropion.

12
Antianxiety Drugs

12.1 INTRODUCTION

Anxiety is the phenomenon that is directly associated with emotional condition. It is an essential part of adaptive response to stressful or threatening stimulation that can heighten to the dangerous or stressful condition. Anxiety is a condition that is clinically difficult to define but may be described as a sense of unpleasant unwanted anticipation and excessive mental tension. However, anxiety is very close to the condition of depression to some extent. Nowadays anxiety is an integral part of our life. It may be chronic, uncontrollable and completely out of proportion to the generating stimulus. There are so many factors such as social factors, pathophysiological factors serve as initiating state of anxiety. Somatic symptoms includes tachycardia, palpitation, sweating, dizziness, nausea, insomnia, irritability and impaired concentration as well.

The anxiety disorders are the following:

1. Generalized anxiety disorders (GAD): Symptoms are restlessness, fatigue, impaired concentration, irritability and muscles tension.
2. Social phobia (SP) or social anxiety disorder (SAD) pertaining social aspects.
3. Obsessive compulsive disorder (OCD).
4. Post-traumatic stress disorder (PTSU): Symptoms includes extreme fear or terror associated with social threatening (war or rape), insomnia, night mares, etc.

According to the clinical studies, monoamine oxidase inhibitors (MAOIs) for illustration iproniazide and phenelzine are the suitable candidates of drugs that are useful in the treatment of anxiety disorders. Another classes of antidepressants drugs are TCAs (tricylic antidepressants) for example, imipramine used to panic disorders of anxiety. Apart from that Benzodiazepines (BZs) discovered in 1950s play a very important role in the treatment of anxieolytic disorders.

12.2 CAUSES OF ANXIETY

Neurotransmitters, neuromodulators and neuropeptides (for illustration, cholecystokinin, corticotropin- releasing factors, neuropeptides Y) are involved in the pathophysiology of anxiety.

12.2.1 Receptor Involve

1. GABA receptors :- $GABA_A$ and $GABA_B$ as well as $GABA_C$ receptors are utilizing the neurotrans-mitters for intracellular communication. $GABA_B$ receptor is (G-protein couple receptor)
2. Serotonergic Receptors:- $5HT_1$, $5HT_2$ and $5HT_3$ receptors
3. Neuropeptides (CCK Receptors) :- CCK-A and CCK-B receptors

12.3 CLASSIFICATION

1. Benzodiazepines

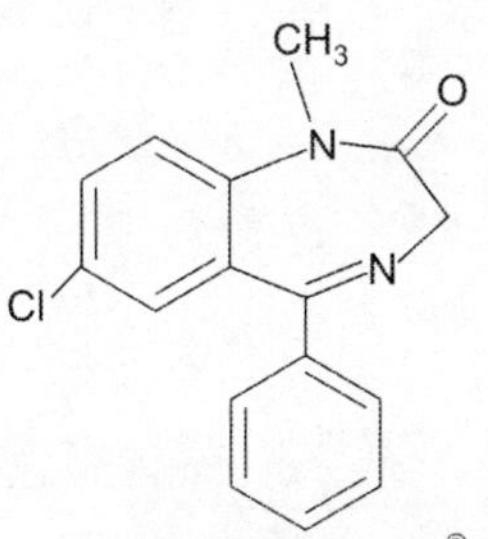

Diazepam (Valium®)

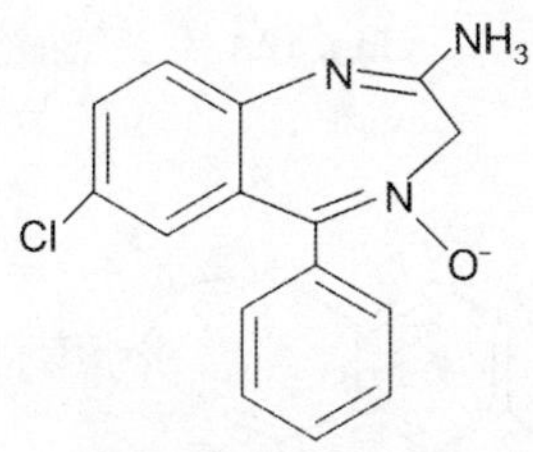

Chlordiazepoxide (Librium®)

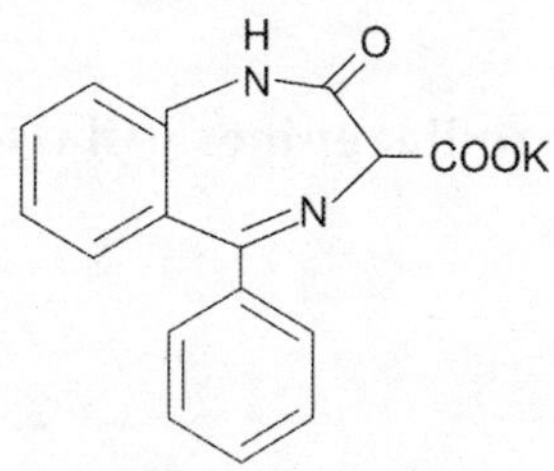

Chlorzepate (Tranxene®)

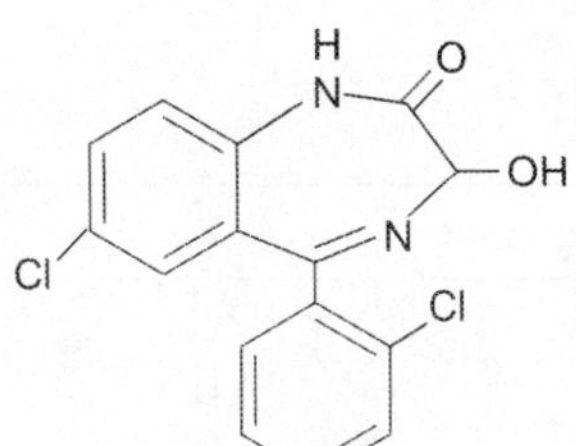

Lorazepam (Ativam®)

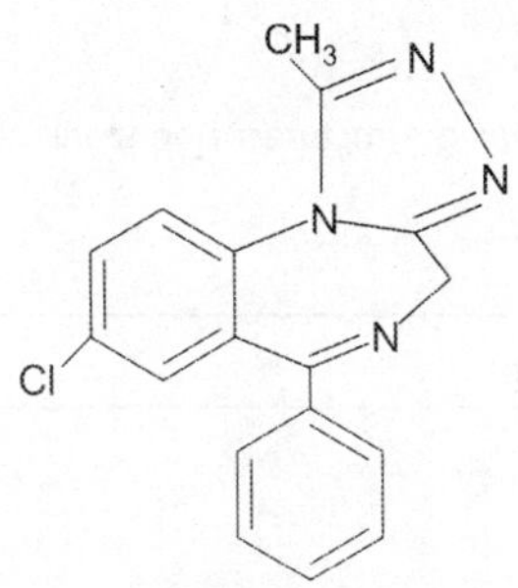

Alprazolam (Xanax®)

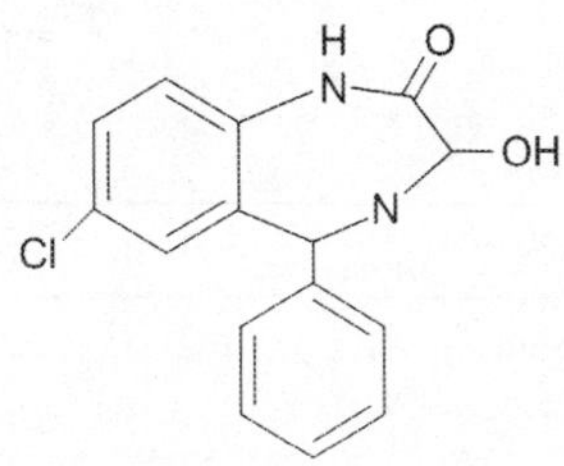

Oxazepam (Serax®)

Fig. 12.1 (Contd.)

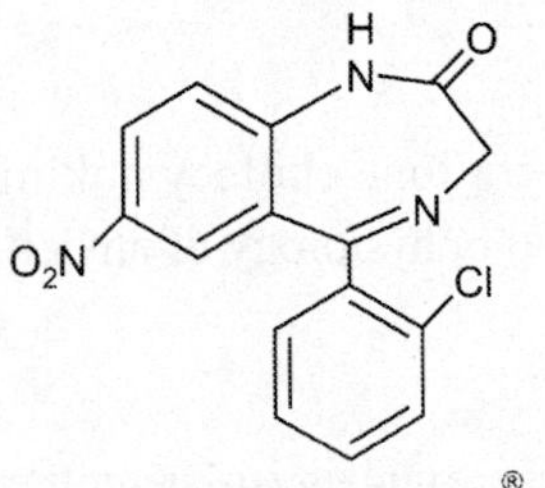

Clonazepam (Klonopin®)

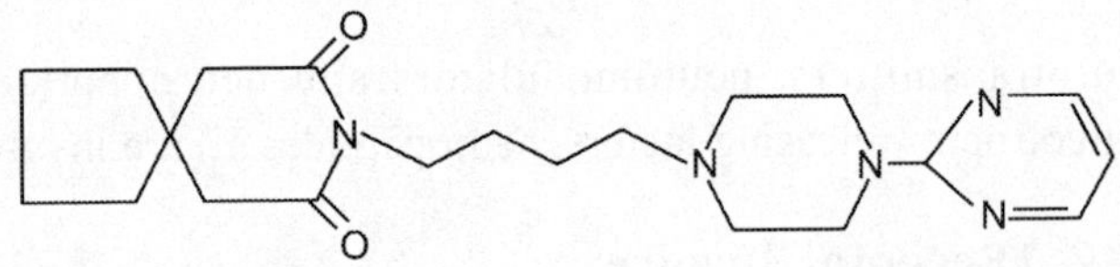

Buspirone (Buspar®)

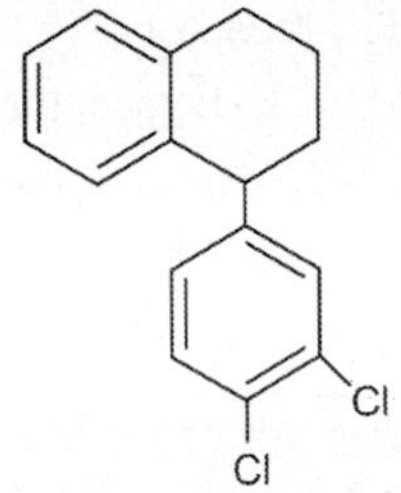

Sertaline (Zoloft®)

Fig. 12.1

Benzodiazepines (Class-A)

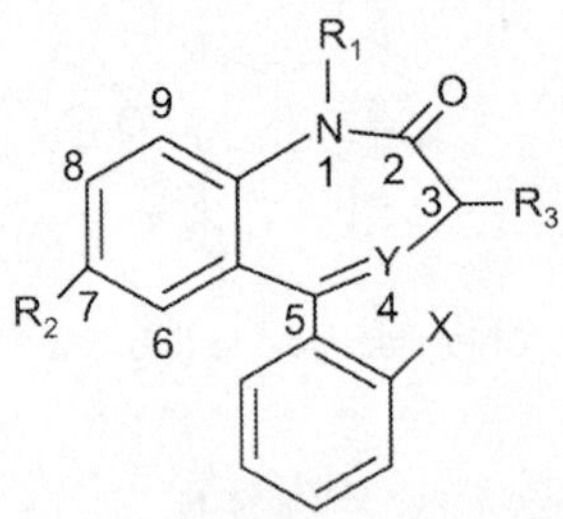

Fig. 12.2: Basic structure of benzodiazepine.

Derivatives	R_1	R_2	R_3	X
Clonazepam	H	NO_2	H	Cl
Diazepam	CH_3	Cl	H	H
Clorazepate	H	Cl	COOK	H
Flurazepam	$(CH_2)_2N(C_2H_5)_2$	Cl	H	F
Halazepam	CH_2CF_3	Cl	H	H
Oxazepam	H	Cl	OH	H
Temazepam	CH_3	Cl	OH	H

2. Triazolobenzodiazepines: (Class-B)

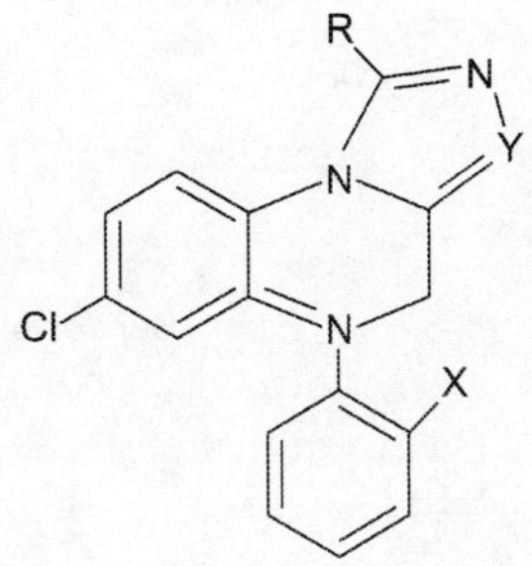

Fig. 12.3: Basic pharmacophore of triazolobenzodiazepines.

Derivatives	R	X	X
Alprazolam	CH_3	H	N
Midaxolam	CH_3	F	Ch
Estazolam	H	H	N
Triazolam	CH_3	Cl	N

3. Non-benzodiazepines Drugs:

A. β Carbolines (9H-Pyrido-[3,4-b]-indole) (Norharmane)

x = COO E + (β-CCE)
x = COO CH_3 (β-CCM)
x = COO–n–Pr
x = Alkoxy

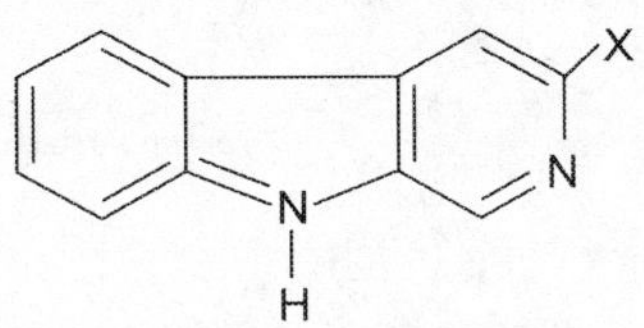

Fig. 12.4

B. Cyclopyrrolones

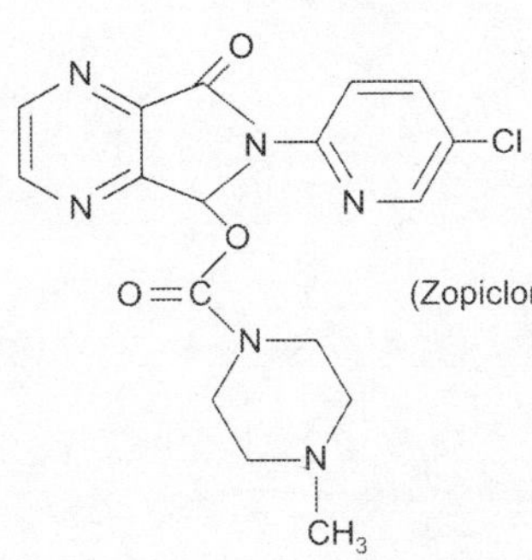

(Zopiclone)

Fig. 12.5 (Contd.)

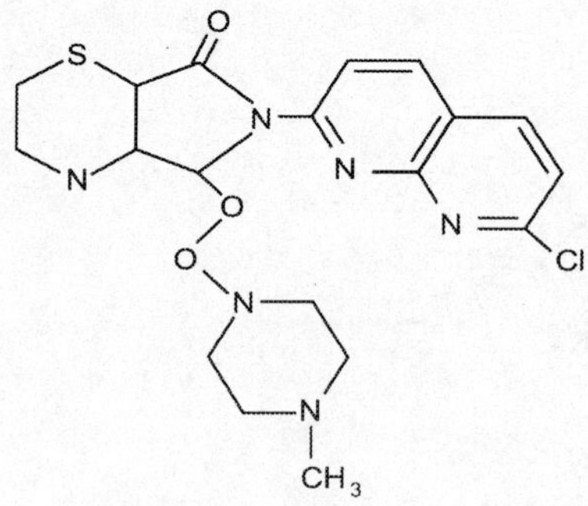

[6-(7-chloro-1,8-Naphthyridin-2yl)-5-oxo-3,7-dihydro-2H-[1,-4]-dithinopyrrol-7-yl]-4-methyl-piperazine-1-carboxylate

Surizoclone

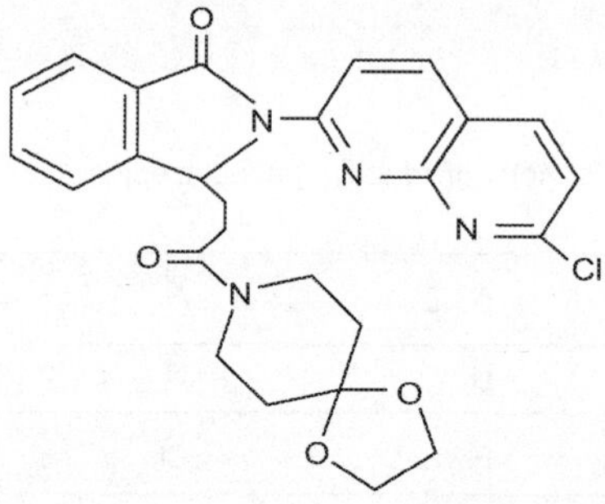

Panizoclone

Fig. 12.5

B. Cyclopyrrolnes

4. Newer Agents

A. Imidazopyridines:

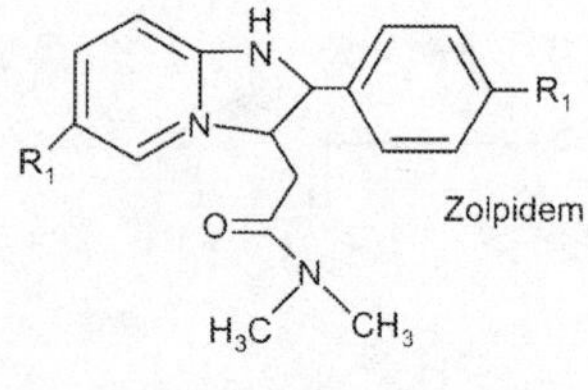

Zolpidem (Ambien®)

Fig. 12.6

B. Pyrazoloquinolines:

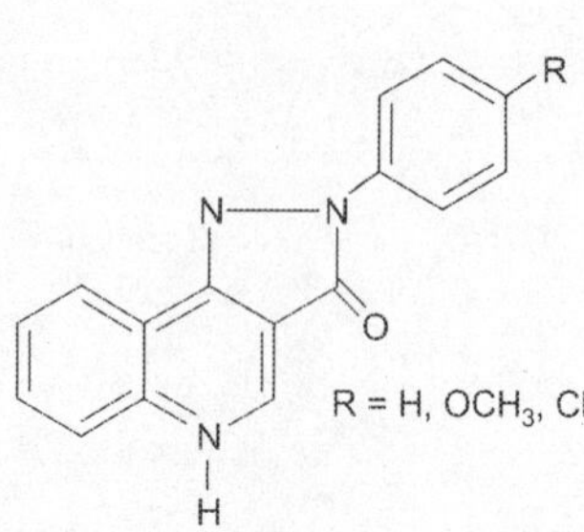

R = H, OCH$_3$, Cl

Fig. 12.7

C. Imidazobenzodiazepines:

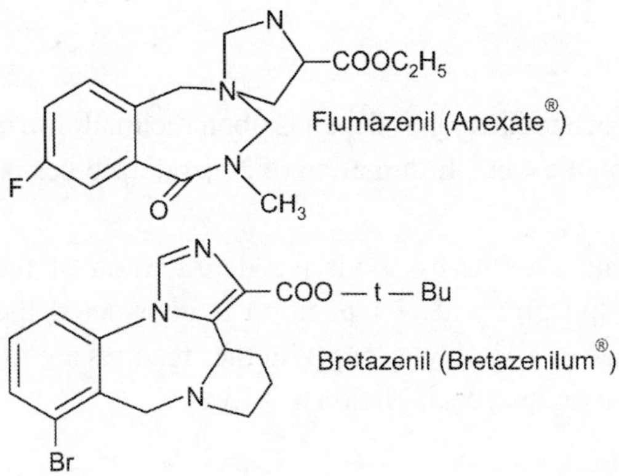

Fig. 12.8

D. Imidazoquinoxalines:

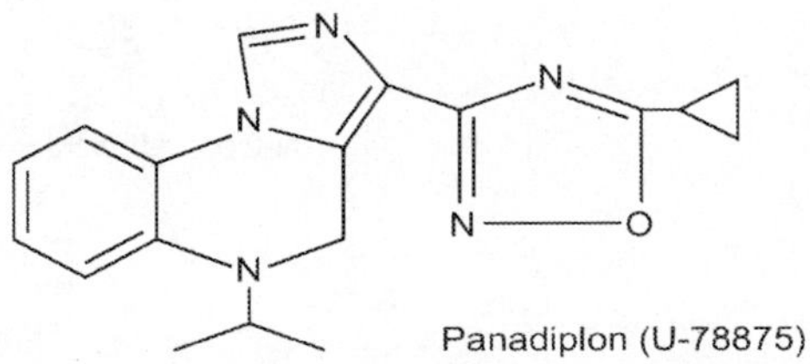

Fig. 12.9

12.4 BENZODIAZEPINES

Benzodiazepines are the most important agents to treat anxiety disorders in the early 1950s. New tranquilizers came in the pipeline; they are barbiturates and phenothiazines. Scientist Dr. Hajiman-La-Roche and Dr. Leo-Sternbach discovered the compound benzoheptox diazines, a synthetic tranquilizers that was structurally similar to that of benzodiazepines. Further studies on SAR proved a new class of compound chlordiazepoxide (Librium®) in 1960. After few years in 1963, diazepam (valium®) was launched and became most popular brand in the treatment of anxiety disorders.

12.4.1 Mechanism of Action

Benzodiazepines exhibit varying degree of pharmacological action like anxiolysis, muscles relaxants, sedation, seizure protection as well as memory impairment. All benzodiazepines act upon benzodiazepines receptors non-selectively. GABA is the major inhibitory neurotransmitter in brain. Many neurotransmitters are affected by the influence of GABA. They include Serotonin, Norepinephrine, CCK as well as Dopamine. Benzodiazepines act on allosteric site of $GABA_A$ receptors and produce their effect by increasing chloride ion conductance followed by opening of channel. All the benzodiazepines produce their effect on the acted upon allosteric subtypes (α_1, α_2, β_1, β_2 and γ_2) and produce the allosteric modulation of GABA receptors.

12.4.2 Metabolic Pattern

Pharmacokinetic profile of benzodiazepines depends upon their half life metabolites which are used as anxiolytics, sedative hypnotics, etc. Elimination of benzodiazepines is by hepatic metabolism in the urine.

1,4-benzodiazepines metabolites by oxidative degradation of the N-1 alkyl substitution. Benzodiazepines with Methyl group at N-1 position is considered the most stable and rate of dealkylation increase the size of N-1 alkyl group. Another features are hydroxylation at Position 3. Hydroxylation half life is longer than dealkylation half lives.

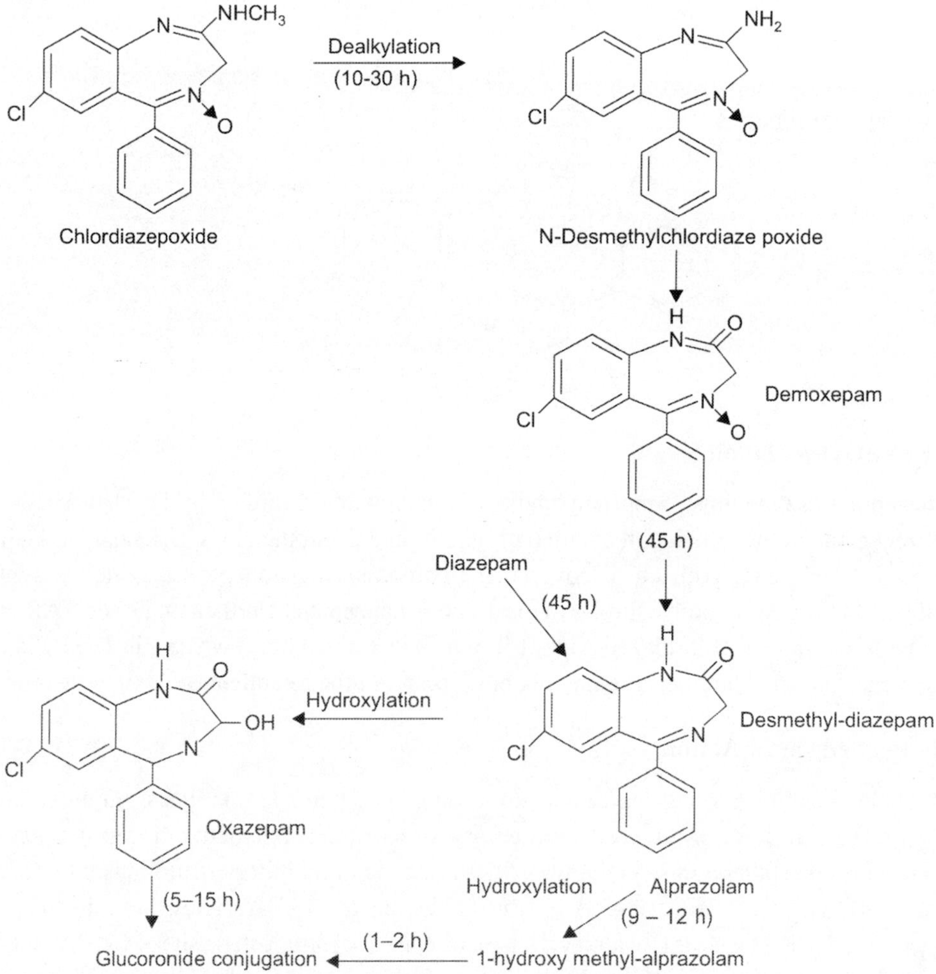

Fig. 12.10: Schematic presentation of benzodiazepine metabolism.

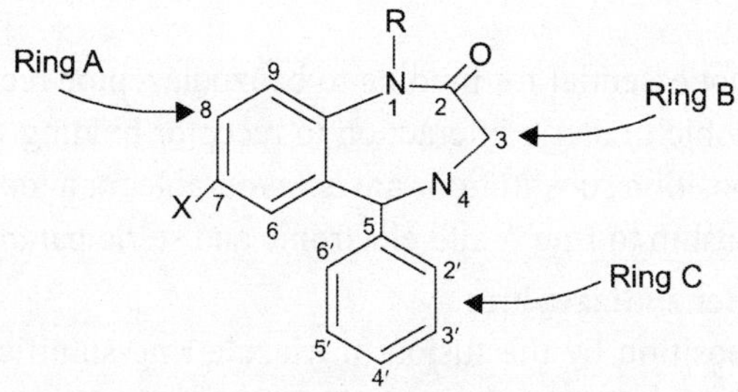

Fig. 12.11

Rate of hydroxylation depends upon the presence of substituents at ring C. 3 hydroxy mealbolities are formed due to glucoronides and excreted in urine. Rate of dealkylation and hydroxylation depends upon the aging, liver function, etc.

Schematic presentation, details of metabolism are due to demethylation and hydroxylation and half-lives given in Fig. 12.11.

12.5 STRUCTURE ACTIVITY RELATIONSHIP (SAR) OF BENZODIAZEPINES

A. 1. 4 – Benzodiazepine - 2 Zone (see in Chapter 5 sedative and hypnotics)
SAR of 5-Phenyl-1,4-benzodiazepine 2-one has been described by scientist Sternbach and other co-workers.

Ring A:

(i) Heteroaromatic ring or aromatic ring is required for activity and essential for receptor interaction by π-π stacking.

(ii) Allotment of electronegative group (halogen or nitro) at position, markedly increase the pharmacological action (anxiolytic activity)

(iii) Introduction of substituents at position 6, 8 or 9 generally decrease the activity.

(iv) Replacement of ring A by other heterocyclic ring results in weak binding properties *in vitro* and have less biological action.

Ring B:

(i) Proton accepting group (carbonyl moeity) at Position 2 of Ring B is the structural requirement for the activity due to positively bind to GABA receptor by interaction with histidine residue of the receptor

(ii) Proton accepting moiety also serve as co-planner spatial orientation with Ring A

(iii) Replacement of oxygen (carbonyl function) moiety by sulphur, maintain the anxiolytic properties

(iv) Substitution of methylene group at position 3 serve as no effect on anxiolytic activity but affect antagonist activity.

(v) Change in 4,5 double bond or 4 position nitrogen atom, results in retaining the anxiolytic activity.

(vi) Esterification of 3-hydroxy moiety resides the anxiolytic activity.

Ring C

(i) 5-phenyl ring C is not essential for binding to benzodiazepine receptors (BZR) but assist as favourable hydrophobic or steric interaction to receptor binding.

(ii) Although 4′ and 2′ position substitutions are unfavourable for agonist activity but they impart the planarity relationship to ring A and electronic and steric parameters.

B. 1,2 Ring Fused 1,4–Benzodiazepines

(i) Extension of 1,2 –position by the fusion of triazole ring significantly enhanced (up to 10 fold) *in vivo* activity, e.g., Alprazolam.

(ii) Introduction of CH_3 group at Position (C_1) of Alprazolam show optimum activity. Replacement with hydrogen or larger alkyl group or aromatic ring moiety is responsible for sudden decrease activity.

(iii) Anxiolytic activity of triazolobenzodiazepine is enhanced by entertaining electron withdrawing groups (Cl, Br, NO_2 and CF_3) at position 8 whereas decrease the activity with presence of electron donating moiety (NH_2, SCH_3, etc).

12.6 BENZODIAZEPINES DERIVATIVES (SPECIFIC DRUGS)

1. Diazepam synthesis given in Chapter-5
2. Chlordiazepoxide synthesis given in Chapter-5
3. Triazolobenzodiazepines see in Chapter-5

1. Oxazepam (Tazepam®): 7-chloro-1,3-dihydro-3-hydroxy-5-phenyl-2H-benzodiazepin-2-one Oxazepam shows significant anxeiolytic activity. It shows much more polarity than Diazepam. It has short duration of action. Oxazepam is used in the treatment of fear, anxiety and stress disorders.

Synthesis:

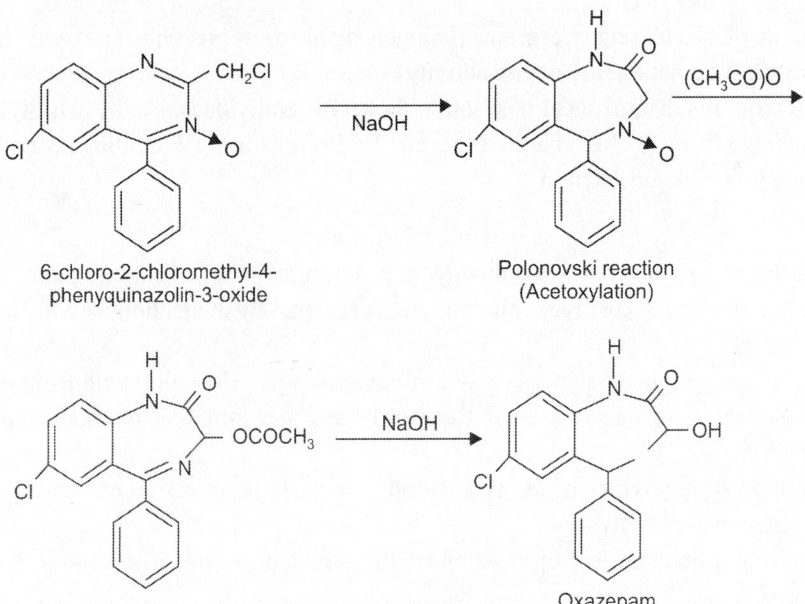

Fig. 12.12

2. Lorazepam (Ativan®): 7-chloro-4-(o-chlorophenyl)-1, 3-dihydro-3-hydroxy-2H-1, 4-benzodiazepin-2-one

Lorazepam is structurally similar to oxazepam. It can be visualized as 2-chloro substituted congener of oxazepam which increases overall activity. Its metabolic activity is rapid and uncomplicated due to presence of 3-hydroxyl group at position 3. Lorazepam is used as tranquilizers and anxieolytic drug. It has short duration of action.

Synthesis:

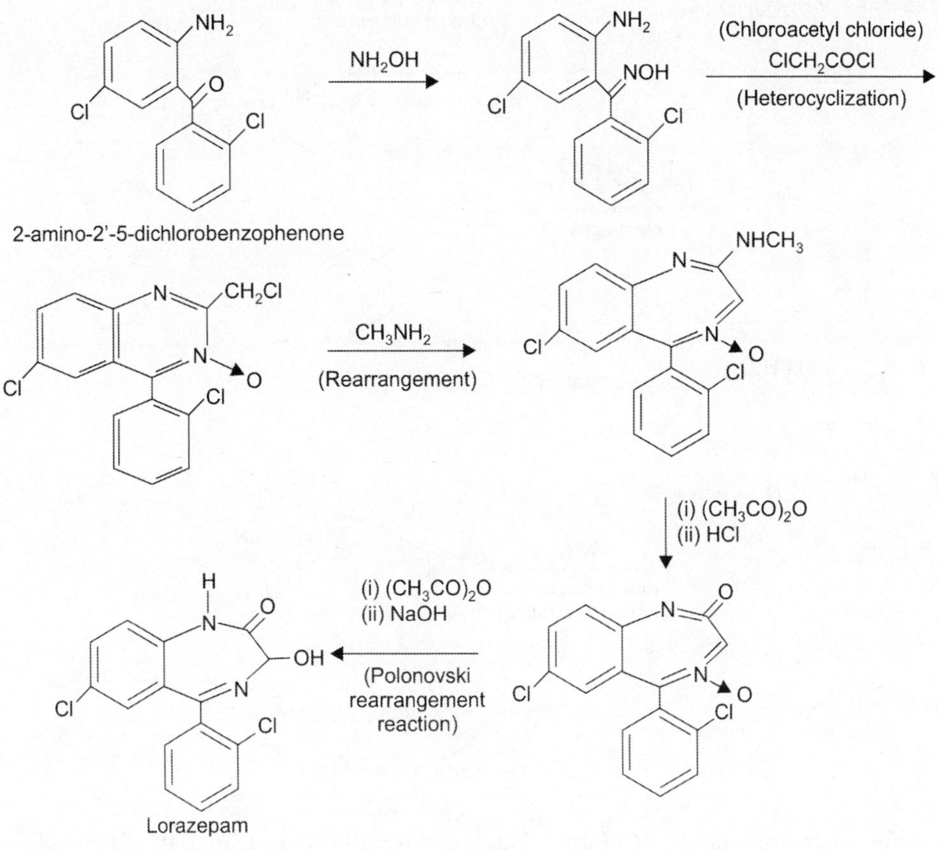

Fig. 12.13

3. Alprazolam (Xanax®): 8-chloro-1-methyl-6-phenyl-4H-S-triazolo [4,3-a]-[1,4]-benzodiazepine. Alprazolam is considered an analogue of triazolam or 1, 2-ring fused, 1-4-benzodiazepine, alprazolam has short duration of action. It is metabolized due to presence of methyl group by oxidative metabolism and followed by conjugation. Alprazolam shows strong anxiolytic action with lower incidence of sedation like side effect. This is because of daytime-tranquilizer. Alprazolam is used in various anxiety disorders.

Synthesis:

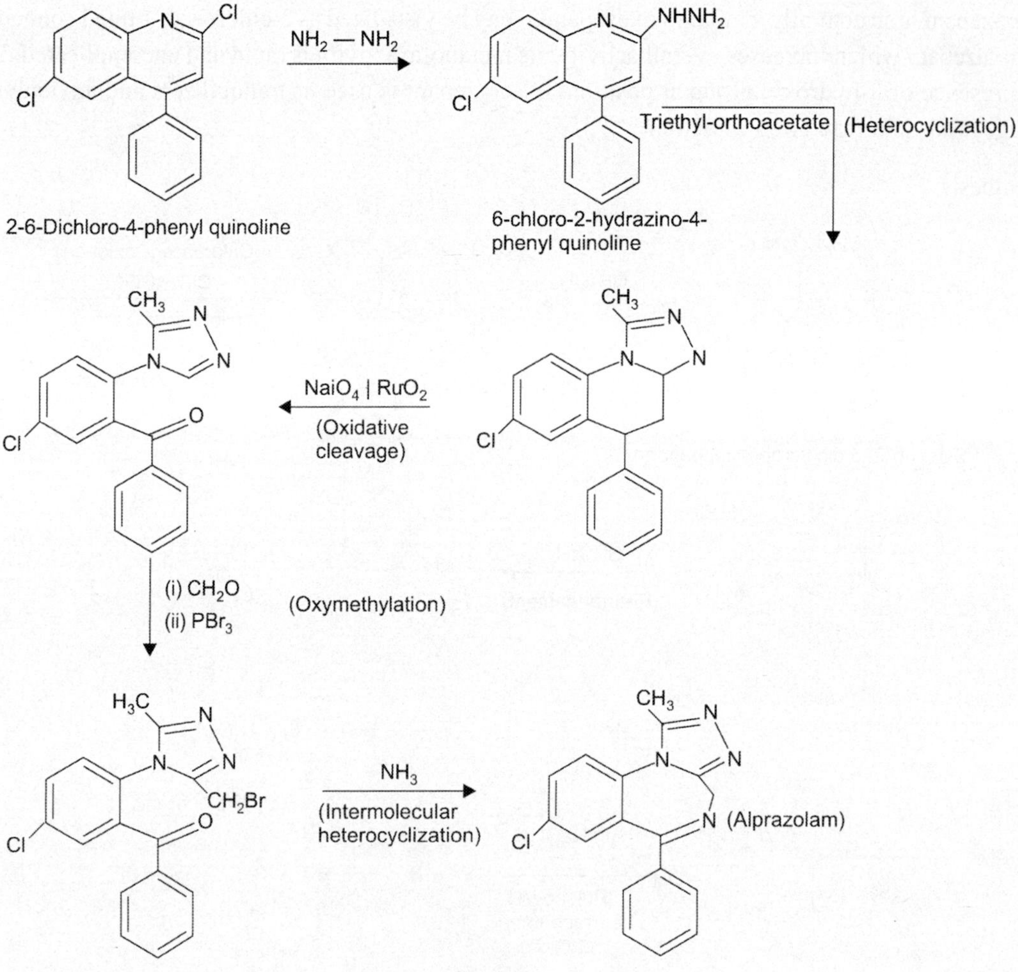

Fig. 12.14

4. **Medazepam (Nobrium®):** 7-Chloro-2, 3-dihydro-1-methyl-5-phenyl-1H-1,4-benzodiazepine Medazepam is a potent day-time tranquilizer. It exhibits anxiolytic effect as well as myorelaxant and anticonvulsant action. Medazepam is suggested for the treatment of neurosis, psychopathy, stress, elevated irritability associated with insomnia symptoms.

Synthesis:

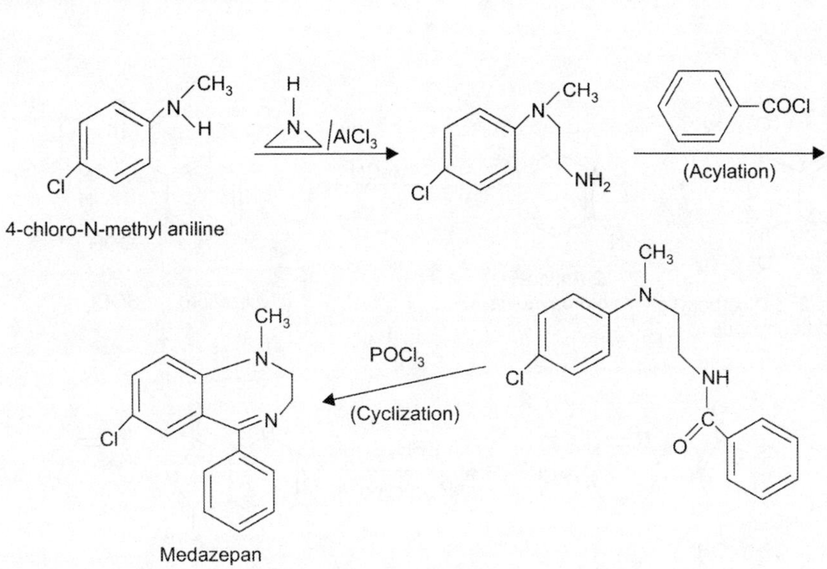

Fig. 12.15

12.7 NON-BENZODIAZEPINE DERIVATIVES (SPECIFIC DRUGS)

1. β-Carbolines: (9H-pyrido-[3,4-b]-indoles)

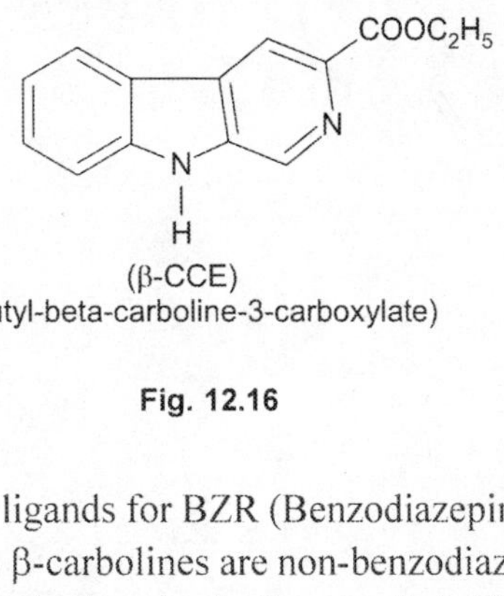

(β-CCE)
(N-Butyl-beta-carboline-3-carboxylate)

Fig. 12.16

β-carbolines are potent endogenous ligands for BZR (Benzodiazepines receptors). In 1980, β-CCE was first isolated from human urine. β-carbolines are non-benzodiazepines anxiolytic agents. They are also known as Norharmane. β-carbolines basically are alkaloids which are found in plant and animals.

2. Zopiclone (Imovane®): 4-methyl-1-piperazine carboxylic acid ester with 6-(5-chloro-2-pyridyl)-6,7-dihydro-7-hydroxy-5H-Pyrrolro (3,4-b)-pyrazin-5-one.

Zopiclone is potent tranquilizer. It is marketed by sleeping pills. Zopiclone is known colloquially as Z-drug include (Zoleplon (Sonata®) and Zolpidem (Ambien®), Zopiclone is less addictive than benzodiazepines. It is recommended as short-term antitranquilizer.

Synthesis:

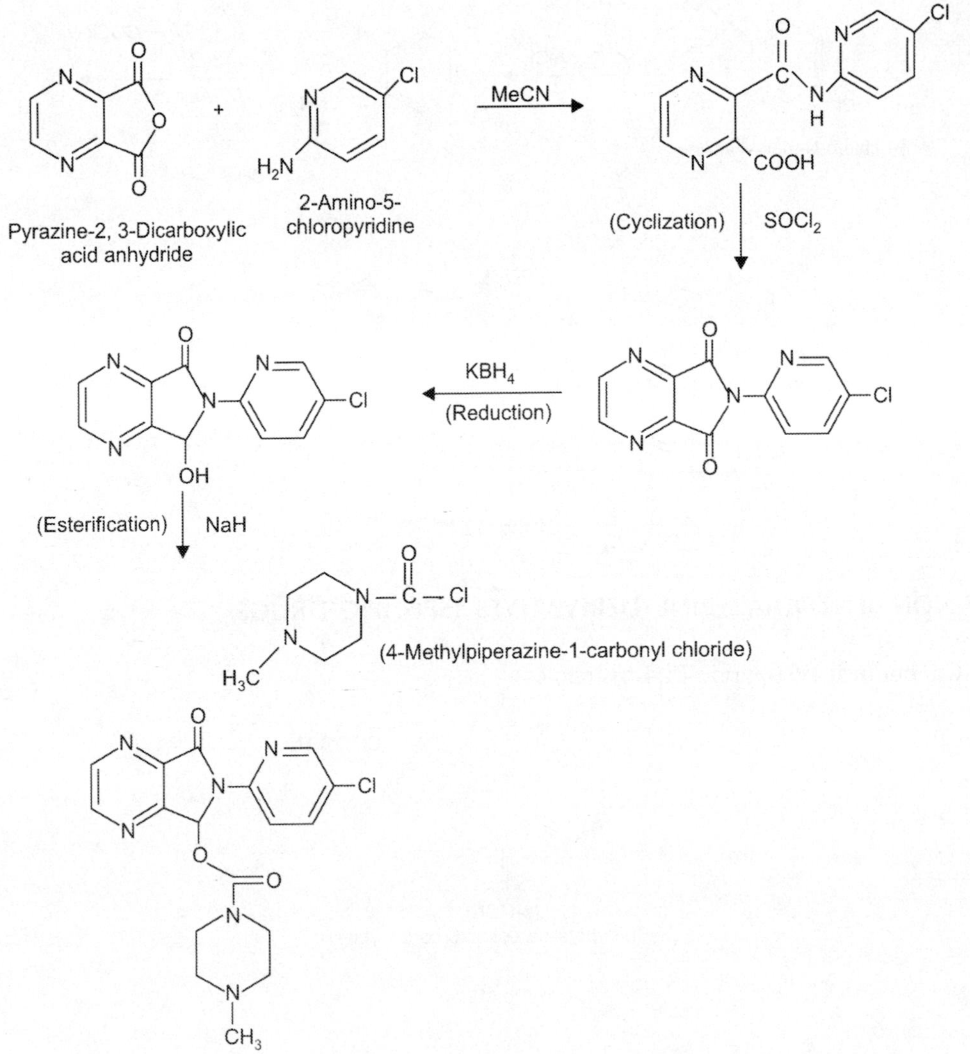

Fig. 12.17

12.8 NEWER AGENTS

1. Zolpidem (Ambiem®):- N,N-dimethyl-2-(6-methyl-2-p-tolylimidazo-[1,2-a]-Pyridin-3yl)-acetamide

Zolpidem is a short acting non-benzodiazepine drug. It has an inhibitory action on benzodiazepine receptors (BZR). It binds to $GABA_A$ receptors as the same location as benzodiazepines.

Zolpidem is used as anxiolytic agent as well as a muscle relaxant and anticonvulsant agent.

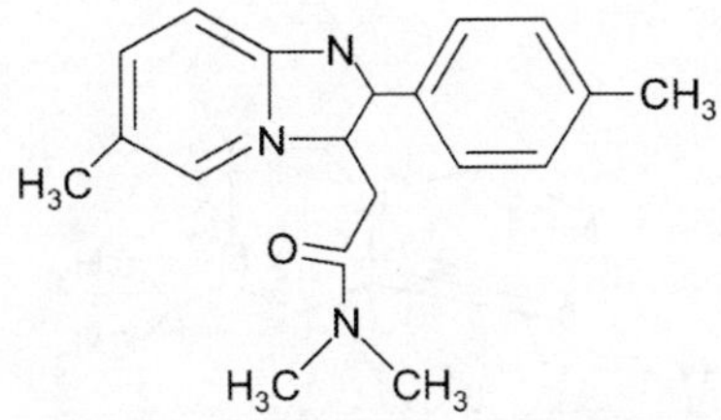

Fig. 12.18

2. Flumazenil (Anexate®):- Ethyl-8-fluoro-5-methyl-6-oxo-5, 6-dihydro-4H-benzo-[f]-imidazo-[1,5-a]-diazepine-3-carboxylate.

Flumazenil is potent anxiolytic Agent. It exhibits excessive drowsiness after benzodiazepines. It may be used as antidotes of benzodiazepines overdose. It is also used in hepatic encephalopathy.

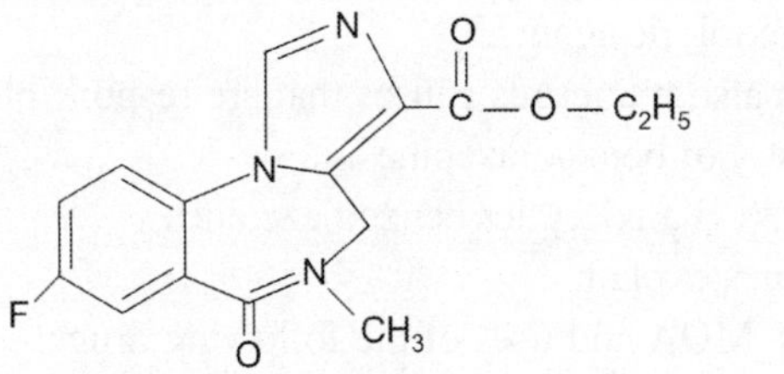

Fig. 12.19

3. Bretazenil (Bretazenilum®) (Ro-16-6028) Code: t-butyl(-8-bromo-11,12,13-13a-tetrahydro-9-oxo-9H-imidazo-pyrrolo-(1,4)-benzodiazepine-1-carboxylate. Bretazenil is the drug belonging to family benzodiazepine. It exhibits high quality binding to BZR and act as partial agonist. It is also used as antidotes of benzodiazepines overdose.

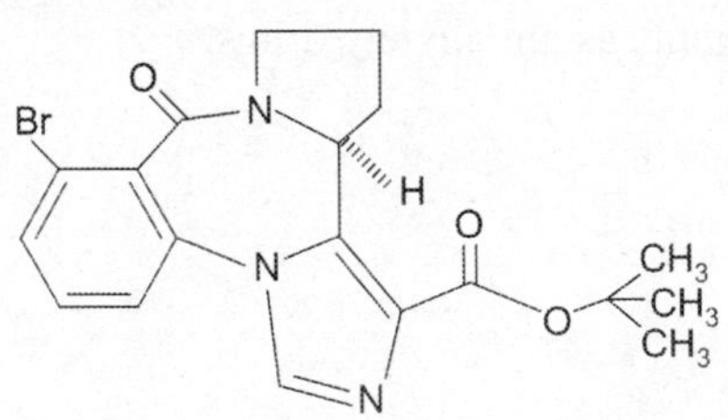

Fig. 12.20

4. Panadiplone (U-78875):- 3-(5-cyclopropyl (-1,2-4-oxadiazol-3yl)-5-propan-2-yl-imidazo[5,1-c]-quinoxalin-4-one

Panadiplone is effective nonbenzodiazepine anxiolytic drug. It acts as high affinity to $GABA_A$ receptor partial agonist. It also shows little sedative effect.

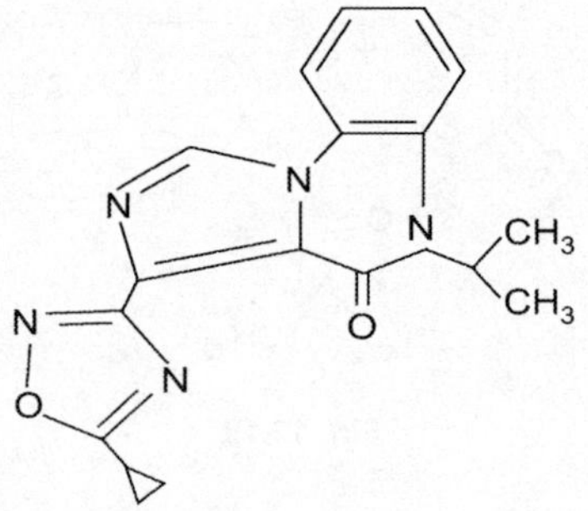

Fig. 12.21

EXERCISE

1. Classify in detail anxiolytic agents.
2. Define anxiety and also numerous causes that are responsible for anxiety state.
3. Explain the chemistry of benzodiazepines.
4. Explain in detail class-B anxiolytics benzodiazepines.
5. What are β-carbolines explain it.
6. Write the synthesis, MOA and uses of the following drugs:
   1. Oxazepam
   2. Alprazolam
   3. Lorazepam
   4. Zopiclone
7. Explain the metabolism of benzodiazepine.
8. Write a note on SAR of benzodiazepines with examples.
9. Give a note on newer agents as antianxiety drugs.

13

Skeletal Muscle Relaxants

13.1 INTRODUCTION

Skeletal muscle relaxants are also known as "spasmolytic" agents. Spasticity of muscles is characterized by the spasms of skeletal muscles in order to increase in tonic stretch reflexes and muscles weakness. It is associated with cerebral palsi, multiple sclerosis, spinal cord injury, etc., results of hyperexcitability of motor neurons. Clinically, spasm is also associated with severe pain in patient. Spasmolytic drugs or skeletal muscles relaxants relax the skeletal muscles and act as analgesia both. The first medicinal agent was recognized to show spasmolytic activity, Antodyne or 3-Phenoxy-1-2-Propaindiol in 1910. Antodyne acts as an analgesic and antipyretic but duration of action of spasmolytic activity was too short to be clinically useful. Later in 1943, glycerylethers related antodynes were made a new series of skeletal muscle in relaxants, mephenesin was discovered in 1946 that had ability to depress postsynaptic spinal cord reflexes.

Many CNS depressant agents cause muscle relaxation in the treatment of sprains, ruptures, arthritis and various muscular disturbances.

13.2 MECHANISM OF ACTION

Skeletal muscle relaxants are functioning to the GABAergic system to depress the skeletal muscles activity. They decrease the tone of spasticiy via hyperpolarization of G Protein activated K^+ channel. Some compounds, e.g., orphenadrine acts on NMDA receptors and exhibit antagonist action (antimuscarinic activity). Cyclobenzaprine drug show muscles relaxant action through antagonism action on Serotonin 5H T_2 receptors in descending neurons of spinal cord.

13.3 CLASSIFICATION

A. Glycerol Monoethers Derivatives:

1. Mephenesin (Tolserol®)

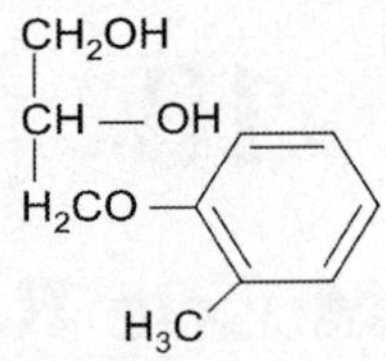

2. Carbamol

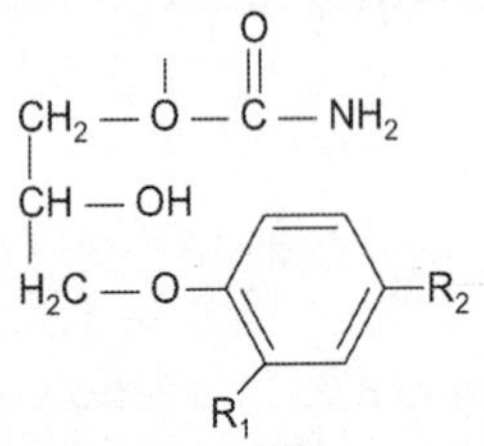

R_1 = H, R_2 = Cl Chlorphenesin (Maolate®)
R_1 = OCH_3, R_2H Methocarbamol (Robaxin®)

Fig. 13.1

B. Alkanediols Derivatives:

1. Meprobomate (Equanil®)

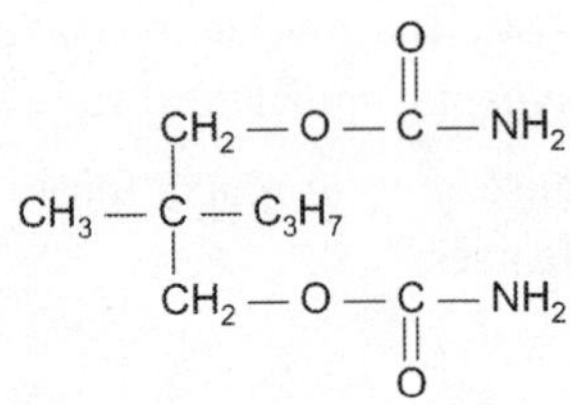

2. Carisoprodol (Rela®)

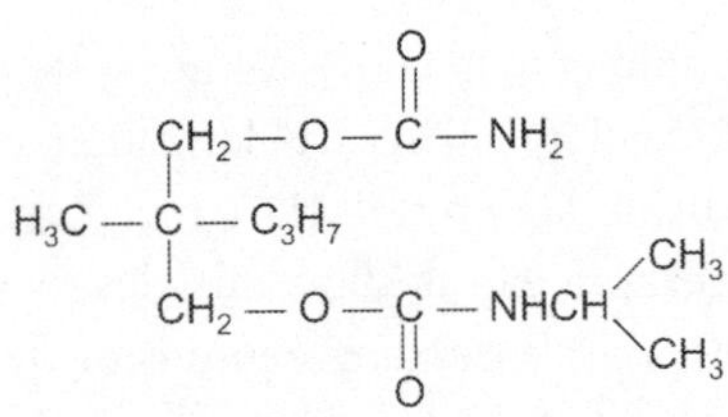

3. Metaxolone (Skelaxin®)

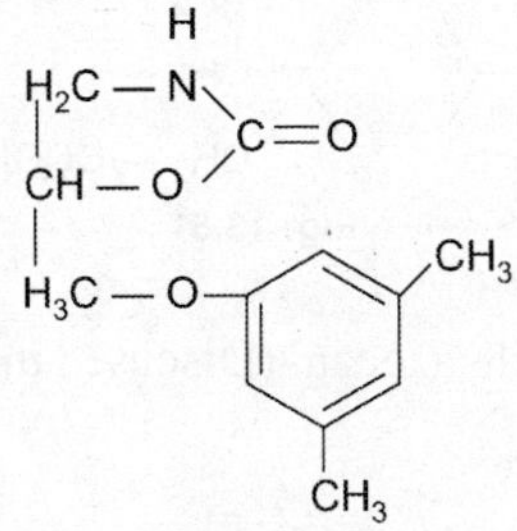

Fig. 13.2

C. Benzazole Derivative:

1. Chlorzoxazone (Paraflex®)

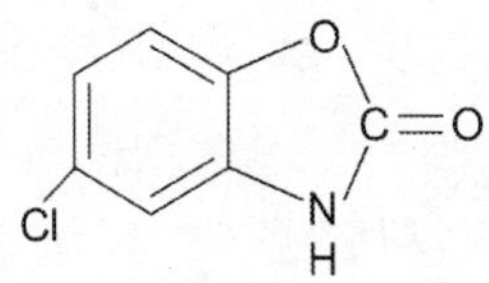

Fig. 13.3

D. Imidazoline Derivative:

1. Dantrolene (Dantrium®)

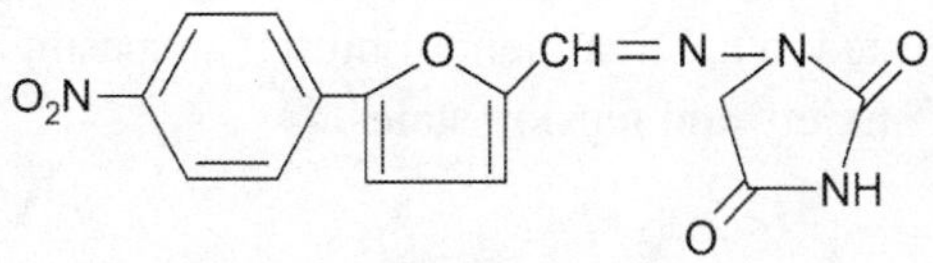

Fig. 13.4

E. Miscellaneous:

1. Orphanadrine citrate (Norflex®)

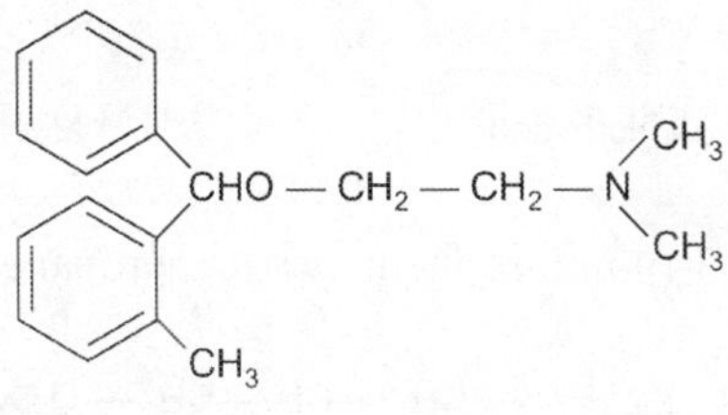

2. Cyclobenzaprine (Flexaril®)

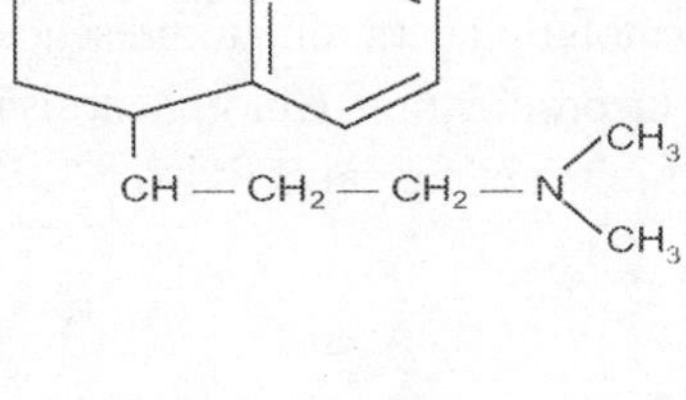

3. Baclofen (Lioresal ®)

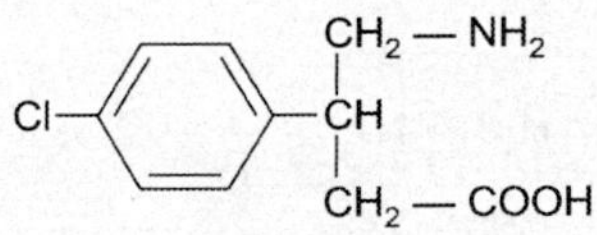

Fig. 13.5

4. Benzodiazepines : Diazepam, Fletazepam (Discused in Chapter 12)

13.4 SPECIFIC DERIVATIVES

13.4.1 Glycerol Monoethers Derivatives

1. Mephenesin (Tolserol®): is 3-(o-methyl phenoxy)-1,2-Propanediol

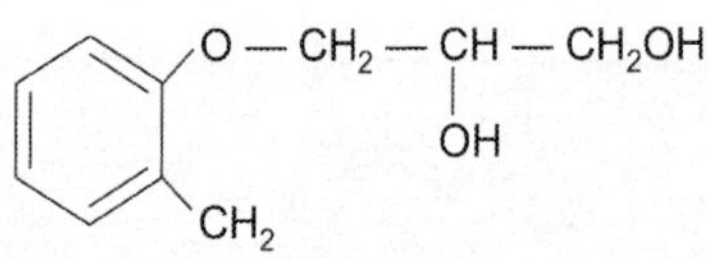

Fig. 13.6

Mephenesin is a potent skeletal muscle relaxant. It cause the skeletal muscle relaxation of hyper tonic muscles to minimize the sensory stimuli. It is also significantly depressed the superficial reflexes. Mephenesis is suggested to the treatment of muscular spasm, hyperkinetic condition. It is also used in the treatment of anxiety and tension states.

Synthesis:

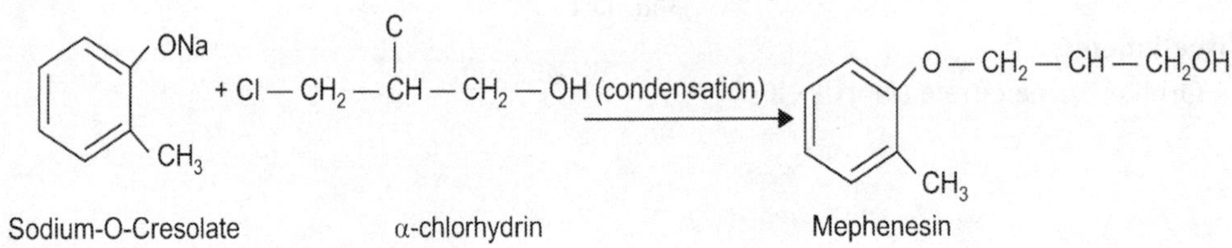

2. Chlorphenesin (Maolate®):- 3-(p-chlorophenoxy)-1, 2-propanediol-1-carbamate

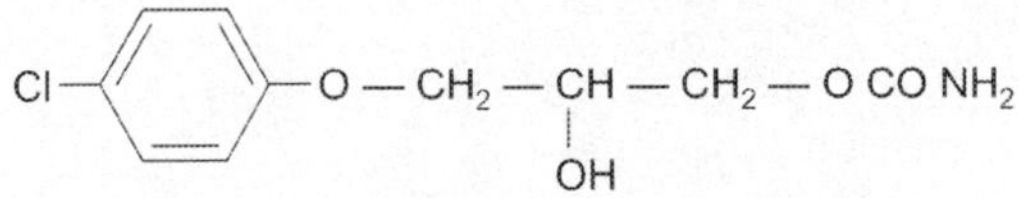

Fig. 13.7

Chlorphenesin is a potent spasmolytic agent, due to presence of secondary hydroxyl group. It undergoes active metabolism via glucoronidation. Chlorphenesis is used for the symptomatic relief of muscular spasm as well as it exhibits antifungal activity.

Synthesis:

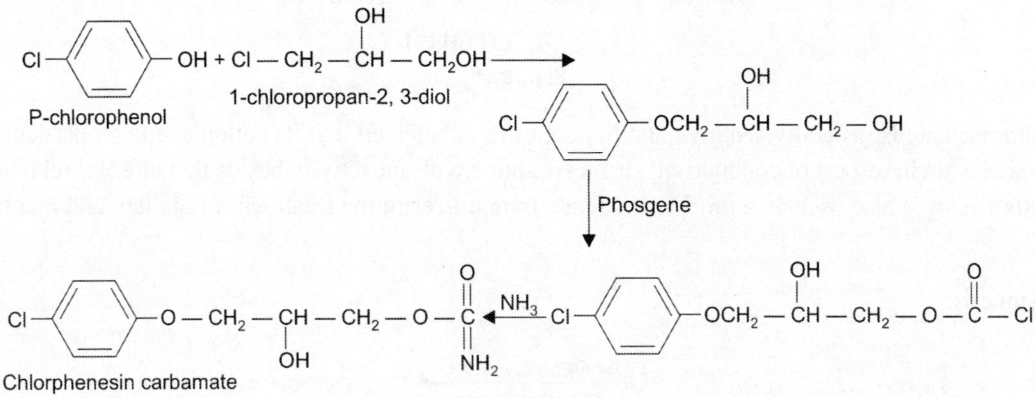

Fig. 13.8

3. Methocarbamol (Robaxin®): 3-(O-methoxy phenoxy)-1, 2-propandiol-1-carbamate

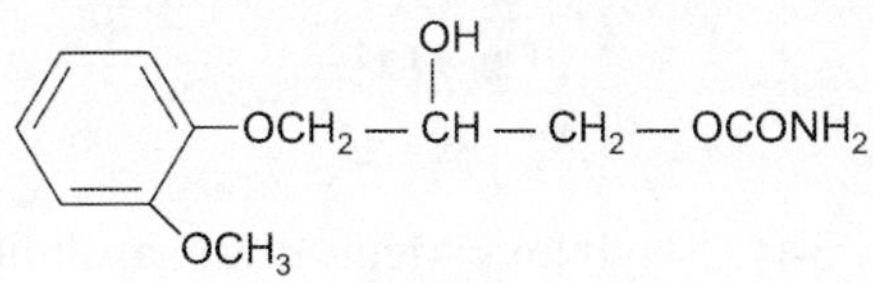

Fig. 13.9

Methocarbamol is potent centrally acting muscles relaxant. It significantly relieves from muscles spasm. It is also used in the treatment of Parkinson's disease.

Synthesis:

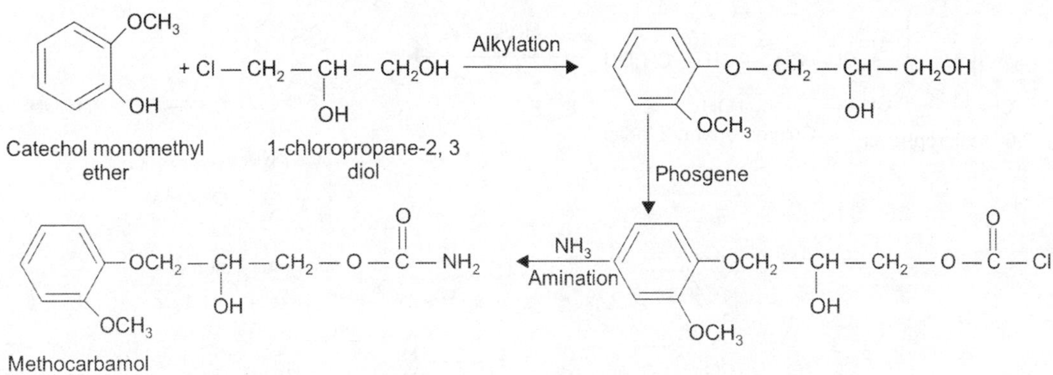

Fig. 13.10

13.4.2 Alkanediols Derivatives

1. Maprobamate (Equanil®): 2-methyl-2-propyl-1,3-propandiol dicarbamate.

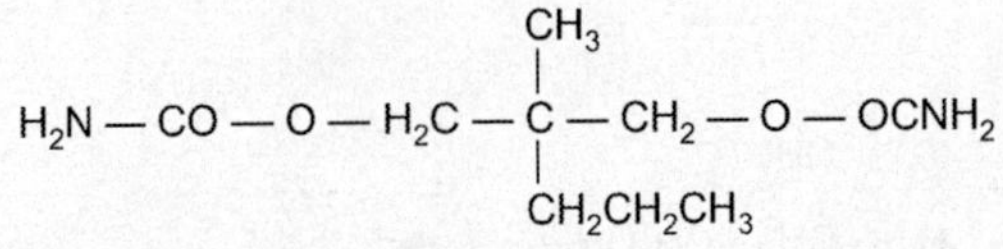

Fig. 13.11

Maprobamate is generally sedative and hypnotics. It is believed that its action exerts on particular area of brain in respect of conductivity. It shows anticonvulsant activity beside that muscles relaxant activities. It is also used as a mild to moderate tranquilizer in the treatment of anxiety and mental disorder.

Synthesis:

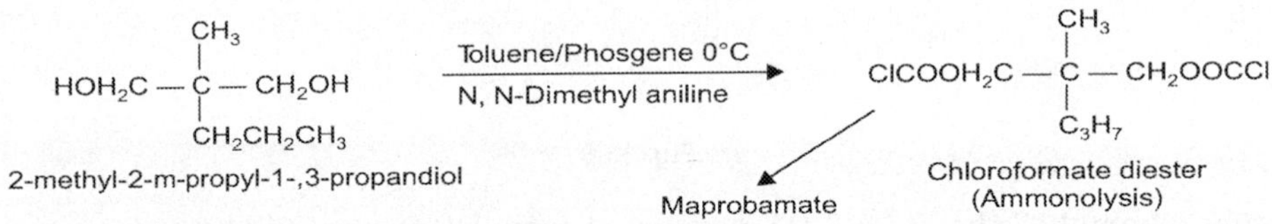

Fig. 13.12

5. Metaxolone (Skelaxin®):- 5-[(3,5-xylyloxy)Methy)]-2-oxazolidinone
Metaxolone significantly reduces the muscle tone. It is used for the relief of acute muscle spasms caused by various injuries or strains.

Synthesis:

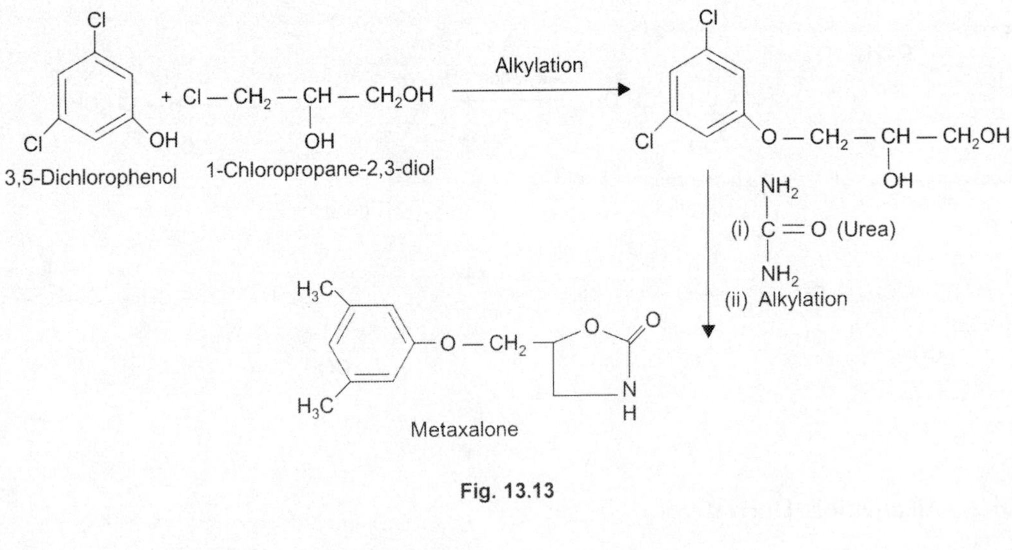

Fig. 13.13

13.4.3 Benzazole Derivative

1. Chlorzoxazone (Paraflex®):- 5-chloro-2-benzoxalinone
Chlorzoxazone is a potent centrally acting skeletal muscle relaxant. It inhibits the postsynaptic reflexes both within spinal cord and subcortical area of brain. It is used in the treatment of severe painful muscle spasm. It is also suggested in the treatment of spondilytis, sprains and muscles strains.

Synthesis:

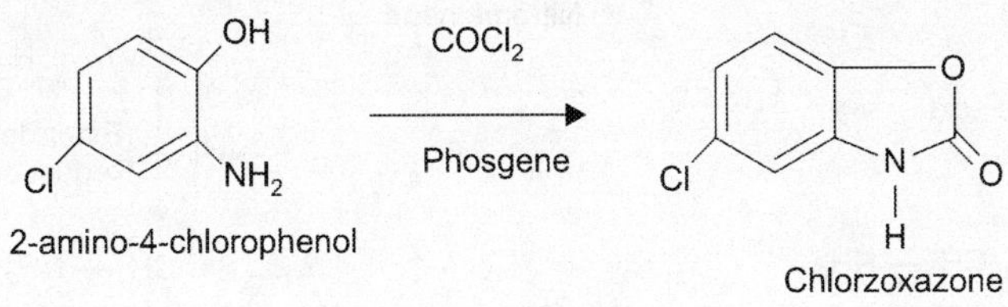

Fig. 13.14

13.4.4 Imidazolin Derivative

2. Dantrolene (Dantrium®): 1-[[5-(p-Nitrophenyl)-furfulidene]-amino]-hydantoin sodium salt hydrate.

Dantrolene is hydantoin derivative which acts centrally and peripherally to reduce spasticity. Dantrolene is used in the treatment of spinal cord injury. Stroke cerebral palsy and multiple sclerosis.

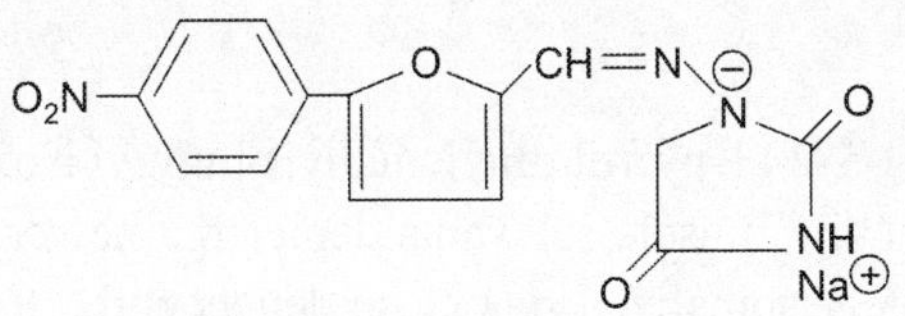

Fig. 13.15

13.4.5 Miscellaneous Agents

Beclofen (±) (Lioresa®): β-(aminomethyl)-p-chlorohydro-cinnamic acid.
Beclofen is the derivative of gamma aminobutyric acid because it is structurally similar to GABA.

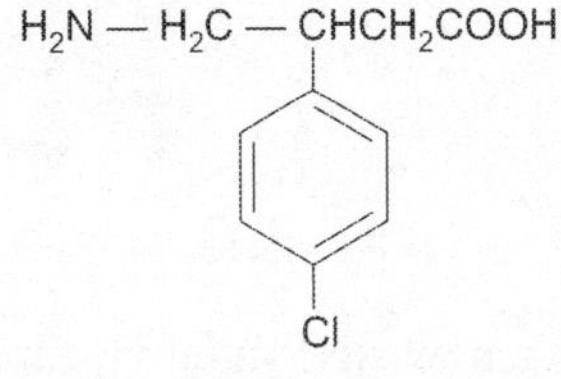

Fig. 13.16

Thus, it shows agonist activity at the size of $GABA_B$ receptor via mechanism G-Protein activated K^Tchannel (hyperpolarization). Beclofen is one of the most common used antispastic drugs. Beclofen is employed by oral administration and has minimum hepatic metabolism. It is also given to the patient via parentral and intrathecal administration.

Synthesis:

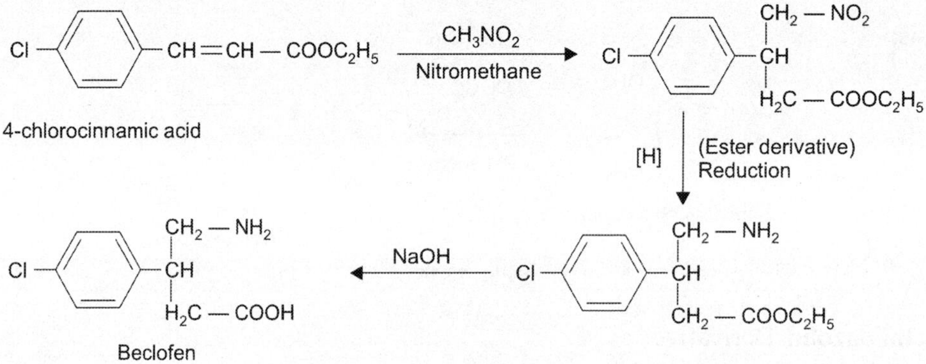

Fig. 13.17

13.5 NEWER AGENTS

1. MS-322 : (R)-(+)-3-phenyl-5-2-(1-pyrrolidinyl methyl)-butyryl isoxazole hydrochloride.
Recent studies on centrally acting muscle relaxants depicted a newer agent MS-322 that reduced spinal reflex potential in acutely spinalized rats close dependently. It shows inhibitory effects on Polysynaptic and dorsal root reflexes were more marked than monosynaptic reflexes therefore it shows better muscle relaxant activity.

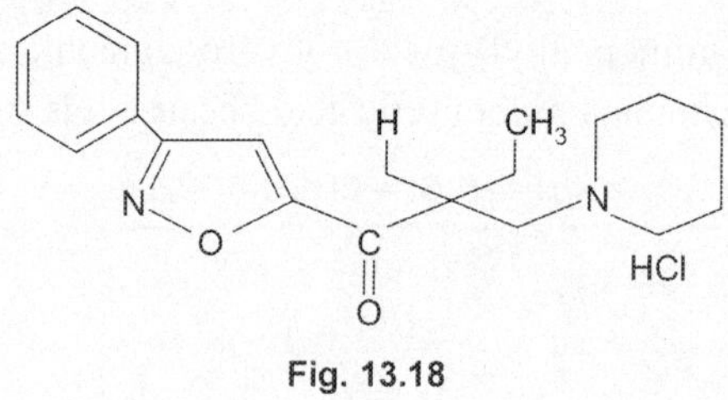

Fig. 13.18

2. Silperisone: N-[(dimethyl-4-fluorobenzyl silyl)-methyl]-piperidine HCl.

Silperisone is organo silicon compound with centrally acting muscles relexant activity. Studies showed that silpersone cause CNS depressant or motorside effect than other drug tolperisone. It is an effective to suppress monosynaptic and postsynaptic spinal reflexes.

[1]General Pharmacology (1998) The Vascular System, Vol. 30 (3), 393-398.

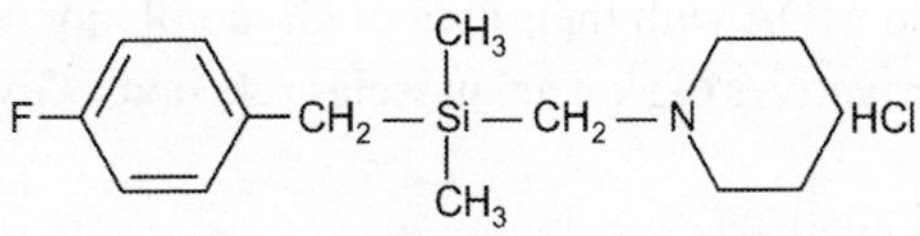

Fig. 13.19

3. Tolperisone: 2-methyl-1-(4-methyl phenyl)-3(1-piperidyl)-propane-1-one

Tolperisone is potent centrally acting muscles relaxant it is much more suitable for cerebral atherosclosis for better movement for treating extra piramidal movement disorders.

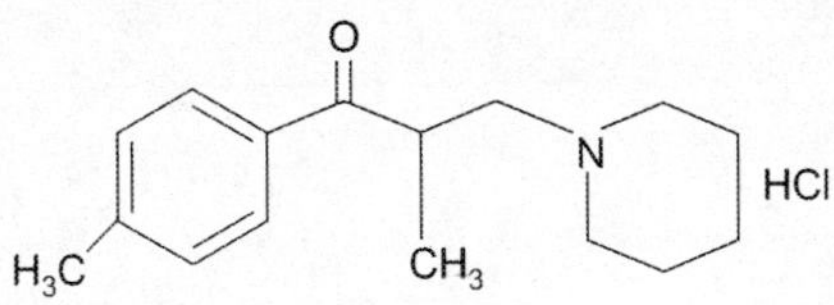

Fig. 13.20

4. Lamperison .HCl (NK-433): (–)-(R)-2-methyl-3-(1-pyrrolidinyl)-4'-tri-fluoromethyl propiophenone monohydrochloride.

It is a novel centrally acting muscle relaxant which inhibits monosynaptic tonic vibration reflex. but it is a weak inhibitor on post-synaptic reflexes.

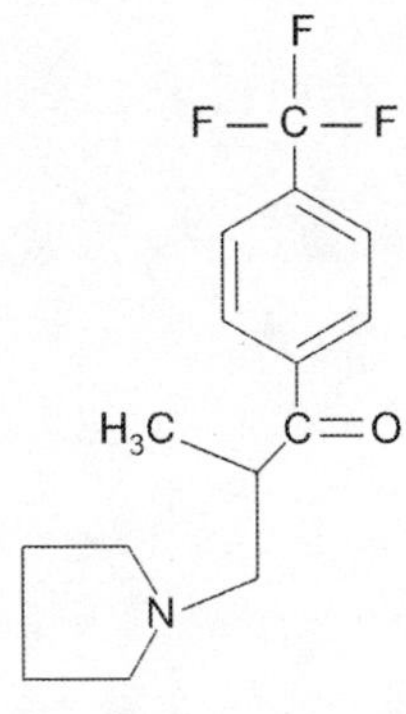

Fig. 13.21

EXERCISE

1. Define spasmolytic agents or centrally acting muscle relaxants.
2. Classify centrally acting muscle relaxants and explain their mechanism of action.

---

[2]CNS—Reviews.(2006), Vol. 12 (3-4), 218–235.

[3]Ozawa et al. (1996) Eur. J. Pharmacol. 298 (1) 57, 62

3. Write the synthesis and MOA with their uses of Glycerol monoether derivatives.
4. Classify alkanediol derivatives of skeletal muscles relaxants. Give their mechanism of action and uses.
5. Write following drugs synthesis.
   1. Beclojen
   2. Mephensin
   3. Chlorzoxazone
   4. Metaxolone
6. What are the recent developments in skeletal muscles relaxants? Explain it.

14

Neuromuscular Blocking Agents

14.1 INTRODUCTION

Neuromuscular blocking agents are also known as nicotinic receptor antagonists or peripherally acting muscles relaxants. These agents are associated with that particular area of neuromuscular contacts with weakened transmission from motor neuron ending to the membrane of skeletal muscles cells. Neuromuscular blocking agents are subdivided into the depolarizing agents and antidepolarizing agents. The neuromuscular blocking action is exhibited earlier in 1510 by the extract of curare d-tubocurarine chloride it was the first neuromuscular blocking agent that had ability of the action against nicotinic receptor antagonists. D-tubocurarine was isolated in 1935 as a pure alkaloid from the tubecurare of unknown botanical region. It is characterized by quaternary ammonium salt compound. Progress of research was of neuromuscular blocking agents are now based on quaternary ammonium compounds.

14.2 MECHANISM OF ACTION

Neuromuscular blocking agents act by transmission blocked at the end of neuromuscular junction by initiating an influx of Ca^{++} that is caused by nerve impulse at the nerve terminal followed by release of Ach from exocytic mechanism in synaptic vescicles. Release of Ach diffuses across the synaptic cleft and binds to the nicotinic receptors located in motor end plates or may be hydrolysed by Ach E enzyme. Binding of Ach to receptor causes opening of Na-K channel and thus cause depolarization of end plates resulting in muscles contraction. Contraction of muscles can be countered by two types of mechanism by non-depolarizing agents (tubocurarine and others) that prevent depolarization by binding to nicotinic receptors. Others are depolarizing agents that mimic the action of Ach and prevent muscle contraction. These types of agents are structurally similar to Ach and cause relaxation of muscles.

14.3 CLASSIFICATION

(I) Non-depolarizing muscle relaxants

1. Tubocurarine (Tubarine®):- (Tetrahydroisoquinoline derivatives) – (+)-Tubocuraranium, 7,12'-dihydroxy-6,6'-methoxy-2-2',-2'-trimethyl-chloride, hydrochloride penta-hydrate.

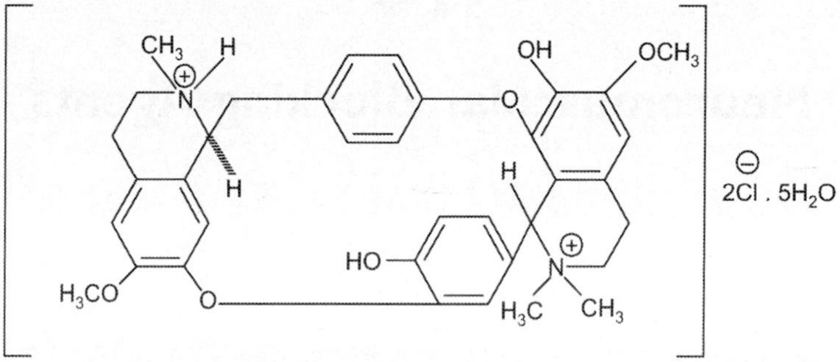

Fig. 14.1

2. Metocurine (Metubine®): (Tetrahydroisoquinoline Derivatives) Tubocuraranium, 6,6',7',12-tetramethoxy-2,2'-2'-2'-tetramethyl-diiodide

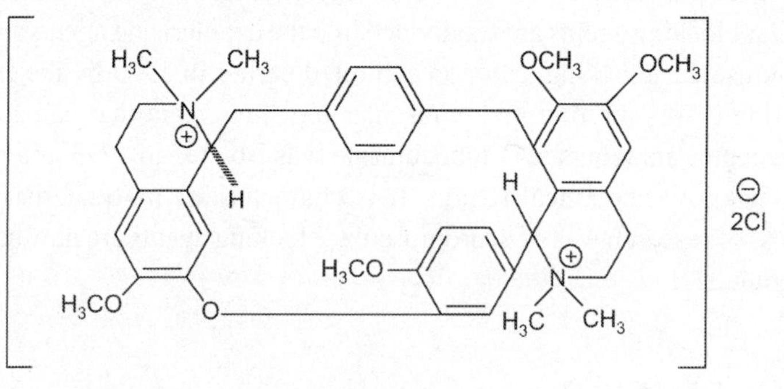

Fig. 14.2

3. Gallamine Triethiodide (Flaxedil®):- [Phenyl-tris-(oxyethylene)]-tris-[triethylammonium]-triiodide (Trisquaternary Ether)

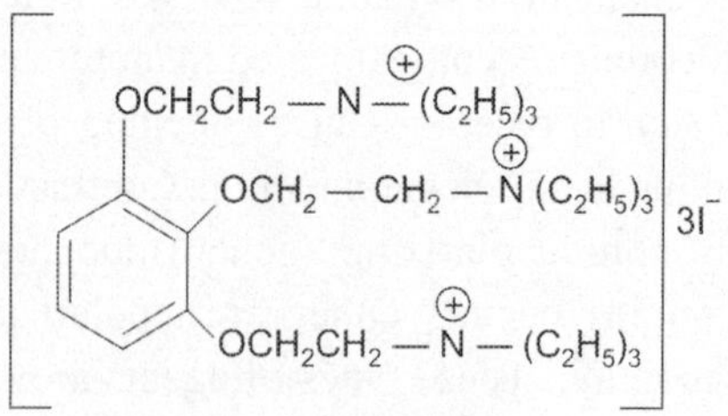

Fig. 14.3

4. Pancuronium Bromide (Pavulon®) :- 1,1'-(3α-17β-dihydroxy-5α-androstane-2β-16β-ylene)-bis-[1-methyl-piperidinium]-dibromidediacetate

(Aminosteroids Derivative)

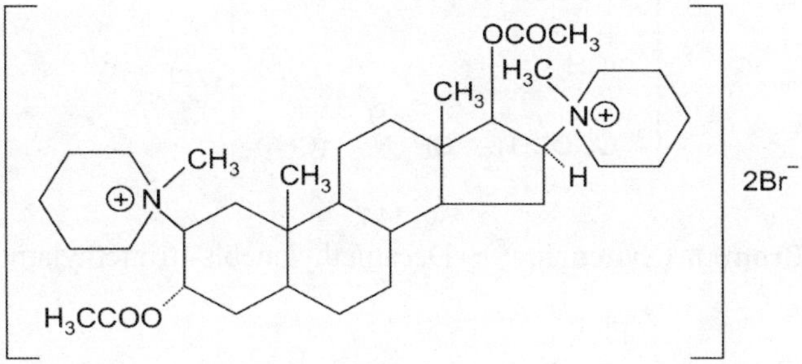

Fig. 14.4

5. Hexafluronium Bromide (Mylaxen®):- Hexamethylene-bis [fluoren-9-yl-dimethyl ammonium]-dibromide

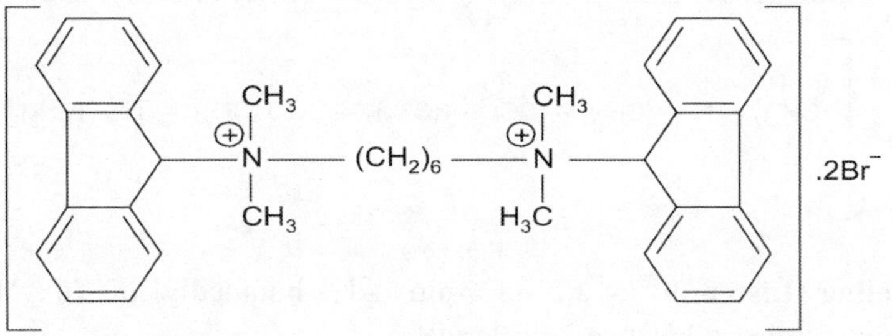

Fig. 14.5

6. Dacuronium Bromide (Organon®):- 3α-17β-dihydroxy-5α-androstane-2β-16β-ylene)-bis-[1-methyl-piperidinium]-dibromidediacetate.

(Aminosteroid Derivative)

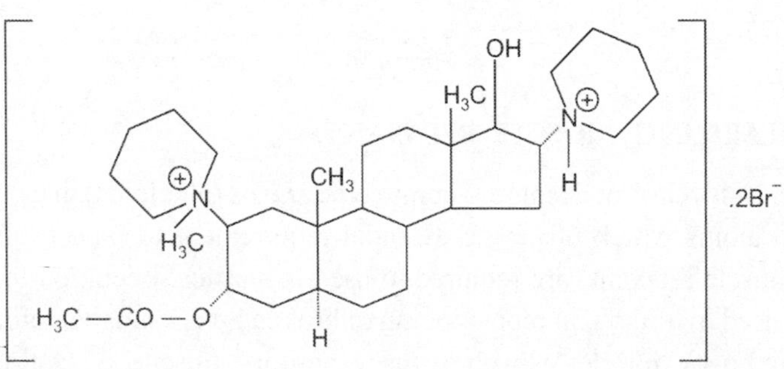

Fig. 14.6

2. Depolarizing Muscles Relaxants

1. **Suxamethonium Chloride (Acectine®):-** Choline-chloride Succinate (2:1).

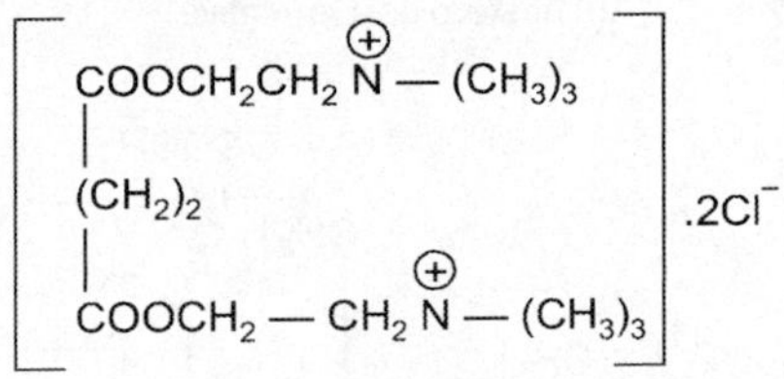

Fig. 14.7

2. **Decamethonium Bromide (Syncurine®):-** Decamethylenebis-[trimethylammonium]-dibromide.

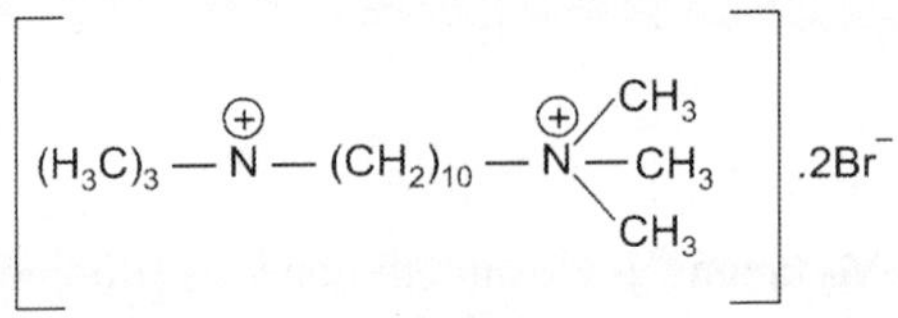

Fig. 14.8

3. **Suxethonium Chloride(Brevidil®):-** Ethyl-(2-hydroxy ethyl)-dimethyl ammonium chloride.

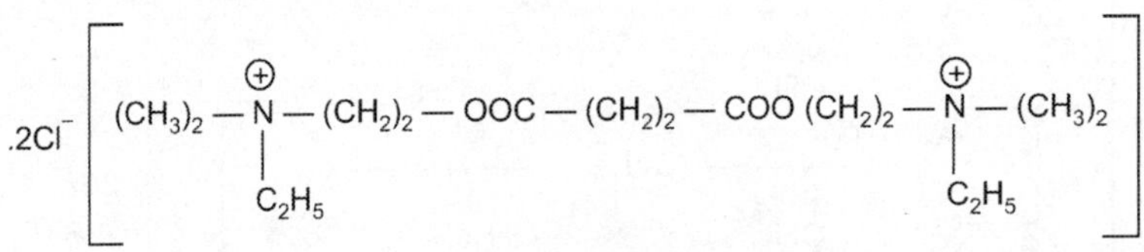

Fig. 14.9

4. **Succinyl Choline (Listenon®):-** 2,2'-[1,4-dioxo-1,4-butanediyl)bis-(oxy)]-bis-[N, N, N-trimethylethanaminum-(diacetylcholine)] dichloride.

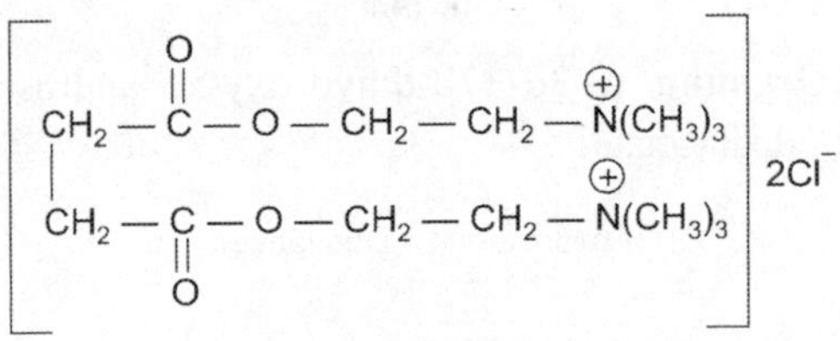

Fig. 14.10

14.4 NON-DEPOLARIZING MUSCLE RELAXANTS

Non-depolarizing muscle relaxant agents are curare type agents (alkaloids) that contain two or more quaternary nitrogen atoms which make the essential requirement to bind to nicotinic receptors. Non-depolarizing muscle relaxants are required to used in surgical operation for skeletal muscles relexants. They are used in trauma, in repositioning of broken bones and their clinical problems, for relaxation of face and neck muscles as well as for respiratory muscles relaxation.

1. **Tubocurarine (Tubarine®):** Tubocurarine is synthesized from plant source extract of "chondodendron".

Tubocurarine is used as myorelaxants, in surgical operation for prolonged muscle relaxation, in endotracheal intubation, in orthopaedic surgery for repositioning of fractured bones.

2. Metocurine (Metubine®): Metocurine iodide is synthesized by the treatment of methyl iodide with d-tubocurarine in the methylation process of two hydroxyl groups. Metocurine has threefold potent action than d-tubocurarine and with longer duration of action.

3. Gallamine (Flaxedil®): Gallamine is used for the similar indication of d-tubocurarine further. It is used in anaesthesiology for the muscle relaxation to facilitate surgical operation.

Synthesis:

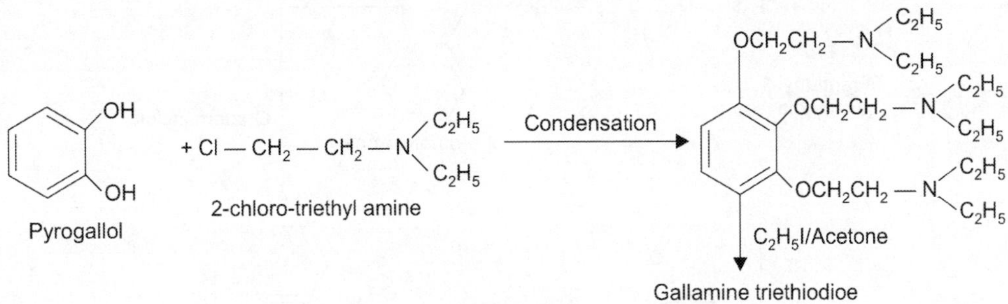

Fig. 14.11

4. Pancuronium (Pavlon®): Pancuronium is steroidal (cyclopetano perhydro phenenthrene ring nucleus) muscle relaxant but it does not have hormonal pharmacological action. Pancuronium is a potent muscle relaxant and is used as a relaxant in anaesthesiology, in cardiovascular disease, in surgical operation for prolonged muscles relaxants, in ophthalmology, orthopaedic and in asthmatic condition.

Synthesis:

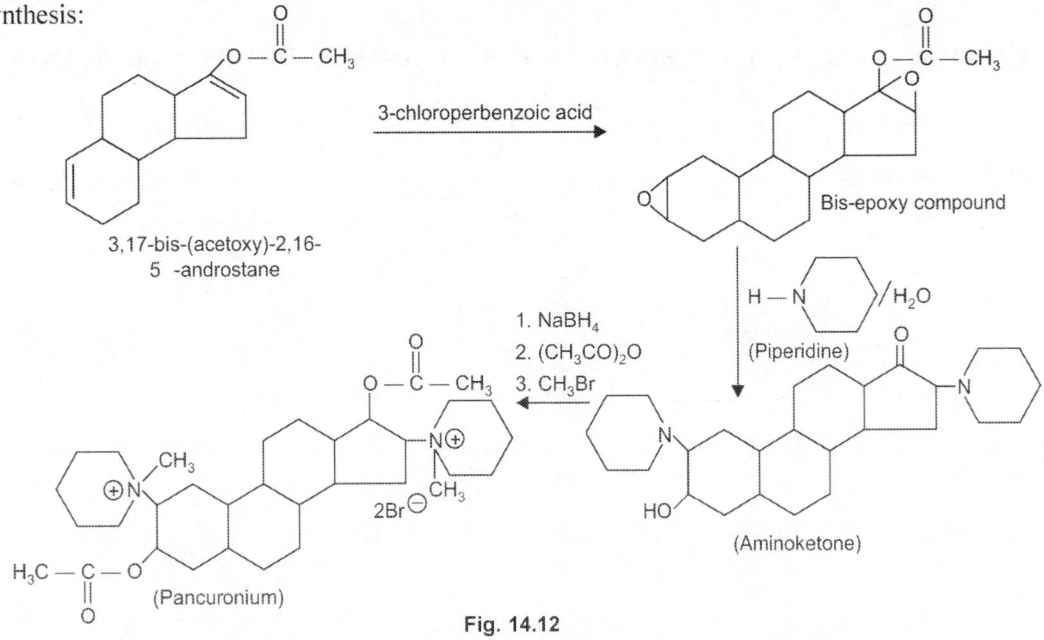

Fig. 14.12

5. Hexafluoronium (Mylaxen®): Hexafluoronium is a weak NMB (neuromuscular blocking agent). It has significant anaesthetic properties and used muscles relaxation. It antagonizes the action of Decamethonium. It exhibits synergistic effect with suxamethonium chloride and prolong the action.

Synthesis:

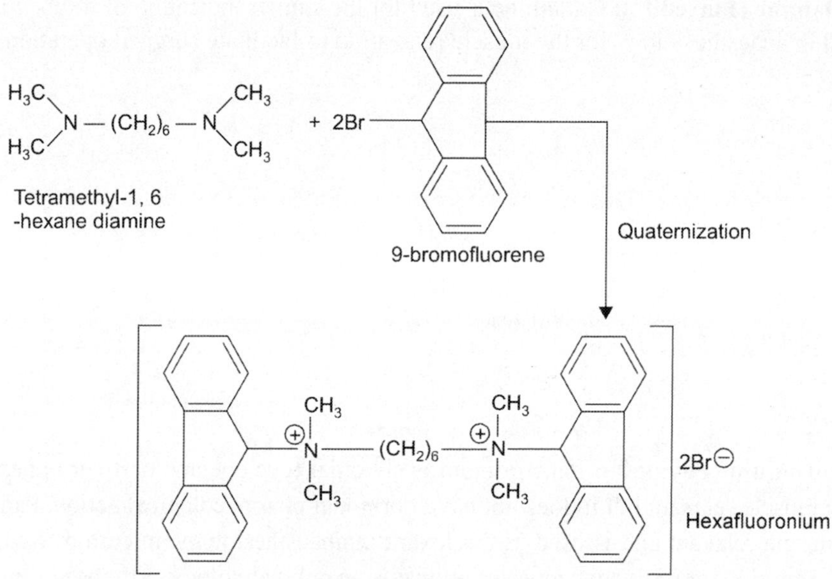

Fig. 14.13

14.5 STRUCTURAL ACTIVITY RELATIONSHIP OF AMINOSTEROIDS NMB AGENTS

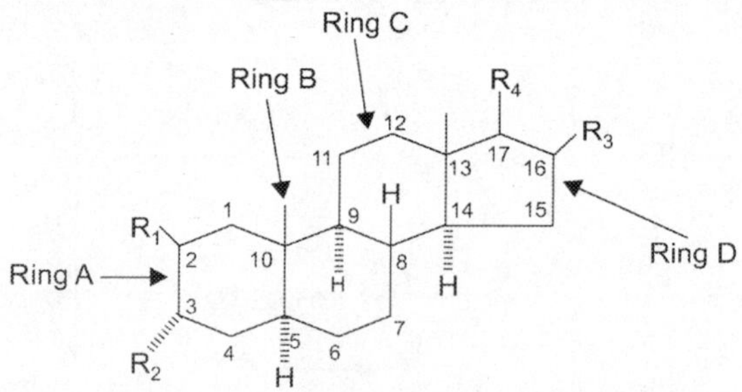

Fig. 14.14

	Derivatives	Substitution R_1	R_2	R_3	R_4
1.	Pancuronium (Pavvlon®)	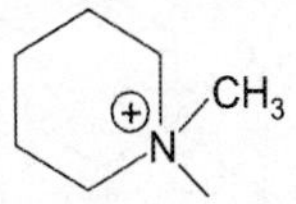	$CH_3COO—$	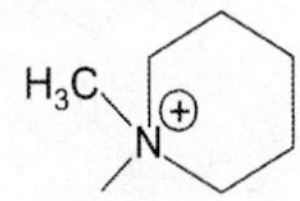	$—OOCCH_3$
2.	Vecuoronium (Norcuron®)	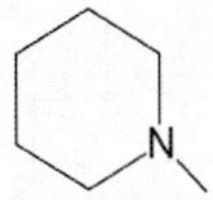	$CH_3COO—$	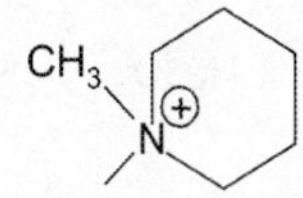	$—OOCCH_3$
3.	Rocuronium (Zemuron®)	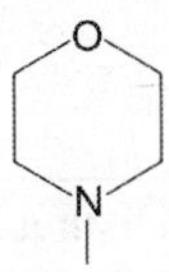	OH—	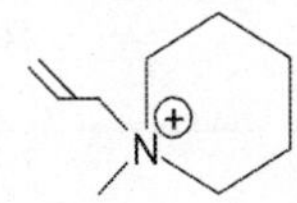	
4.	Repacuronium (Newer)	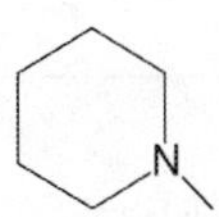	CH_3COO	═ (double band)	$—OOCCH_2CH_3$
5.	Pipecuronium (Arduan®)	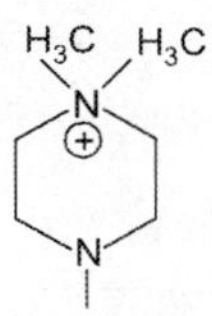	$CH_3COO—$	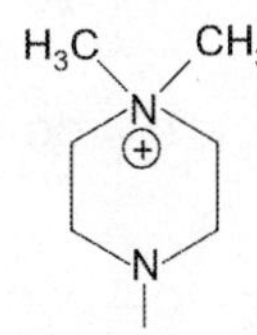	$—OOCCH_3$

1. Presence of steroid muscles provides a rigidity of bulky structural base.
2. Presence of two Ach moiety of Pancuronium at Ring A make the compound trans in geometry while ring D is Cis in nature. Hence, Pancuronium and Vancuronium are equipotent, their NMB action may be attributed to their common Cis-D-ring Ach moiety.
3. Beta surface of the molecules is crowded with substituents (R_1 and two methyl groups, R_3 and R_4) are essential for NMB activity.
4. Replacement by larger groups with R_3 or R_4 in place of methyl groups decrease the potency.
5. Introduction of bis quaternary with dimethypiperazinium moiety at ring A and D makes the compound more potent, e.g., Pipecuronium.
6. Repacuronium is the newer agent it has 16-N-propenyl and 17-propionyloxy (C_2H_5—COO —) groups. It is more than 10 times weaker than vecuronium.
7. 10 Atoms-rules: Ten atoms (C or O) separates the quaternary N-atoms of C_{10} with onion heads are interconnected. This imparts the greatest NMB activity, e.g., in Pancuronium, 10 atoms along with one path or another from N to the other atoms therefore is called rules for NMB activity.

14.6 SAR OF TETRAHYDROISOQUINOLINE DERIVATIVES

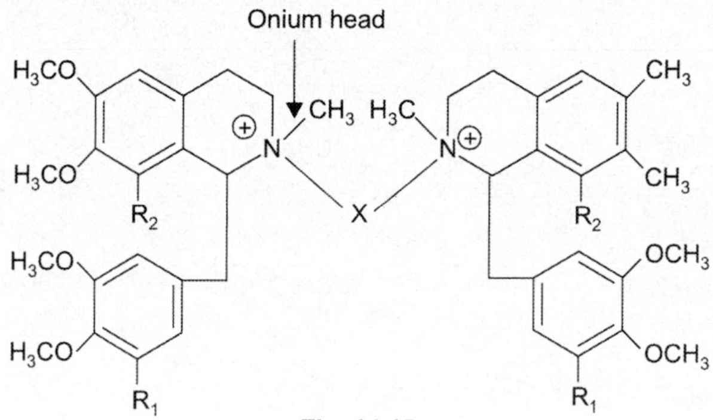

Fig. 14.15

Derivatives	Substitution		
	Y	R_1	R_2
1. Atracurium (Tracrium®)	$-(CH_2)_2-\overset{O}{\overset{\|}{C}}-O-(CH_2)_5-\overset{O}{\overset{\|}{C}}-O-(CH_2)_2-$	H	H
2. Mivacurium (Mivacron®)	$-(CH_2)_3-O-OC-(CH_2)_2-CH=CH-(CH_2)_2-\overset{O}{\overset{\|}{C}}-O-(CH_2)_3-$	$-OCH_3$	H
3. Doxacurim (Nuromax®)	$-(CH_2)_3-O-OCCH_2-CH_2-\overset{O}{\overset{\|}{C}}-O-(CH_2)_3-$	$-OCH_3$	$-OCH_3$

1. Presence of methoxymoeity at position 4,5,6 on benzylisoquinolium heads are essential for NMB activity and reduce the side effects.
2. Presence of onium heads connecting chain(X) exhibit increase in NMB activity as indicated in the same order.
3. α-tubocurarine is not like the atracurium, mivacurium or doxacurium. However, it is benzyl isoquinolinium compound. Its onium heads are interconnected with ether linkage.
4. Both hydroxyl groups of d-tubocurarine (d-tc) are methylated in metocurine (mtc) while metocurine has four methoxy groups. Mtc has 3-fold activity than d-tubocurarine.

14.7 DEPOLARIZING MUSCLE RELAXANTS (DPR)

Depolarizing muscle relaxant are not competitive antagonists like undepolarizing agents. They are based on structurally similarity of Ach and are considered more stable than Ach. DPR have prolonged duration of action than Ach. DPR acts on same receptors as Ach with longer action in the synapse and cause more stable depolarization that is positive symptom of muscles relaxants by blocking of receptor repolarization in order to accept new stimuli.

1. **Suxamethonium Chloride (Acectine®):** Ethanaminium, 2,2'-[1,4-dioxo-1,4-butanediyl)-bis-(oxy]-bis [N,N,N-trimethyl]-dichloride

Suxamethonium chloride is a potent depolarizing neuromusclar blocking agent. Because of its rapid hydrolysis by cholinesterase, its blocking effect is very limited. It has significant properties to relax the muscles. It is used in orthopaedic problems, endotracheal intubation, in laryngospasm, in electroconvulsion therapy.

Synthesis:

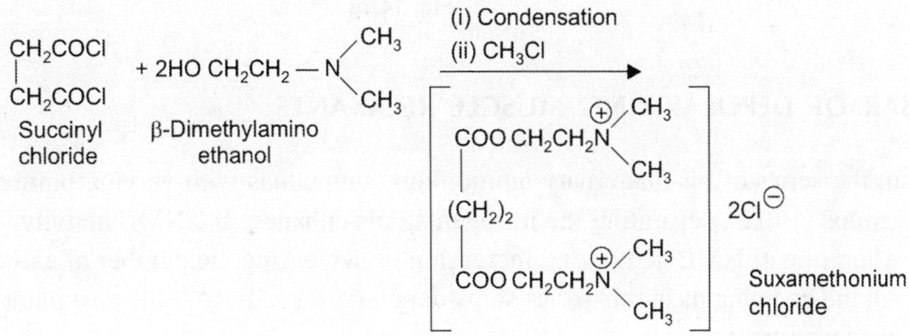

Fig. 14.16

2. Succinylcholine (Listenone®):

Succinylcholine is a potent depolarizing muscle relaxant. It has longer duration of action than Ach and it is more stable agonist rather than Ach. It causes more stable depolarization and blocks repolarization state, results in relaxation of skeletal muscles. Succinylcholine is therapeutically used in the treatment of muscles spasm, in operation, intubation of trachea, etc.

Synthesis:

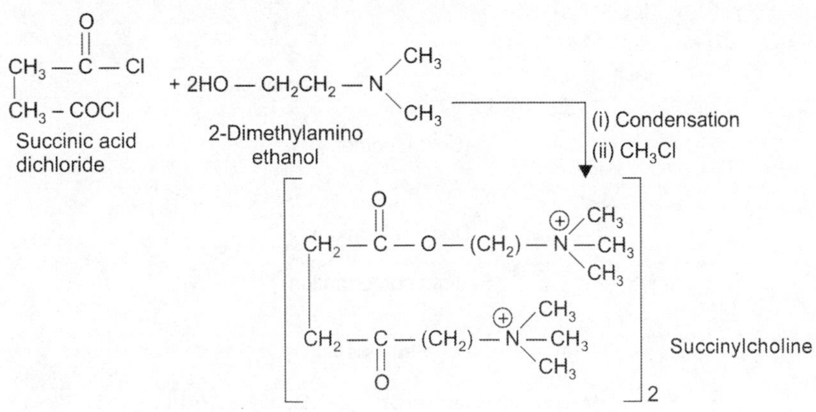

Fig. 14.17

3. Decamethonium Bromide (Syncurine®):

Decamethonium was one of the first NMB agents which was synthesized. Decamethonium exhibits the primary action on nicotinic receptors as depolarization at the motor end plate membrane. It is used in the surgical operation for relaxation of skeletal muscles.

Synthesis:

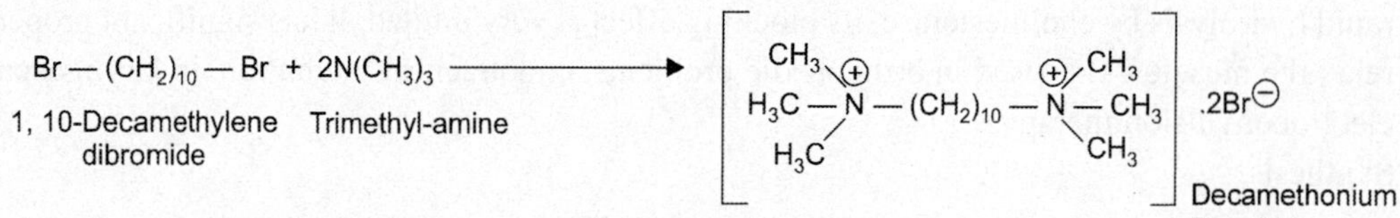

Fig. 14.18

14.8 SAR OF DEPOLARIZING MUSCLE RELAXANTS

(1) In the series of bis-quaternary ammonium compounds with varying number of methylene groups (10,12), separating the nitrogen atoms enhances the NMB activity.

(2) Alteration of NMB activity by increasing or decreasing the number of carbon atoms.

(3) All the compounds in this series show depolarizing activity at the post junctional motor end plate membrane.

(4) Decamethonium prefers the straight conformation while Ach and succinyl choline prefers the bent geometry due to electrostatic interaction between the functional groups. (Methonium head, carbonyl group and estero atom.)

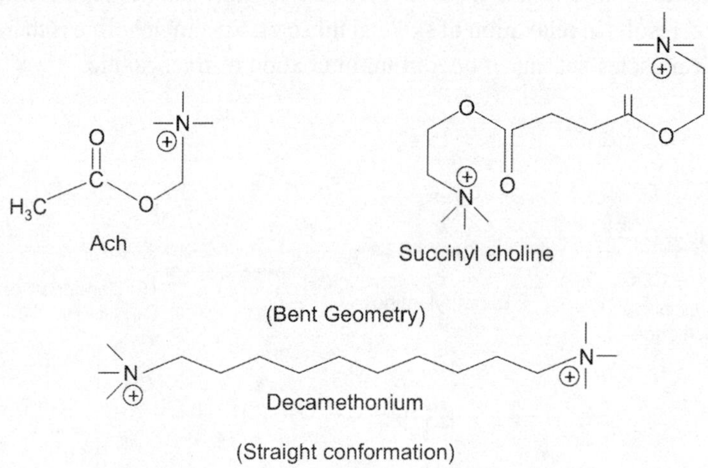

Fig. 14.19

Methonium Head — $(CH_2)_n$ — Methonium Head

Where n = 10 to 12

14.9 NEWER AGENTS

1. Atracurium Besylate: 2,2'-(3,11-dioxo-4,10-dioxatridecylene)-bis-[6,7-dimethoxy-benzyl)-2-methyl-1,2,3,4 tetra hydro Isoquinolinum]. Dibenzene sulfonate is the new series of competitive neuromuscular blocking agents. Atracurium besylate is used in anaesthesiology for muscle relaxants.

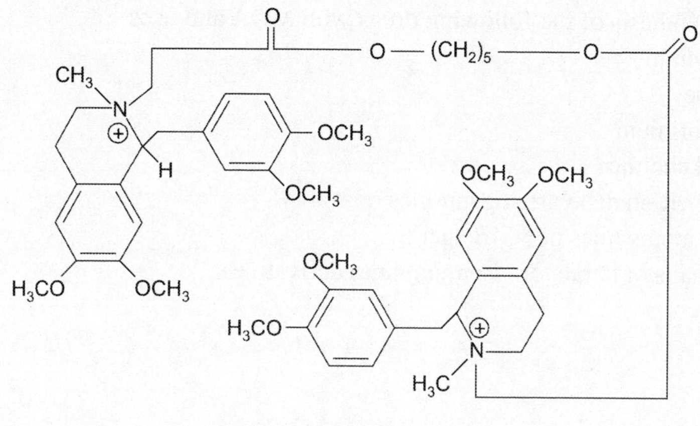

Fig. 14.20

2. Mivacurium:

Mivacurium is a short acting non-depolarizing muscle relaxant (NDPMR) with benzyl-isoquinoline structure. Neuromuscular blocking activity may be prolonged due to low plasma choline sterase activity.

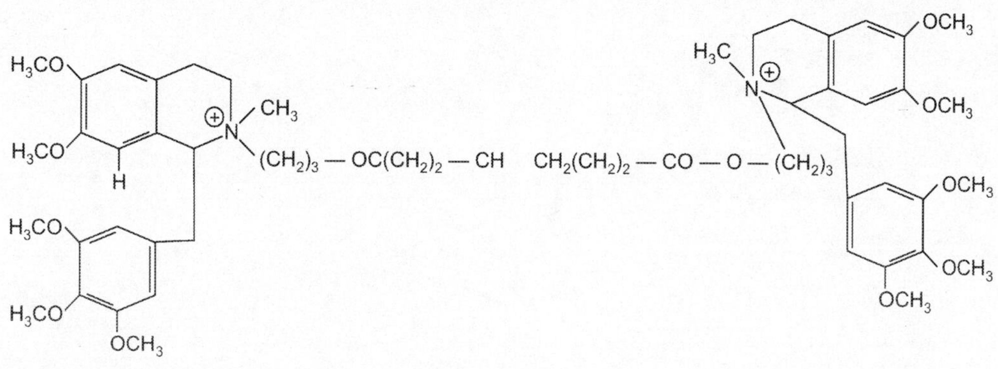

Fig. 14.21

2, 2-(1,8–dioxo-4-dtene 1,8-diyl)-bis-(oxy-3,1-Propanediyl)-bis (1,2,3,4-tetrahydro-6,7-dimethoxy-2-methyl-1-(3,4,5-trimethoxyphenyl)-methyl] isoquinoinium) dichloride.

C. Lee (2001) "British-J. of Anaesthesia, 87(5): 755-69(2001) Structure, conformation and action of neuromuscular blocking drugs.

EXERCISE

1. Define neuromuscular blocking agents.
2. Explain non-depolarizing and depolarizing muscle relaxant with their structures.
3. Classify NMB agents and give their mechanism of action.
4. Explain aminosteroids derivatives. Give the SAR of aminosteroids agents.
5. Write the synthesis of the following drugs with MOA and uses
   1. Pancuronium
   2. Gallamine
   3. Hexa fluorinium
   4. Succinyl choline
6. Write the SAR on depolarizing muscles relaxants.
7. Explain 10 atoms roles in NMB agents.
8. Write in details of tetrahydro isoquino lines derivatives.

15

Antibiotics

15.1 INTRODUCTION

An **antibiotic** is a type of antimicrobial substance active against bacteria. It is the most important type of antibacterial agent for fighting bacterial infections, and antibiotic medications are widely used in the treatment and prevention of such infections. They may either kill or inhibit the growth of bacteria. A limited number of antibiotics also show antiprotozoal activity. The term *antibiotic* takes Greek roots ἀντι (*anti*, "against") and Bio (*bios* "life") and is broadly used to refer to any substance used against microbes. Antibiotics (such as penicillin) are those produced naturally (by one micro-organism fighting another), on the other hand, nonantibiotic antibacterials (such as sulfonamides and antiseptics) are fully synthetic. Antibiotics have been used since ancient times.

15.2 CLINICAL USES

Antibiotics are used to treat or prevent bacterial infections, and sometimes protozoan infections. (Metronidazole is effective against a number of parasitic diseases.) When an infection is suspected of being responsible for an illness but the responsible pathogen has not been identified, an empiric therapy is adopted. This involves the administration of a broad-spectrum antibiotic based on the signs and symptoms presented and are initiated pending laboratory results that can take several days.

When the responsible pathogenic microorganism is already known or has been identified, definitive therapy can be started. This will usually involve the use of a narrow-spectrum antibiotic. The choice of antibiotic given is also be based on its cost. Identification is critically important as it can reduce the cost and toxicity of the antibiotic therapy and also reduce the possibility of the emergence of antimicrobial resistance. To avoid surgery, antibiotics may be given for non-complicated acute appendicitis.

Antibiotics may be given as a preventive measure and this is usually limited to at-risk populations such as those with a weak immune system (particularly in HIV cases to prevent pneumonia), those taking immunosuppressive drugs, cancer patients, and those having surgery. Their use in surgical procedures is to help prevent infection of incisions. They have an important role in dental antibiotic prophylaxis where their use may prevent bacteraemia and consequent infective endocarditis. Antibiotics are also used to prevent infection in cases of neutropenia particularly related to cancer.

15.3 ADMINISTRATION

There are many routes of administration for antibiotic treatment. Antibiotics are usually taken orally. In more severe systemic infections, antibiotics can be given intravenously or by injection. Where the site of infection is easily accessed, antibiotics are also topically used in the form of eye drops onto the conjunctiva for conjunctivitis or ear drops for ear infections and for the treatment of some skin disease conditions including acne and cellulitis.

15.4 CLASSIFICATION ON THE BASIS OF MECHANISM OF ACTION

Table 15.1: Method of administration of local anaesthetics

Mode of Action	Antibiotics
Cell wall synthesis inhibitors	Penicillins Cephalosporins Vancomycin Beta-lactamase Inhibitors Carbapenems Aztreonam Polymycin Bacitracin
Protein synthesis inhibitors	<u>Inhibit 30s Subunit</u> Aminoglycosides (gentamicin) Tetracyclines <u>Inhibit 50s Subunit</u> Macrolides Chloramphenicol Clindamycin Linezolid Streptogramins
DNA synthesis inhibitors	Fluoroquinolones Metronidazole
RNA synthesis inhibitors	Rifampin
Mycolic acid synthesis inhibitors	Isoniazid
Folic acid synthesis inhibitors	Sulfonamides Trimethoprim

15.5 β-LACTAM ANTIBIOTICS

β-lactam antibiotics (beta-lactam antibiotics) are antibiotics that contain a beta-lactam ring in their molecular structure. This includes penicillin derivatives (penams), cephalosporins (cephems), monobactams, carbapenems and carbacephems. Most β-lactam antibiotics work by inhibiting cell wall biosynthesis in the bacterial organism and are the most widely used group of antibiotics. Bacteria often develop resistance to β-lactam antibiotics by synthesizing a β-lactamase, an enzyme that attacks the β-lactam ring. To overcome this resistance, β-lactam antibiotics can be given with β-lactamase inhibitors such as clavulanic acid.

15.6 PENICILLINS

Penicillins: The first β-lactam antibiotic discovered, penicillin, was isolated from a rare variant of *Penicillium notatum* (since renamed *Penicillium chrysogenum).* In the structure of penicillins carboxylic group is present at C-3 position. β-lactam carbonyl functional centre is present at position 7, while germinal dimethyl substitution is located at C-2 position.

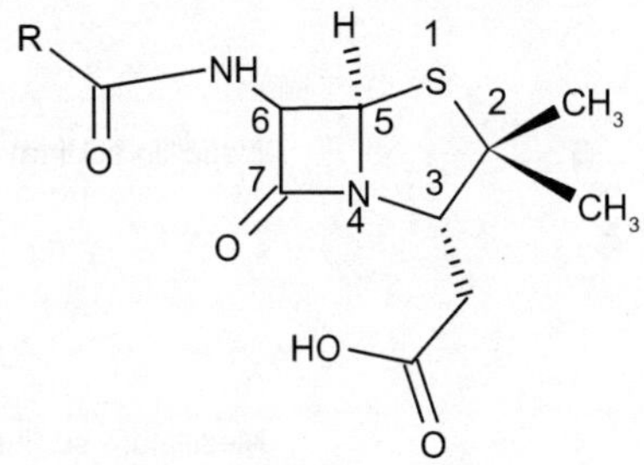

Fig. 15.1 Penicillin scaffold

15.6.1 Classification of Penicillins

Table 15.2: Naturally occurring penicillins

S.No.	Penicillins	Substitution R	Chemical name
1	Penicillin I or F	$-CH_2{\cdot}CH{=}CH{\cdot}CH_2{\cdot}CH_3$	Pent-2-enylpenicillin
2	Penicillin H or G	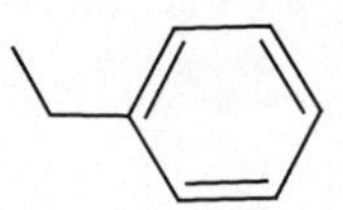	Benzylpenicillin
3	Penicillin X or III		*p*-Hydroxybenzylpenicillin
4	Penicillin K or IV	$-(CH_2)_6{\cdot}CH_3$	n-Heptylpenicillin
5	Dihydro-F-penicillin	$-(CH_2)_4{\cdot}CH_3$	n-Amylpenicillin
6.	Penicillin V	$-CH_2\text{-}O\text{-}C_6H_5$	Phenoxymethylpenicillin

Table 15.3: Clinically Relevant Penicillins

S.No.	Penicillins	Substitution R
1	Cloxacillin sodium	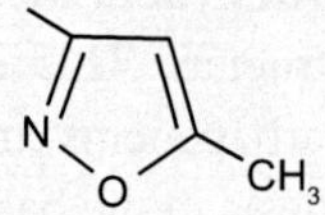
2	Methicillin sodium	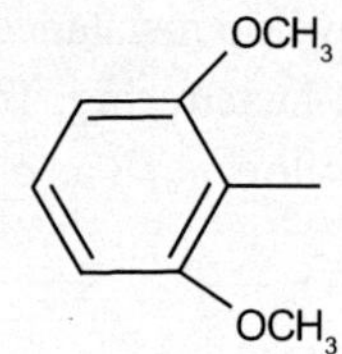
3	Naficillin sodium	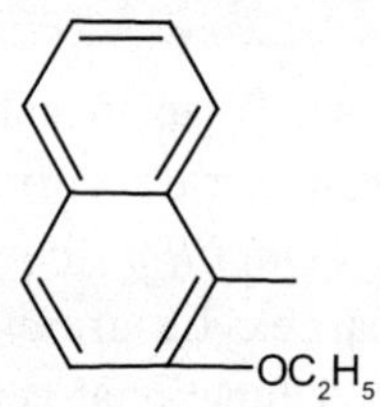
4	Mecillinam sodium	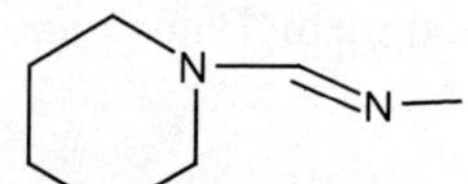
5	Ampicillin	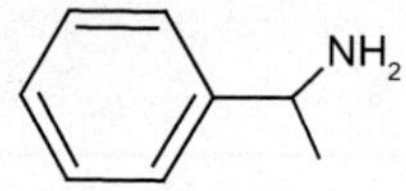
6	Amoxicillin dihydrate	
7	Ciclacillin	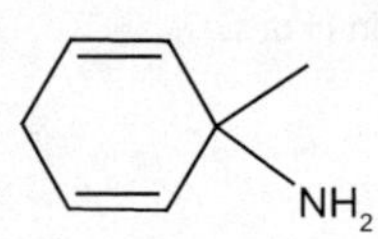

15.6.2 General Features of Penicillins

1. Penicillins are monobasic acids, i.e., they form salts.
2. Penicillins are hydrolysed in the presence of dilute inorganic acids which form two forms penicillamine (amine) and penniloaldehyde (aldehyde).
3. Penicillamin gives indigo colour reaction with ferric chloride which is a characteristic of cysteine.
4. Penilloaldehyde gives substituted acetic acid and aminoacetaldehyde on vigorous hydrolysis.

Varients of penicillins

1. Naturally occurring (from streptococcal/narrow spectrum).
2. Penicillinase-resistant penicillins.
3. Aminopenicillins (active against Gram-negative bacteria: *Enterococcus, Salmonella, Shigella*).
4. Extended spectrum (antipseudomonal).
5. β-lactamase combination (β-lactamase producers).

15.6.3 Biosynthesis

Penicillin G biosynthesis

Important steps are involved in biosynthesis of penicillin G (benzylpenicillin).

- **First step** is the condensation of three amino acids: L-α-aminoadipic acid, L-cysteine and L-valine, into a tripeptide. Before condensing into the tripeptide, the amino acid L-valine undergo epimerization to become D-valine. The condensed tripeptide is named α-(L-α-aminoadipyl)-L-cysteine-D-valine (ACV).
- **Second step** is the biosynthesis of penicillin G is the oxidative conversion.
- **Third step** is transamidation by isopenicillin N N-acyltransferase, in which the α-aminoadipyl side-chain of isopenicillin N is removed and exchanged for a phenylacetyl side chain.

15.6.4 Structure Activity Relationship (SAR) of Penicillins

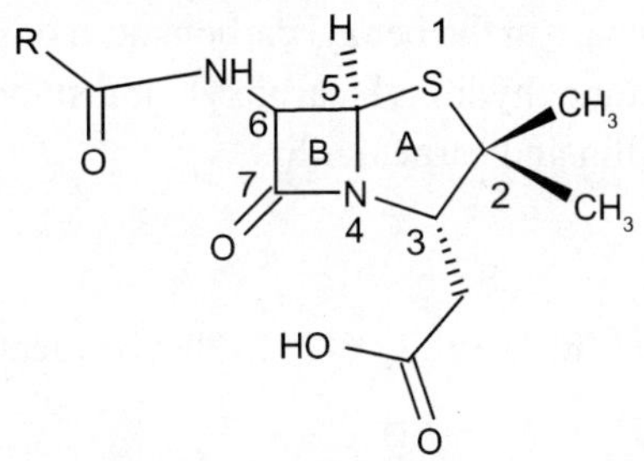

Fig. 15.2 Penicillin

1. **Position 1:** The sulphur of the thiazolidine ring with O, CH , and CH-•-CH_3 provides broad-spectrum antibacterial activity.
2. **Position 2:** The geminal dimethyl group at C-2 position is a characteristic of the penicillin.
3. **Position 3:** Many esters of the carboxyl group attached to C-3 have been prepared as prodrugs to increase lipophilicity and acid stability. For illustration, Acetoxymethyl ester derivatives are used for preparing prodrugs.
   Derivatization of the C-3 carboxylic acid functionality should not be tolerated unless the free penicillin carboxylic acid can be generated in vivo. Doubly activated penicillin esters undergo rapid cleavage in vivo to generate active penicillin. Example: pivampicillin and becampicillin.
4. **Position 4:** The antibacterial activity is evidenced by N-4 atom at ring junction.

5. **Position 6: 6-Acyl side chain**
   (a) The substitution of R on the primary amine with an electron withdrawing group resultant decreases the electron density on the side chain and protects from acid degradation.
   (b) Substitution on the •-carbon of the side chain, such as amino (ampicillin), chloro, and guanidine exhibit good resistance to inactivation by acids. Benzyl penicillin undergoes acid and alkali degradation and is susceptible to all known •-lactamase.
   (c) Increased length of acyl amino side chain through acylation of 6APA resultant with tremendous increase in biological activity. Substitution of •-aryl of the alkyl group in the side chain gives increased stability and oral absorption.
   (d) Substitution of bulky groups on •-carbon of the side chain confers •-lactamase resistance activity. In all the penicillins (methicillin, nafcillin, oxacillin, etc.), an aromatic ring is attached directly to the side chain amide carbonyl atom as well as substitution at both positions of ortho to the point of attachment. The size of the ring systems plays an important role in determining the ability of the ortho substitutent to show penicillinase resistance activity.
   (e) Replacement of acyl side chain with other hydroxymethyl groups confers improved Gram-negative activity. Introduction of •-methoxy group on C-6 produces greater stability against •-lactamase.
6. The isomeric forms of penicillins differ in their activity. Example: D-isomer is 2–8 times more active than L-isomer of amoxicillin. The introduction of polar group or ionized molecule into the •-position of the side chain in the benzyl carbon atom of penicillin-G confers against the Gram-negative bacilli. Amino, hydroxyl, carboxyl, and sulphonyl increases Gram-negative activity. Example: ampicillin and carbenicillin.

15.6.5 Specific Drugs

Phenoxymethylpenicillin (penicillin-V): (6R)-6-2-(2-Phenoxyacetamido) penicillanic acid

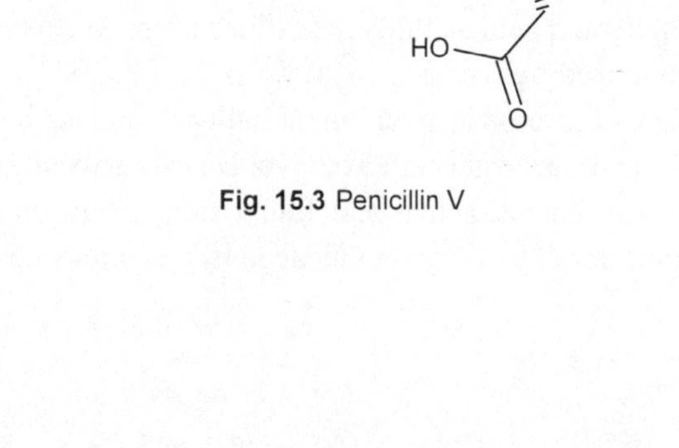

Fig. 15.3 Penicillin V

Synthesis:

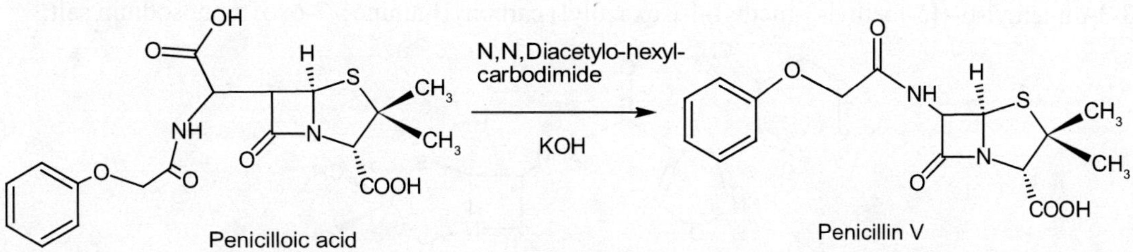

Fig. 15.4

Penicillin V is used in the management and control of Gram-positive bacterial infectious disease. It is active against *Streptococcal, Staphylococcal, Pneumococcal* infections. It is also used in the treatment of *Gonococcal* and *Meningococcal* infections.

15.6.5.1 Penicillinase-Resistant Penicillins

Cloxacillin sodium: Cloxapen (R) 4-thia-1-azabicyclo-[3,2,0] heptanes-2-carboxyllic acid-6-[[[3-(2—chlorophenyl)-5-methyl-4-isoxazolyl] carbonyl]amino]-3,3-dimethyl-7-oxo-monosodium salt.

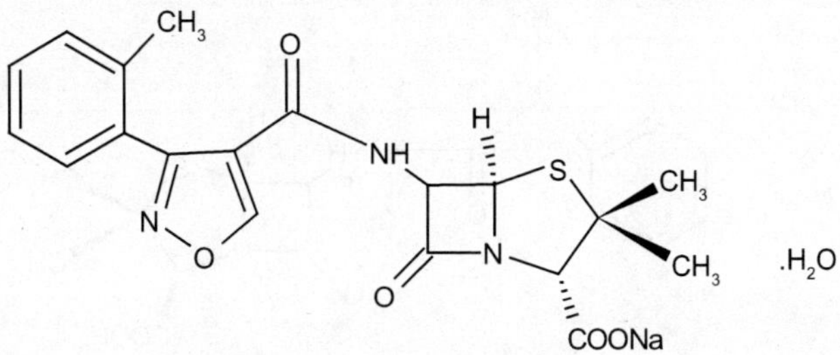

Fig. 15.5 Cloxacillin sodium

Cloxacillin sodium is a penicillinase-resistant penicillin. It is active against *staphylococcal* infections. It is administered orally.

Methicillin sodium: Staphcillin(R) 4-thia-1-azabicyclo-[3,2,0] heptanes-2-carboxyllic acid, 6-[(2,6-dimethoxybenzoyl)-amino]-3,3-dimethyl-7-oxo-monosodium salt.

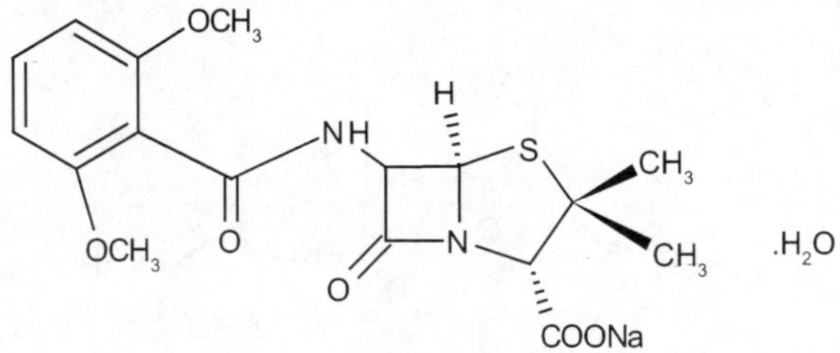

Fig. 15.6 Methicillin sodium

Methicillin sodium is used in the treatment of staphylococcal infections caused by strain resistant to other penicillins.

Oxacillin sodium: Prostaphlin(R) 4-thia-1-azabicyclo-[3,2,0] heptanes-2-carboxyllic acid, 3-3-dimethyl-6-{[5-methyl-3-methyl-4-isoxazolyl) carbonyl]-amino}-7-oxo, monosodium salt.

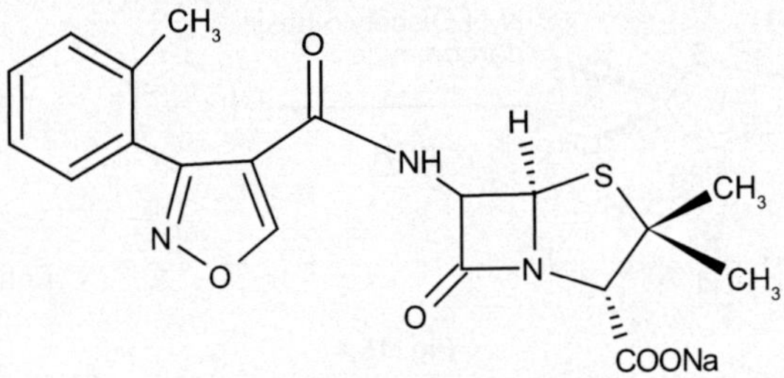

Fig. 15.7

Oxacillin sodium is used in the treatment of staphylococcal infections caused by strain resistant to other penicillins.

15.6.5.2 Aminopenicillins

Ampicillin: Polycillin(R) 4-thia-1-azabicyclo-[3,2,0] heptanes-2-carboxyllic acid, 6-(aminophenylacetyl)amino-3-3-dimethyl-7-oxo.

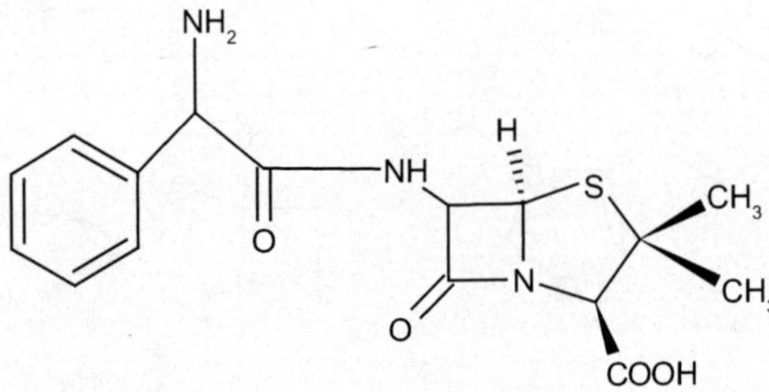

Fig. 15.8 Ampicillin

Ampicillin is a broad-spectrum antibiotic. It is active against Gram-positive organism which is susceptible to other penicillins. It is used in the treatment of the infections caused by Gram-negative organisms.

Becampicillin hydrochloride: Spectrobid(R) 4-thia-1-azabicyclo-[3,2,0] heptanes-2-carboxyllic acid, 6-[(aminophenylacetyl) amino]-3,3-dimethyl-7-oxo-1-[(ethoxycarbonyl)-oxyethyl] ester monohydrochloride.

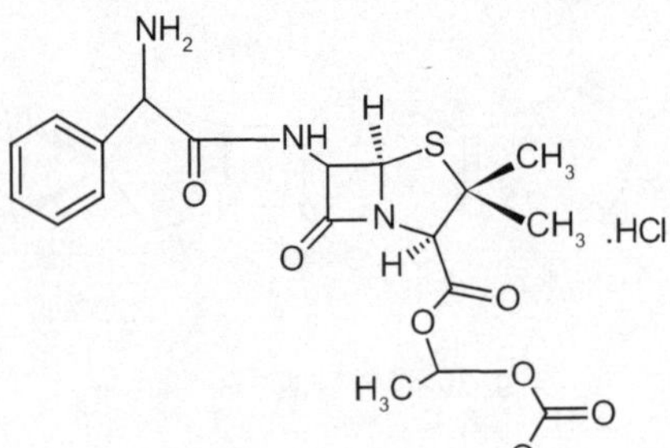

Fig. 15.9 Becampicillin

Becampicillin hydrochloride is a prodrug which is converted to ampicillin in vivo after undergoing hydrolysis in the presence of esterase in plasma.

15.6.5.3 Extended-Spectrum Penicillins

Piperacillin sodium: Pipracil [(R)] 4-thia-1-azabicyclo-[3,2,0] heptanes-2-carboxyllic acid, 6-[[[(4-ethyl-2,3-dioxo-1-piperazinyl) carbonyl]-amino]-phenylacetyl]-amino]-3,3-dimethyl-7-oxo-monosodium.

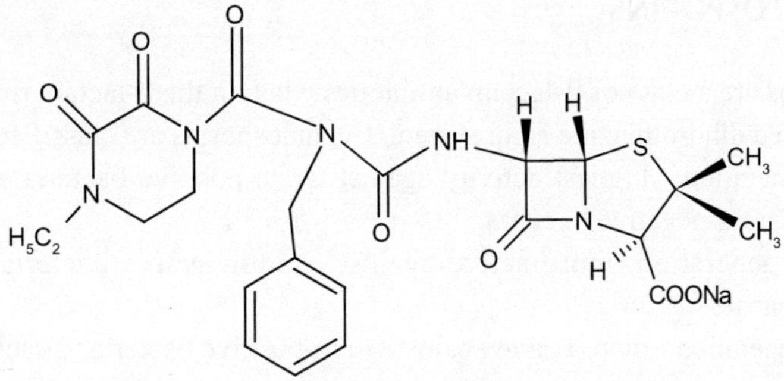

Fig. 15.10 Piperacillin

Piparacillin is mostly active against *P. aurigenosa.* It is also used in the treatment of infection caused by enteric bacilli. It shows low efficacy against penicillinase and other members of beta-lactamase producing organism.

15.6.6 β-lactam Inhibitors

Generally, penicillanase resistant penicillins bound to enzyme (drug-enzyme complex), therefore dissociation of the complex gets rapidly. To overcome this problem some β-lactam inhibitors (Clavulanic acid, Salbactum, Tazobactum) *acylated* by forming the double bond, the resulting dissociation would be slow and enhances the potency of penicillins. Therefore, β-lactam inhibitors with other antibiotics help to expand the spectrum of antibiotics.

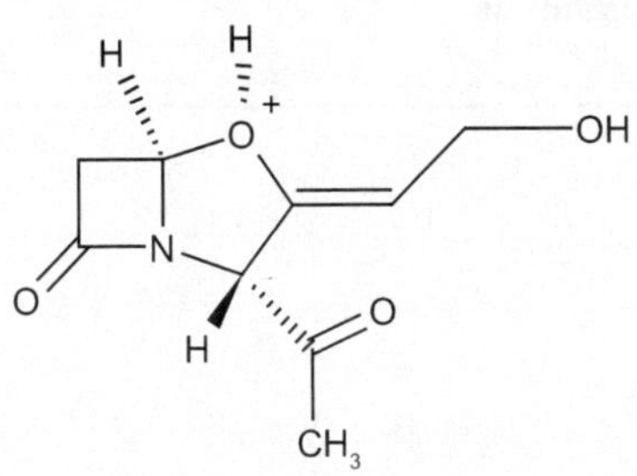

Fig. 15.11 Clavulanic acid

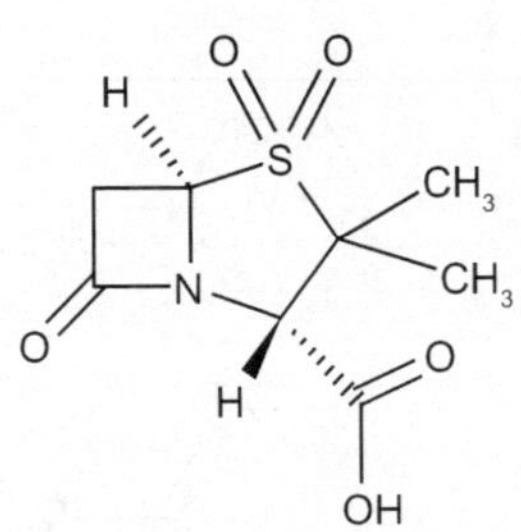

Fig. 15.12 Salbactum

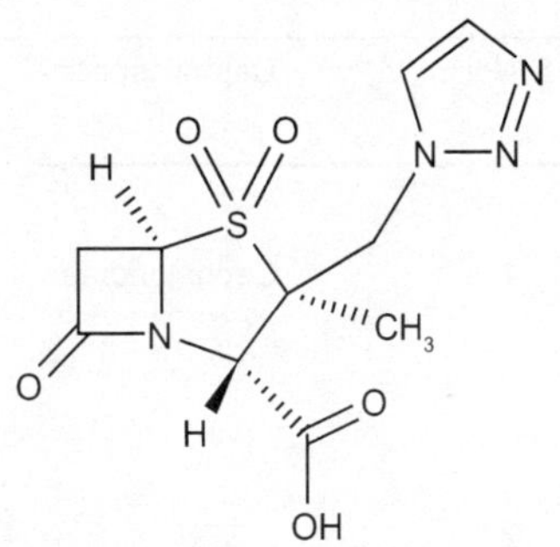

Fig. 15.13 Tazobactum

15.6.7 Clinically Useful Combinations

Clavulanate- Amoxicillin: Used in treatment in diarrhoea, more potent than Amoxicillin.
Clavulanate--Ticarcillin: This combination is more potent for Gram-negative bacilli.
Salbactum--Ampicillin: This combination is more potent for *Staphylococcus* infections.
Tazobactum-- Piperracillin: This combination is more potent for Gram-negative bacilli.

15.7 CEPHALOSPORINS

Cephalosporins are a class of β-lactam antibiotics wherein the β-lactam ring is strategically fused with 6-membered dihdrothiazine rimg system. Cephalosporins are classified in four categories.

1. First generation : highest activity against Gram-positive bacteria and the lowest activity against Gram-negative bacteria.
2. Second generation: more active against Gram-negative bacteria, some active against *H. influenzae.*
3. Third generation : more active against Gram-positive bacteria, β-lactamase resistant
4. Fourth generation.

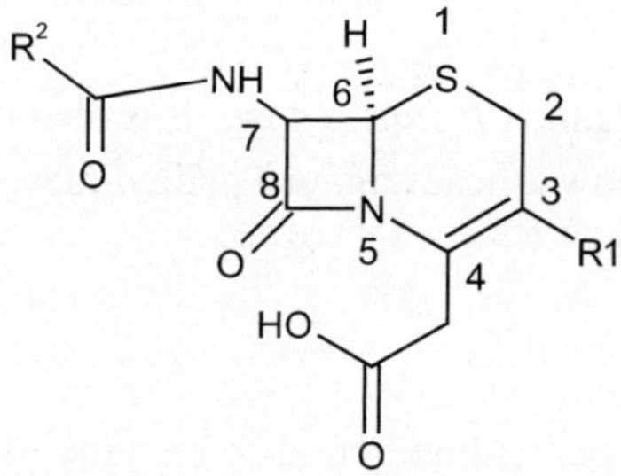

Fig. 15.14 Cephalosporin scafold

15.7.1 Classification

Table 15.4: First generation cephalosporins

S.No.	Cephalosporin	Substitutions R_1	Substitutions R_2
1	Cephaloridine	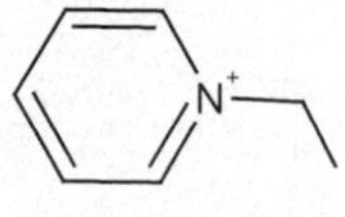	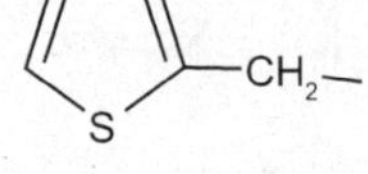
2	Cephalothin	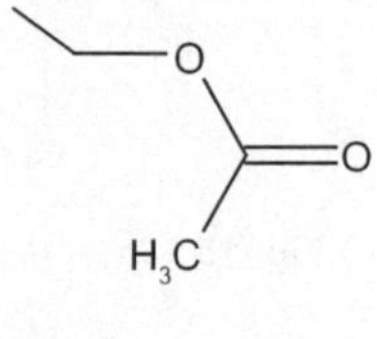	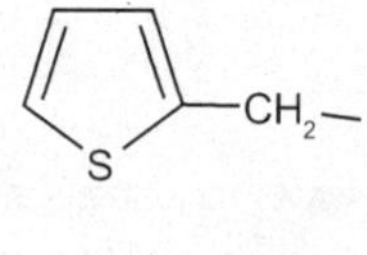

3	Cephapirin	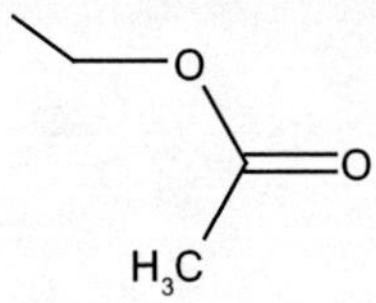	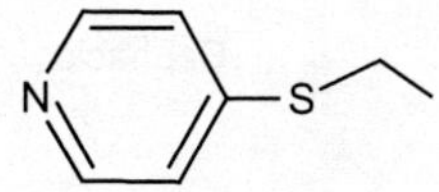
4	Cephalexin	-CH_3	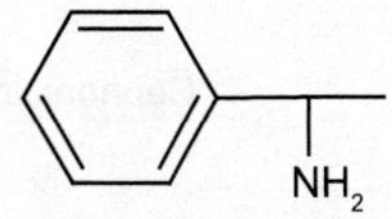
5	Cephaloglycine	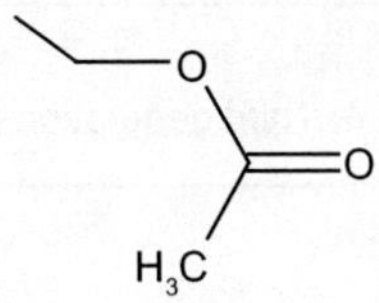	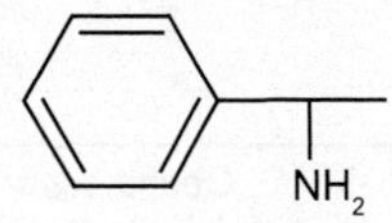
6	Cephadroxil	-CH_3	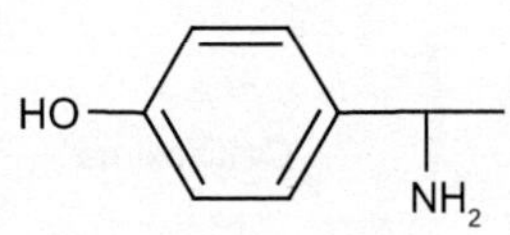
7	Cephradine	-CH_3	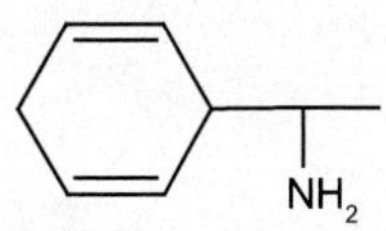
8	Cephazoline	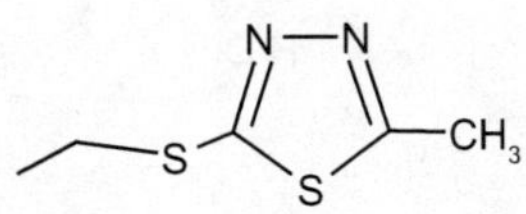	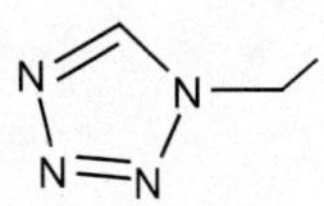

Table 15.5: Second generation cephalosporins

S.No.	Cephalosporin	Substitutions	
		R_1	R_2
1	Cephamandole	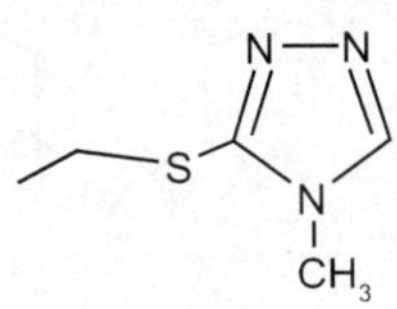	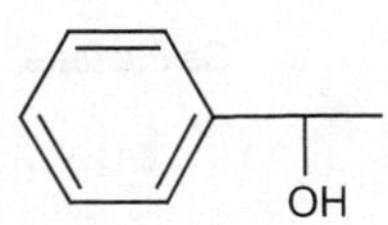
2	Cephuroxim	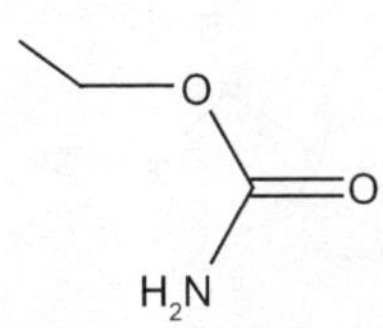	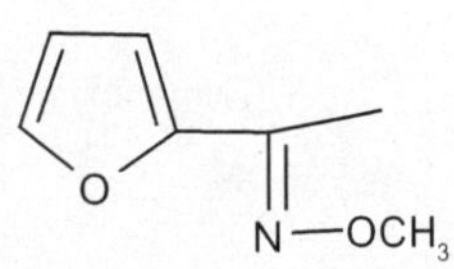

3	Cephaclor	-Cl	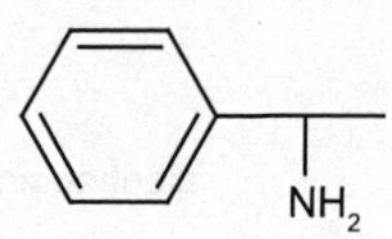
4	Cephonicid	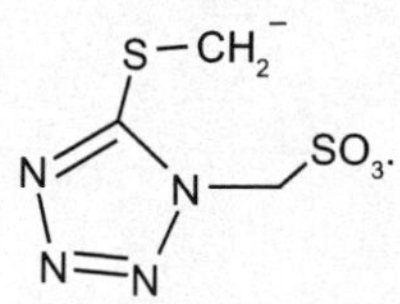	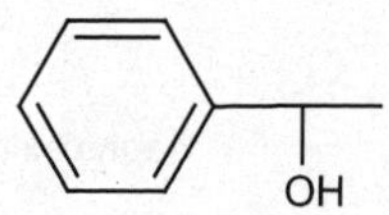

Table 15.6: Third generation cephalosporins

S.No.	Cephalosporin	Substitutions	
		R_1	R_2
1	Ceftizoxime	H	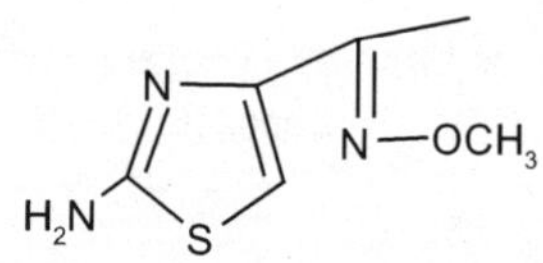
2	Cefotoxime	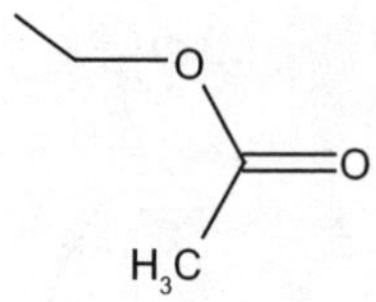	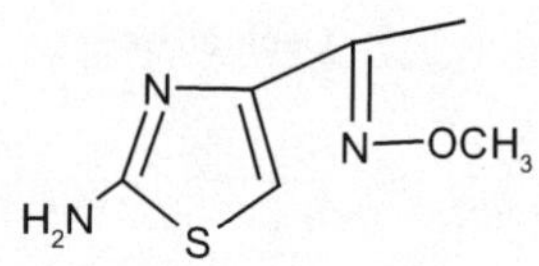
3	Ceftazidime	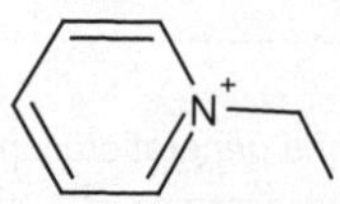	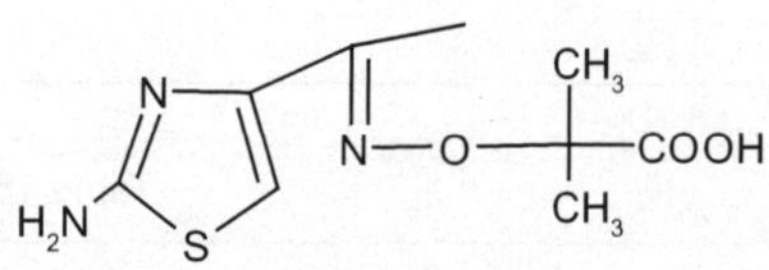
4	Ceftriaxone	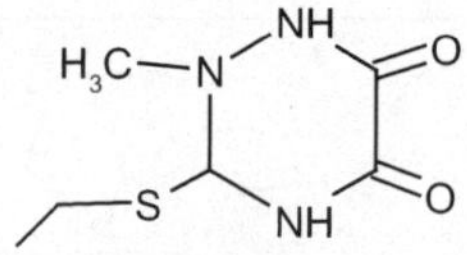	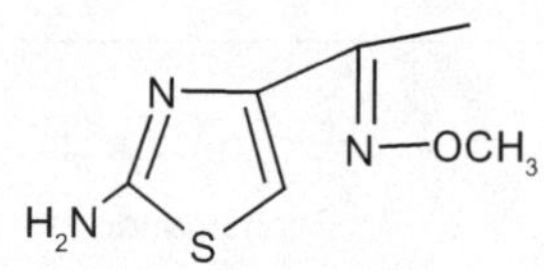
5	Cefmenoxime	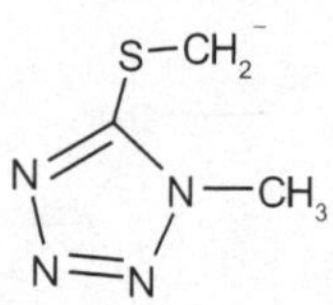	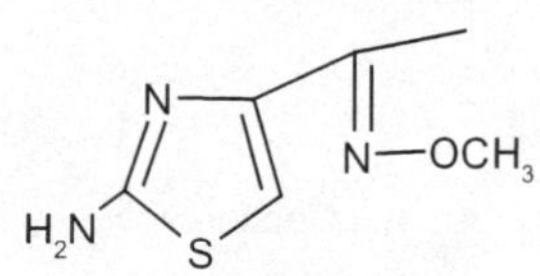

Table 15.7: Fourth generation cephalosporins

S.No.	Cephalosporin	Substitutions R_1	R_2
1	Cefepime	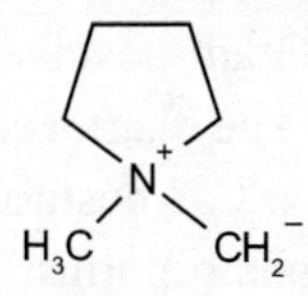	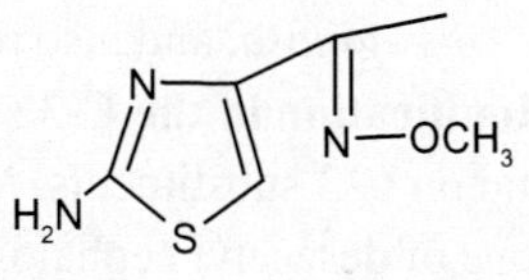
2	Cefpirome	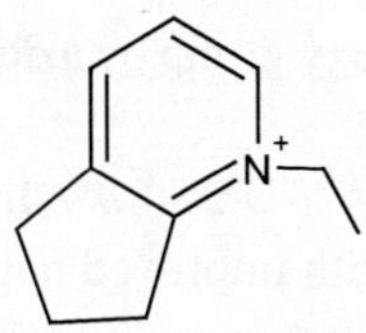	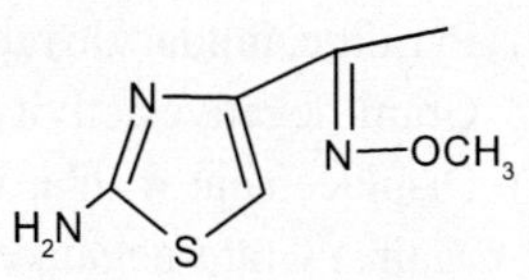

15.7.2 Structure Activity Relationship (SAR) of Cephalosporins

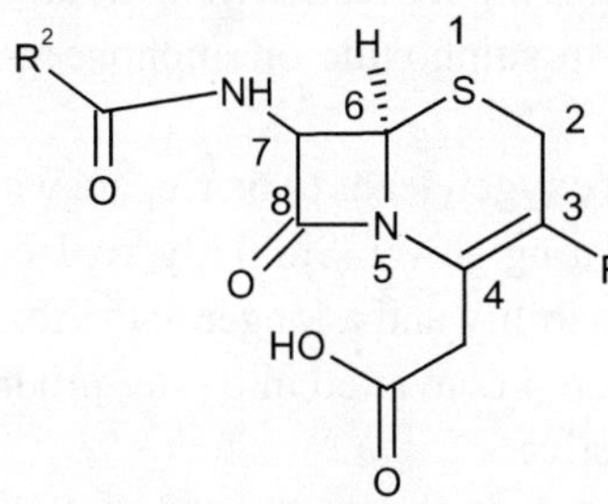

Fig. 15.15 Cephalosporins

1. 7-Acylamino substitution

(a) The addition of amino group and a hydrogen to • and •$_1$ positions on R^2 produces basic compound, which is protonated under acidic conditions in vivo. The ammonium ions improve the stability of •-lactum of cephalosporins and make the drug orally active. Activity against Gram-positive bacteria is increased and for Gram-negative bacterial species is decreased by acylation of amino group.

(b) When the new acyl groups are derived from carboxylic acids, they exhibits a good spectrum of antibacterial action against Gram-positive bacteria.

(c) Substitutions on the aromatic ring phenyl attached with R^2 increase lipophilicity, provide higher Gram-positive activity and lower Gram-negative activity.

(d) Replacement of phenyl ring in the side chain with other heterocycles rings (thiophene, tetrazole, furan, pyridine, and aminothiazoles with improved spectrum of activity as well as pharmacokinetic properties.

(e) The L-isomer of •-amino •$_1$ -hydrogen derivative of cephalosphorins shows 30–40-fold stable than D-isomer. Addition of methoxy oxime to • and •$_1$ on R^2 increases the stability to nearly 100 times.

(f) Introduction of catechol nucleus enhances activity, particularly, against *Pseudomonas aeruginosa*, and also retains some Gram-positive activity.

2. Modification in the C-3 substitution: The pharmacokinetic and pharmacodynamics properties depend on C-3 substituents. Modification at C-3 position has been done to reduce the degradation (lactone of desacetyl cephalosporin) of cephalosporins.

(a) The benzoyl ester displayers improved Gram-positive activity, but lowered Gram-negative activity.

(b) Pyridine, imidaozle replaced acetoxy group by azide ion, yields derivative with relatively low Gram-negative activity.

(c) Displacement with aromatic thiols of 3-acetoxy group results in an enhancement of activity against Gram-negative bacteria with improved pharmacokinetic properties.

(d) Orally active compounds can be produced by replacement of acetoxy group at C-3 position with CH_3 and Cl.

3. Other modifications

(a) Methoxy groups at C-7, exhibit higher resistance to hydrolysis by •-lactamase.

(b) Oxidation of ring spectrum to sulphoxide or sulphone greatly diminishes or destroys the antibacterial activity.

(c) Replacement of sulphur with oxygen leads to oxacepam with increased antibacterial activity, because of its enhanced acylating power. Similarly, replacement of sulphur with methylene group has greater chemical stability and a longer half-life.

(d) The carboxyl group at position-4 converted into ester prodrugs to increase bioavailability of cephalosporins, and orally active.

(e) The antibacterial activity also depends on the olefinic linkage at C-3 and C-4 position and their activity is lost due to the ionization of double bond to 2nd and 3rd positions.

15.7.3 Specific Clinically Useful Drugs

15.7.3.1 First Generation Cephalosporins

Cephalexin: Keflex(R) 5-Thia-1-azabicyclo-[4,2,0]-oct-2-ene-2-carboxylic acid-7-[(aminophenylacetyl)-amino]-3-methyl-8-oxo, monohydrate.

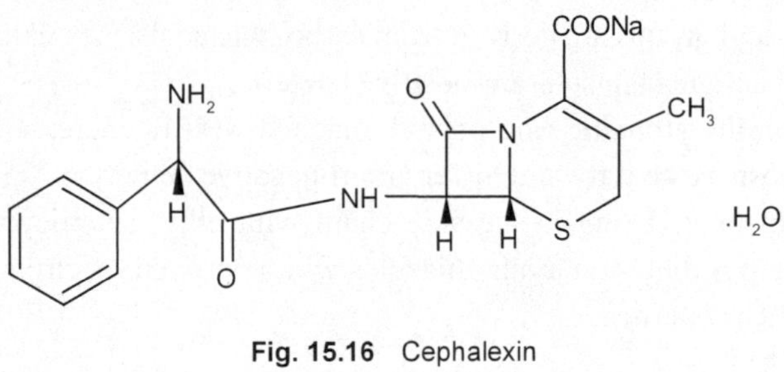

Fig. 15.16 Cephalexin

Synthesis

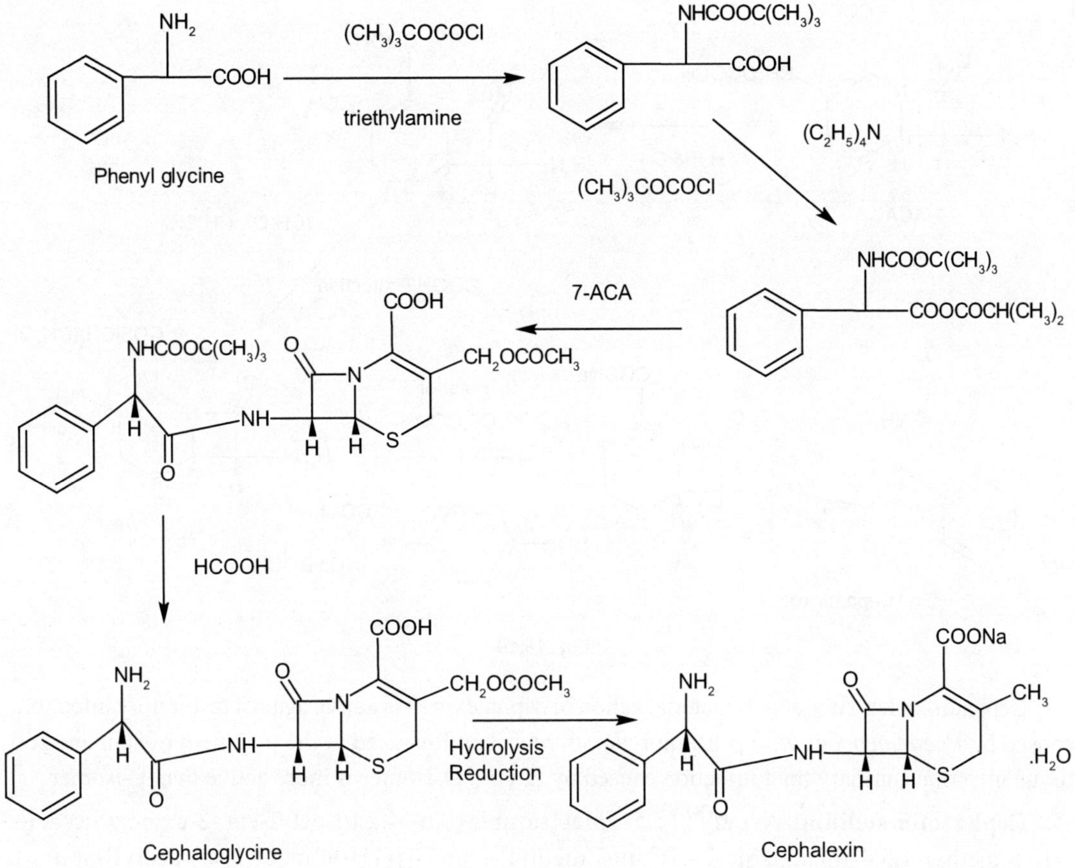

Fig. 15.17

Cephalexin is active against respiratory infections caused by *Pneumococcus* with β-haemolytic *Streptococci*. It is used in the treatment of skin and soft tissue infections, and urinary tract infection caused by *E. coli*.

Cephadroxil: Cephadrox[(R)] 5-Thia-1-azabicyclo-[4,2,0]-oct-2-ene-2-carboxylic acid-7-[(amino-4-hydroxyphenylacetyl)-amino]-3-methyl-8-oxo, monohydrate.

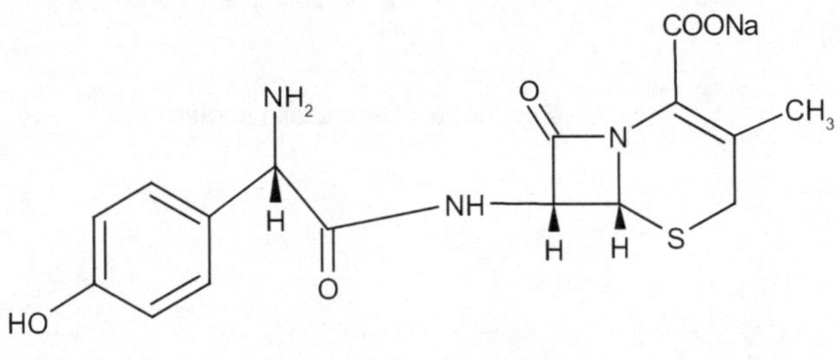

Fig. 15.18 Cephadroxil

Synthesis

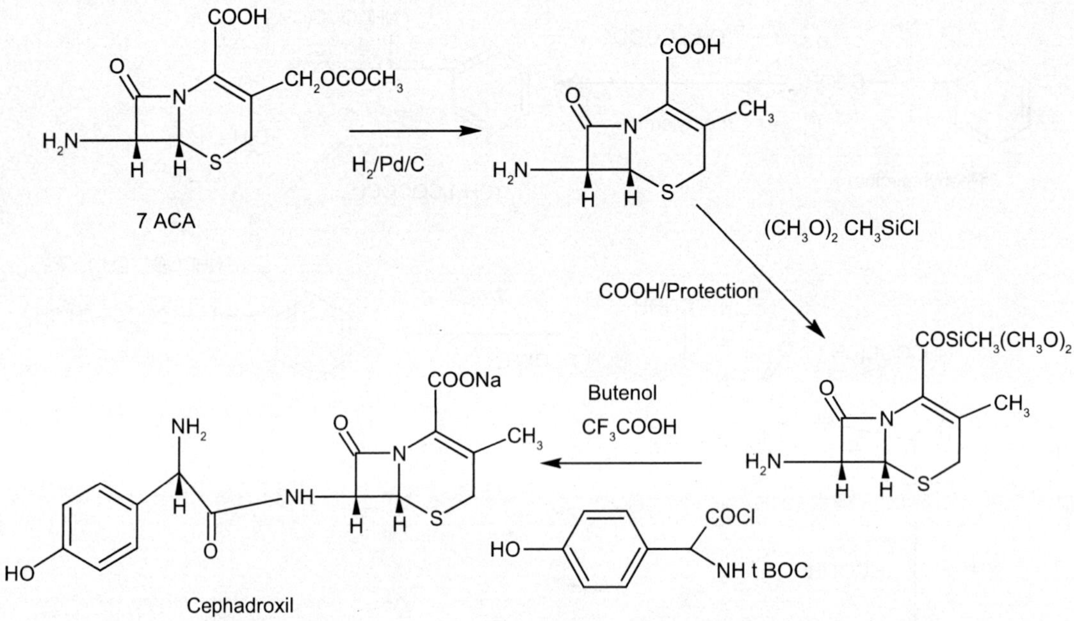

Fig. 15.19

Cephadroxil has similar antibacterial action of cephalexin. It is active against respiratory infections caused by *Pneumococcus* with β-haemolytic *Streptococci*. It is used in the treatment of skin and soft tissue infections, urinary tract infection caused by *E. coli*. D-isomer is more active than L-isomer.

Cephazolin sodium: Ancef[(R)] : 5-Thia-1-azabicyclo-[4,2,0]-oct-2-ene-2-carboxylic acid, 3-[[(5-methyl-1,3,4-thiadiazole-2-yl) thio] methyl]-oxo-7-[[(1H-tetrazol-1-yl) acetyl]-amino], monosodium salt.

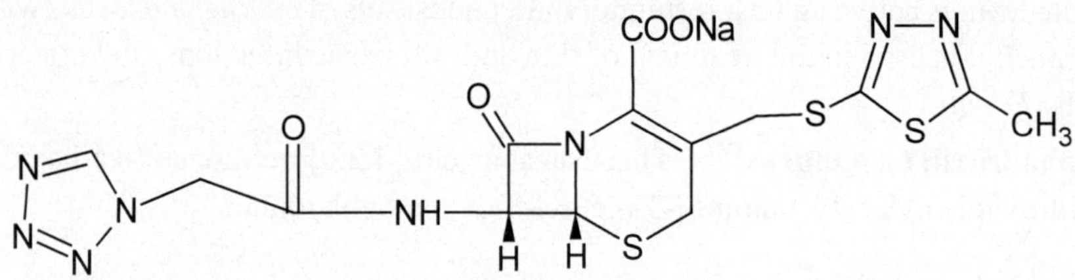

Fig. 15.20 Cephazolin sodium

Synthesis

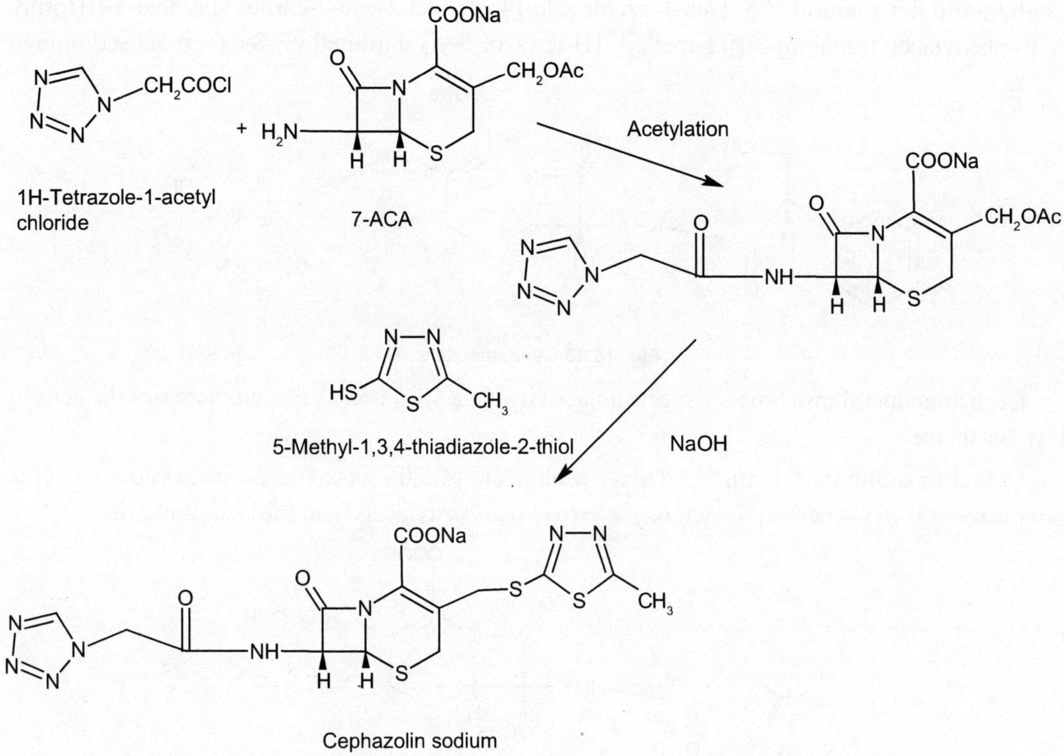

Fig. 15.21

Cephazolin is used in the treatment of infectious disease caused by Gram-positive bacteria. It shows narrow spectrum activity against Gram-negative bacterial species like *E. coli*, *Klebsiella*, etc.
Ceptradine: Anspor[(R)] 5-Thia-1-azabicyclo-[4,2,0]-oct-2-ene-2-carboxylic acid-7-[(amino 1,4-cyclohexadien-1-ylacetyl)-amino]-3-methyl-8-oxo.

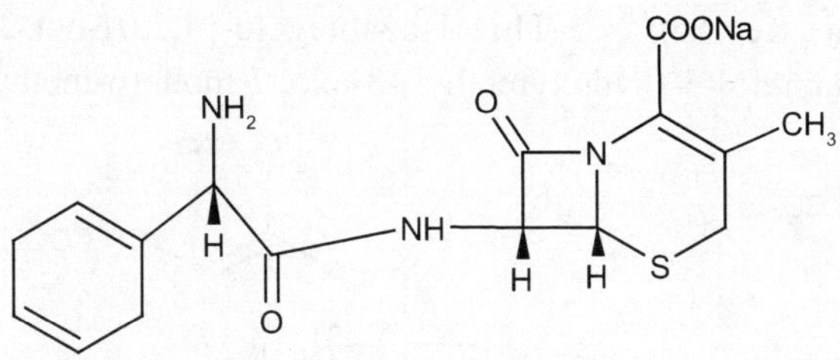

Fig. 15.22 Ceptradine

Ceptradine is active against respiratory infections caused by *Pneumococcus*. It is used in the treatment of skin and soft tissue infections, urinary tract infection caused by *E. coli*.

15.7.3.2 Second Generation Cephalosporins

Cephamandole: mandol(R) 5-Thia-1-azabicyclo-[4,2,0]-oct-2-ene-2-carboxylic acid-7-[[(formyloxy)-phenylacetyl]-amino]-3-[[(1-methyl, 1H-tetrazol-5-yl)-thio]methyl]-8-oxo, monosodium salt.

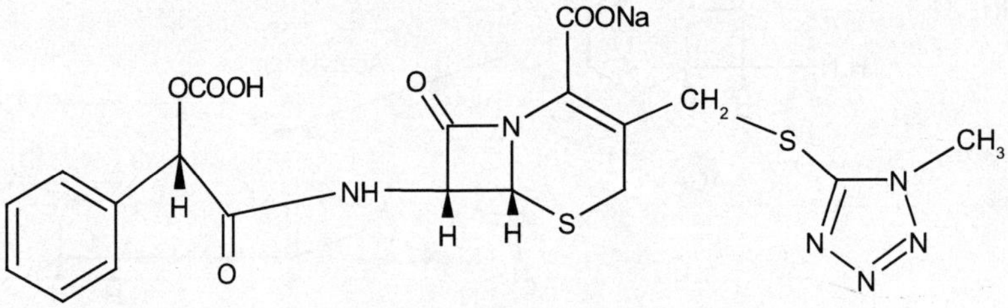

Fig. 15.23 Cephamandole

Cephamandole shows broader spectrum activity angaist *H. ifluenzae*, and increases the activity of β-lactamase.

Cefoxitin sodium: Mefoxin(R) 5-Thia-1-azabicyclo-[4,2,0]-oct-2-ene-2-carboxylic acid,3-[[(a-monocarbonyl)-oxy]-methyl]-7-methoxy-8-oxo-7-[(2-thienylacetyl) amino], sodium salt.

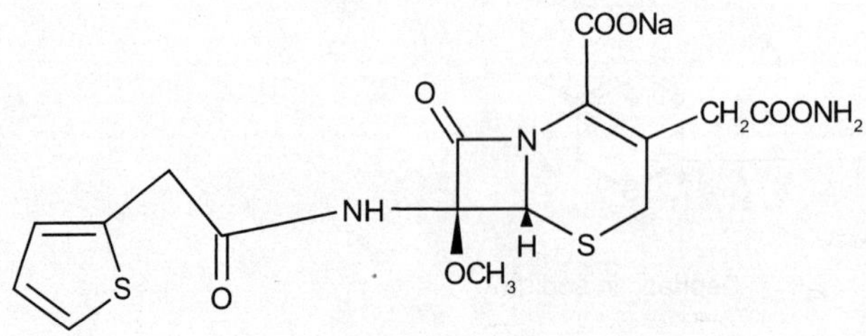

Fig. 15.24 Cefoxitin sodium

Cefoxitin sodium is used in the treatment of intra-abdominal infection and lower respiratory infection. It is active against enteric anaerobes, Gram-negative bacilli infections caused by *E. coli, H. influenzae, Klebsiella* sp., *S. aureus*.

Cefuroxim sodium: Kefurox(R) 5-Thia-1-azabicyclo-[4,2,0]-oct-2-ene-2-carboxylic acid-7-[2-(2-furyl)-glyoxylamido]-3-(hydoxymethyl)-8-oxo, 7-mono(o-methyloxime)-cabamates.

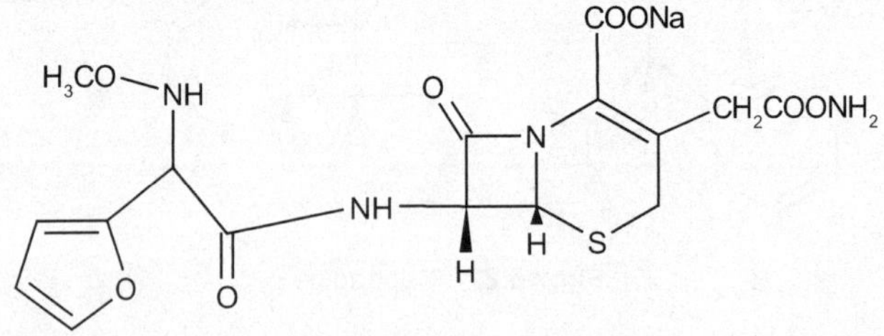

Fig. 15.25 Cefuroxim sodium

Cefuroxim sodium is used in the treatment of meningitis caused by *Sptreptococcus pneumonia*. It is also used in the treatment of lower respiratory tract infections caused by *H. influenzae, Kleb-*

siella sp., *E. coli.* Cefuroxim sodium is also recommended in the treatment of urinary tract infections caused by *Klebsiella* sp., *E. coli.*

15.7.3.3 Third Generation Cephalosporins

Cefixim: Suprax^(R) 5-Thia-1-azabicyclo-[4,2,0]-oct-2-ene-2-carboxylic acid-7-[[(2-amino-4-thiazoyl)-[(carboxymethyoxy) imino]-acetyl] amino]-3-ethenyl-8-oxo.

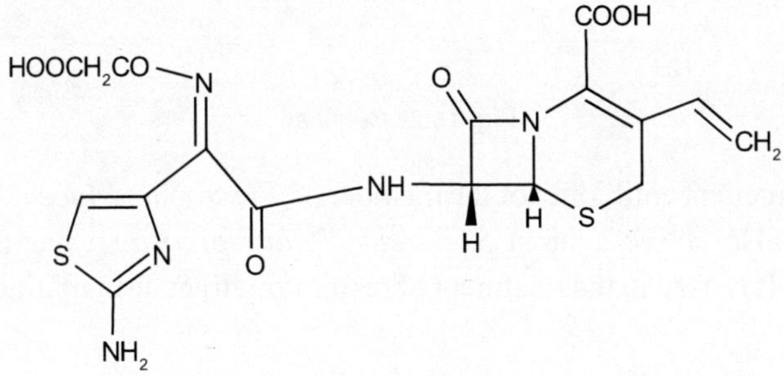

Fig. 15.26 Cefixim

Cefixim is a well known third generation cephalosporin. Cefixim has excellent therapeutic effect against *H.influenzae, Klebsiella* sp., *E. coli, N. gonorrhoeae, meningitides* as well as *β-lactamase* producing bacterial strains. It is also used in the treatment of UTI.

Ceftazidime sodium: Fortaz^(R) Pyridinium-1-[[7-[[(2-amino)-4-thiazolyl)-1-(1-carboxy-1-methylethoxy)-imino]acetyl]amino]-2-carboxy-8-oxo-5-thia-1-azbicyclo[4,2,0]-oct-2-ene-3yl], hydroxide.

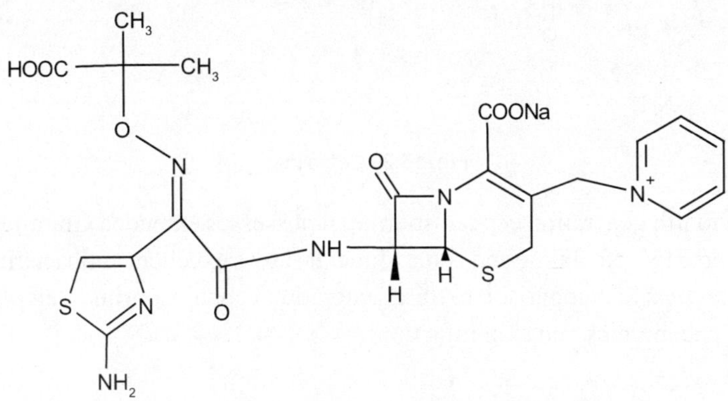

Fig. 15.27 Ceftazidime sodium

Ceftazidime sodium is specially used in the treatment of infections caused by *Pseudomonas* and *Enterobacteriaceae.* It is used in the treatment of bone and joint infections. Ceftazidime is also used in the treatment of CNS and gynaecological infections.

Ceftibuten: Cedax^(R) 5-thia-1-azbicyclo[4,2,0]-oct-2-ene-2-carboxylic acid-7-[[(2-amino)-4-thiazolyl)-4-carboxy-1-oxo-2-butenyl] amino]-8-oxo, dehydrate.

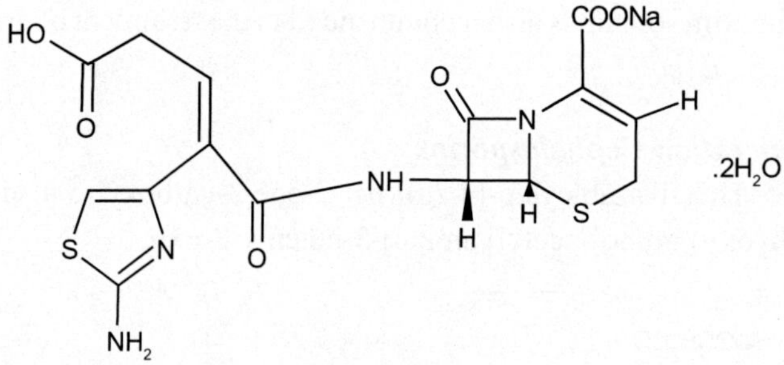

Fig. 15.28 Ceftuben

Ceftibuten is an excellent antibiotic for the member of *Enterobacteriaceae* family, *H. influenzae*, *M. catarrhalis*. It is also active against *S. aureus*, *P. aureginosa*. It has mild activity against *streptococcal* activity. It is used in the treatment of respiratory tract, urinary tract and gynaecological infections.

15.7.3.4 Fourth Generation Cephalosporins

Cefepime: Maxipime$^{(R)}$ Pyrolidinium-1-[[7-[[(2-amino)-4-thiazolyl) (methoxy-imino) acetyl]-amino-2-carboxy-8-oxo-5-thia-1-azbicyclo[4,2,0]-oct-2-ene-3yl]-methyl]-1-methyl, hydroxide.

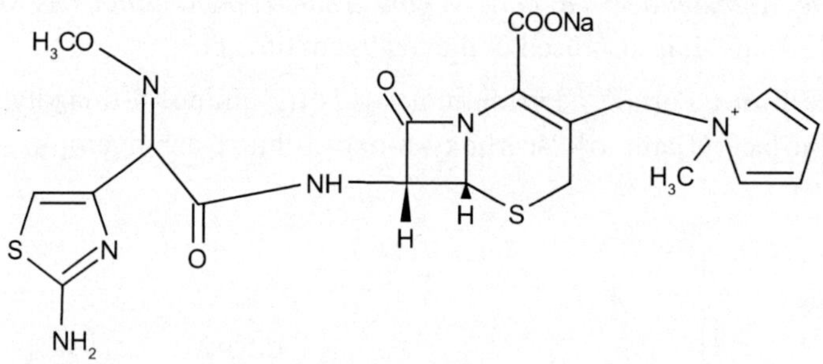

Fig. 15.29 Cefepime

Cefepime is a fourth generation cephalosporine. It possesses extended Gram-negative spectrum against *Bacilli* covered by cefotaxime and ceftazidime. It shows excellent antibacterial activity against *S. pneumoniae*, *S. aureus* in comparison to third generation cephalosporins. Cefepime is used in the treatment of UTI, pneumonias and skin infections.

15.8 AMINOGLYCOSIDES ANTIBIOTICS

In the concern of the development of streptomycin, the first antibiotic of this group, was a well-planned work of Waksman (1944) and his associates, who isolated it from a strain of Streptomyces griseus. The aminoglycoside antibiotics consist of one or more amino sugars linked to an aminocytitol ring by glycosidic bonds. Aminoglycoside antibiotics are broad-spectrum antibiotics, generally, they have greater activity against Gram-negative than Gram-positive bacteria. The aminoglycoside can produces severe adverse effects, which include nephrotoxity, ototoxicity, and neuro effects. These properties have limited use of aminoglycoside chemotherapy to serious systemic indications. Some aminoglycosides can be administered for ophthalmic and topical purposes.

Mode of action: The aminoglycosides exhibit bactericidal effects as a result of several phenomena. Ribosomal binding on 30s and 50s subunits as well as the interface produces misreading; this disturbs the normal protein synthesis. Cell membrane damage also plays an integral part in ensuring bacterial cell death.

Table 15.8: Commercially available aminoglycosides

Name	Source
Gentamycin	*Micromonospora purpurea*
Neomycin	*Streptomyces fradiae*
Streptomycin	*Streptomyces griseus*
Tobramycin	*Streptomyces tenebrarius*
Framycetin	*Streptomyces decaris*
Kanamycin	*Streptomyces kanamycetius*
Amikacin	1-L-(-)-4-amino-2-hydrobutyryl kanamycin
Capreomycin	*Streptomyces capreolus*

Toxicity: Toxicity depends on the dose of antibiotics so that the every individual could have side effects provided the dose is sufficiently high enough. Ototoxicity, nephrotoxicity, vestibular damage, hearing loss and tinnitus are the observed side effects of aminoglycosides.

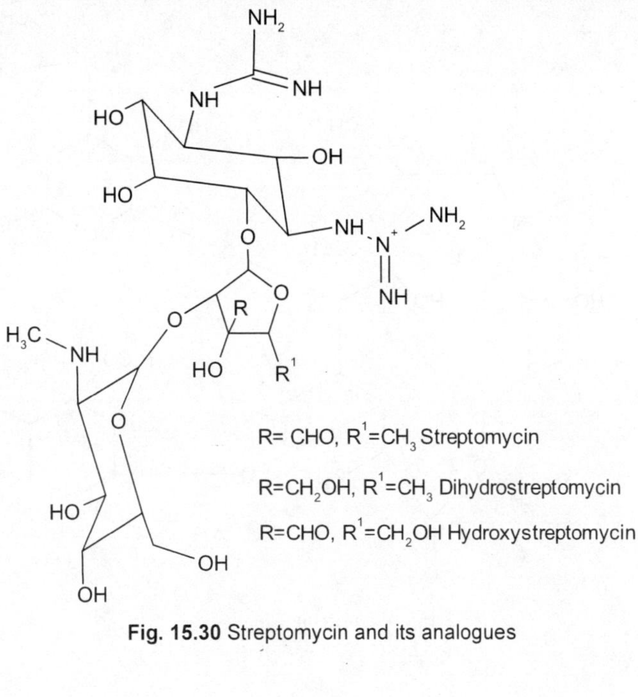

Fig. 15.30 Streptomycin and its analogues

15.8.1 Structure Activity Relationship (SAR) of Aminoglycosides

1. Aminoglycosides is a triacidic base due to the presence of two chemical scaffolds:
   a. Two strong basic guanidine moieties.
   b. Weak basic methylamino function.
2. Hydroxystreptomycin differs from streptomycin in essentially having a strategically positioned OH fuctional group in place of H atom of streptosemethyl group.
3. Manniside streptomycin (streptomycin B) consists of mannose residue attached to glycosidic linkage through OH group at C-4 position of N-methyl-L-glucosamine group.
4. In kanamycin sugar residue is linked to glycosidic oxygen at C-4 position of central deoxy streptomine

15.9 MACROLIDE ANTIBIOTICS

Macrolide antibiotics are remarkable chemotherapeutic agents for the treatment of a variety of infectious disorders caused by Gram-positive bacterial pathogens. Erythromycin is a common antibacterial agent of this category, which is generally effective against *Streptococci*, *Staphylococci*, *Chlamydia*, *Legionela* and *Mycoplasm* pathogens. Macrolide antibiotics are commonly administered for respiratory, skin, tissue and genitourinary infections.

Chemical Structure: Macrolide antibiotics have three common chemical characteristics.

1. Lactone ring (ring has 12,14 or 16 atoms with partially unsaturated)
2. Ketone group
3. Glycosidically liked amino sugar

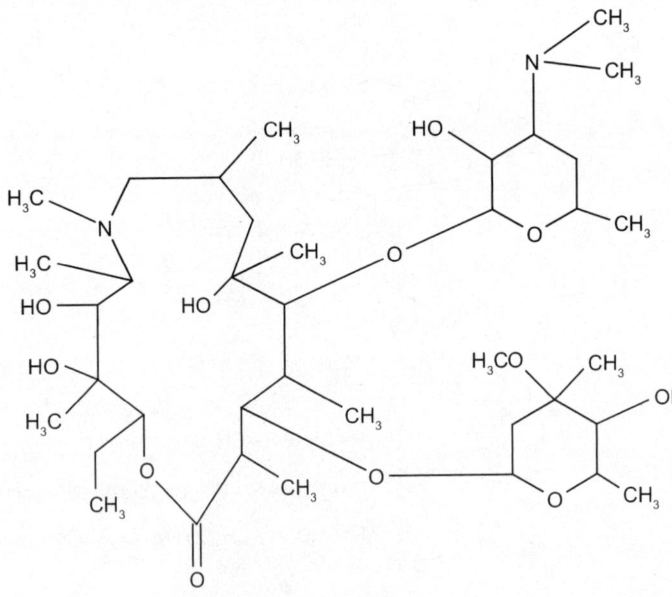

Fig. 15.31 Azithromycin

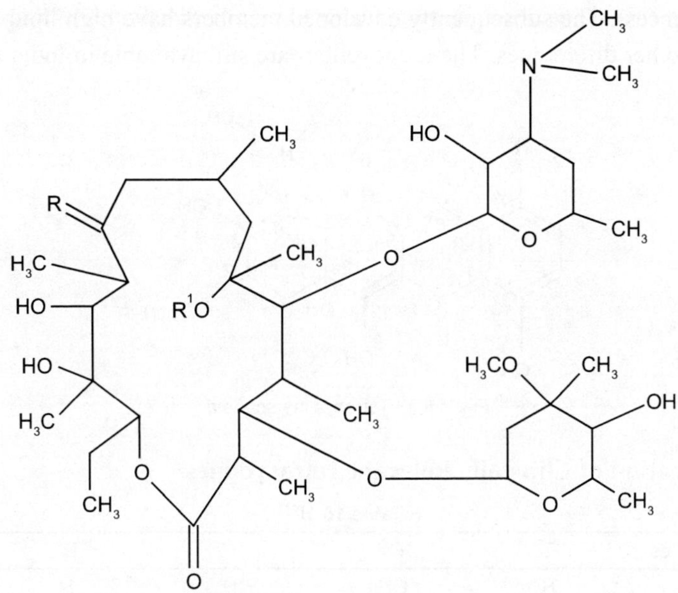

Fig. 15.32 Macrolide scaffold

Table 15.9: Commercially available macrolides

Derivatives	R	R1
Erythromycin	O	H
Roxithromycin	$CH_3OCH_2CH_2OCH_2ON=$	H
Clarithromycin	O	CH_3

Mechanism of action: Macrolide antibiotics are bacteriostatics. They inhibit bacterial protein biosynthesis, specifically; 50S ribosomal subunits are the targets of macrolides where they stop protein synthesis at different stages of peptide chain elongation thus inhibit peptidyl transferase activity. Some of macrolides inhibit translocation of the ribosome along with m-RNA template.

15.10 TETRACYCLINE ANTIBIOTICS

Tetracycline antibiotics are obtained from soil actinomycetes. The first tetracycline antibiotics was introduced as chlortetracycline in 1948 under the name Aureomycin (because of the golden yellow colour of *S. aureofaciens* colonies). These are different from penicillin and streptomycin (the other two antibiotics available at that time) in being active orally and in affecting a wide range of microorganisms, hence called broad spectrum antibiotics. Oxytetracycline and others were produced later, either from mutant strains or semisynthetically.

All tetracyclines are slightly bitter solids which are weakly water soluble, but their hydrochlorides are more soluble. Aqueous solutions are unstable. All have practically the same antimicrobial activity (with minor differences). The subsequently developed members have high lipid solubility, greater potency and some other differences. The tetracyclines are still available in India for clinical use.

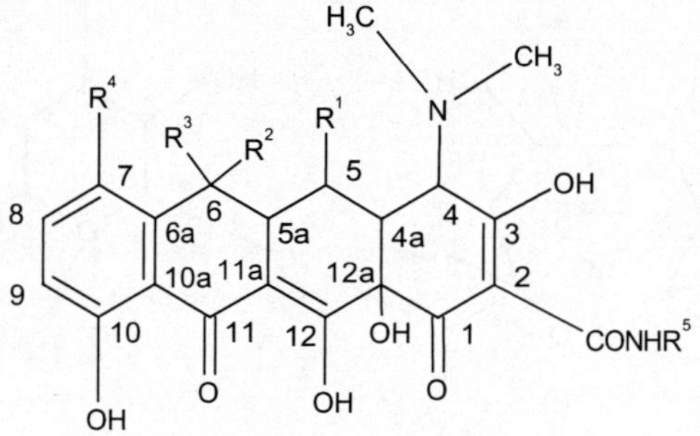

Fig. 15.33 Tetracyclins scaffold

15.10.1 Classification of Clinically Relevant Tetracyclines

Table 15.10

Tetracyclines derivatives	R^1	R^2	R^3	R^4	R^5
Tetracycline	H	OH	CH_3	H	H
Oxytetracycline	OH	OH	CH_3	H	H
Chlortetracycline	H	OH	CH_3	Cl	H
Demeclocycline	H	OH	H	Cl	H
Methacycline	OH	=	CH_3	H	H
Doxycycline	OH	H	CH_3	H	H
Rolitetracycline	H	OH	CH_3	H	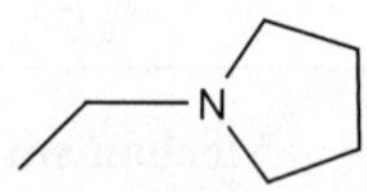
Minocycline	$-N(CH_3)_2$	H	H	H	H

15.10.2 Structural Features of Tetracyclines

A. **Effect of pH:** Tetracyclines has the ability to undergo epimerization at C-4 in the solution of intermediate pH range, called *epitetracyclines* which show less activity than natural isomers. Acidic condition influences and equilibrium set up in one day consists of equal amount of isomers.

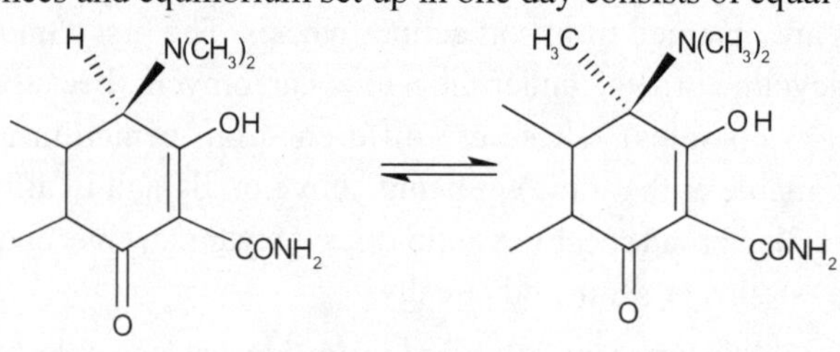

Epi Natural

Fig. 15.34

Strong acids and bases attacks on hydroxyl group at C-6 position resultant loss in antibacterial activity in the modification of C-ring. Strong acids produce dehydration, double bond, formed between C-5a and C-6 position, induce a shift of double bond between C-11a and C-12 to the position between C-11 and C-11a, producing efficient resonance of the naphthalene ring found in inactive anhydrotetracyclines. However, strong bases propagate a reaction between C-6 hydroxyl group and ketonic group at C-11 position resultant cleavage of bond between C-11 and C-11a and developing the lactone ring found in inactive isotetracyclines.

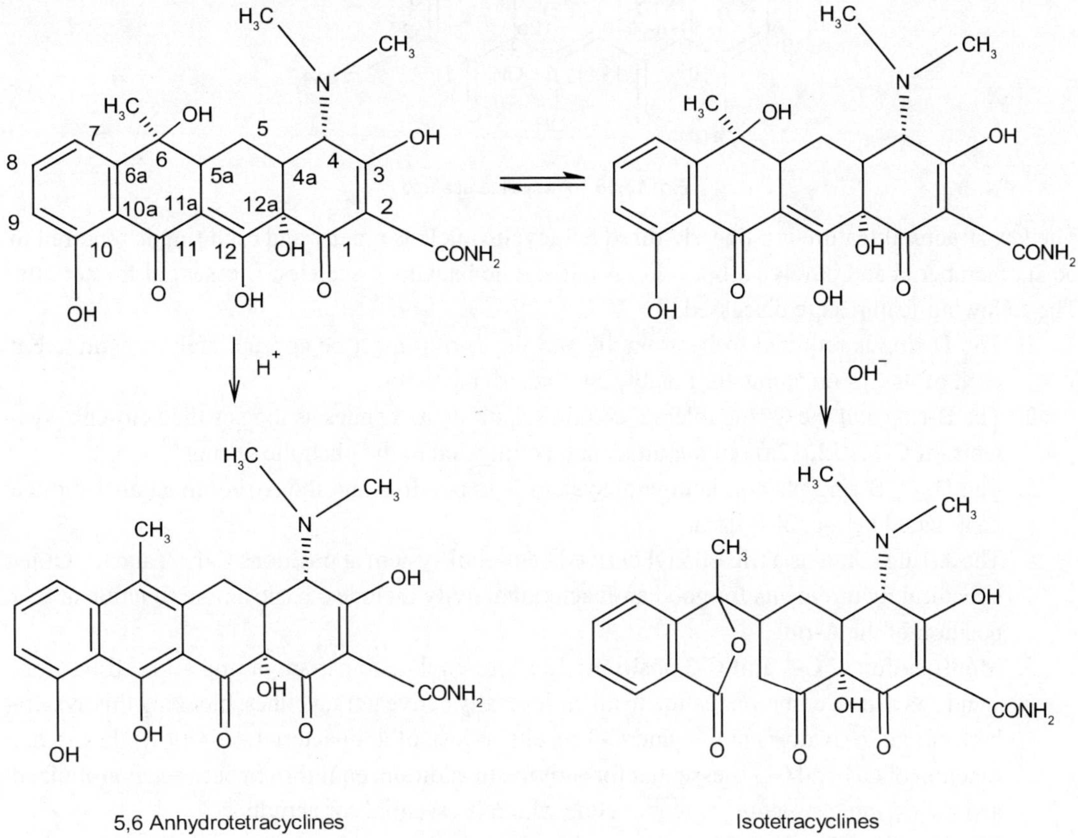

Fig. 15.35

B. Effect of metals on tetracyclines: Tetracyclines form stable chelate complexes with many metals, including calcium, magnesium, iron, etc. The chelates formation is insoluble in water accounting stop absorption of most of the tetracyclines in the presence of milk, calcium, magnesium, and aluminium containing antacids and iron salts.

The affinity of tetracyclines for calcium ions causes them to be laid down in newly formed bones and teeth as tetracycline calcium orthophosphate complexes respectively. Deposition of tetracyclines in tooth causes yellow discolouration that darkens due to photochemical reaction. Tetracyclines are distributed into the milk of lactating mothers and also across the placenta into the foetus. The possible effect of these agents on bones and teeth of the child should be taken as precaution before they are used during pregnancy or in children under eight years of age.

15.10.3 Structure Activity Relationship (SAR) of Tetracyclines

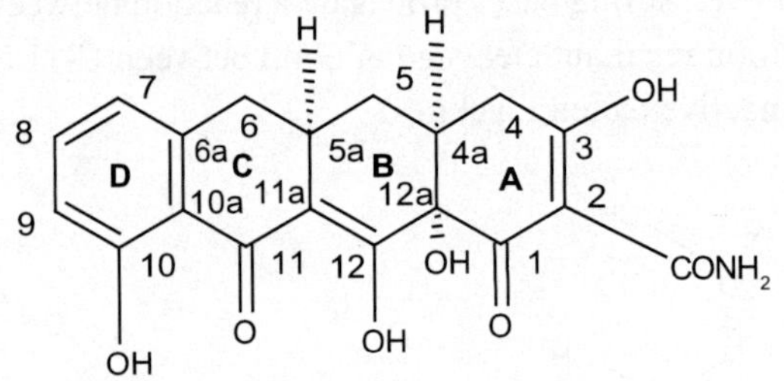

Fig. 15.36 Tetracycline scaffold

The key structural feature is a linearly fused tetracyclic nucleus moeity and each ring is required to be six membered and purely carbocyclic. A tetracyclic backbone skeleton is essential for activity. The following features are discussed.

1. The D-ring is required to be aromatic and the A-ring must be appropriately substituted at each of its carbon atoms for notable antibacterial activity.
2. The B-ring and the C-ring tolerate certain substitutent changes as long as the keto-enol systems (at C-11, 12, 12a) remain intact and conjugated to the phenolic D-ring.
3. The D, C, B-ring phenol, keto-enol system is imperative and the A-ring must also show a conjugated keto-enol system.
4. The A-ring contains a tricarbonyl derived keto-enol system at positions C-1, 2, and -3. Other structural requirements for good antibacterial activity include a basic amine function at C-4 position of the A-ring.
5. **Modification of C-1 and C-3 position:** The keto-enol tautomerism of ring A in carbon atom 1 and 3 is a most common feature to all biologically active tetracyclines, blocking this system by forming derivatives at C-1 and C-3 results in loss of antibacterial activity while C=O, a function of C-1 and C-3 is essential for activity. In addition, equilibrium between non-ionized and zwitterionic structure of tetracycline which is essential for activity.
6. **Modification of C-2 position:** The antibacterial activity resides on the carboxamide moiety. Unsubstituted amide or monosubstitution is acceptable in the form of activated alkylaminomethyl amide (Mannich bases). An example includes rolitetracycline has large alkyl group on the carboxamide that may change the normal keto-enol equilibrium of the C-1, 2, and 3 conjugated systems and diminishes inherent antibacterial activity. The replacement of carboxamide group or dehydration of carboxamide to the corresponding nitrile group, results in a loss of activity.
7. **Modification of C-4 position:** The keto-enolic character of the A-ring is due to the dimethyl amino substituent at C-4. Tremendous depletion of activity is exerted when dimethyl amino group is replaced with hydrazone oxime or hydroxyl group.

8. **Modifcation of C-4a position:** The hydrogen atom at C-4a position of tetracyclines is necessary for useful antibacterial activity.
9. **Modification of the C-5 and C-5a positions:** Alkylation of the C-5 hydroxyl group results in loss of activity. Naturally occurring antibacterial tetracyclines have an unsubstituted methylene moiety at the C-5 position. However, oxytetracycline contains C-5 hydroxyl group, was found to be a potent antibacterial compound, and has been modified chemically to some semisynthetic tetracyclines. However esterification is only acceptable if the free oxytetracycline can be liberated in vivo; only small alkyl esters are useful. Epimerization is detrimental to antibacterial activity.
10. **Modification at the C-6 position:** The C-6 methyl group attributed little to the activity of tetracycline. The C-6 position is tolerant to a variety of substituents. The majority of tetracyclines have CH_3 and OH group at this position. Demeclocyclin is a naturally occurring C-6 demethylated chlortetracycline with an excellent activity. However, removal of C-6 hydroxyl group affords doxycycline, which exerts good antibacterial activity.
11. **C-7 and C-9 substituents:** The nature of the aromatic D-ring is vulnerable C-7 position to electrophilic substitution. Substitution with electron withdrawing group such as nitro and halogen groups are introduced in some C-7 tetracyclines, which produces the most potent of all the tetracyclines in vitro, but leads to potentially toxic and carcinogenic compounds. The C-7 acetoxy, azido, and hydroxyl tetracyclines are inferior in terms of antibacterial activity.
12. **C-10 substituents:** The C-10 phenolic moiety is necessary for antibacterial activity. C-10 substitution with para or ortho hydrogen group activates the C-9 and C-7.
13. **C-11 substituents:** The C-11 carbonyl moiety is a part of one of the conjugated keto-enol systems required for antibacterial activity.
14. **C-11a substituents:** Modification at the C-11a position leads to no stable tetracyclines.
15. **C-12/12a substituents:** Esterification of the hydroxyl group leads to the incorporation of drug with the tissues due to the enhanced lipophilicity and it should undergo hydrolysis to leave the active tetracycline with hydroxyl group at 12a position, which is necessary to produce good antibacterial action. The transport and binding of these drugs depend on keto-enol system.

15.10.4 Specific Clinically Useful Tetracyclines

Oxytetracycline: 4-dimethylamino-1,4,4a,5,5a,6,11,12a-octahydro-3,5,6,10,12,12a-hexahydroxy-6-methyl-1,11-dioxo-2-naphthacencarboxamide.

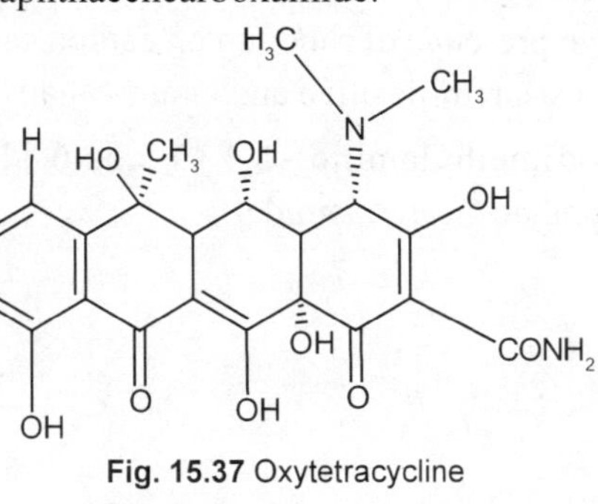

Fig. 15.37 Oxytetracycline

Oxytetracycline is obtained from fermentation using *Streptomyces rimosus*. Its free base is less bitter in taste than hydrochloride salt. Therefore, free base is preferred for oral use. oxytetracycline is indicated in the use of infections caused by Gram-positive and Gram-negative bacteria.

Doxycycline: 4-dimethylamino-1,4,4a,5,5a,6,11,12a-octahydro-3,5,10,12,12a-pentahydroxy-6-methyl-1,11-dioxo-2-naphthacencarboxamide.

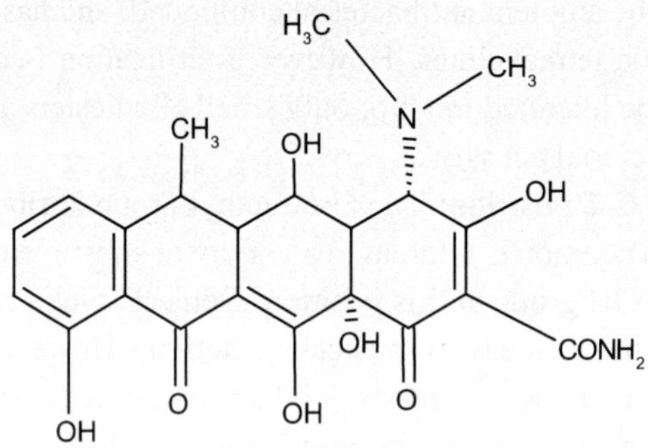

Fig. 15.38 Doxycycline

Synthesis

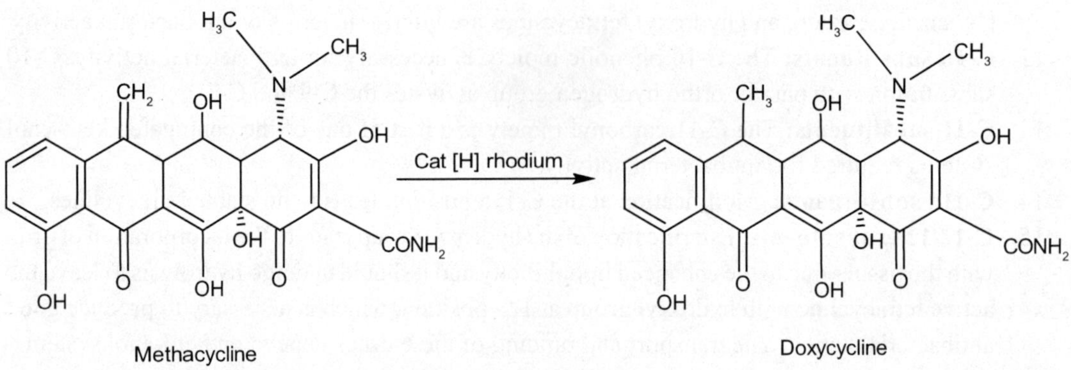

Fig. 15.39

Doxycycline is isomer of tetracycline. Structural feature of doxycycline differs only in the placement of –OH group and transferring the C-6 hydroxyl group to C-5 position. It is synthesized from reduction of methacycline in the presence of rhodium on carbon catalyst. Doxycycline is indicated in the use of infections caused by Gram-positive and Gram-negative bacteria.

Minocycline: 4,7-*bis*(4-dimethylamino)-1,4,4a,5,5a,6,11,12a-octahydro-3,10,12,12a-tetrahydroxy-1,11-dioxo-2-naphthacencarboxamide.

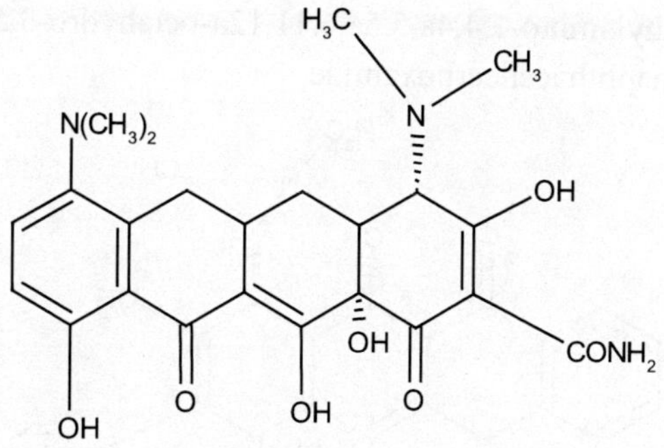

Fig. 15.40 Minocycline

Synthesis

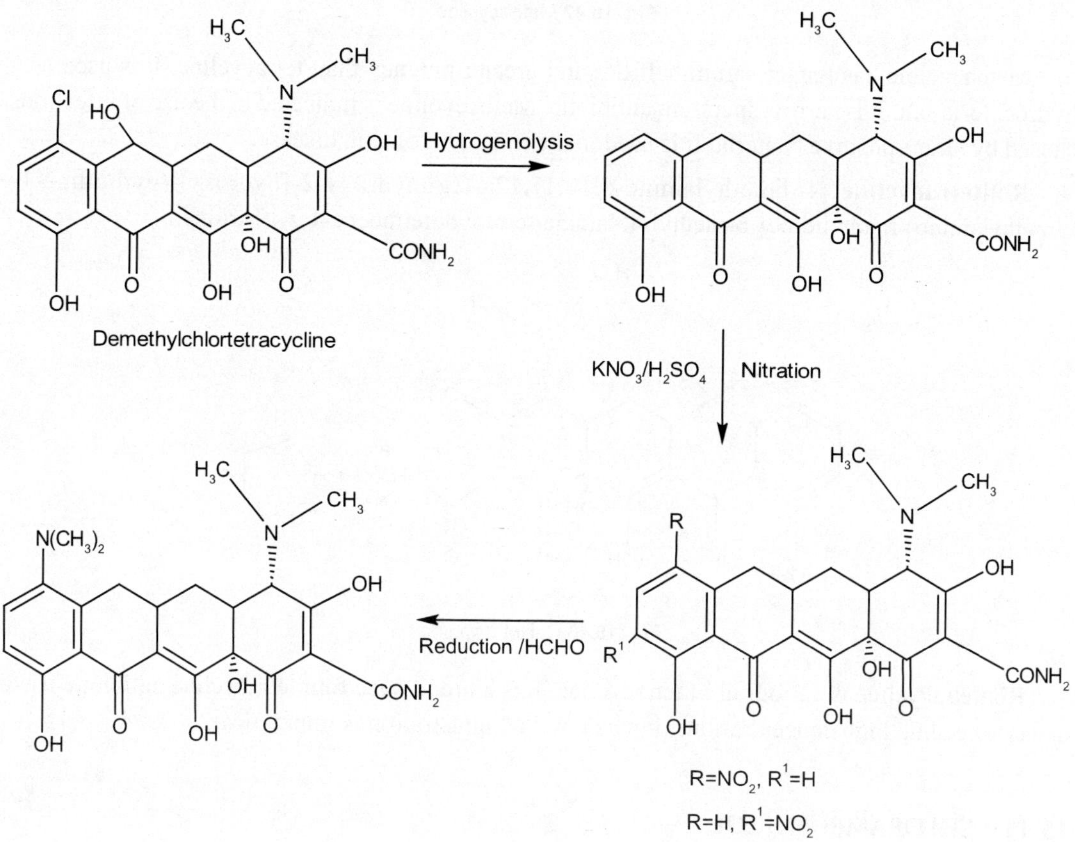

Fig. 15.41

Minocycline is an important tetracycline with appreciable potency. Minocycline lacks 6-OH group in the structure and stable to acids. It is given orally to give high plasma level. It is used in the treatment of chronic bronchitis and other respiratory infections. It is also used in the treatment of UTI.

Methacycline: 4-dimethylamino-1,4,4a,5,5a,6,11,12a-octahydro-3,5,10,12,12a-pentahydroxy-6-methylene-1,11-dioxo-2-naphthacencarboxamide.

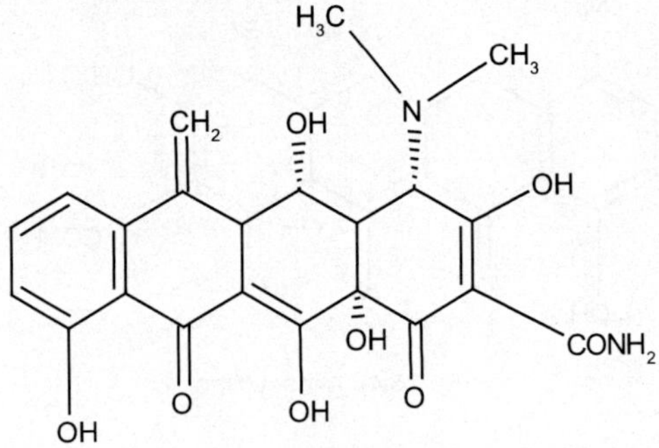

Fig. 15.42 Methacycline

Methacycline has longer serum half life and greater potency than tetracycline. It is used as a hydrochloric salt. It is narrow spectrum antibiotic. Methacycline is indicated in the use of infections caused by Gram-positive bacteria. It is used in the treatment of skin disease.

Rolitetracycline: 4-dimethylamino-6,10,11,12a-tetrahydroxy-2-[hydroxyl-(pyrolidine-1-ylmethylamino)-methylidine]-6-methyl-4,4a,5,5a-tetrahydotetracene-1,3,12-trione.

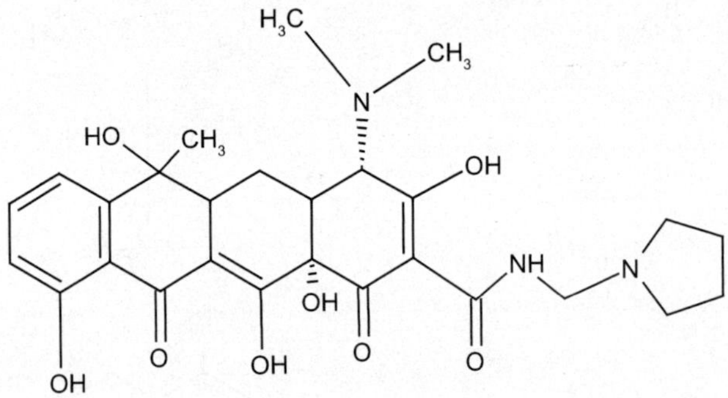

Fig. 15.43 Rolitetracycline

Rolitetracycline is a prodrug of tetracycline. It is a broad-spectrum tetracycline antibiotic used in cases needing high concentrations or when oral administration is impractical.

15.11 CHLORAMPHENICOL

Chloramphenicol is D-threo-2,2-dichloro-N-[B-hydroxy-o-(hydroxymethyl)]-N-nitrophenyl-acetamide. It was initially obtained from *Streptomyces venezuelae* in the year of 1947. Chloramphenicol was soon synthesized chemically and the commercial available products are synthetic. It is a yellowish white crystalline solid, aqueous solution is quite stable, stands boiling, but needs protection from

UV light. Chloramphenicol has a nitrobenzene substitution, which is probably responsible for the antibacterial activity and bitter taste.

Mechanism of Action: Chloramphenicol inhibits bacterial protein synthesis It specifically binds reversibly to the 50S ribosome subunit which prevents amino acid ending of t-RNA and thus may hinder the access of aminoacyl-RNA to the acceptor site for amino acid incorporation probably by acting as a peptide analogue, it prevents formation of peptide bonds. 30S ribosomal subunit which is present in mammalian cells is not disturbed.

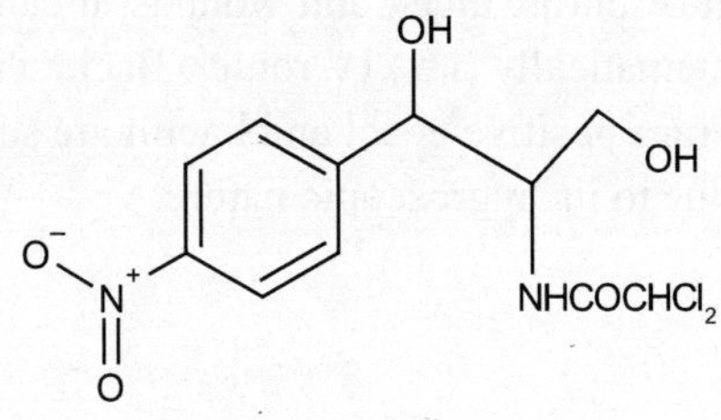

Fig. 15.44 Chloramphenicol

Synthesis:

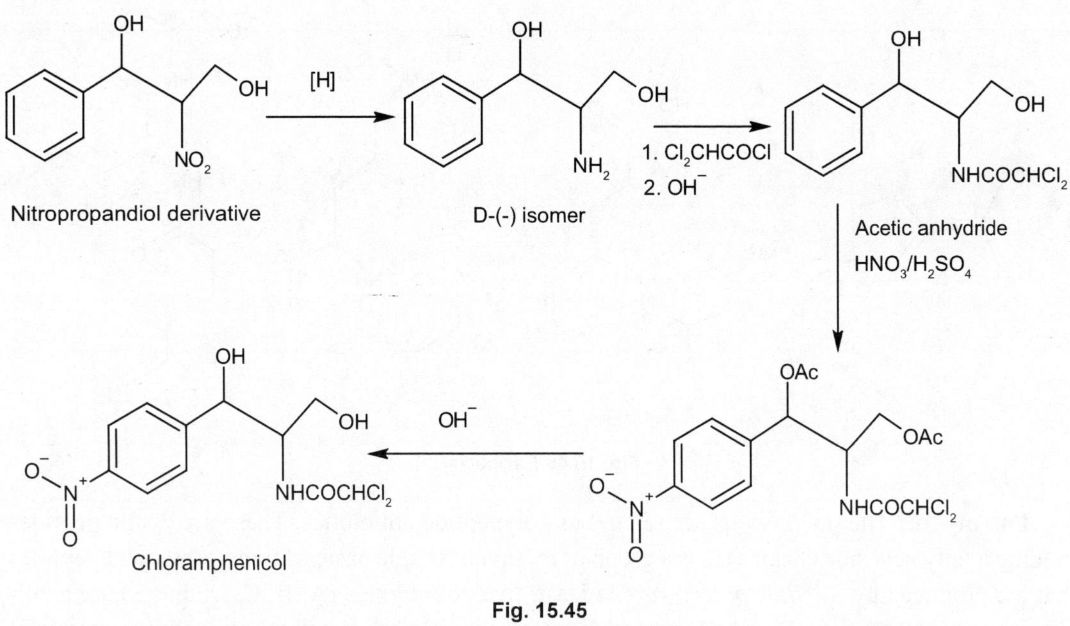

Fig. 15.45

15.12 POLYPEPTIDE ANTIBIOTICS

Polypeptide antibiotics have complex polypeptide structure. These are resistant to animal and plant proteases, respectively. These antibiotics contain lipid moieties besides amino acids that are not found in peptides of animal and plant origins. Bacitracin, Polymycin, Amphomycin, Tyrothricin, and Vancomycin are the commercially available polypeptide antibiotics.

Bacitracin: Bacitracin antibiotic is isolated from the fermentation broth of a culture of tracyl-1 strain of *Bacillus subtilis*. It is a complex mixture of at least 10 polypeptides (A, A1, B, C, D, E, F1, F2, F3, and G), of which bacitracin A fraction is believed to be the most abundant and the most potent. Presence of a divalent ion Zn++ enhances its antibacterial activity. Although bacitracin is occasionally employed for topical application (often in combination with neomycin, polymycin, and tyrothicin) for the treatment of burns, ulcer, and wounds, it can cause serious necrosis of the kidney tubules; if it is given systematically (i.e., IV route). Bacitracin does not work orally due to weak absorption. A variety of Gram-positive cocci and bacilli are sensitive to bacitracin. It should be stored in airtight containers due to its hygroscopic nature.

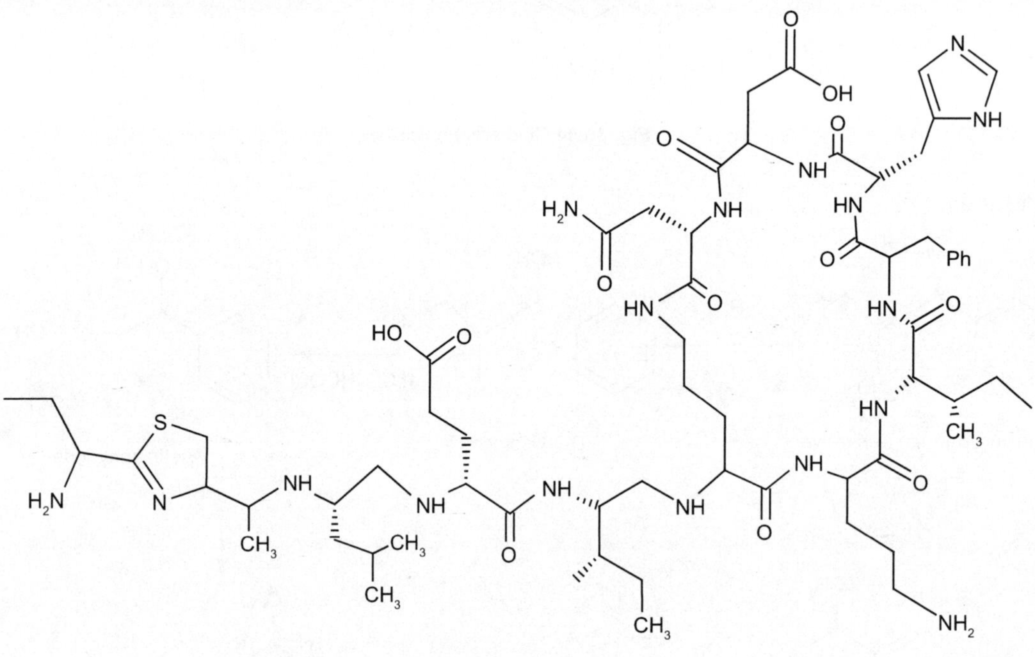

Fig. 15.46 Bacitracin

Polymyxin: The polymyxins are related to polypeptide antibiotics. These are cyclic peptides holding a fatty acid side chain. This is a group of relatively simple basic, cationic, detergent peptides that are produced by *Bacillus polymyxia*. At least, five polymyxins (A, B, C, D, and E) are available, but only polymyxin B and polymyxin E are of clinical useful antibiotics. Both polymyxin B and polymyxin E (colistin) are mixtures of two components and is used in the treatment of various bacterial infections, bacterial meningitis, urinary tract infection, burns, wounds, and gastroenteritis. On the other hand, Polymyxin may affect renal tubules and central nervous system (CNS), and because of their nephrotoxicity associated with their systemic use, they are primarily employed to treat topical infections.

Cycloserine: Cycloserine is an antibiotic obtained from *Streptomyces orchidaceus*, it is a chemical analogue of Dalanine. It inhibits bacterial cell wall synthesis by inactivating the enzymes which recemize L-alanine and link two D alanine residues. Cycloserine acts as tuberculostatic and inhibits some other Gram-positive bacteria, *E. coli*, and *Chlamydia* also.

Clindamycin: Clindamycin is a potent lincosamide antibiotic. Its mechanism of action (inhibits protein synthesis by binding to 50S ribosome) and spectrum of activity resembles to erythromycin with which it exhibits partial cross resistance. Modification of the ribosomal binding site by constitutive methylase enzyme leads resistance to both. Clindamycin inhibits most Gram-positive cocci (including penicillinase producing Staphylococcus), *C. diphtheriae, Nocardia, Actinomyces,* and *Toxoplasma,*

15.13 NEWER ANTIBIOTIC AGENTS

Newer cephalosporins

Ceftaroline: Tafralo(R)

Ceftaroline was developed by modifying the structure of the fourth-generation cephalosporin cefozopran. Ceftaroline fosamil is considered a prodrug of Ceftaroline. It is a novel broad-spectrum antibiotic effective against methicillin resistant *Staphylococcus aureus* (MRSA), penicillin and cephalosporin resistant *S. pneumoniae*, vancomycin-intermediate *S. aureus* (VISA), and vancomycin-resistant *S. aureus* (VRSA). It is also pathogenic against many Gram-negative bacteria but inactive against extended-spectrum β-lactamase producing bacteria. It has been approved in the treatment of pneumonia and other bacterial infections.

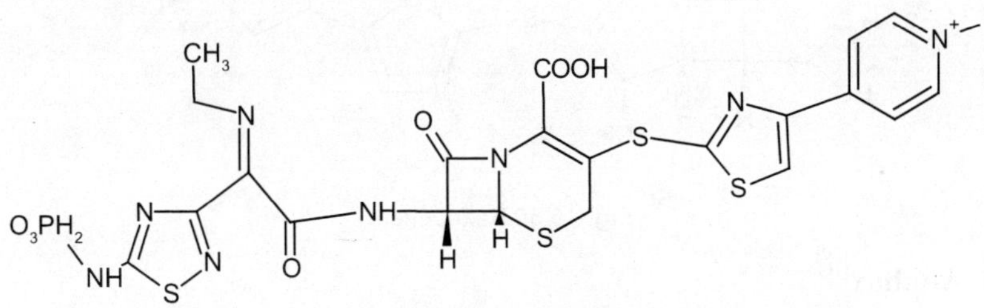

Fig. 15.47 Ceftaroline

Ceftobiprole: Zevtera(R)

Ceftobiprole is a newer cephalosporin. It is a broad-spectrum antibiotic which shows good spectrum of activity against Methicillin Resistant *Staphylococcus aureus*, penicillin-resistant *S. pneumoniae, P. aeruginosa* and *Enterococci*. Ceftobiprole shows strong affinity for of *S. pneumoniae*.

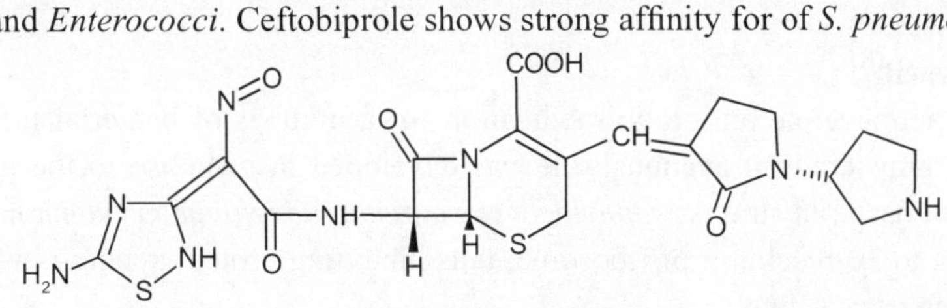

Fig. 15.48 Ceftobiprole

Newer Carbapenems

Doripenem: Doribax[(R)]

Doripenem is a newer parenteral carbapenem approved for the treatment of severe urinary tract infections and intra-abdominal infections. Doripenem has a unique spectrum of activity. It shows activity against Gram-positive cocci like imipenem and activity against Gram-negative bacilli like meropenem.

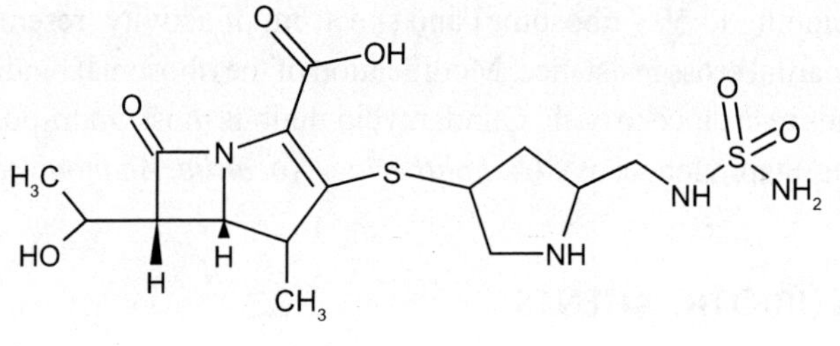

Fig. 15.49 Doripenem

Razupenem (PZ-601) [(R)]

Razupenem is a broad spectrum injectable antibiotic, similar to carbapenem subgroup of beta-lactam antibiotics. It is another novel carbapenem active against multi-drug-resistant Gram-positive and Gram-negative (ESBL producers) bacteria.

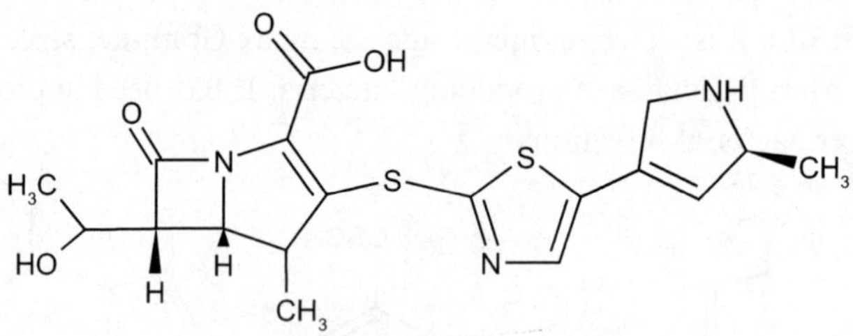

Fig. 15.50 Razupenem

Retapamulin: Altabax[(R)]

It is a novel topical antibiotic and the first approved member in this new class. It is approved for the treatment of skin and soft tissue infections caused by *S. pyogenes* and *S. aureus* which are resistant to the most commonly used topical antibiotics. It is ineffective against Gram-negative organisms. Retapamulin is a semisynthetic pleuromutilin derivative isolated from *Clitopilus scyphoides* (an edible mushroom). It is a protein synthesis inhibitor which acts by binding to 50-S subunit of bacterial ribosomes.

Tigecycline: Tygacil[(R)]

Tigecycline is a tetracycline antibiotic medication for a number of bacterial infections. It is a glycylcycline administered intravenously. It was developed in response to the growing rate of antibiotic resistant bacterial strains *Staphylococcus aureus*, *Acinetobacter baumannii*, and *E. coli*. It acts by binding to 30-S subunit of ribosome, thus inhibiting protein synthesis which is 20-fold more efficient than tetracycline.

EXERCISE

1. Classify β-lactam antibiotics with their chemical structures.
2. Write the synthesis, mechanism of action and uses of the following drugs.
   (a) Penicillin V
   (b) Cephalexin
   (c) Minocyclin
   (d) Doxycelin
3. Explain SAR of penicillins with examples and chemical structures.
4. Explain SAR of cephalosporins with examples and chemical structures.
5. Explain medicinal chemistry of tetracyclines.
6. Write a note on chemistry of chloramphenicol.
7. Write a note on aminoglycosides with examples.
8. Give an explanatory note on effect of pH on tetracyclines.
9. Explain SAR on tetracyclines with examples.

16

Quinolones Antibacterials

16.1 INTRODUCTION

Nalidixic acid is a drug which belongs to quinolones. It was developed as a treatment for urinary tract infections. Other first-generation quinolones were produced in the 1960s and 1970s. Quinolones had remarkable antibacterial spectrum that was largely restricted to *E. coli* and other *Enterobacteriaceae*. All the quinolones compounds are bactericidal, and act by inhibiting the bacterial enzymes that are responsible for coiling of DNA. Quinolones are all orally active, and mostly are available for intravenously administration. The first-generation drugs were only used for the treatment of urinary infections, but the much broader spectrum derivatives of those that followed permitted their use for respiratory, soft tissue, bone and joint, gastrointestinal and sexually transmitted infections as well.

16.2 MECHANISM OF ACTION

Quinolones/fluoroquinolones antibacterial agents are bactericidal. They inhibit DNA gyrase and topoisomerase therefore leads to irreversible conformation of bacterial DNA that is responsible for modifying the topology of DNA via supercoiling and twisting the macromolecules to permit the replication.

16.3 THERAPEUTICS APPLICATIONS

Second generation antibiotics (Ciprofloxacin, Norfloxacin) have broad spectrum activities and are used for the treatment of urinary tract infections, gonorrhea, and upper respiratory tract infections. While third generation quinolones (Levofloxacin, Gatifloxacin) are used in the treatment of chronic bronchitis, pneumonia.

16.4 CLASSIFICATION OF QUINOLONE AGENTS

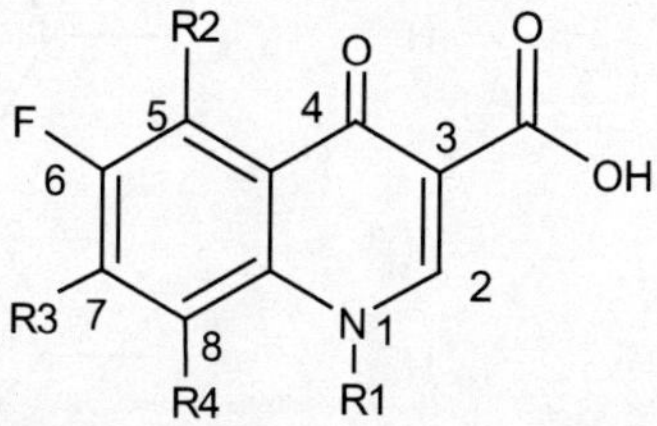

Fig. 16.1 Basic structure of quinolones

Table 16.1: Classification of clinically relevant quinolones

Quinolones	R_1	R_2	R_3	R_4
Norfloxacin	C_2H_5	H	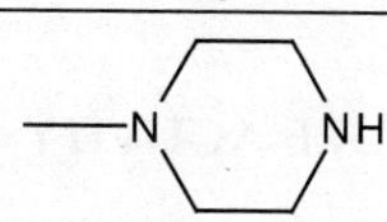	H
Ciprofloxacin	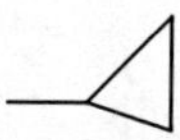	H	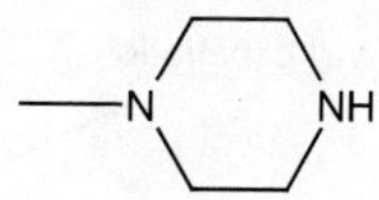	H
Lomefloxacin	C_2H_5	H	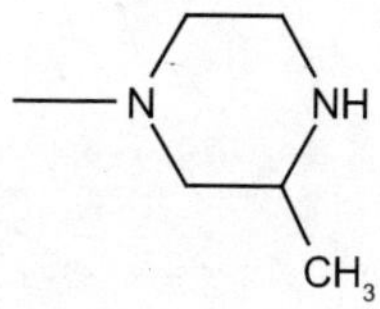	F
Pefloxacin	C_2H_5	H	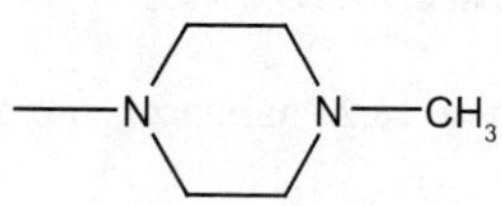	H
Gatifloxacin	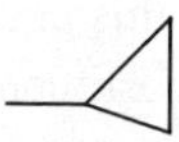	H	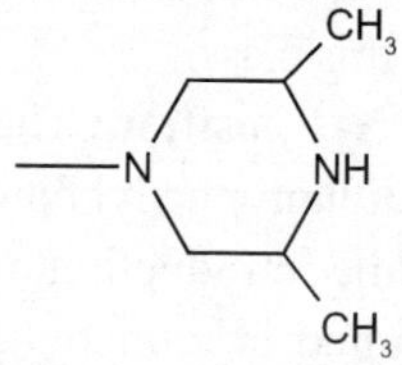	OCH_3
Sparfloxacin	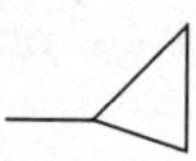	NH_2	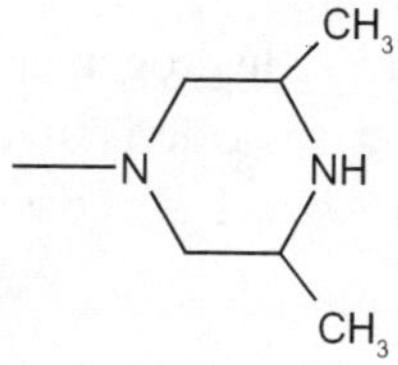	F

Ofloxacin	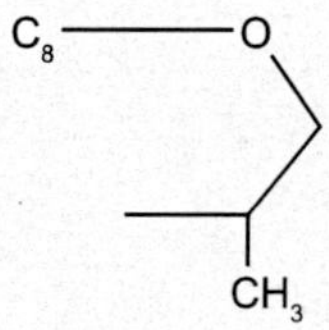	H	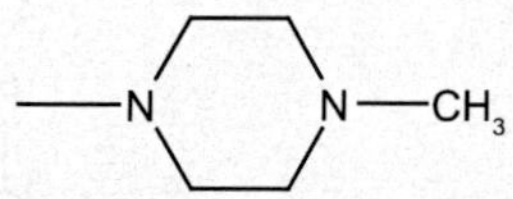	1,8 bridge
Moxifloxacin	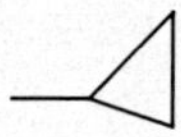	H	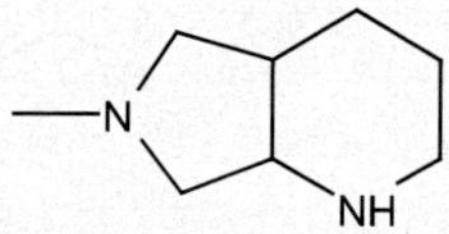	OCH_3

First generation: Nalidixic acid
Second generation: Ciprofloxacin, lomefloxacin. norfloxacin, ofloxacin, pefloxacin
Third generation: Levofloxacin, sparfloxacin, temafloxacin
Fourth generation: Moxifloxacin, gatifloxacin, sitafloxacin

16.5 SAR (STRUCTURE ACTIVITY RELATIONSHIP)

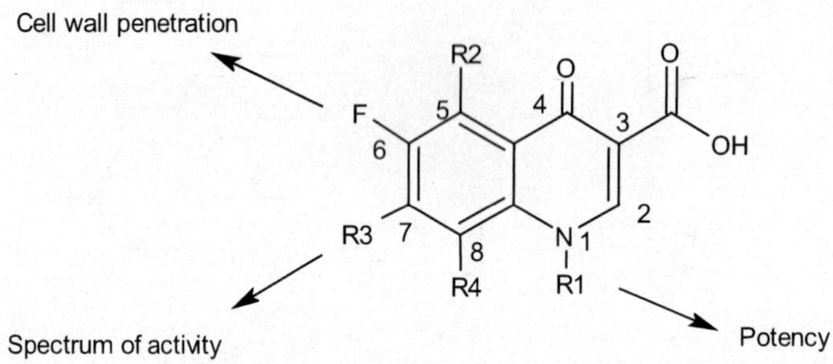

Fig. 16.2 Pharmacophore of quinolones (carboxy-4-pyridone nucleus)

The structural features of quinolones influence the antimicrobial and pharmacokinetics properties of the drugs belong to this class.

1. **Substituent at N-1 position:** The optimum substituents at this position by ethyl, butyl, cyclopropyl, and difluorophenyl functional groups and these substituents have resulted in potent derivatives. While introduction of an F atom into the N-1 cyclopropyl group or the 1-butyl substituent resulted in compounds with overall improved activity against Gram-positive bacteria.
2. Replacement of C-2 hydrogen generally disfavour for activity (e.g., C-2 methyl or hydroxy groups); however, some derivatives containing a suitable C-1, C-2 ring have shown to possess notable activity.

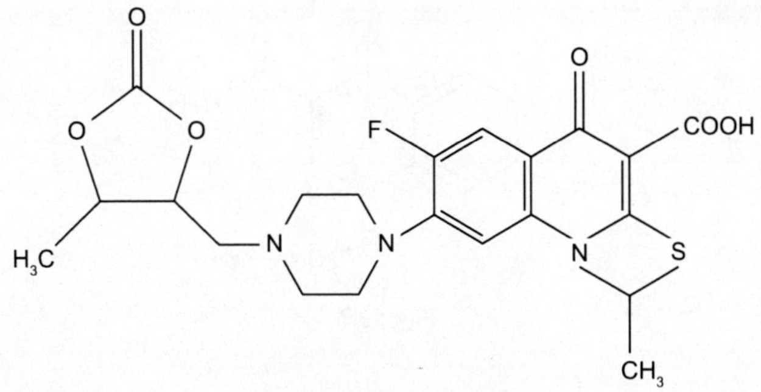

Fig. 16.3 Prulifloxacin

3. **Importance of carboxy functions at position-3:** Replacement or modification of C-3 carboxylic acid group generally leads to decrease in antibacterial activity. However, replacement of C-3 carboxylic group with isothiazolo group afforded most active isothiazolo quinolone, which has been 4–10-fold greater in vitro antibacterial activity than ciprofloxacin. The isothiazolo system possesses aromatic character and the nitrogen proton is very acidic and mimics the carboxylic group function whereas other groups, such as sulphonic acid, phosphonic acid, tetrazole as well as derivatization, as an ester lead to loss of antibacterial activity.
4. 4-oxo group at position-4 of the quinolone nucleus found to be essential for antibacterial activity while replacement with 4-thioxo or sulphonyl group leads to a loss of activity.
5. The incorporation of a group at the C-5 position has proven beneficial in terms of antibacterial activity. The order of activity is NH_2: CH_3>F, H>OH or SH, SR.
6. The incorporation of an F atom at the C-6 position of the quinolone is monumental. The order of activity is F>Cl, Br, CH_3>CN.
7. The introduction of a piperazine moiety at the C-7 position is fruitful. Other aminopyrrolidines also are compatible for activity.

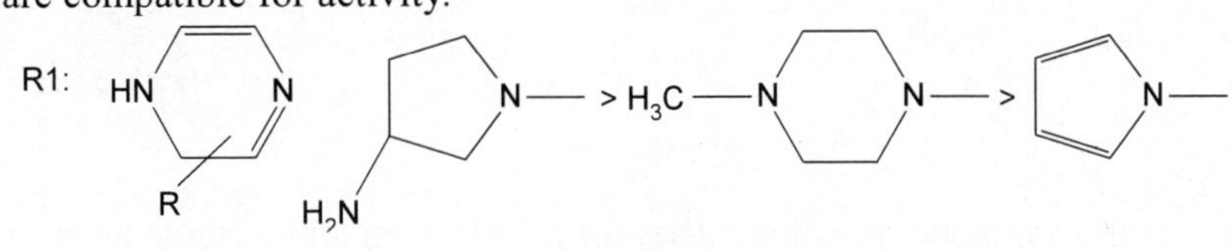

Fig. 16.4

8. In general, a C-8 fluoro substituent offers good potency against gram-negative pathogens, while a C-8 methoxy moiety offers activity against gram-positive bacteria. The order of activity is F, Cl, OCH_3>H, CF_3>methyl, vinyl, propargyl.
9. A halogen (F or Cl) at the C-8 position improves oral absorption properties.
10. Linking of N-1 group to the C-8 position with oxazine ring leads to active oflaxacin compound.

16.6 RELEVANT COMPOUNDS OF QUINOLONES

Nalidixic acid: Navigramon(R), 1-ethyl-7-methyl-4-oxo-[1,8] naphthyridine-3-carboxylic acid

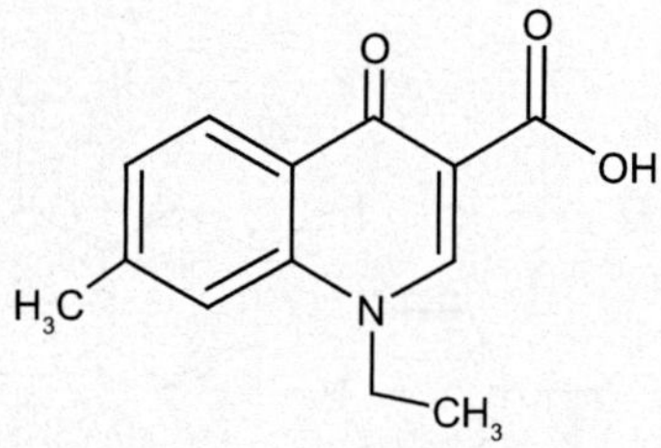

Fig. 16.5

Synthesis:

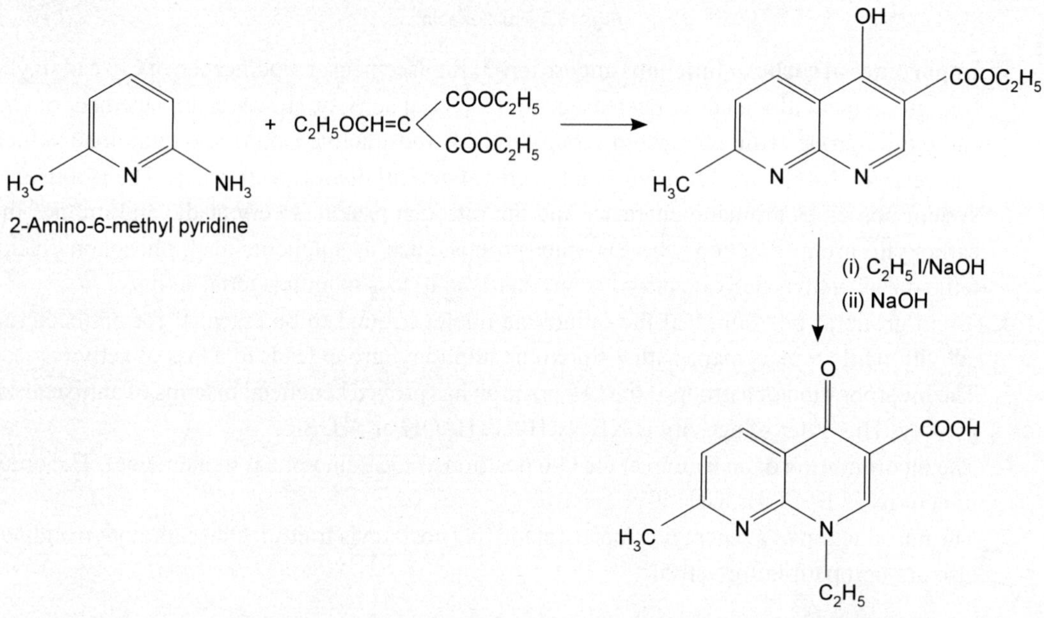

Fig. 16.6

Nalidixic acid is a synthetic quinolone derivative. This is an antibacterial agent with urinary tract antiseptic activity by interference of DNA gyrase activity thereby inhibiting DNA synthesis during bacterial replication in dose dependent manner.

Norfloxacine: Noroxin(R, 1-ethyl-6-fluoro-4-oxo-7-piperazin-1-yl-1H—quinolne-3-carboxylic acid

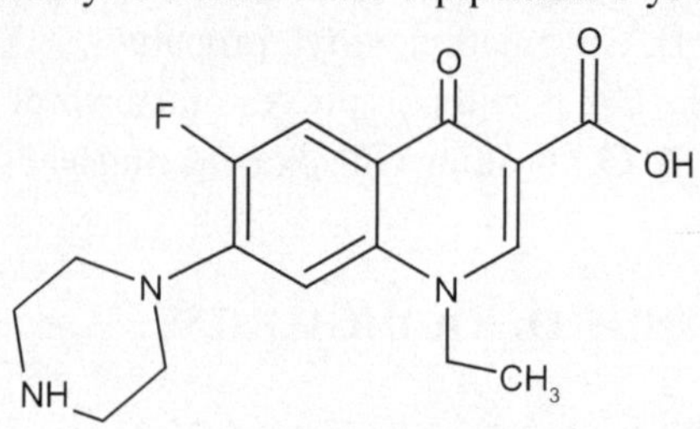

Fig. 16.7

Norfloxacin belongs to the class of fluoroquinolones antibiotics. It is frequently used in the treatment of UTI infections as well as gynaecological infections. It is also used in the treatment of gonorrhea and bladder infections.

Ciprofloxacin: Cipro(R), 1-cyclopropyl-6-fluoro-1,4-dihydro-4-oxo-7-[piperazine-1-yl]-quinoline-3-carboxylic acid

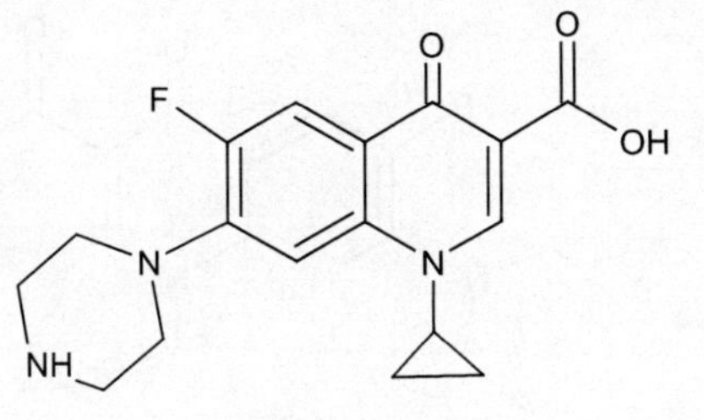

Fig. 16.8

Synthesis:

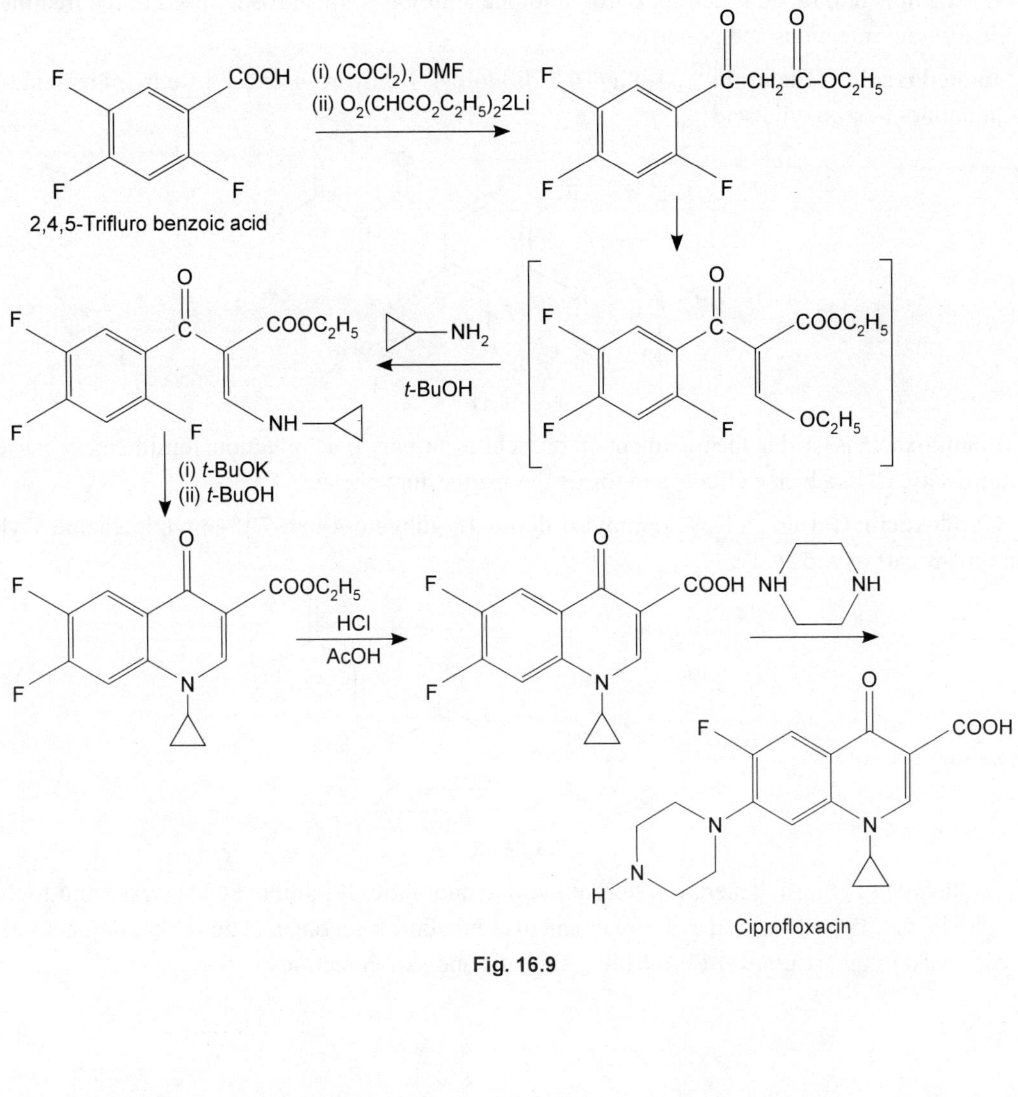

Fig. 16.9

Ciprofloxacin is listed in essential medicines of the World Health Organization. It is another derivative of fluroquinolone. Ciprofloxacin is used in several bacterial infections, includes joint infection, intra-abdominal infections, respiratory tract infections, UTI, and typhoid fever also.

Enoxacin: enroxil(R), 1-ethyl-6-fluoro-1,4-dihydro-4-oxo-7-[piperazine-1-yl]-quinoline-3-carboxylic acid

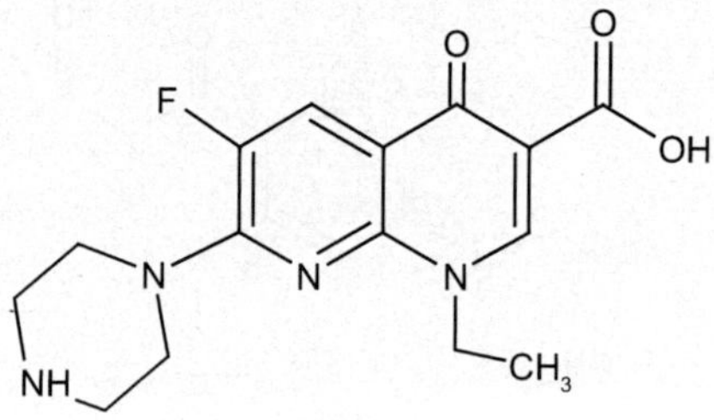

Fig. 16.10

Enoxacin is oral broad spectrum fluroquinolone antibiotics. It is broadly used in the treatment of urinary tract infections and gonorrhea.

Lomefloxacin: Maxaquin(R), 1-ethyl-6,8-difluoro-1,4-dihydro-4-oxo-7-[3-ethylpiperazine-1-yl]-quinoline-3-carboxylic acid.

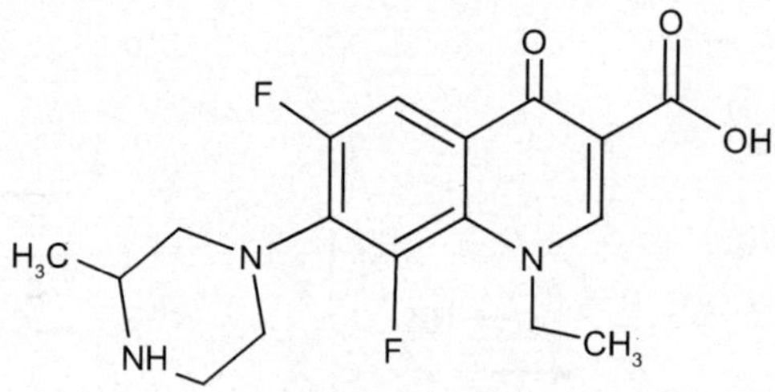

Fig. 16.11

Lomifloxacin is used in the treatment of bronchitis, urinary tract infection. lomifloxacin caused phototoxicity, CNS adverse effects and forms the magnesium chelate.

Gatifloxacin: Gatiflo(R), 1-cyclopropyl-6-fluoro-1,4-dihydro-4-oxo-7-[3-ethylpiperazine-1-yl]-quinoline-3-carboxylic acid.

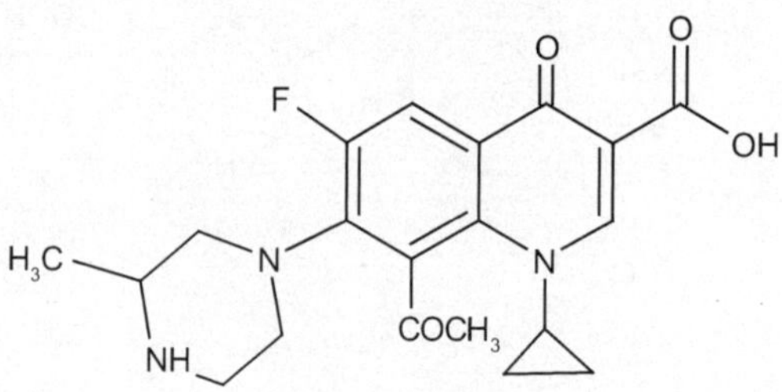

Fig. 16.12

Gatifloxacin is fourth generation fluroquinolone antibiotic. It inhibits DNA gyrase and topoisomerase IV. Gatifloxacin is used in the treatment of ophthalmic bacterial infections like conjunctivitis. It is also used in the treatment of bronchitis, sinusitis and skin infections.

Sparfloxacin: Spacin[(R)], 5-amino-1-cyclopropyl-6,8-difluoro-1,4-dihydro-4-oxo-7-[3,5-dimethylpiperazine-1-yl]-quinoline-3-carboxylic acid.

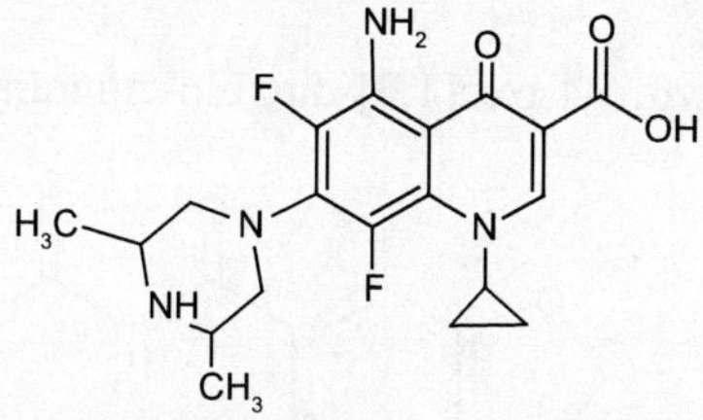

Fig. 16.13

Sparfloxacin is basically used in the treatment of commonly acquired lower repiratory tract infections like bronchitis, sinusitis and commonly acquired pneumonia.

Ofloxacin: Ocuflox[(R)], 7-fluoro-2-methyl-6-(4-methylpiperazin-1-yl)-10-oxo-4-oxa-1-azatri-cyclo-trideca-5,6,8,11-tetraene-11-carboxylic acid

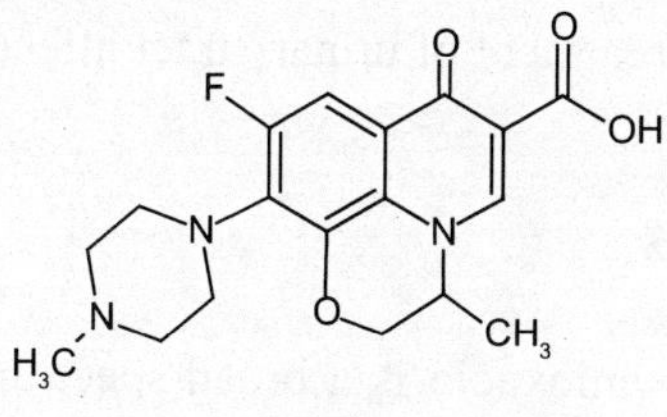

Fig. 16.14

Synthesis:

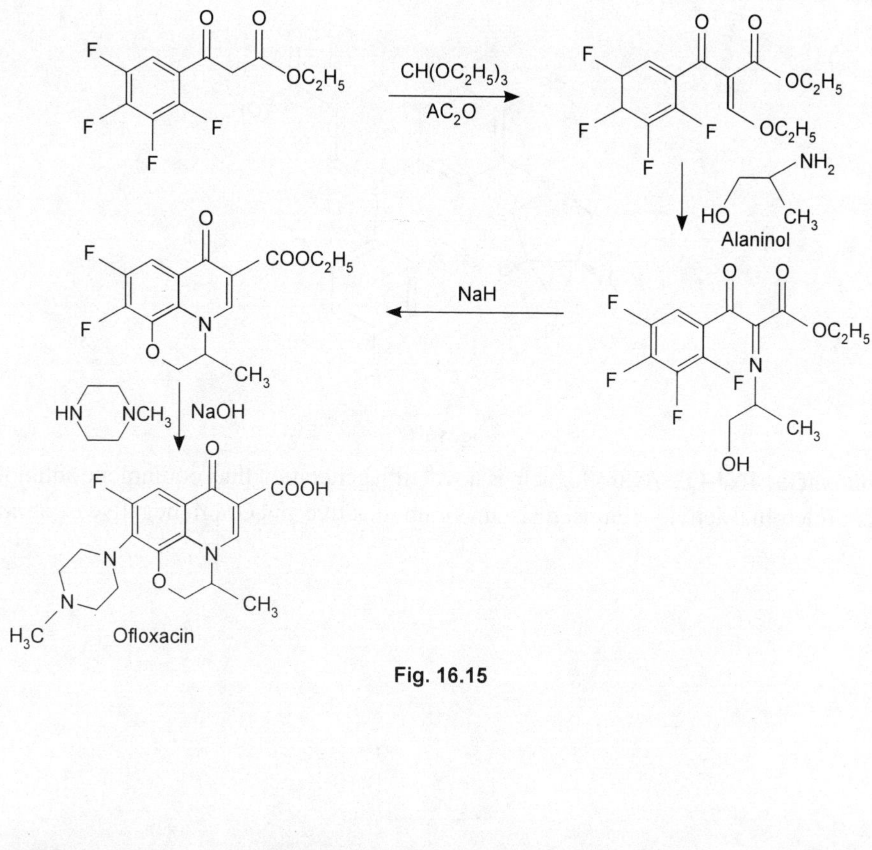

Fig. 16.15

Ofloxacin is common quinolone antibiotic and is used in the treatment of various bacterial infections such as urinary tract infection, respiratory tract infections, prostatitis, plague, eye infections, ear infections, etc.

Cinoxacin: 1-ethyl-1,4-dihydro-4-oxo[1,3]-dioxolo-cinnoline-3-carboxylic acid.

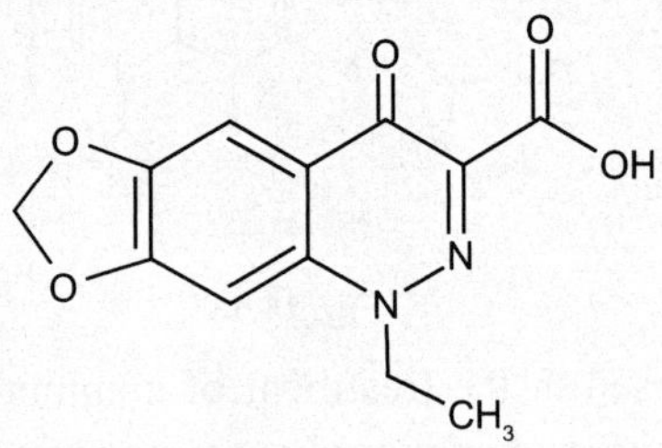

Fig. 16.16

Cinnoxacin is an older synthetic antimicrobial agent with similar oxolinic acid and nalidixic acid activity. It was used in the treatment of urinary tract infections.

16.7 NEWER COMPOUNDS

Trovafloxacin: Trovan^(R)^: Trovafloxacin is a broad spectrum antibiotic. It was withdrawn due to severe hepatotoxicity but it shows excellent antibacterial activity against Gram-positive and Gram-negatives bacterial strains.

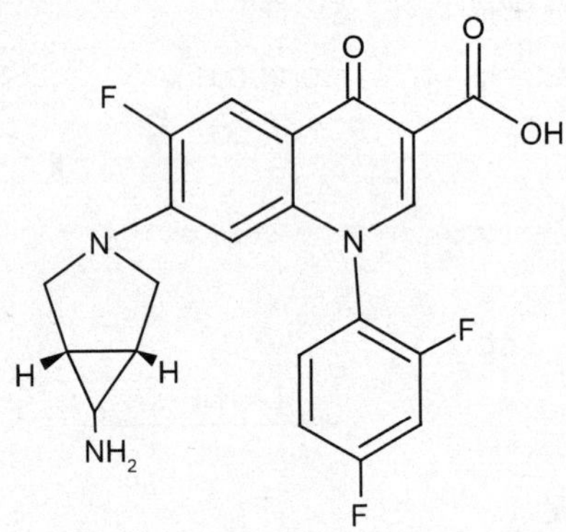

Fig. 16.17

Avarofloxacin: JNJ-Q2: Avarofloxacin is novel fifth generation fluoroquinolone antibiotic. It has excellent antimicrobial activity against resistant Gram -positive and Gram- negative bacterial strains.

EXERCISE

1. Classify quinolone antibiotics with suitable structures.
2. Give chemistry of fluoroquinolones.
3. Explain SAR of quinolones with suitable examples.
4. Write synthesis, MOA and uses of the following drugs.
   (a) Ofloxacin
   (b) Nalidixic acid
   (c) Ciprofloxacin

17
Sulphonamides

17.1 INTRODUCTION

Sulphonamides were the first antimicrobial agents effective against pyogenic bacterial infections. Girhard Domagk discovered Sulphonamido-chrysoidine also referred to as prontosil rubrum (red dye), to treat experimental streptococcal infection in mice and found it to be highly effective. He subsequently cured his daughter of streptococcal septicaemia by prontosil. By 1937, a large number of Sulphonamides were produced and used extensively in the subsequent years, but because of rapid emergence of bacterial resistance and the availability of many safer and more effective antibiotics, their current utility is limited, except in combination with trimethoprim (as cotrimoxazole) or pyrimethamine (for malaria).

All Sulphonamides considered to be derivatives of sulphanilamide (p-aminobenzene sulphonamide). Individual members differ in the nature of N_1 (Sulphonamido N) substitution, which governs solubility, potency and pharmacokinetic property. A free amino group in the para position (N_4) is required for antibacterial activity.

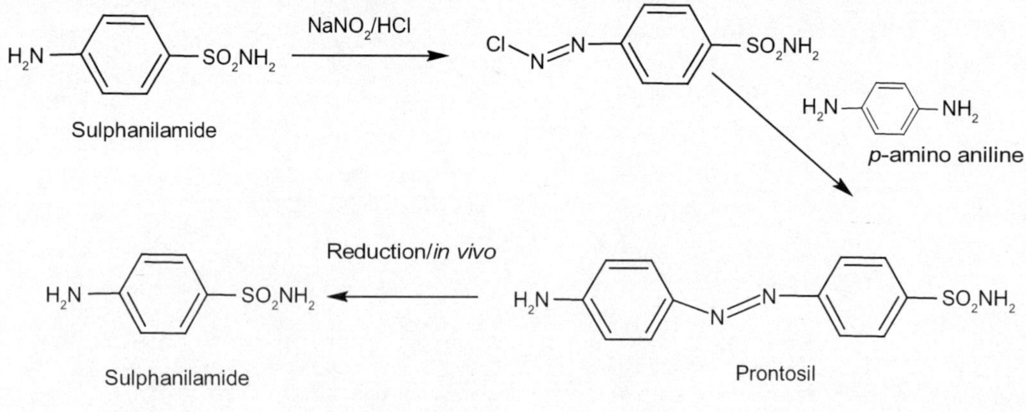

Fig. 17.1 Synthesis of prontosil

17.2 THERAPEUTIC APPLICATIONS

Sulphonamides are primarily bacteriostatic against many Gram-positive and Gram-negative bacteria starins *Strept. pyogenes, Haemophilus influenzae, H. ducreyi, Calymmatobacterium granulomatis, Vibrio cholerae*. Only a few *Staphalococcus aureus, gonococci, meningococci, pneumococci, Escherichia coli,* and *Shigella.*

17.3 MECHANISM OF ACTION

Many bacteria synthesize their own folic acid (FA) of which p-aminobenzoic acid (PABA) is a constituent, and is taken up from the medium. Sulphonamides inhibit bacterial folate synthase therefore, FA is not formed and a number of essential metabolic reactions suffer. Sulphonamides competitively inhibit the union of PABA with pteridine residue to form dihydropteroic acid which conjugates with glutamic acid and to produce dihydrofolic acid. However, Pus and tissue extracts contain purines and thymidine which decrease bacterial requirement for FA and antagonize sulphonamide action.

17.4 SIDE EFFECTS

Allergic reactions are most common adverse effects which takes the form of rashes, photosensitivity and fever while less common effects are kidney and liver damage, haemolytic anaemia and other blood related problems. The most serious side effect is Stevens-Jonhnson syndrome characterized by ulceration of mucous membranes of eyes, mouth and urethra.

17.5 CLASSIFICATION OF SULPHONAMIDES

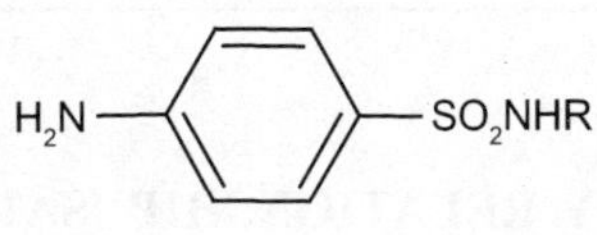

Fig. 17.2 General structure of sulphonamide

Table 17.1: Clinically useful sulphonamides

Sulphonamides	Substitution R
Sulphanilamide	H
Sulphapyridine	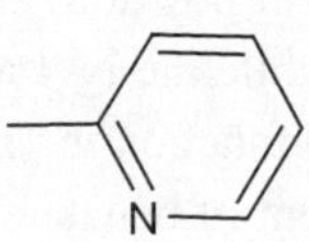

Sulphathiazole

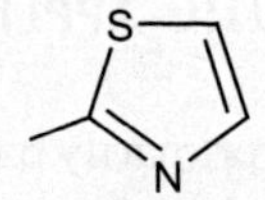

Sulphadiazine

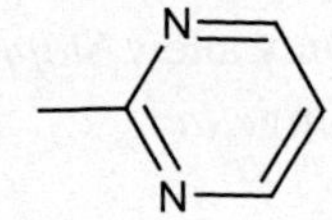

Sulphamerazine

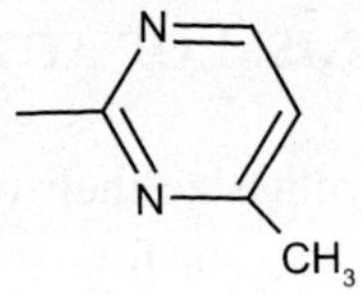

Sulphadimidine

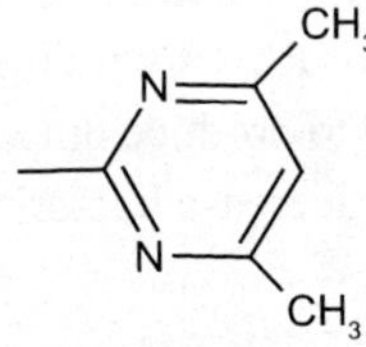

Sulphalene

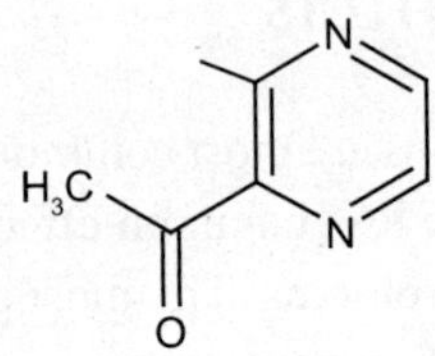

Sulphamethiazole

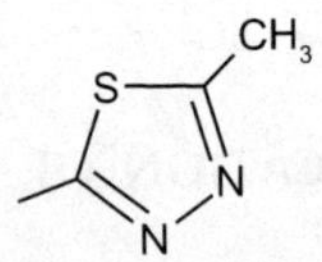

17.6 STRUCTURE ACTIVITY RELATIONSHIP (SAR)

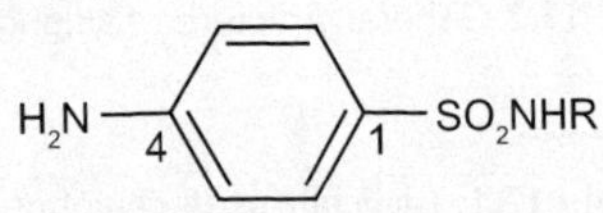

Fig. 17.3

1. Sulphanilamide lead molecule is the minimum structural requirement for antibacterial activity.
2. The amino (1 position) and sulphonyl (4 position) functional groups on the benzene ring are essential while the N-4 amino group could be modified (prodrugs) which are converted to free amino function in vivo.
3. Replacement of benzene ring by other ring systems or additional substituents resultant decreases or abolishes antibacterial activity.

4. Replacement of the $-SO_2NH$ group by –CONH reduces the activity.
5. On N-1-substituted sulphonamides, activity varies with the nature of the substituent at the amino group. With substituents imparting electron-rich characters to SO_2 group, consequently increases in bacteriostatic activity.
6. Heterocyclic substituents lead to highly potent derivatives, while sulphonamides, which contain a single benzene ring at N-1 position, are considerably more toxic than heterocyclic ring analogues.
7. The free aromatic amino groups should reside para to the sulphonamide group. Its replacement at ortho or meta position results in compounds devoid of antibacterial activity.
8. Ionization of sulphonamide is active at pKa values 6.6–7.4.
9. Substitutions in the benzene ring of sulphonamides produced inactive compounds.
10. Substitution of free sulphonic acid ($-SO_3H$) group for sulphonamido function intolerates the activity, but replacement by a sulphinic acid group ($-SO_2H$) and acetylation of N-4 position retains the activity.
11. Sulphonamides bind to the basic centres of arginine, histidine, and lysine sites of proteins. The binding groups are alkyl, alkoxy, and halides. The binding affects the activity of sulphonamides; protein binding appears to modulate the availability of the drug and its half-life.
12. The lipid solubility influences the pharmacokinetic and antibacterial activity therefore increases the half-life and antibacterial activity in vitro.

17.7 CLINICALLY USEFUL SULPHONAMIDES

Sulphisoxazole: Gantrisin(R), N-(3,4-Dimethyl-5-isoxazolyl) sulphanilamide

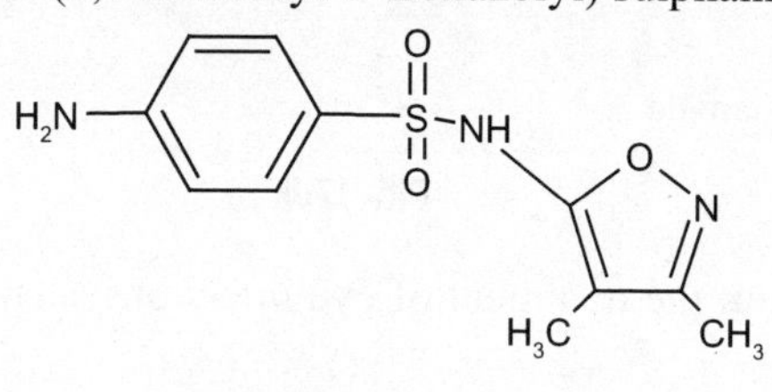

Fig. 17.4

Sulphisoxazole is also known as sulphafurazole. Sulphafurazole is a sulphonamide antibacterial with dimethyl-isoxazole substituent. It possesses antibiotic activity against a wide range of Gram-negative and Gram-positive organisms.

Sulphamethoxazole: Bactrim(R), *N1*-(5-Methyl-3-isoxazolyl)-Sulphanilamide

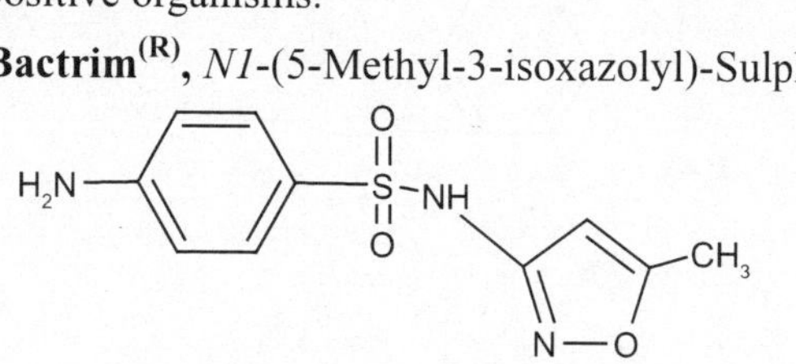

Fig. 17.5

Synthesis:

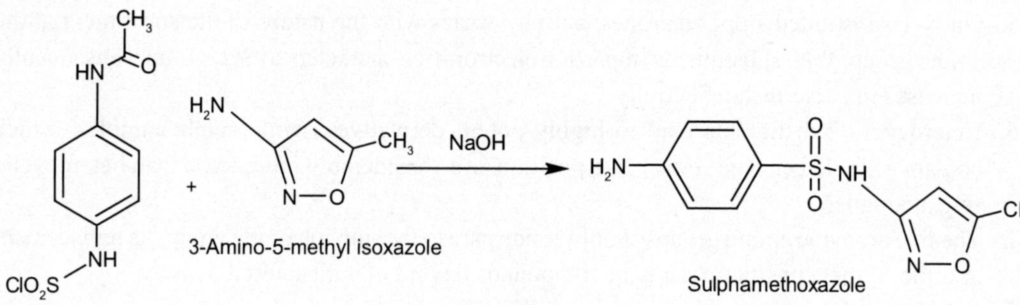

Fig. 17.6

Sulphamethoxazole is antibacterial drug and used in the combination with trimethoprime. It is used to treat a number of bacterial infection such as respiratory, intestinal and urinary tract infections.

Sulphacetamide: Ovace[(R)], *N*-Acetylsulphacetamide

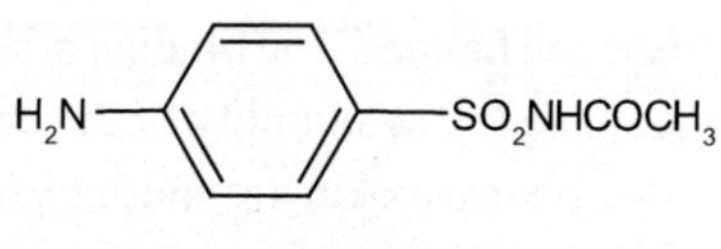

Fig. 17.7

Synthesis:

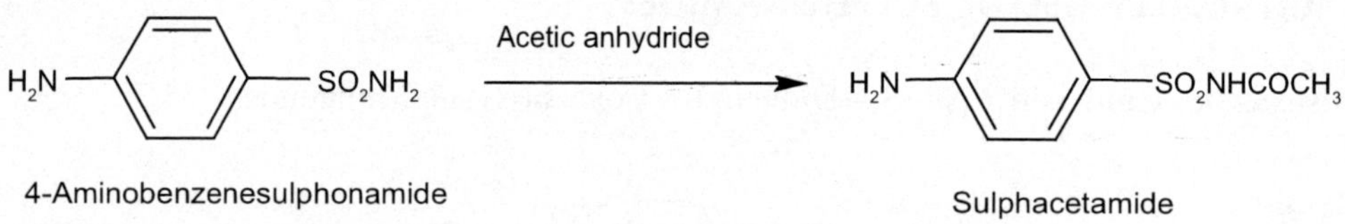

Fig. 17.8

Sulphacetamide is used in the treatment of eye infections such as conjunctivitis, keratitis and blepharitis.

Sulphadiazine: Silvadene[(R)], N-2-pyrimidyl sulphanilamide

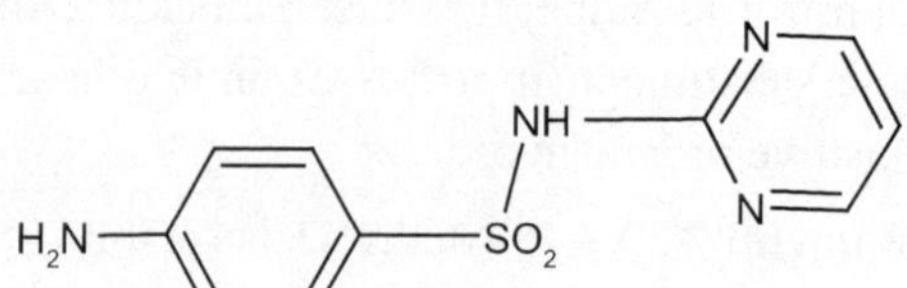

Fig. 17.9

Synthesis:

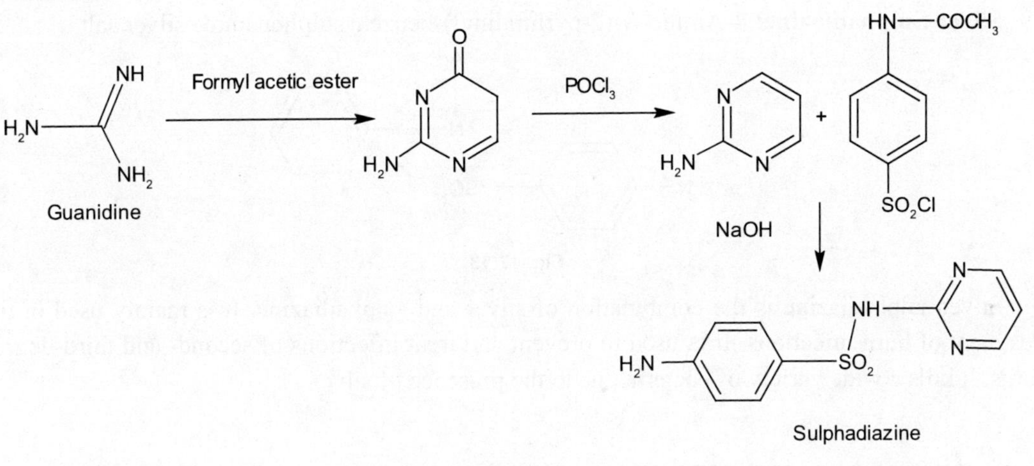

Fig. 17.10

Sulphadiazine is used in the treatment of UTI, ear infections, meningitis, malaria and other infections.

Suphasalazine: Azulfidine[(R)], 2-Hydroxy-5-[[4-[(2-pyridinyl amino) sulphamoyl] phenyl] azo] benzoic acid

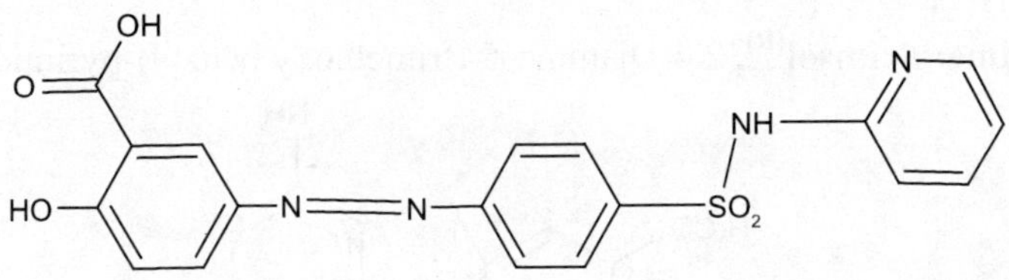

Fig. 17.11

Synthesis:

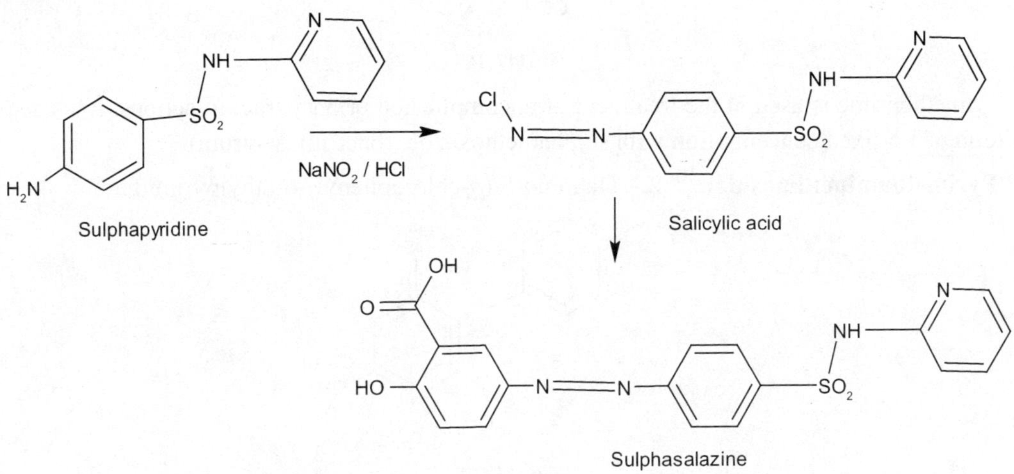

Fig. 17.12

Sulphasalazine is used in the treatment of bowl inflammation, diarrhoea, ulcerative colitis, etc.

Silver-Sulphadiazine: 4-Amino-*N*-(2-pyrimidinyl)-benzenesulphonamide silver salt

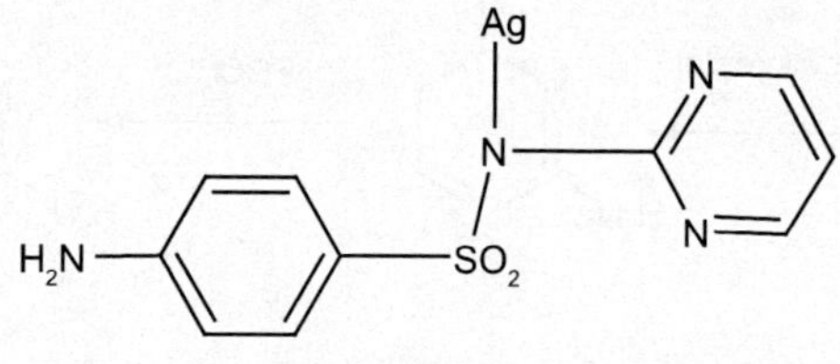

Fig. 17.13

Silver-sulphadiazine is the combination of silver and sulphadiazine. It is mainly used in the treatment of burn infections. It is used to prevent and treat infections of second- and third-degree burns. It kills a wide variety of bacteria due to the presence of silver.

17.8 DIHYDROFOLATE REDUCTASE (DHFR) INHIBITORS

These inhibitors are used to prevent tetra-hydrofolic acid biosynthesis results in bacteriostatics. tetra-hydrofolic acid is an important co-factor essential for supplying cabon unit in thymidine biosynthesis for DNA synthesis. Trimethoprime and pyrimethamine are the examples of dihyrofolate reductase inhibitors.

Trimethoprime: Primsol[(R)], 2,4-Diamino-5-(trimethoxy benzyl)-pyrimidine

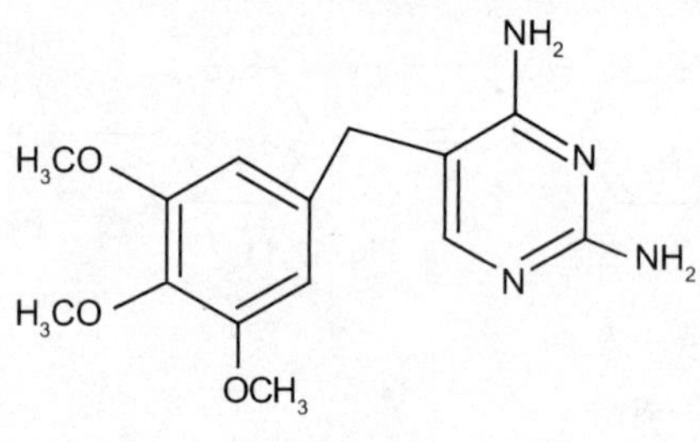

Fig. 17.14

Trimethoprime is used in the treatment of uncomplicated urinary tract infactions. It is used in the ratio of 1:5 fixed concentration with sulphamethoxazole (bactrim, septrum).

Pyrimethamine: Fansidar[(R)], 2,4-Diamino-5-(*p*-chlorophenyl)-6-ethylpyrimidine

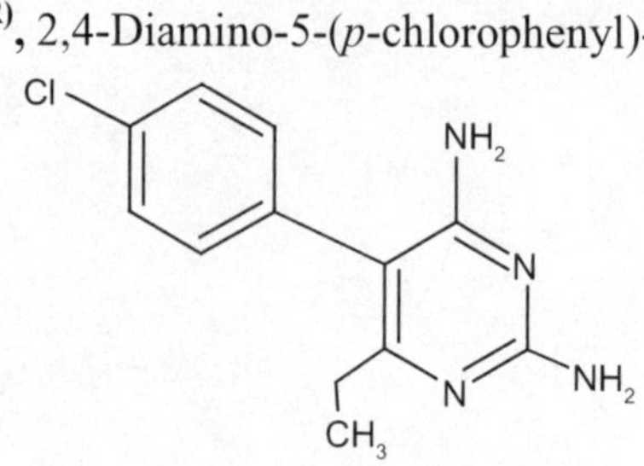

Fig. 17.15

Pyrimethamine is used in the treatment of malaria in the combination of sulphadoxim.

17.9 NEWER AGENTS

Pazopanib: 5-[[4-[(2,3-dimethylindazol-6-yl)-methylamino]pyrimidin-2-yl]amino]-2-methylbenzenesulphonamide.

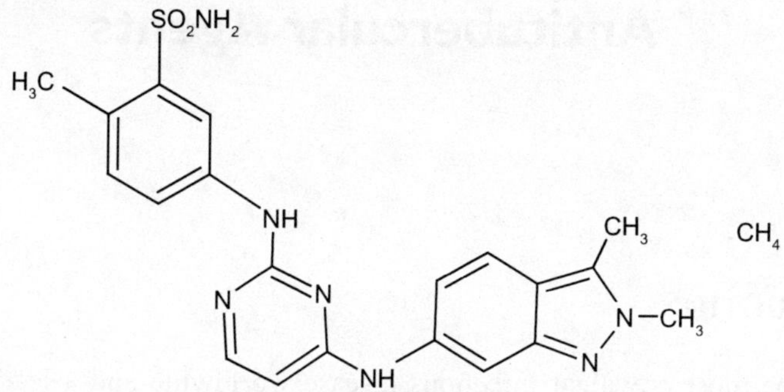

Fig. 17.16

Pazopanib (Votrient), is an anti-cancer medication drug. It is a potent and selective multi-targeted receptor tyrosine kinase inhibitor that blocks tumour growth and inhibits angiogenesis. It has been approved for renal cell carcinoma and soft tissue sarcoma by numerous regulatory administrations worldwide.

EXERCISE

1. Classify sulphonamides with suitable structures.
2. Write the synthesis, MOA and uses of sulphamethoxazole and suphadiazine.
3. Explain structure activity relationship of sulphadrugs.
4. Explain mechanism of action of sulphonamides.

18

Antitubercular Agents

18.1 INTRODUCTION

Tuberculosis is the most prevalent infectious disease worldwide and a leading killer caused by *Mycobacterium tuberculosis*. According to the World Health Organization (WHO) report, billions of peoples worldwide are tubercular patients, with 30 million new cases reported every year. This intracellular infection accounts for at least 3 million deaths annually. Common infection sites of the tuberculosis are lungs (primary site), brain, bone, liver, and kidney. The main symptoms are cough, tachycardia, cyanosis, and respiratory failure.

Drugs used in the treatment of tuberculosis can be divided into two major categories.

1. **First-line drugs:** Isoniazid, streptomycin, rifampicin, ethambutol, and pyrazinamide.
2. **Second-line drugs:** Ethionamide, *p*-amino salicylic acid, ofloxacin, ciprofloxacin, cycloserine, amikacin, kanamycin, viomycin, and capreomycin.

Anti-tubercular medications are FDA approved for the treatment of Mycobacterium tuberculosis infections. The combination and duration on which medications to use for therapy, depending on severity of patient. A feared complication of tuberculosis therapy, multi-drug-resistant tuberculosis (MDR-TB), is distinguished from its resistance to first-line medications isoniazid and rifampin. Second-line drugs that are in common use for MDR-TB are Kanamycin, Capreomycin, and Amikacin via injections. Fluoroquinolones such as Levofloxacin, Moxifloxacin, and Gatifloxacin and are also amongst the common second-line agents used when drug resistance develops to the first-line agents.

18.2 CLINICALLY USEFUL ANTITUBERCULAR AGENTS

Isoniazid: Tubizid (R), Isonicotinic acid hydrazide

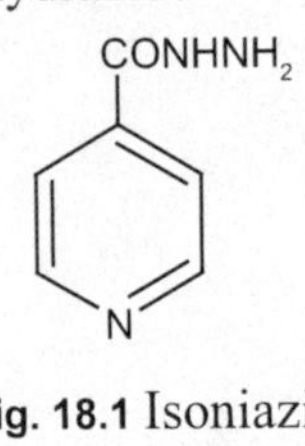

Fig. 18.1 Isoniazide

Isoniazid is bacetriostatic and bactericidal. It is a prodrug of isonicotinic acid. It is converted by KatG enzyme to isonicotinic acid which inhibits bacterial cell wall mycolic acid thereby make *M. tuberculosis* susceptible to reactive oxygen redicals during cell division.

Synthesis:

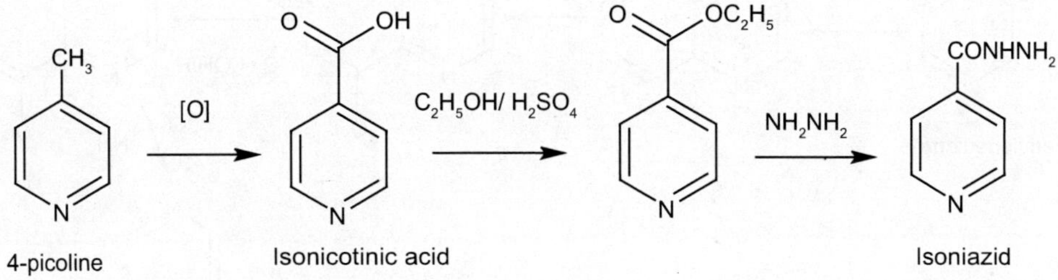

Fig. 18.2

SAR of Nicotinamides derivatives (Isoniazide)

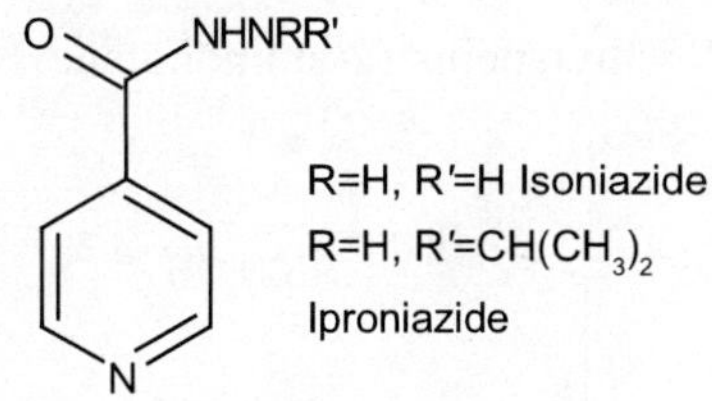

Fig. 18.3 Basic structure

1. About 100 analogues of isoniazide have been synthesized and clinically tested for antimicrobial activity results in reduction in activity.
2. N2-isopropyl derivative (iproniazide) was found to be active in vitro. It is used in the treatment of tuberculosis It also causes reduction in monoamine oxidase enzymes therefore used in the treatment of psychotic condition.

Pyrazinamide: Pyrazine-2-carboxamide

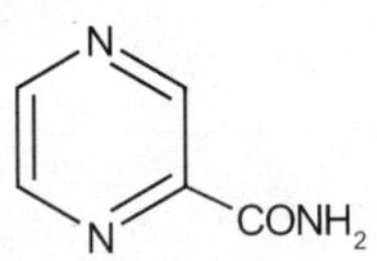

Fig. 18.4 Pyrazinamide

SAR of pyrazinamide

1. Compound with other substitution in pyrazine nucleus were found to be inactive or loss of activity then pyrazinamide.
2. Only active analogues is Morphazimamide which is potentially active over pyrazinamide

Pyrazinamide is also a prodrug. It shows bactericidal activity by inhibiting fatty acid synthetase I in bacterial cell.

Synthesis:

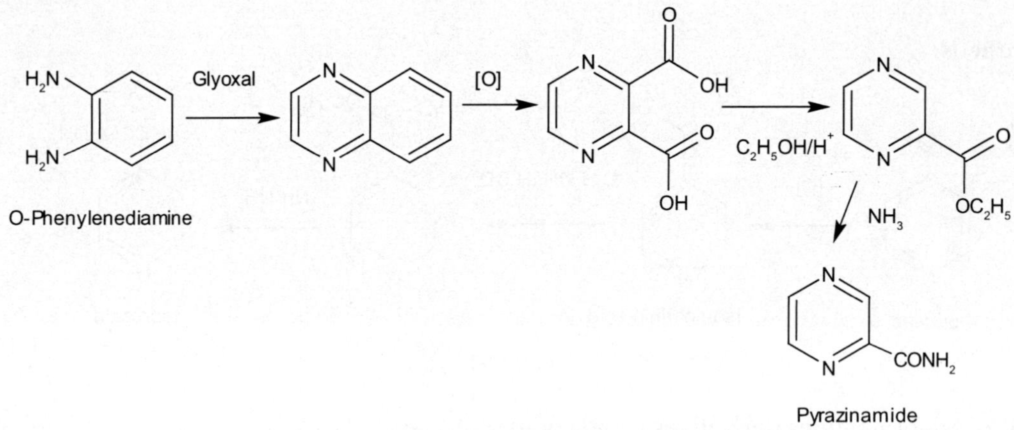

Fig. 18.5

Ethambutol: (**Myabutol**)[(R)], *N,N*-Ethylenebis-(2-aminobutane-1-ol)

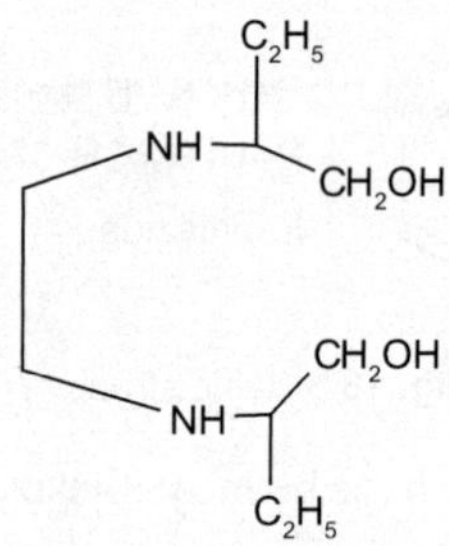

Fig. 18.6 Ethambutol

Ethambutol is a bacteriostatic drug that inhibits the incorporation of mycolic acid into the mycobacterium cell wall.

Synthesis

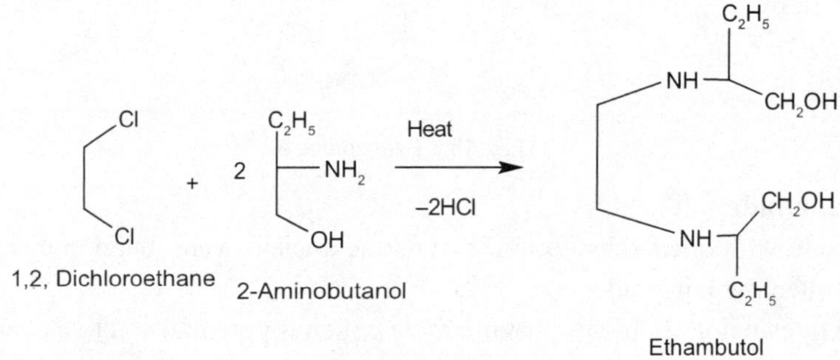

Fig. 18.7

P-Aminosalicylic acid: Mycopas(R), 5-amino-2-hydroxybenzoic acid

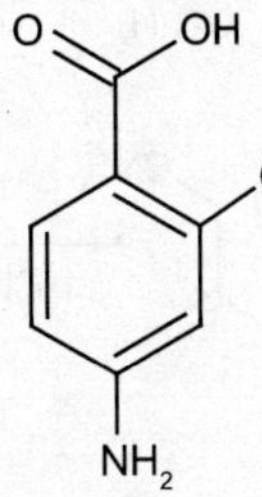

Fig. 18.8

P-Aminosalicylic acid (PAS) is a bacterial folate metabolism inhibitor. It is used in case of resistance, retreatment and intolerance of other antimycobacterial agents.

Synthesis:

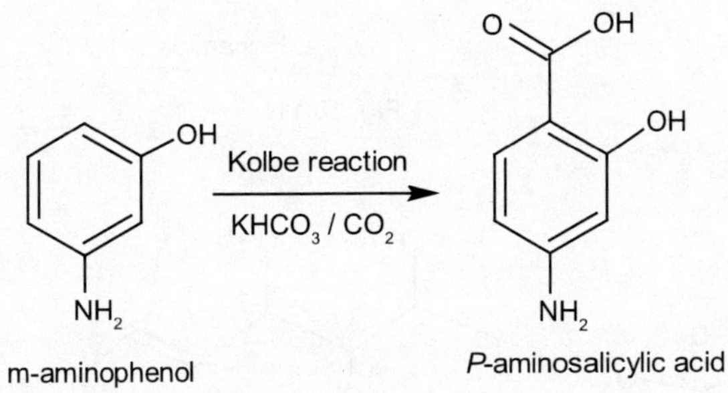

Fig. 18.9

Ethionamide: Trecator-SC(R), 2-(ethyl)-isonicotinthioamide

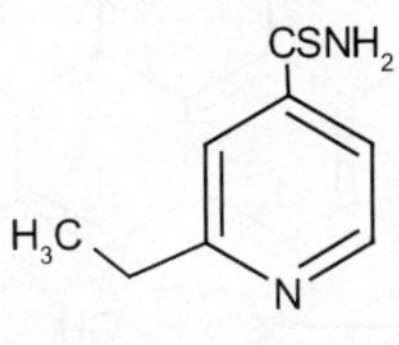

Fig. 18.10 Ethionamide

Ethionamide is bactericidal. It is considered a prodrug that is converted via oxidation by catalase peroxidise enzyme to active acetylating agent.

Synthesis:

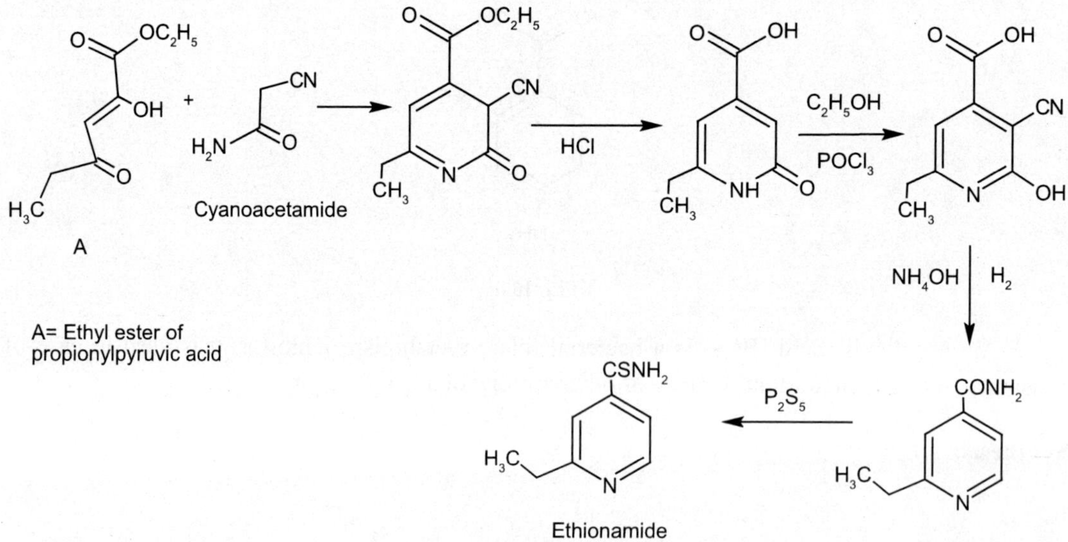

Fig. 18.11

Rifampicin: Rifadin[(R)]

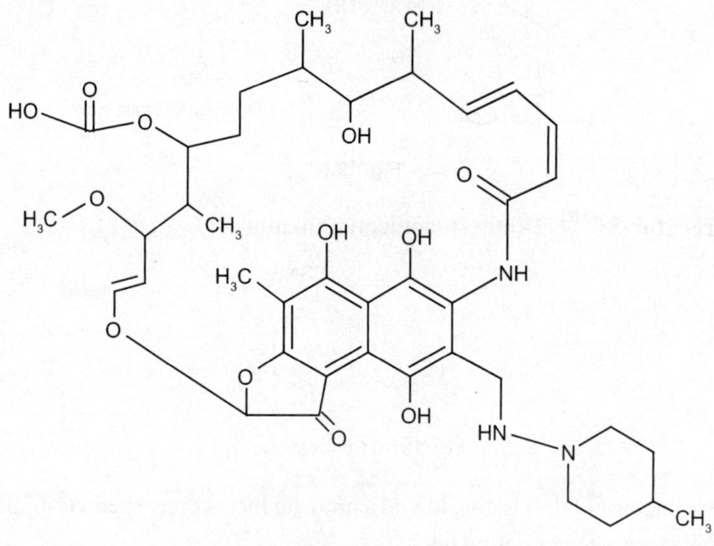

Fig. 18.12

Rifampicin is an antibiotic obtained from *Streptomyces mediterranei* spp. Rifampicin inhibits DNA-dependent RNA polymerase of mycobacteria by forming a stable drug enzyme complex resulting in suppression of initiation of chain formation in RNA synthesis and acts as a bactericidal drug.

Rifabutin: Mycobutin(R)

Rifabutin is a semisynthetic antibiotic rifamycin used in 1992, structurally close to rifampicin. It is used against *M. avium.* Rifabutin's lipophilic nature attributed its ability to penetrate the cell wall of the organism more effectively than other agents.

Cycloserine: Seromycin(R)

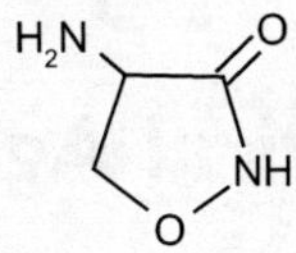

Fig. 18.13 Cycloserine

Cycloserin is naturally isolated from *Streptomyces orchidaceus*. Its dextro- isomer D-cycloserin is active against tubercular infections.

EXERCISE

1. Classify antitubercular dugs with suitable structures.
2. Write an explanatory note on antitubercular agents.
3. Write the synthesis of isoniazid, pyrimethamine and ethionamide.
4. Write a note on antitubercular antibiotics.
5. Define multidrug-resistant TB.

19
Antiviral Agents

19.1 INTRODUCTION

Viruses, smallest of the human infections, cause the ultimate expression of parasitism: take nutrition from the host cell and direct its metabolic machinery to synthesize new virus particles. Viruses contain one of the nucleic acid either RNA or DNA. Viral chemotherapy, therefore, is difficult, as it would require interference with cellular metabolism in the host. However, virus directed enzymes have been identified in the infected cell and some viruses have a few enzymes of their own which may have higher affinities for some anti-metabolites or inhibitors than the regular cellular enzymes. Drugs could also target virus specific steps like cell penetration, un-coating, reverse transcription, virus assembly or maturation. Another stumbling block is that in majority of acute infections viral replication is already at its peak when symptoms appear. To be effective, therefore, therapy has to be started in the incubation period, i.e., has to be prophylactic.

DNA Viruses: Many viruses replicate vary widely and are characterized by transcribe mRNA for translation of viral proteins. Most of the DNA viruses contain double stranded DNA as their genome and replicate in host cells thus produce mRNA.

RNA Viruses: RNA viruses are single stranded gemone. They are divided in three groups. First group included *piconaviruses, flaviviruses and togaviruses* where mRNA directly used. Second group includes *orthomyxoviruses, bunyaviruses* and *paramyxoviruses*, where, antisense RNA genome are used for transcription of new mRNA. Third group consists of retroviruses, in which single stranded RNA exist as dimer and antisense strands.

Viral Disease includes HIV, Herpesvirus, Hepatitis, Hepatitis C Carcinoma, influenza, Tumour viruses, Prion disease.

Chemotherapy: Antiviral compounds designed in the general area of chemotherapy.

1. Disruption of virus and host cell attachment.
2. Inhibition of virus associated enzymes.
3. Inhibition of virus transcription.

4. Inhibition of viral translation.
5. Interference with viral regulatory proteins.
6. Interference with virus glycosylation, phosphorylation, sulfation, etc.
7. Interference with viral protein assembly.
8. Prohibit the release of viruses from cell surface membrane.

19.2 CLINICALLY USEFUL ANTIVIRAL AGENTS

1. Agent interfering with viral nucleic acid replication (Acyclic nucleoside analogues)

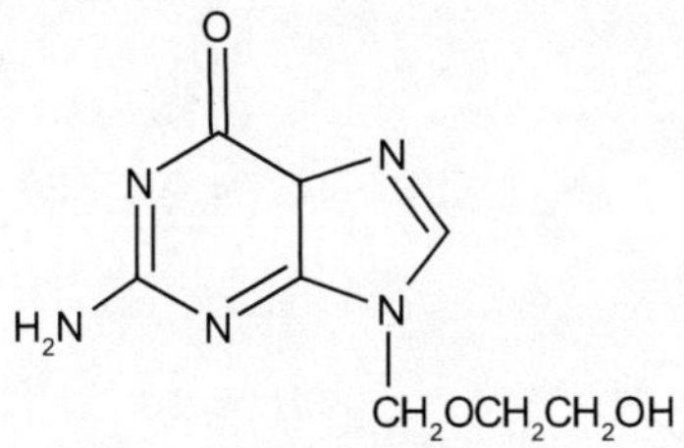

Fig. 19.1 Acyclovir

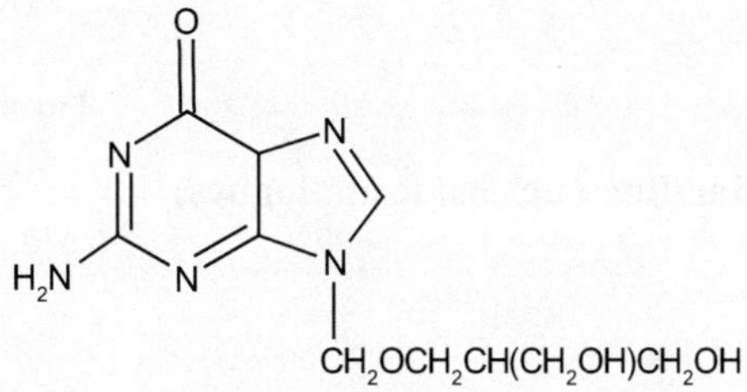

Fig. 19.2 Ganciclovir

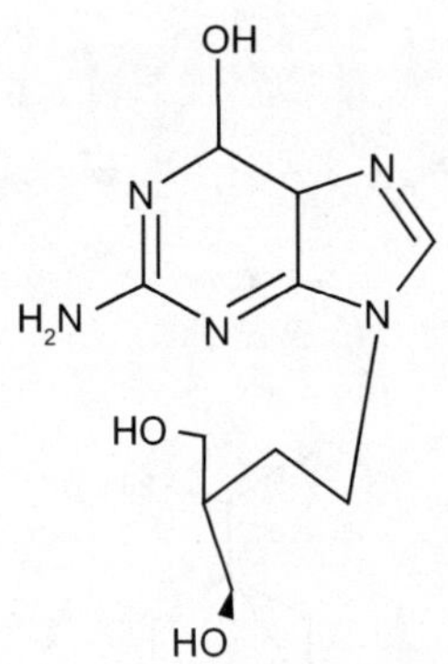

Fig. 19.3 Penciclovir

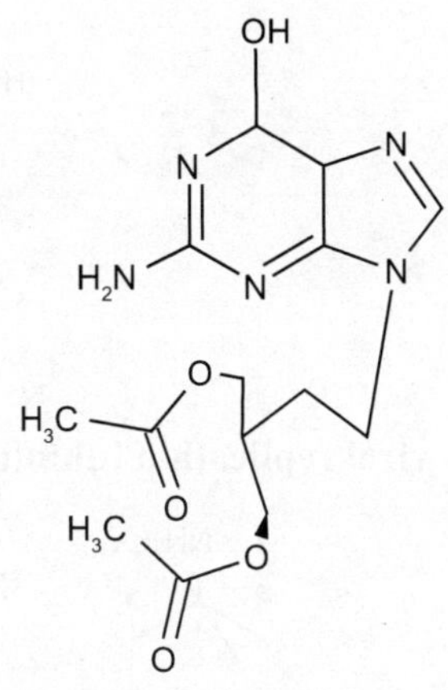

Fig. 19.4 Fenciclovir

(Purine nucleoside analogues)

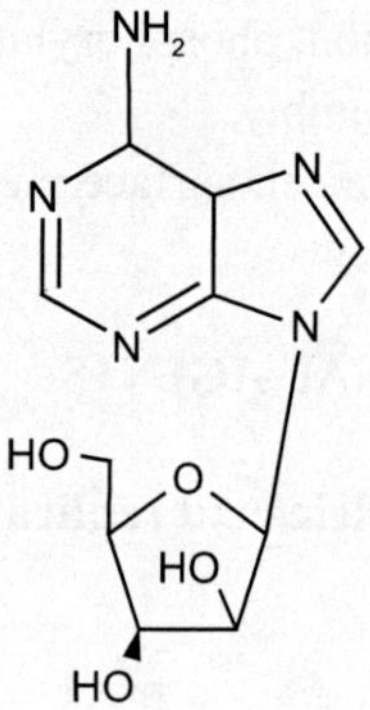

Fig. 19.5 Vidarabin

(Pyrimidine nucleoside analogues)

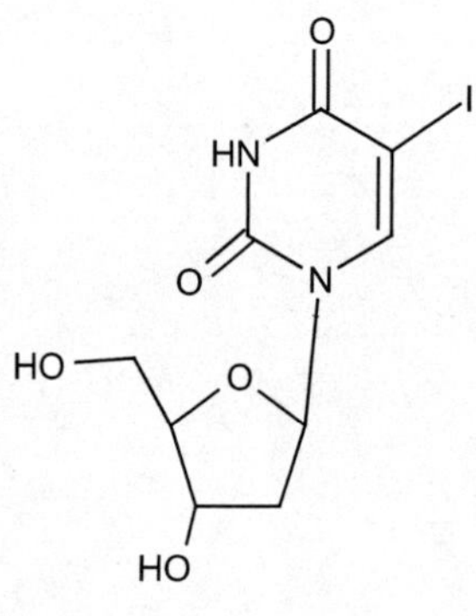

Fig. 19.6 Idoxurideine

2. Early viral replication inhibitors

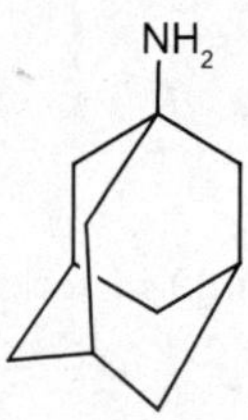

Fig. 19.7 Amantadine

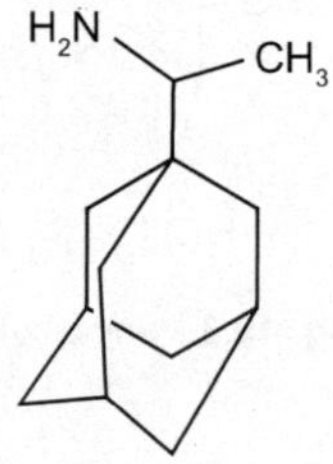

Fig. 19.8 Rimantadine

3. Neuraminidase inhibitors

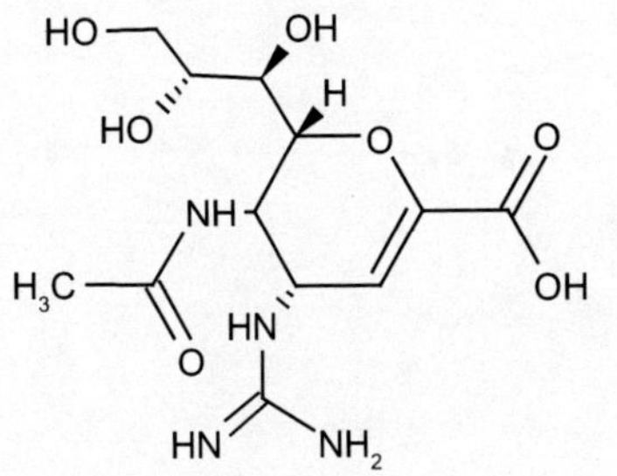

Fig. 19.9 Zanamivir

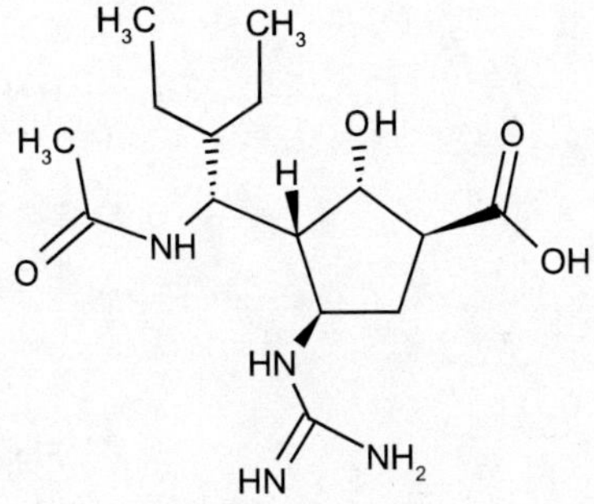

Fig. 19.10 Peramivir

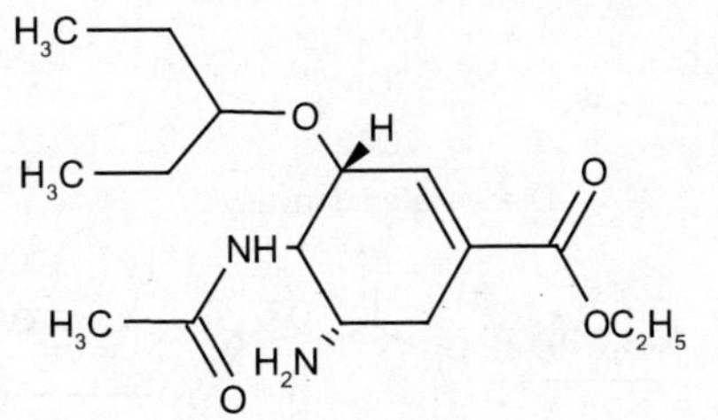

Fig. 19.11 Oseltamivir

4. Entry (fusion) inhibitor

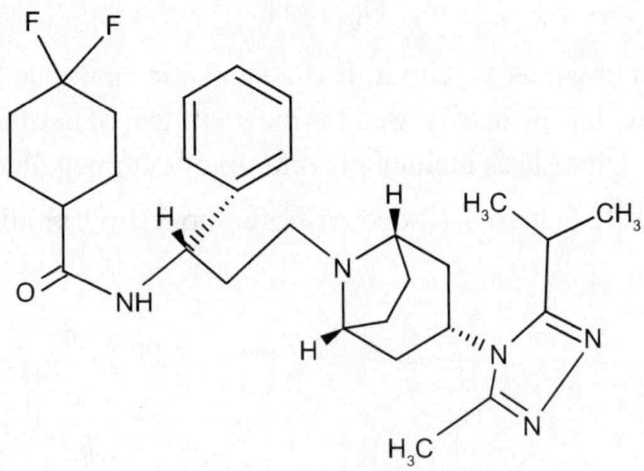

Fig. 19.12 Maraviroc

5. Agent interfering with viral nucleic acid replication

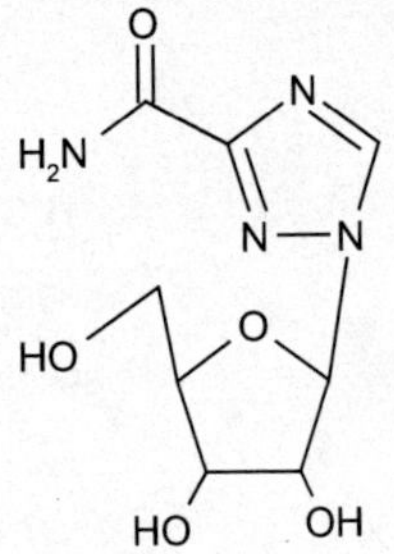

Fig. 19.13 Ribavirin

Acyclovir: Zovirax[(R)] (2-Amino-1,9-dihydro-9-[(2-hydroxyethoxy)methyl]-6*H*-purin-6-one)

Synthesis:

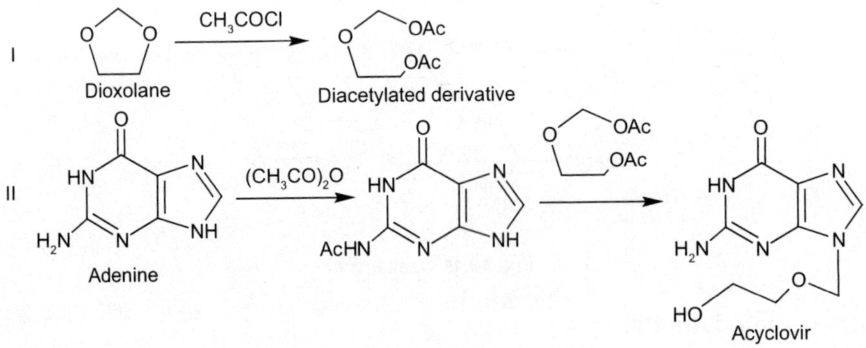

Fig. 19.14

Aciclovir Commonly known as acyclovir. It is a synthetic analogue of deoxyguanosine where carboxylic moiety is acyclic. It is primarily used for the treatment of herpes simplex virus infections, chicken pox, and shingles. Other uses include prevention of cytomegalovirus infections.

Idoxuridine: Herplex[(R)], 5-Indo-1-(2-deoxyribofuranosyl)-pyrimidin-2,4-(*1H.3H*)-dione

Synthesis:

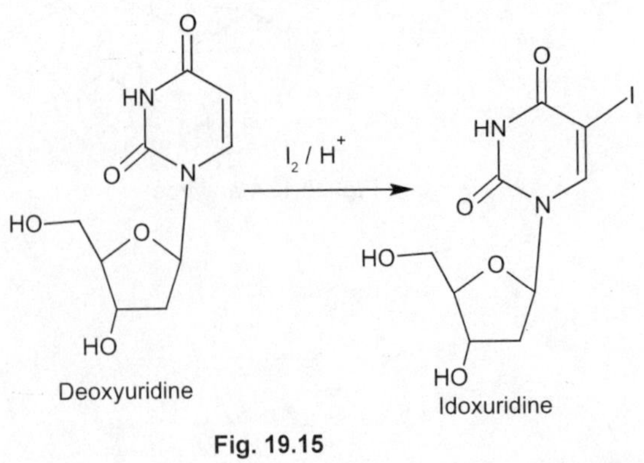

Fig. 19.15

Idoxuridine is used to treat anti-herpes virus infection. It is a nucleoside analogue, a modified form of deoxyuridine, similar enough to be incorporated into viral DNA replication, but the iodine atom added to the uracil component blocks base pairing. It is used only topically due to cardiotoxicity.

6. Antiretroviral Agents (Anti-HIV) Agents

1. Nucleoside reverse transcriptase inhibitors (NRTI)

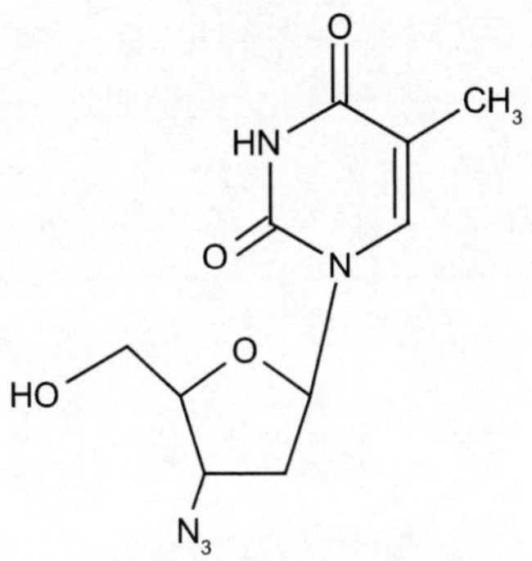

Fig. 19.16 Zidovidine

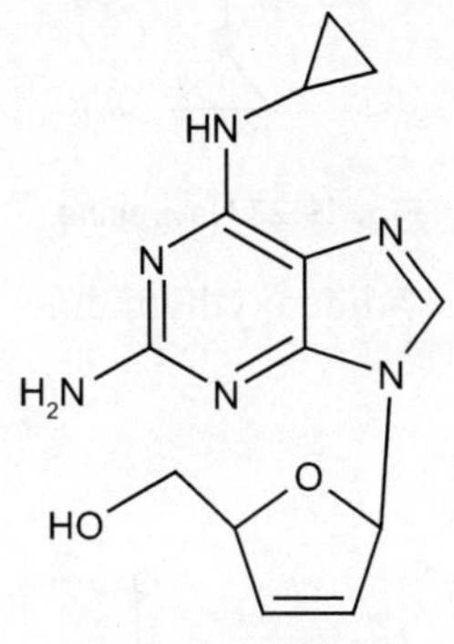

Fig. 19.17 Abacavir

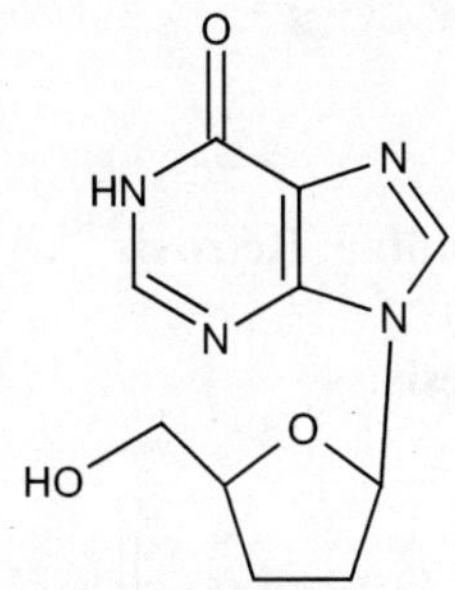

Fig. 19.18 Didanosine

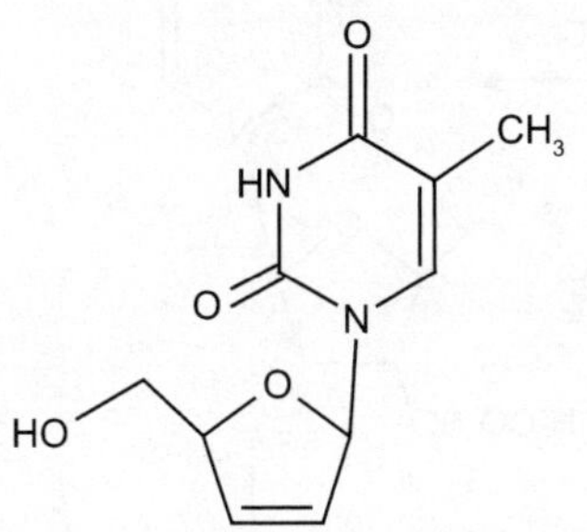

Fig. 19.19 stavudine

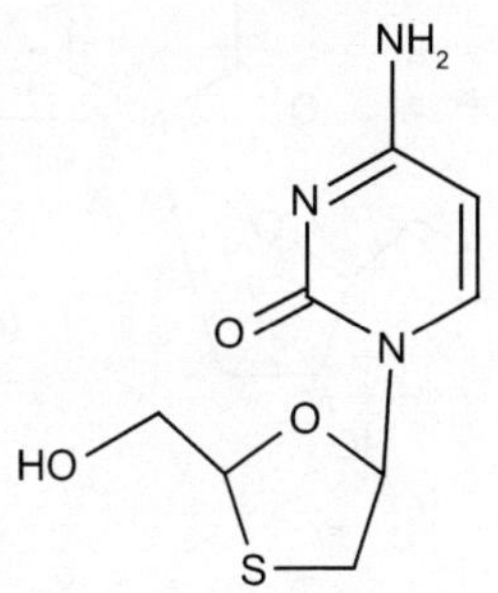

Fig. 19.20 Lamivudine

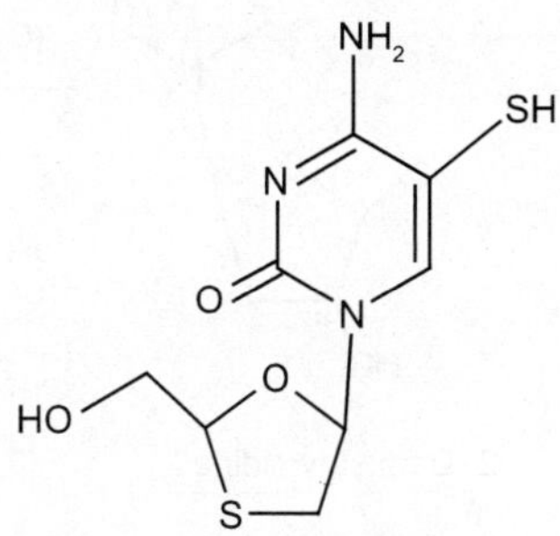

Fig. 19.21 Emtrictavine

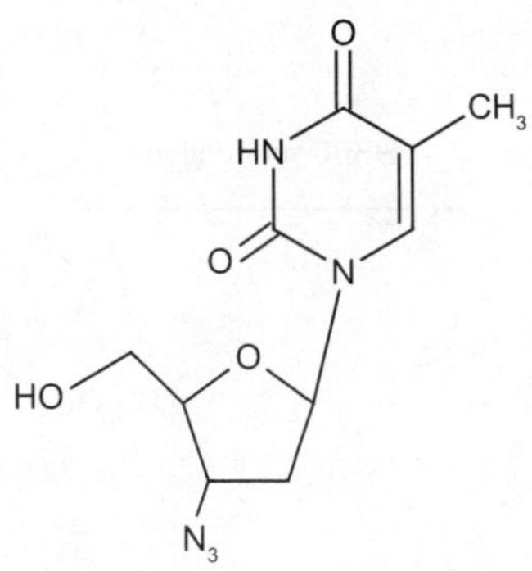

Fig. 19.22 Zalcitabin

2. Non-Nucleoside RT inhibitors (NNRTI)-Anti HIV agents

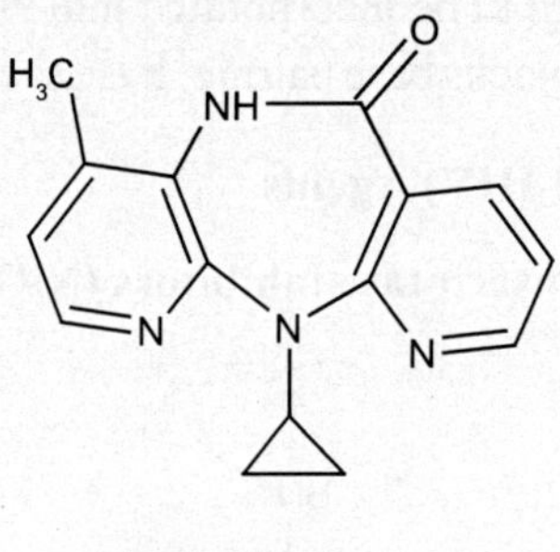

Fig. 19.23 Navirapine

Zidovudine: Retrovir(R), 3'-Azido-2',3'-dideoxythymidine

Synthesis:

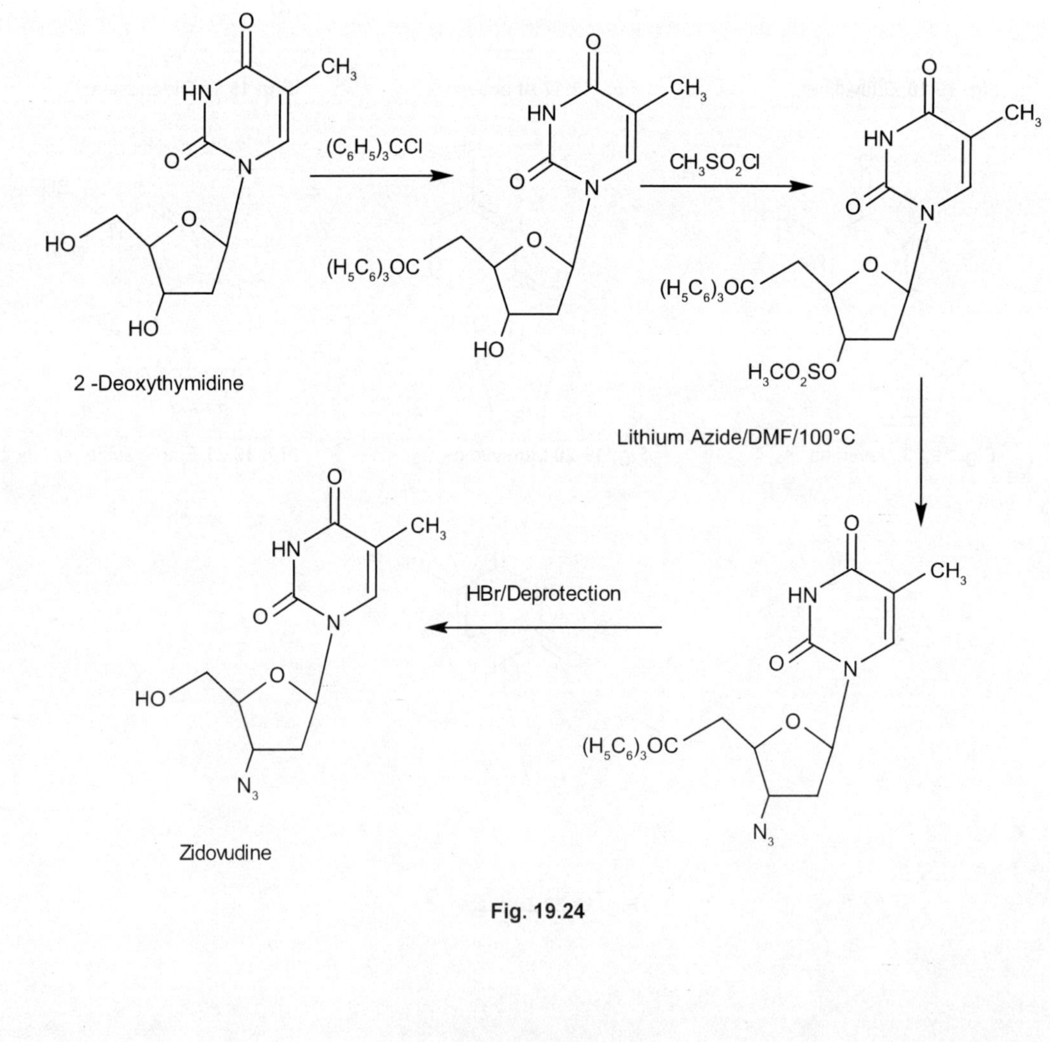

Fig. 19.24

Zidovudine (Azidothymidine). It is an antiretroviral agent used to prevent and treat HIV/AIDS. It is generally recommended in combination with other antiretroviral agents. It may be clinically used to prevent newborn spread during birth or after a needle stick injury or other potential exposure.

Zalcitabine: Hivid(R), 2', 3'-dideoxycytidine

Synthesis:

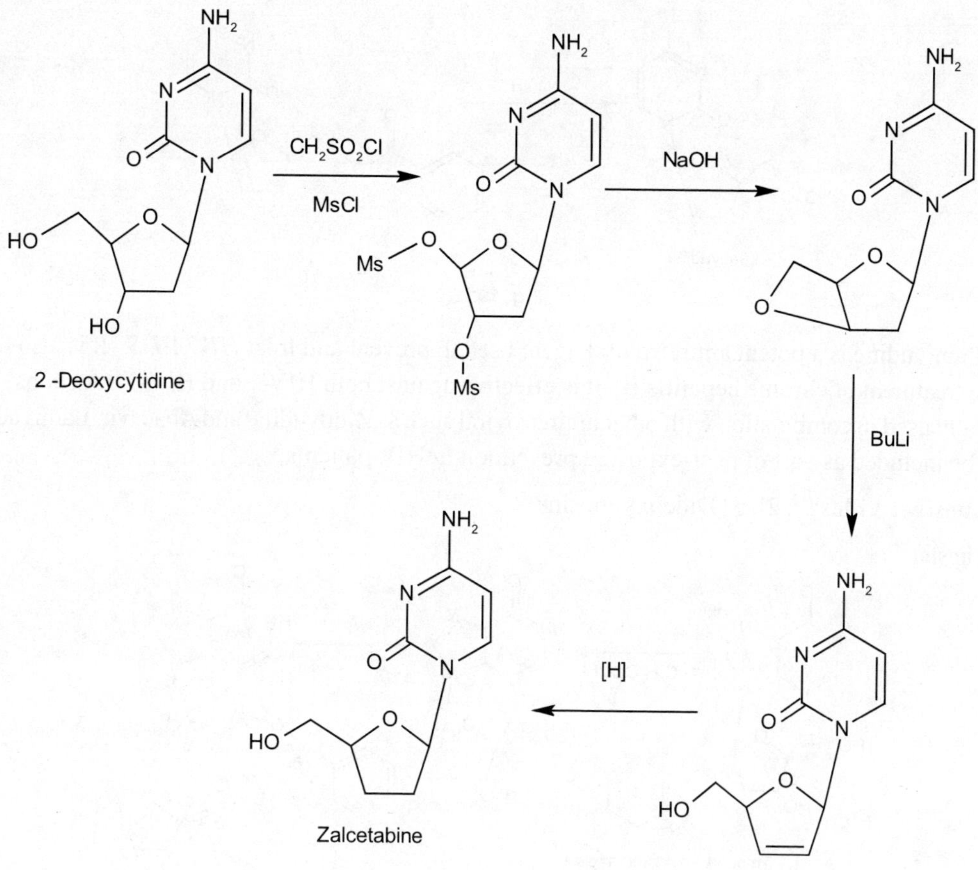

Fig. 19.25

Zalcitabine is dideoxycytidine. It is a potent nucleoside reverse-transcriptase inhibitor (NRTI). Zalcitabine was the third antiretroviral agent approved by the Food and Drug Administration (FDA) for the treatment of HIV/AIDS.

Lamivudine: Epivir(R), 3'-Thia-2, 3'-Dideoxy cytidine

Synthesis:

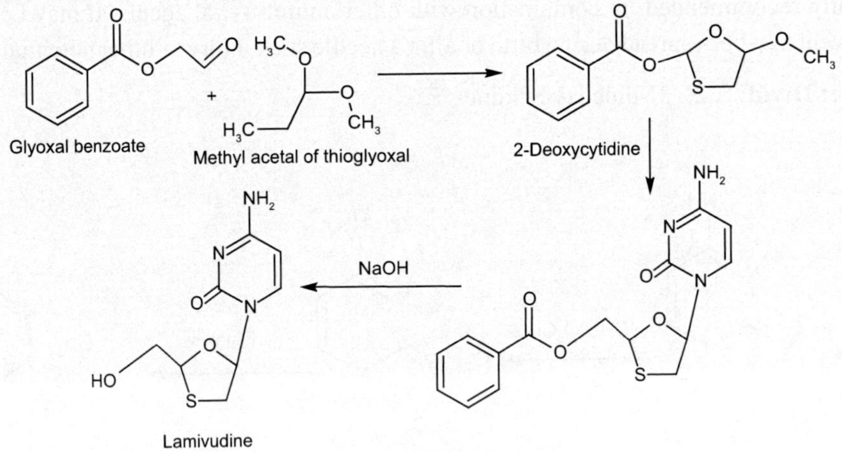

Fig. 19.26

Lamivudine is a potent antiretroviral agent used to prevent and treat *HIV/AIDS*. It is also used in the treatment of chronic hepatitis B. It is effective against both HIV-1 and HIV-2 infections. It is typically used in combination with other antiretroviral such as Zidovudine and Abacavir. Lamivudine may be included as part of post-exposure prevention in HIV patients.

Didanosine: Videx$^{(R)}$, 2', 3' Dideoxy inosine

Synthesis:

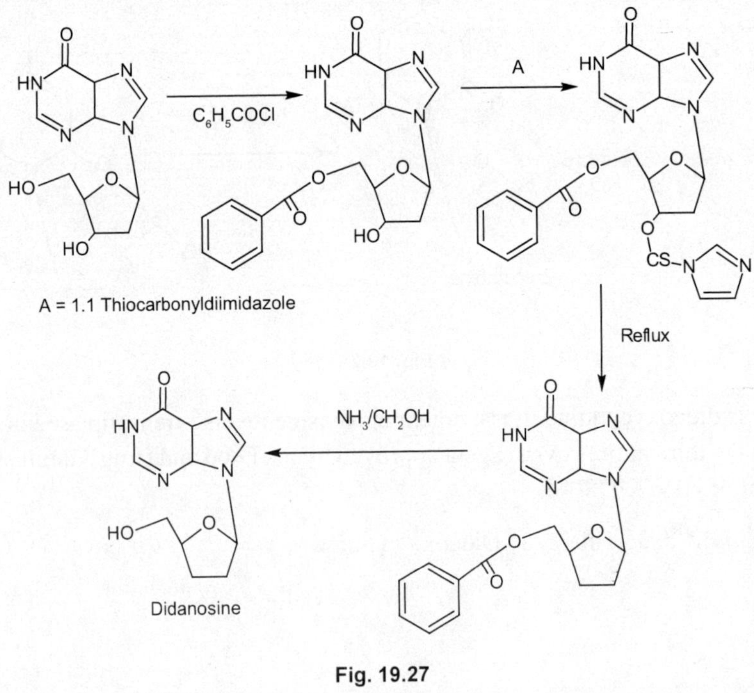

Fig. 19.27

Didanosine is used to treat HIV/AIDS. It is reverse-transcriptase inhibitor. It is used in combination with other medications as part of highly active antiretroviral therapy.

19.3 NEWER AGENTS

Emtricitabine: Emtriva$^{(R)}$, 4-amino-5-fluoro-1-[2-(hydroxymethyl)-1,3-oxathiolan-5-yl]-pyrimidine-2-one

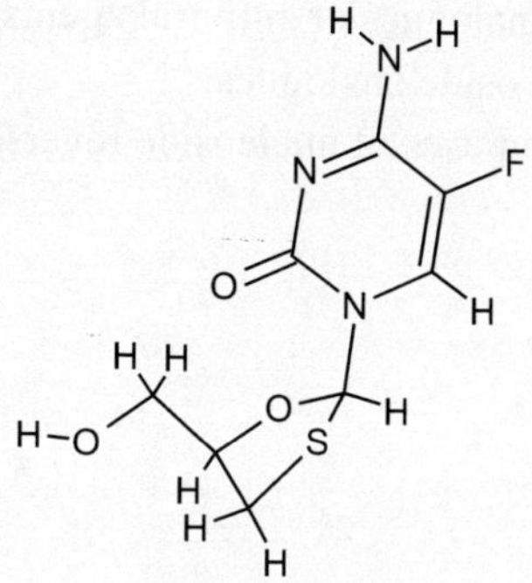

Emtricitabine

Fig. 19.28

Emtricitabine is used as HIV reverse transcriptase inhibitor.

Tenofovir: Viread$^{(R)}$, 1-(6-aminopurin-9-yl)-propane-2-yloxymethylphosphoric acid.

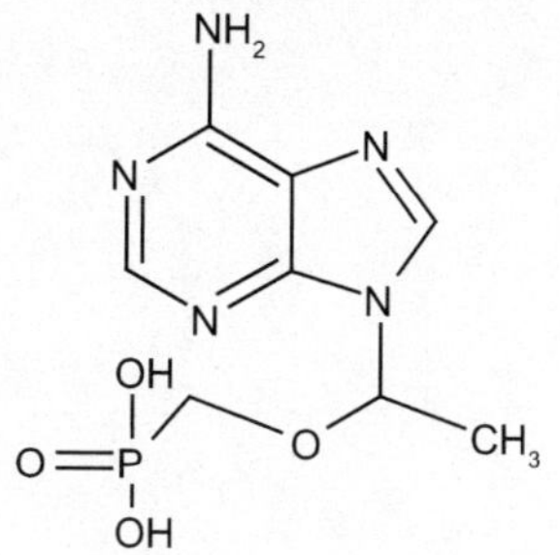

Fig. 19.29 Tenofovir

Tenofovir is used as antiretroviral drugs.

EXERCISE

1. Classify antiviral agents with their chemical structures.
2. Write the synthesis of the following drugs.
   (a) Zudovudine
   (b) Acyclovir
   (c) Idoxuridine
3. Explain anti-HIV agents with examples.
4. Explain acyclic nucleoside analogues of antiviral agents.
5. Write a note on acyclic nucleoside analogues.
6. Write the synthesis, MOA and uses of nucleoside reverse transcriptase inhibitor.

20

Antifungal Agents

20.1 INTRODUCTION

Human-fungi-parasitic are found in mycotic illnesses. Most fungal infections (mycoses) occur as superficial invasion of the skin or mucous membrane of the body orifices. These diseases can usually be treated by local application of the antifungal agents. However, fungi have different shapes and sizes. Some are large while others are minute, parasitic, and saprophytic cells, respectively. The fungal kingdom includes yeast, molds, rust and mushrooms. Mostly fungi are saprophytics (lives on dead organic matters in the soil or decaying leaves or woods), therefore cause of limited to superficial topical infections, systemic, or both. Some of these fungi can cause opportunistic infections by wounds or by inhalation. Some of these are fatal.

Fungal Pathogens

***Candida spp.*:** Invasive candidiasis is most common cause mycosis. *C. glabratta*, *C. krusei*, *C. tropicali* has emerged in recent years as infectious organism. While, *C. albicans* is most common infectious organism cause fungal infection. Most common manifestation found in AIDS patients, in oral fungal infections, catether related candida infections.

Cryptococcus neoformans: This is survives in pegion feces and transferred through Aerosolization and inhalation in human beings. It is responsible for *Cryptococcal meningitis*.

Opportunistic filamental fungi: *Aspergillus* spp. is most common fungal pathogens in this category. This causes for pulmonary *Aspergillus fungal* infections.

20.2 CLASSIFICATION

1. Antibiotics: Amphotericin B, Nystatin, Griseofulvin
2. Azoles: Triazoles (Fluconazole, Itraconzole, Terconazole) Imidazoles (Clotrimazole, Ketoconazole, Miconazole, Bifonazole, Butoconazole, and Zinoconazole)

3. Fluorinated pyrimidines: Flucytosine
4. Chitin synthetase inhibitors: Nikomycin Z
5. Peptides/proteins: Cispentacin
6. Miscellaneous: Ciclopirox, Tolnaftate, Naftifine, and Terbinafine

20.3 CLASSIFICATION BASED ON THE ROUTE OF ADMINISTRATION

1. Drugs for subcutaneous and systemic mycoses: Amphotericin B, Fluconazole, Flucytosine, Itraconazole, Ketoconazole.
2. Drugs for superficial mycoses: Clotrimazole, Econazole, Griseofluvin, Miconazole, Nystatin.

20.4 BIOCHEMICAL TARGET OF ANTIFUNGAL CHEMOTHERAPY

As per cellular level status, the major difference between fungal cells and mammalian cells is that fungal cell contains cell wall therefore fungal cell wall is the target site for antifungal agents. Potent antifungal agents have some toxicity towards human cells. However, targeting of antifungal agents includes inhibition of DNA biosynthesis, disruption of mitotic spindles and interference with intermediary metabolism. Another difference between fungal and mammalian cell walls is mammalian cell wall contains cholesterol while fungal cell wall contains ergesterol both sterols are slightly same on stereochemistry model and construction which provide selective toxicity of antifungal agents.

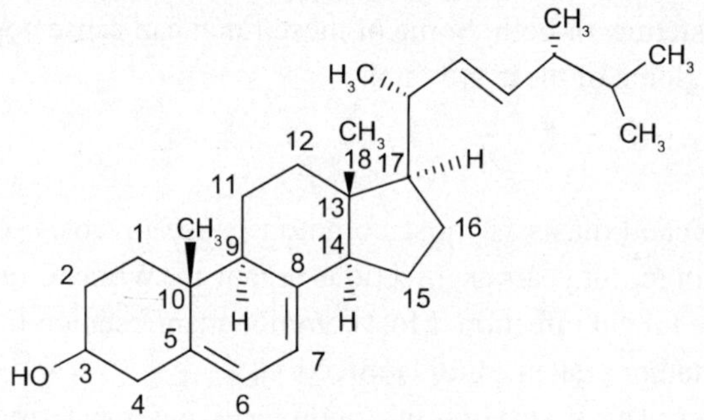

Fig. 20.1 Ergosterol

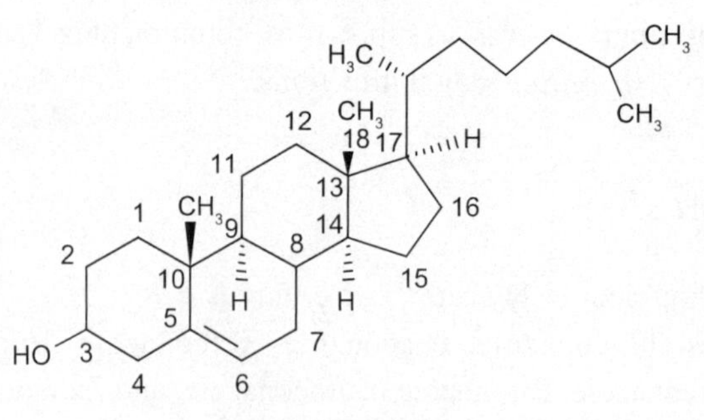

Fig. 20.2 Cholesterol

20.5 ERGASTEROLE BIOSYNTHESIS INHIBITORS

Azole Analogues (Imidazole and Triazole) as Antifungal Agents

Azole antifungal agents are the largest class of antimicotics today. Topically used antifungal derivatives are imidazole based antifungal agents are used to treat systemic fungal infections.

Mode of action: They inhibit sterols-14-α-demethylase, cytochrome P450 dependent enzyme system thus inhibit the biosynthesis of ergosterol for cytoplasmic membrane. So accumulation of methylsterols disrupt the close packing of acyl chains of phospholipids which impaired the membrane bound enzyme system and thus inhibiting the growth of fungi.

Structure activity relationship (SAR) of Azole antifungal agents

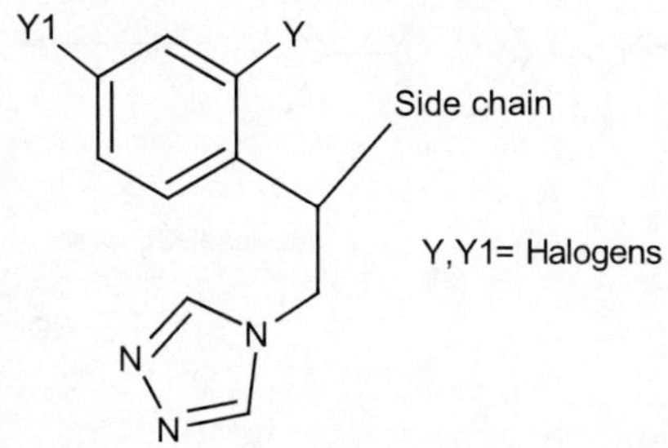

Fig. 20.3 Basic structure of Azole antifungal agents

1. Azoles have been reported by several diverse groups in the structural features, several groups emerge: imidazole or triazole, a halo substituted aryl ring separated from azole moiety by two atoms and a side chain represent the diversity across the azoles family.
2. Halo substituted (Y.Y1) aromatic (2 and 2,4 substituted) residue bind in the pocket occupied by six member rings and yields potent antifungal agents.
3. 1,2,4-triazole ring bonded by nitrogen carbon linkage to rest of the structure is required for antifungal activity.
4. Nitrogen atom of imidazole or triazole is believed to bind the heme iron enzyme bound CYT P450 to inhibit molecular oxygen and prevent the steroidal substrate.
5. Incorporation of methyl group β to the aromatic ring boost up the potency because of methyl group occupies the binding region on the β phase of lamosterole which favour the conformation in which the side chain is anti-periplanar to aromatic ring.

20.6 CLINICALLY USEFUL AZOLE DERIVATIVES

Ketoconazole: Nizoral$^{(R)}$ (cis-1-acetyl-4-[4-[2-(2,4-dichlorophenyl)-2-(1H-imidazole-1-ylmethyl)-1,3-dioxolane-4-ylmethyl]phenyl]piperazine)

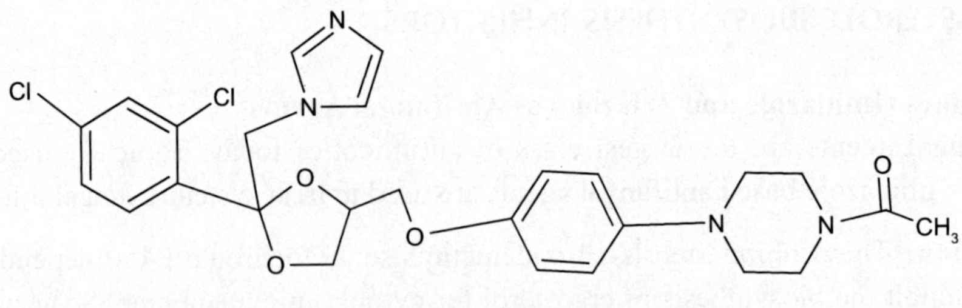

Fig. 20.4 Ketoconazole

Synthesis

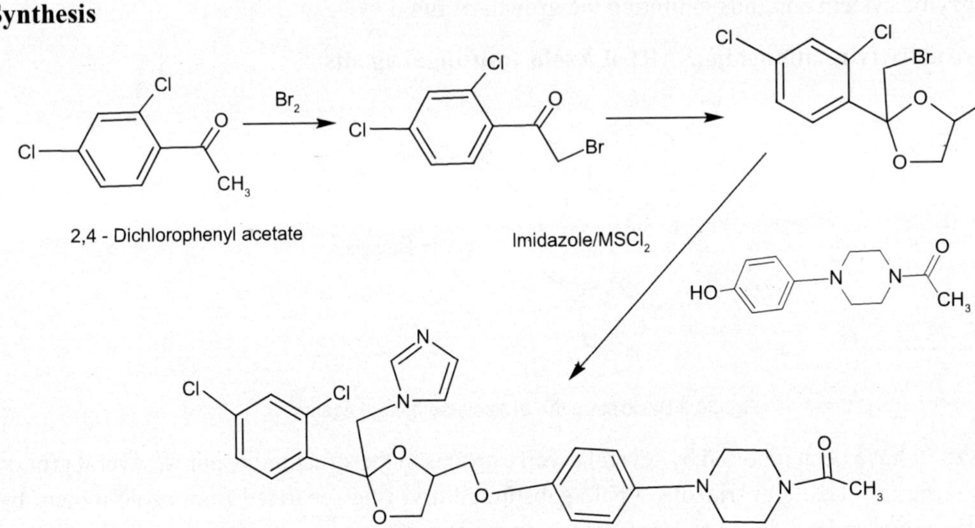

Ketoconazole

Fig. 20.5

Ketoconazole is a potent antifungal agent. It is highly dependent on low stomach pH for absorption. Ketoconazole exhibit its high concentration in fatty tissues due to highly lipophilicity. It is active against *Candida* sp. and *Chryptococcus neoformance*.

Fluconazole: Difulcan$^{(R)}$, 2-(2,4-diphluorophenyl)-1,3-bis-(1H-1,2,4-triazol-1-yl)-propane-2-ol

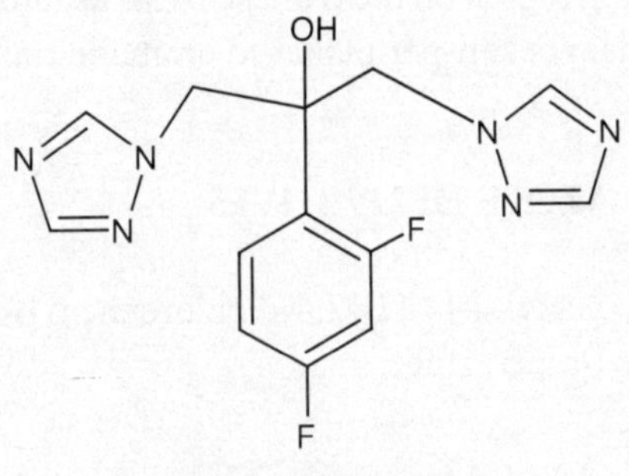

Fig. 20.6 Fluconazole

Fluconazole is used as bis-triazole antifungal agents. Fluconazole is used against *Candida* sp. as well as *Cryptococcus* sp. It is also used in infection caused by *Coccidioides immitis*.

Synthesis:

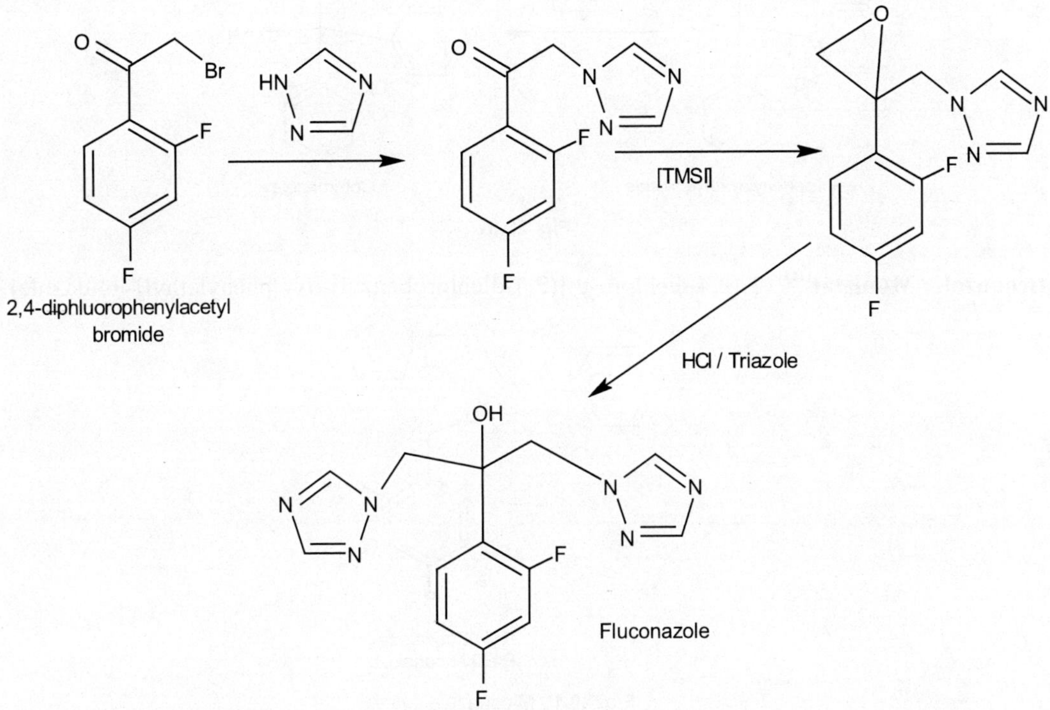

Fig. 20.7

Cotrimazole: Lotrimin(R), -1-(o-chloro-α, α-diphenylbenzyl)-imidazole

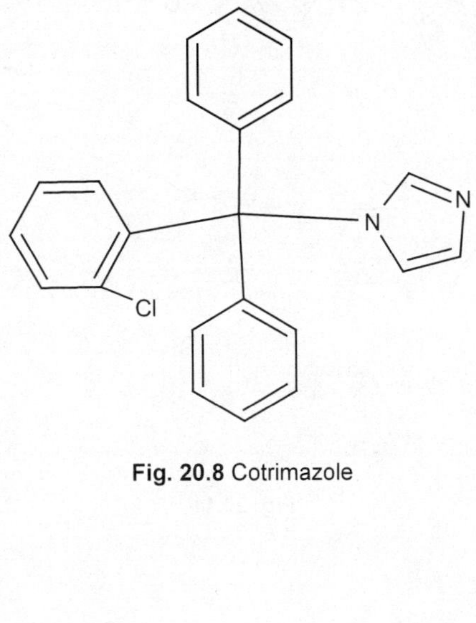

Fig. 20.8 Cotrimazole

Synthesis:

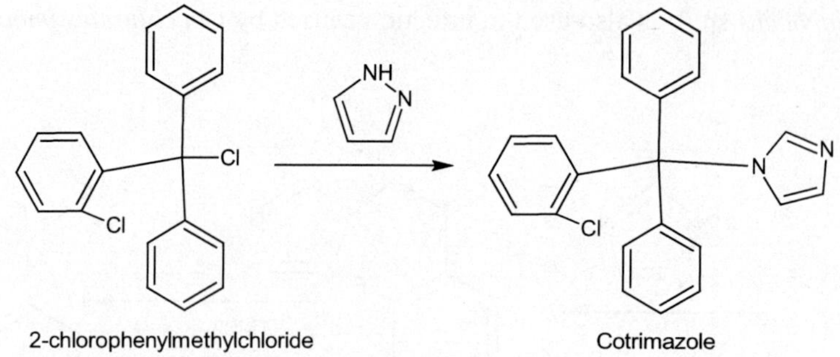

Fig. 20.9

Miconazole: Monistat(R), (1-[2,4-dichloro-b-[(2,4-dichlorobenzyl)-oxy]phenylethyl]-imidazole)

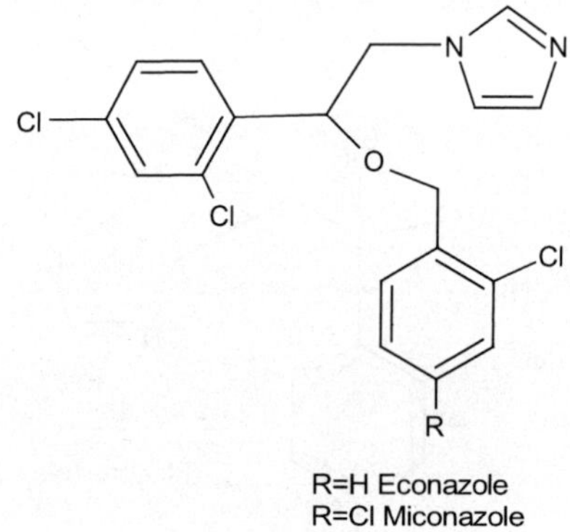

Fig. 20.10 Miconazole

Synthesis:

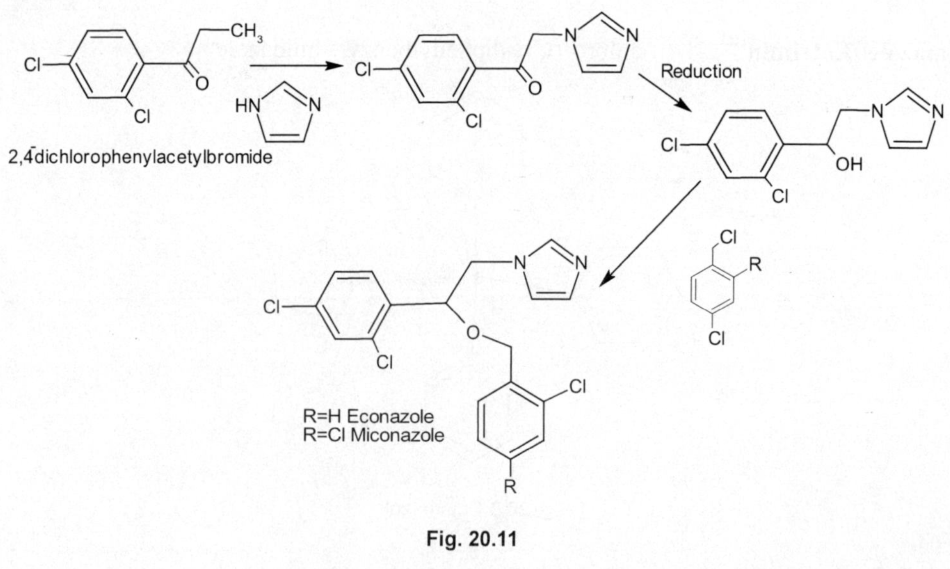

Fig. 20.11

Miconazole and Econazole are potent antifungal agents. Miconazole is used to treat *Tinea corporis* (ringworm: fungal infection). It is also used to treat athlete's foot fungal infection.

Butoconazole: Gynazole-1(R), 1-[4-(4-chlorophenyl)-2-[(2,6-dichlorophenyl)-thio)-butyl]-1H-imidazole

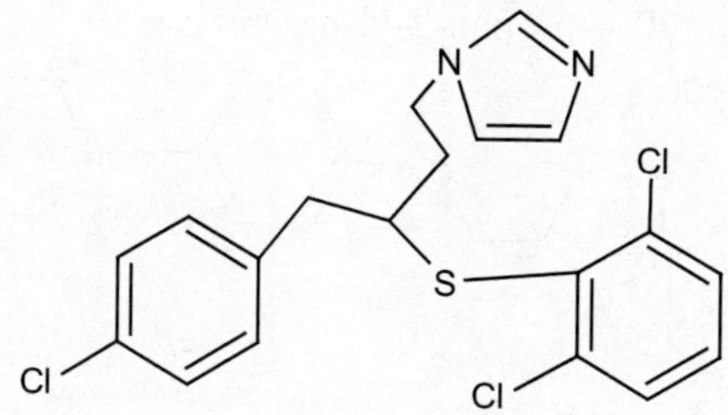

Fig. 20.12 Butoconazole

Synthesis:

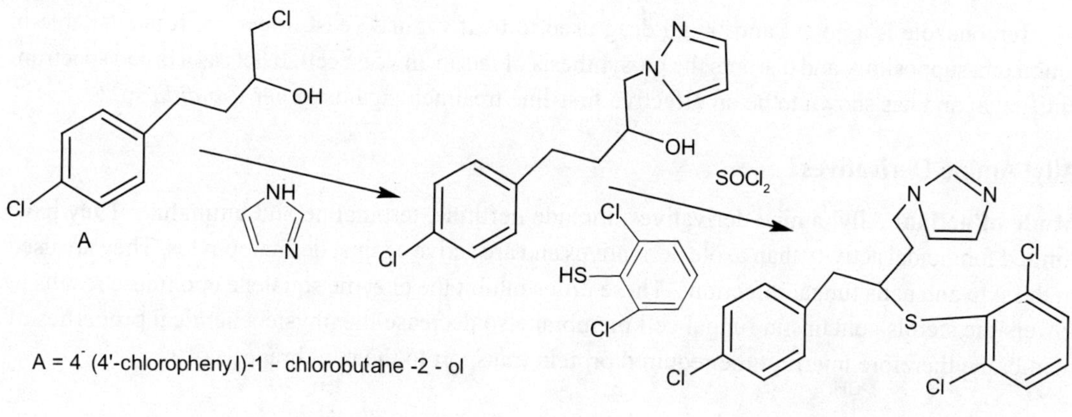

Fig. 20.13

Butoconazole is a potent antifungal agent used in the treatment of yeast infection in vagina.

Oxiconazole: Oxistat(R) [1-(2,4-Dichlorophenyl)-2-(1H-imidazol-1-yl)ethylidene]-[(2,4-dichlorophenyl)-methoxy]amine.

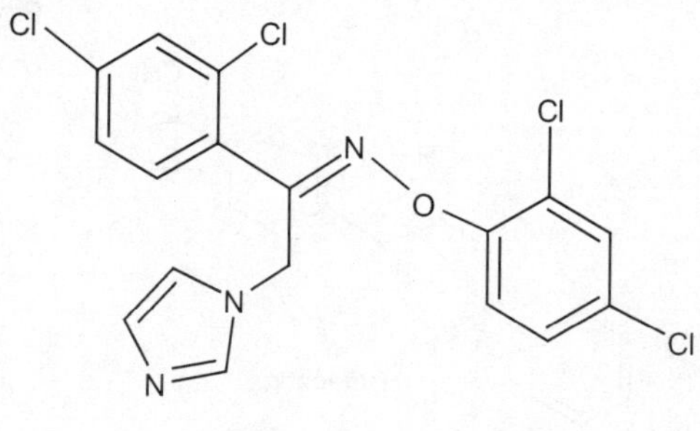

Fig. 20.14 Oxyconazole

Oxiconazole is an antifungal medication typically administered in a cream or lotion to treat skin infections, such as athlete's foot, jock itch and ringworm. It can also be used to treat the skin rash known as tinea versicolour, caused by systemic yeast overgrowth (Candida spp.).

Terconazole: Terazol$^{(R)}$ 1-[4-[[(2S,4S)-2-(2,4-Dichlorophenyl)-2-(1,2,4-triazol-1-ylmethyl)-1,3-dioxolan-4-yl]methoxy]phenyl]-4-propan-2-yl-piperazine.

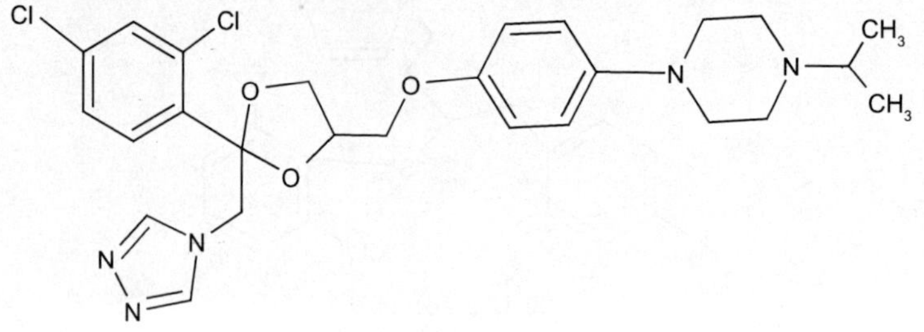

Fig. 20.15 Terconazole

Terconazole is a potent antifungal drug used to treat vaginal yeast infection. It is available in lotion or a suppository and disrupts the biosynthesis of fats in an yeast cell. It acts as a broad spectrum antifungal and has shown to be an effective first-line treatment against other *Candida* sp.

Allyl Amine Derivatives

Mode of action: Allyl amine derivatives, include neftifine, terbinafine and butinafine, They have limited fungicidal activity than azole derivatives and are active against dermatophytes. They are used in the skin and nails fungal infections. These drugs inhibit the enzyme squalene epoxidase results in lowers the sterols contains in fungal cell membrane so decrease the physicochemical properties of membrane therefore interrupt the required protein transport to the membrane.

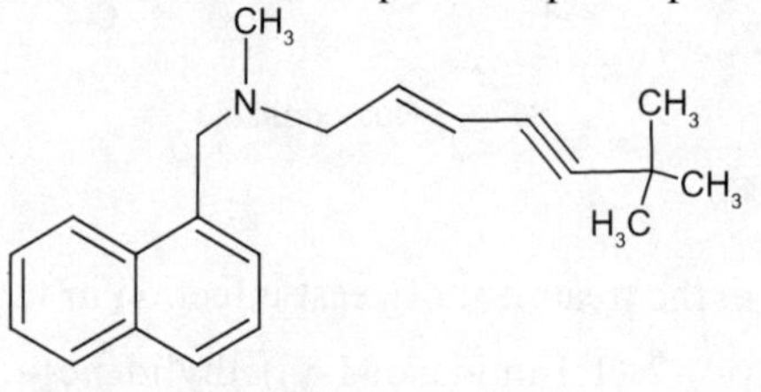

Terbinafine

Fig. 20.16

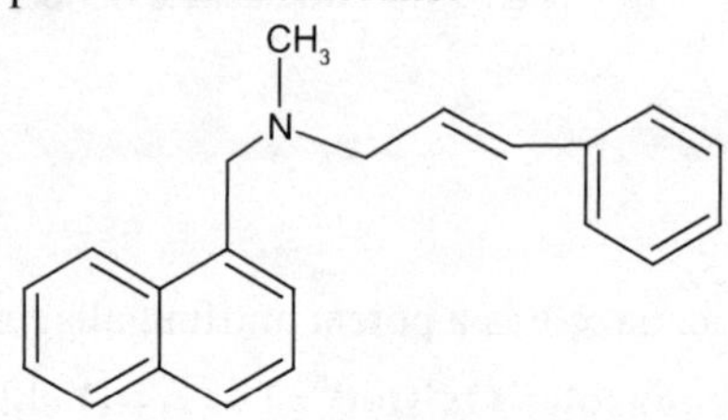

Naftifine

Fig. 20.17

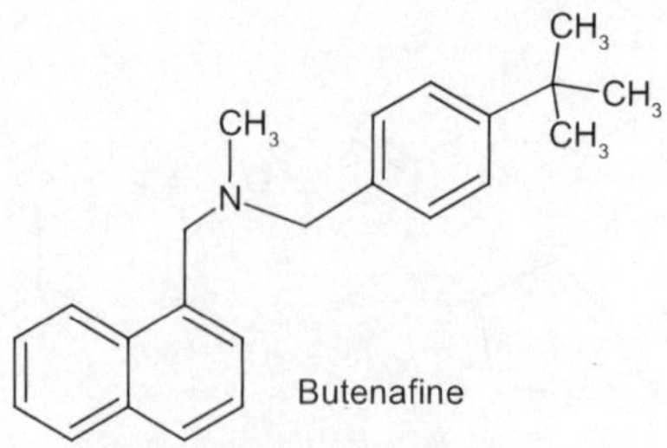

Butenafine

Fig. 20.18 Allyl amines derivatives

20.7 ANTIFUNGAL ANTIBIOTICS: (POLYENE MEMBRANE DISRUPTORS)

Mode of action: Polyene are basically macrocyclic lacnones with hydrophilic (contains alcohols, carboxylic acid and sugars) and lopophilic portions (contains chromophores of 4-7 conjugated double bonds). Conjugated double bonds play antifungal activity in vitro while cause toxicity in mammalian cells. Polyenes have higher affinity for membrane containing ergosterole over membrane containing cholesterol therefore greater toxicity to fungal cells.

20.8 CLINICALLY USEFUL ANTIFUNGAL ANTIBIOTICS

Amphotericin B: Fungizone (R)

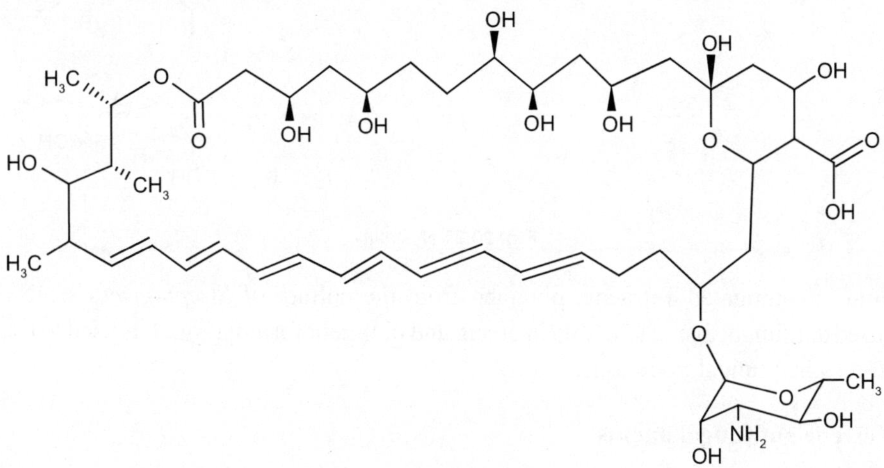

Fig. 20.19 Amphotericin

Amphotericin B is heptaene. It is toxic drug but show low toxicity to mammalian cells. It is used to treat life threatening fungal infections. This cannot cross the blood brain barrier so administered intrathecally for CNS fungal infections. Amphotericin B is 10 folds more toxic to fungal cells.

Gresiofulvin: Gris-peg(R)

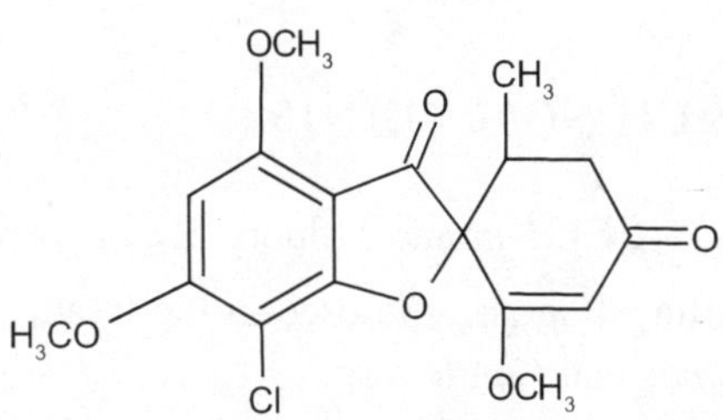

Fig. 20.20 Gresiofulvin

Gresiofulvin is potent antifungal agent is produced by strains of *Penicillium*. It is used orally as well as for superficial fungal infections. It may also interfere with DNA replication.

Nystatin: Mycostatin[(R)]

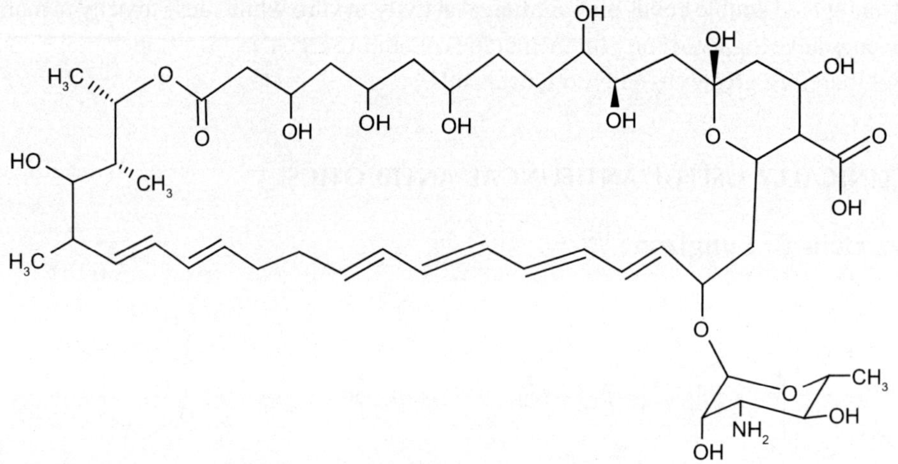

Fig. 20.21 Nystatin

Nystatin is conjugated Tetraene, obtained from the culture of *Streptomyces* sp. Nystatin is topically used antifungal agent available in cream and ointment formulation. It is used to treat mouth and gastrointestinal fungal infections.

SAR of Polyene antifungal agents

1. Polyenes are amphoteric, poorly water soluble and unstable due to the presence of higher unsaturation.
2. They consist of large macrocyclic core: 35 membered Amphotericin B1 which is hydrophilic, have polyhydroxylated chain (C_1-C_{35} IN Amphotericin B1) which is bound in lipophilic polyene fragments (C_{20}-C_{35}) by two alkyl linkage.
3. Diversity develops in polyene from a number and configurational arrangement of hydroxylic group and double bonds essential for antifungal activity.

20.9 MISCELLANEOUS ANTIFUNGAL AGENTS

Flucytosine: 5-fluorocytosine (5-FC), 4-amino-5-fluoro-1,2-dihydropyrimidin-2-one

Flucytosin is a potent antifungal agent. It is used in the treatment of serious systemic fungal infections caused by *Cryptococcus* and *Candida* sp.

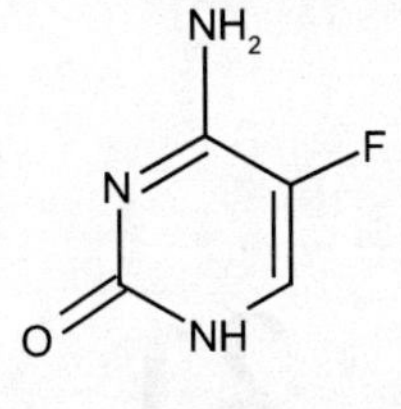

Fig. 20.22 Flucytosine

20.10 NEWER AGENTS

Voriconazole: vfend[(R)]

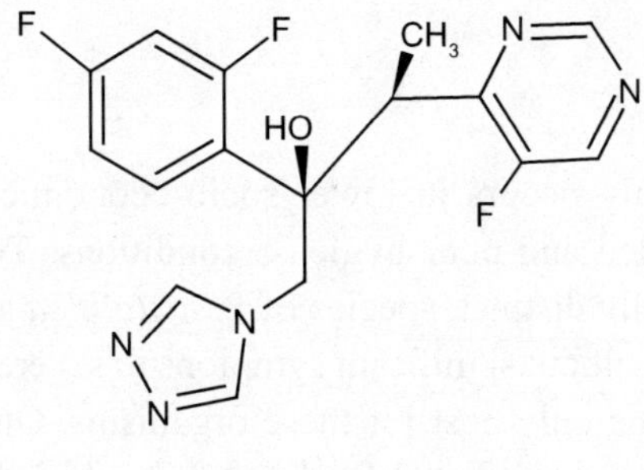

Fig. 20.23 Voriconazole

Voriconazole is suggested for serious fungal or yeast infections, such as aspergillosis (fungal infection in the lungs), candidemia (fungal infection in the blood), esophageal candidiasis (candida esophagitis), it is also used to treat fungal infections in the skin, stomach, kidney, bladder, or wounds.

Rezafungin (Cidara)
Fosmanogepix (Amplyx)

EXERCISE

1. Classify antifungal agents with chemical structures.
2. Explain Azole antifungal agents with their MOA.
3. Explain SAR of flucanazole antifungal agent.
4. Write a note on antifungal antibiotics.
5. Write synthesis, MOA and uses of the following drugs.
    1. Fluconazole
    2. Cotrimazole
    3. Butoconazole.

21
Antiprotozoal Agents

21.1 INTRODUCTION

Amoebiasis infection commonly occurs in lower socio-economic groups and institutionalized individuals living under crowded and poor hygienic conditions. Two morphologically identical, but genetically and biochemically distinct, species of *E. histolitica* and *E. dispar* are the causative organisms. Amoebiasis can be without significant symptons to severe condition and life threatening dysentery. Human beings are the only host for these organisms. Other factors are transformation of ingested amoebic cysts from contaminated food or water which form acid gastric contents and transform them into trophozoites causing colitis. In some cases, targeted organs are the brain and the liver producing abscesses and systemic diseases.

Other Protozoal diseases are:

Giardiasis: It is caused by *Giardia lamblia* found in duodenum and jejunum. It is a common disease among people who drink contaminated water. Abdominal cramps, diarrhoeal, flatulence and anoxia are the main symptoms of this disease.
Trichomoniasis: It is caused by *Trichomonas vaginalis*. It causes infection in vagina, urethra and prostate.
Pneumocystis: This is commonly known as PCP. It is caused by *Pneumocystis carini*. This protozoal infection is generally caused by pneumocystis.
Tricryps: Three pathogens belong to this family *Trypanosomatidae, Kineroplastida* and *Trypanosome* (*African trypanosomiasis*).

21.2 CLASSIFICATION OF AMOEBICIDES

1. **Luminal amoebicides:** It is active only against intestinal forms of amoeba.
   Example: Diloxanide furoate.

2. **Systemic amoebicides:** These compounds have been employed primarily to treat severe amoebic dysentery or hepatic abscesses. Example: Dihydroemetin, Choroquine.
3. **Mixed amoebicides:** These agents are active against both intestinal and systemic forms of amoeba. Examples: Metronidazole, tinidazole, and ornidazole.

21.3 CLINICALLY USEFUL AMOEBICIDES

1. **Metronidazole: Flagyl(R)** 2-methyl-5-nitroimidazole-1-ethanol

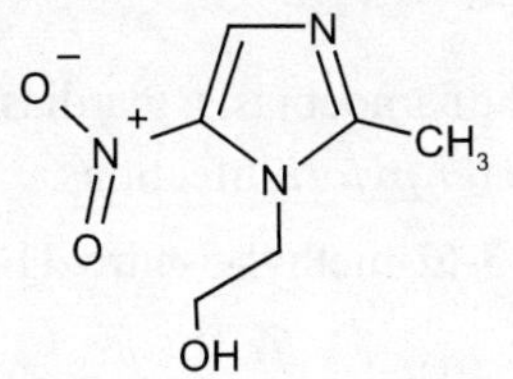

Fig. 21.1 Metronidazole

Synthesis:

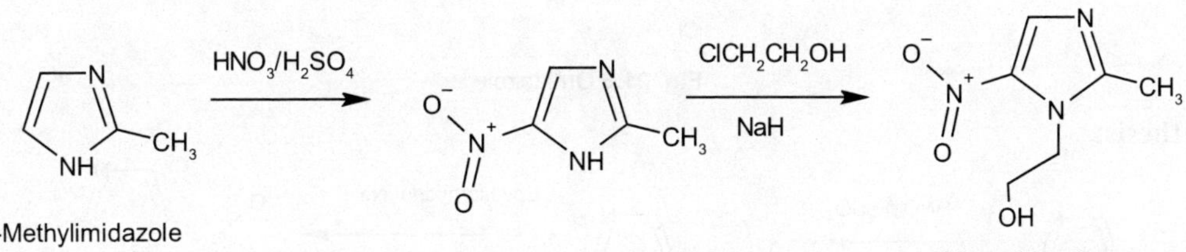

Fig. 21.2

Metronidazole is a prodrug and anaerobic organism which reduces nitro group into hydroxylamine during which reactive species produce cause destructive effect on cell component (DNA, protein and membrane) of organism. Metronidazole was initially used in the treatment of vaginal infection. It has also been exhibited to be effective for the treatment of amoebiasis, giardiasis and anaerobic bacterial infections.

Tinidazole: Tindamax(R), 1,2-(Ethylsuphonyl)-ethyl-2-methyl-5-nitroimidazole

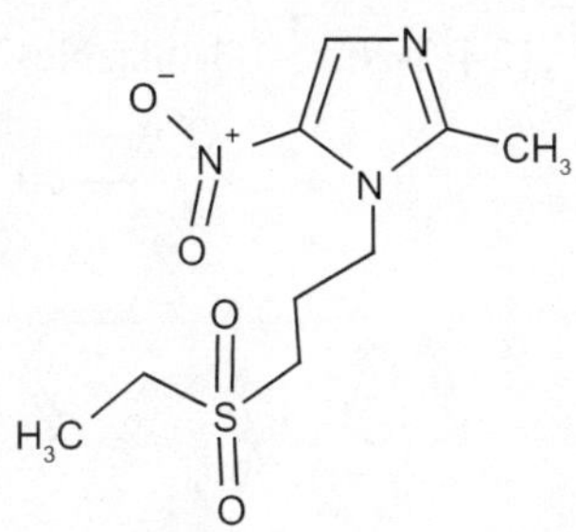

Fig. 21.3 Tinidazole

Synthesis:

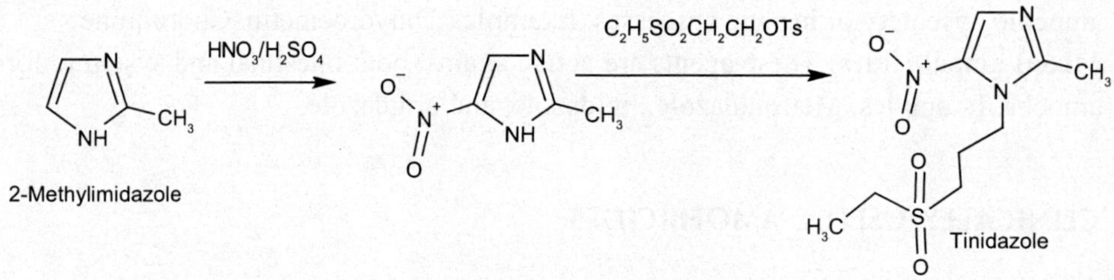

Fig. 21.4

Tinidazole is used in the treatment of amoebiasis, giardiasis and trichomoniasis. It is also highly effective in the treatment of *Hylicobacter pylori* infections.

Ornidazole: Xynor(R), 1-Chloro-3-(2-methyl-5-nitro-1H-1-imidazolyl)-2-propanol

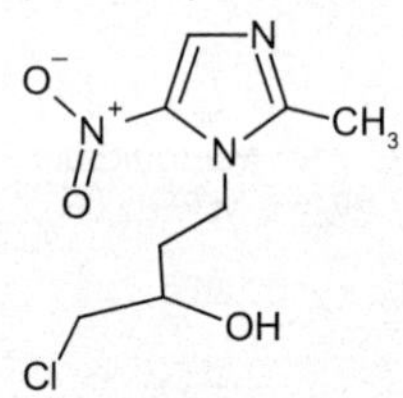

Fig. 21.5 Ornidazole

Synthesis:

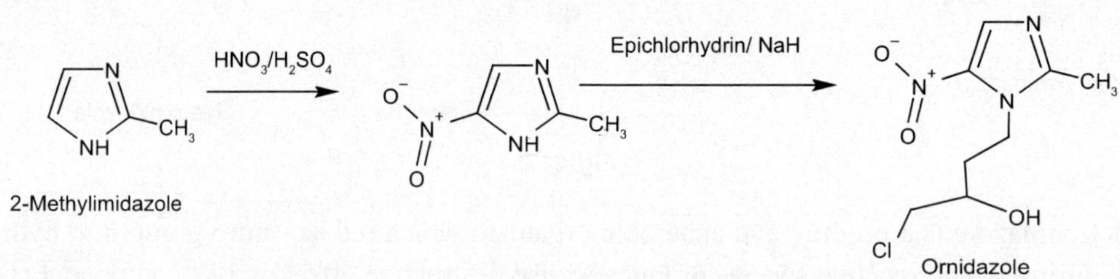

Fig. 21.6

Ornidazole was initially used for treating trichomoniasis before being recognized for its broad anti-protozoan and anti-anaerobic-bacterial properties. It has also been investigated for use in Crohn's disease after bowel resection.

Nitrazoxamide (NTZ): Alinia(R), [2-[(3-Nitro-1,3-thiazole-2-yl)-carbamoyl)]phenyl]-ethanoate

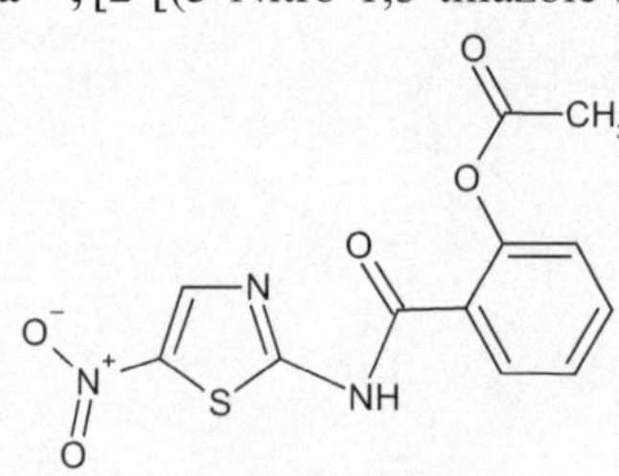

Fig. 21.7 NTZ

Synthesis:

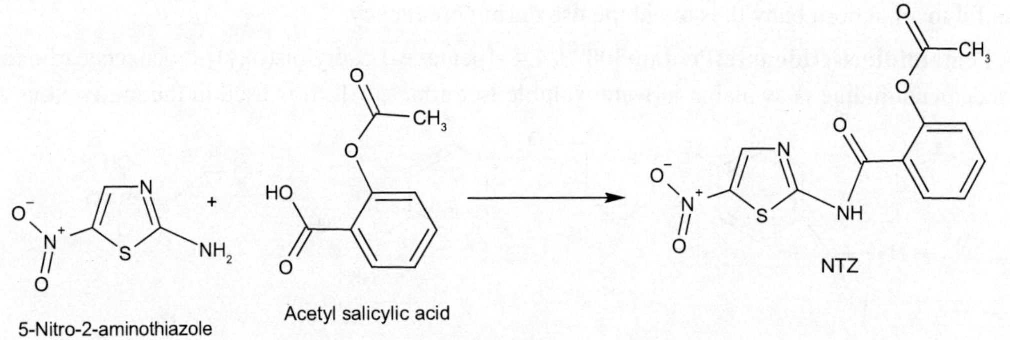

Fig. 21.8

Nitrazoxamide is approved orphan drug used in the treatment of diarrhoea and giardiasis in children. It is used in the treatment of cryptosporidiosis protozoal infection. NTZ is inhibit pyruvate farridoxin oxydoreductase in *Trichomonas vaginalis, Entamoaeba histolica* and *Clostridum perfingens*. Therefore, disruption of bioenergetics in the organism.

Diloxanide furoate: Furamide(R), 2,2-Dichloro-N-(4-furoyloxyphenyl)-N-methylacetamide

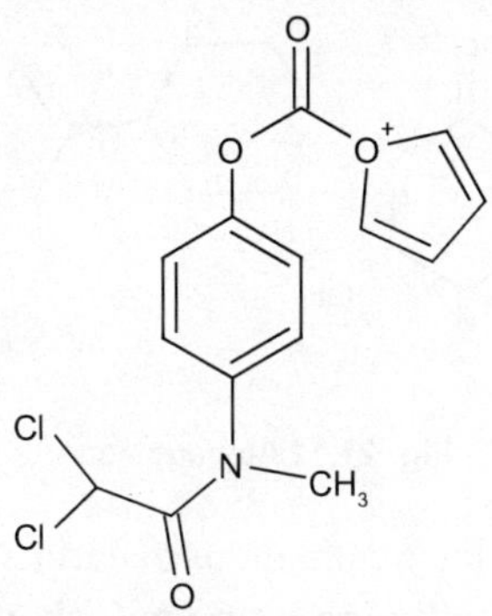

Fig. 21.9 Diloxanide furoate

Synthesis:

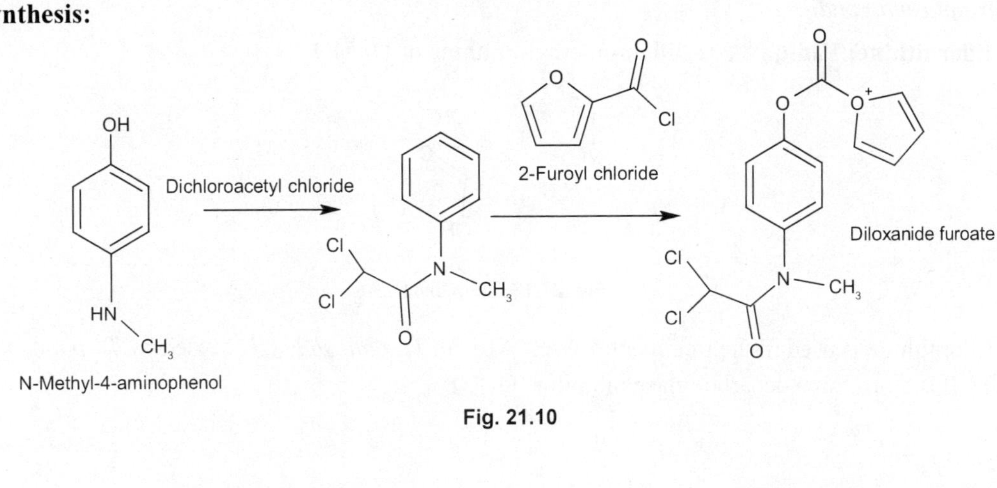

Fig. 21.10

Diloxanide is used in the treatment of worm infections and intestinal amoebiasis. It may be harmful to an unborn baby thus avoid the use during pregnancy.

Pentamidineisethionate:Pentam300^(R), 4,4'-[pentane-1,5-diylbis(oxy)]-dibenzenecarboximidamide pentamidine is available in water soluble isethionate salt. It is used in the intravenous and

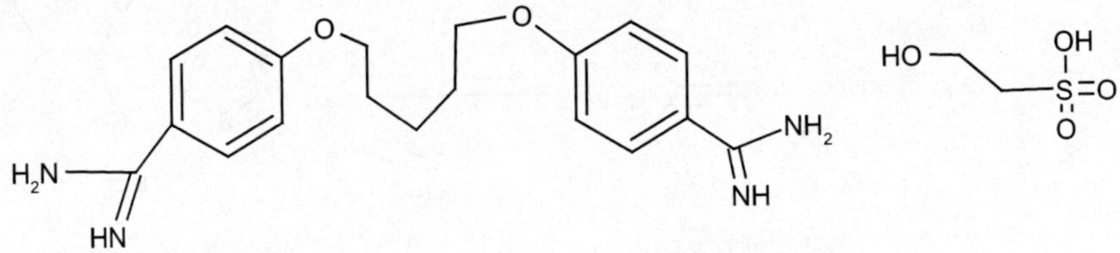

Fig. 21.11 Pentamidine isethionate

as aerosol formulation. It impart the cleavage process of DNA in the trypanosome parasite and inhibit topoisomerase which leads to double strand cleavage of DNA. Pentamidine is used as second line therapeutic agent in the treatment and prophylaxis of PCP (*Pneumocystis pneumonia*).

Atovaquinone: Mepran^(R), trans-2-[4-(4-chlorophenyl)cyclohexyl]-3-hydroxy-1,4-naphthalenedione.

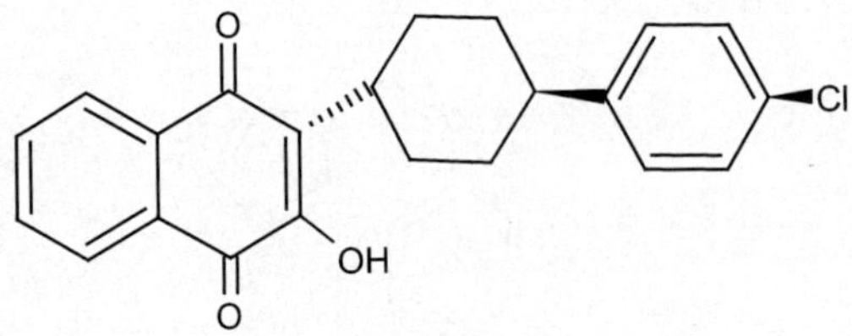

Fig. 21.12 Atovaquinone

Atovaquinone is initially used in the treatment of malaria. Nowadays, it is used in the treatment of PCP. It exhibited antiparasite action by inhibiting mitochondrial respiratory chain and collapse mitochondrial membrane. Atovaquinone also reported for the treatment of Toxoplasmosis caused by *Toxoplasma gondii*.

Eflornithine: Vaniqa^(R), α-difluoromethylornithine or DFMO

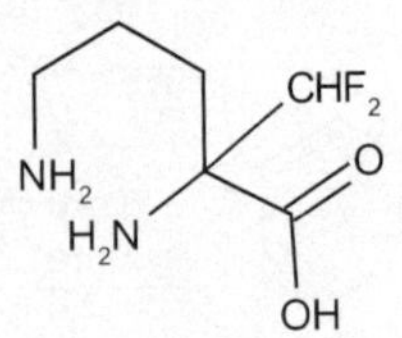

Fig. 21.13 Eflornithin

Eflornithine is used in the treatment of West African *Trypanosomiasis* caused by *Trypanosome brucei*. It is a ornithine decarboxylase inhibitor (ODC)

Iodoquinol: iodoxin$^{(R)}$, 5,7-diiodoquinolin-8-ol

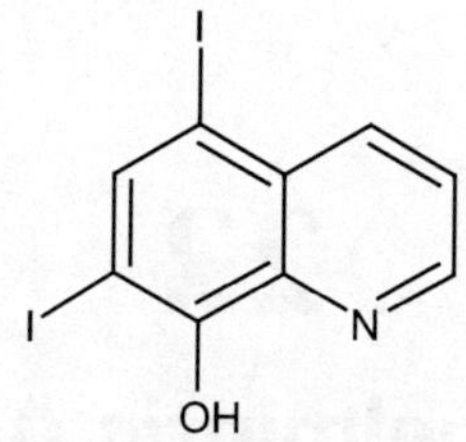

Fig. 21.14 Iodoquinol

Iodoquinol is used against *Entamoeba histolytica*, and it is effective against both *cyst* and *trophozoites* that are localized in the lumen of the intestine. It is considered the drug of choice for treating asymptomatic or moderate forms of amebiasis.

21.4 MISCELLANEOUS AGENTS

Melarsoprol: Arsobal$^{(R)}$

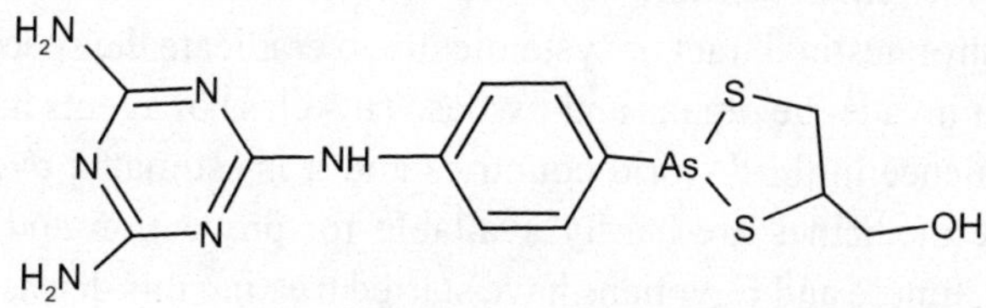

Fig. 21.15 Melarsoprol

Melasoprol is arsenic containing drug (Organoarsenical). It is used in the treatment of *African trypanosomiasis.* It is also used in the treatment of *Meningoencephalitic trypanosomiasis.*

EXERCISE

1. Explain Amoebicides with their chemical structures.
2. Write the synthesis, MOA and uses of
   (a) Tinidazole
   (b) Ornidazole
   (c) Diloxanide furoate
3. Give the example of arsenic containing anti-amoebic agent.
4. Write a note on Atovaquinone and Iodoquinol.

22
Antithelmintic Agents

22.1 INTRODUCTION

Anthelmintics are medicinal agents, used to treat parasitic infections caused by pathogenic worms referred to as metazoan which are conventionally classified into roundworms (nematodes) and two types of flatworms, flukes (trematodes) and tapeworms (cestodes). Anthelmintics act locally either to out worms from the gastrointestinal tract or systemically to eradicate the species and the developing forms of helmintics that invade the organs and tissues. This class of agents have added significance due to their wide prevalence in third world countries and it is estimated over 2 billion people are affected. Cost effective medicines are hardly available for prevention and treatment, even now, although the primitive Chinese and Egyptians have started treating this disease thousands years ago.

22.2 BENZIMIDAZOLE DERIVATIVES

Benzimidazoles are clinically broad-spectrum nematocides with some agents also having activity against tapeworms and flukes. Fenbendazole is effective against the protozoan parasite *Giardia*. Many BZs kill larval stages of nematodes, as well as the adults, but usually larval efficacy is lower than for adults. In some cases increasing the dose and/or duration of treatment (e.g., treat for multiple days) enhance the efficacy, particularly against larval stages.

Mode of action: These drugs act by blocking the glucose transportation in the parasites and lead to the depletion of glycogen storage of the intracellular microtubules in the cells of the worms, thereby arresting the nematodes and cell division in the metaphase. The major site of action is microtubular protein β tubilin of the parasite. These drugs are bound with β tubilin and inhibit the polymerization.

22.3 CLASSIFICATION OF BENZIMIDAZOLES

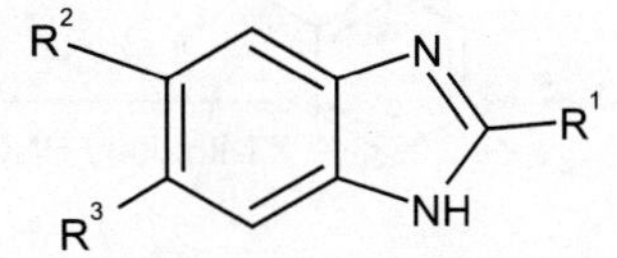

Fig. 22.1 Basic structure of benzimidazole nucleus

Derivatives	R^1	R^2	R^3
Mebendazole	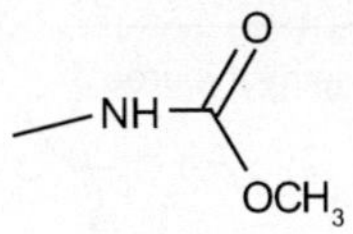	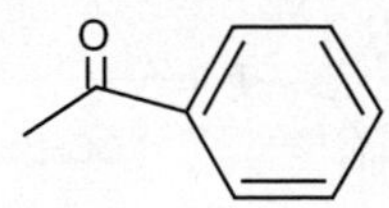	H
Albendazole	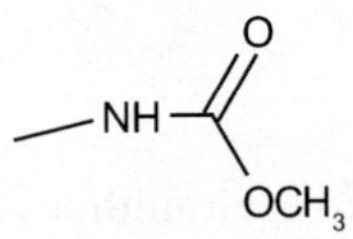	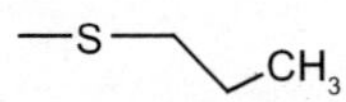	H
Triclabendazole	—S—CH_3	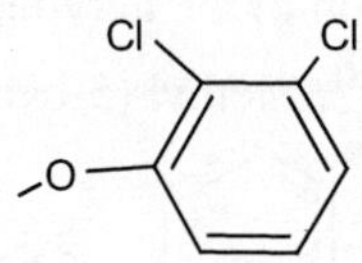	Cl
Fenbendazole	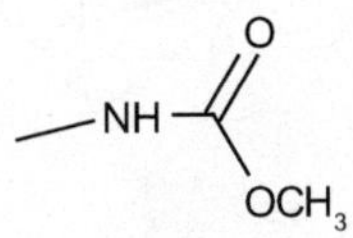	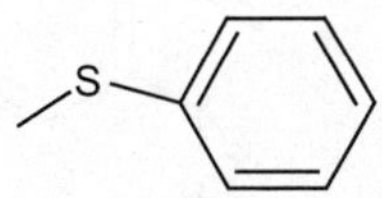	H
Flubendazole	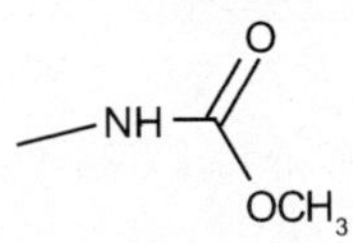	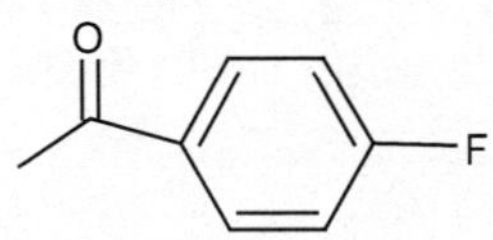	H

22.4 CLINICALLY USEFUL BENZIMIDAZOLE ANTHELMINTIC AGENTS

Albendazole: Albenza$^{(R)}$, Methyl-[5-(propylthio)-1H-benzoimidazol-2-yl]-carbamate

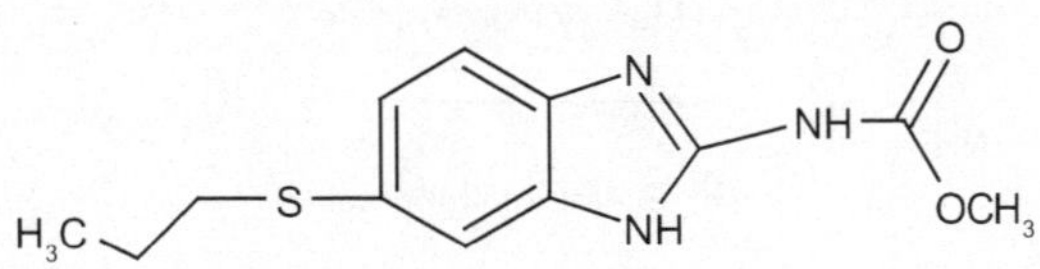

Fig. 22.2 Albendazole

Synthesis:

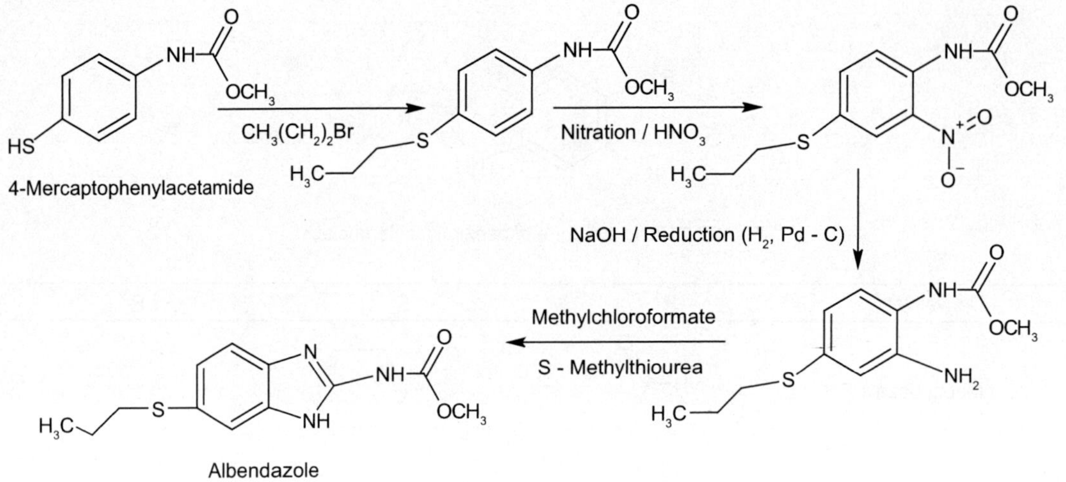

Fig. 22.3

Albendazole is clinically useful in the treatment of Helminthiasis infection. It is used to treat neurocysticercosis, an infection of the nervous system caused by pork tapeworms.

Mebendazole: Vermox(R), Methyl (5-benzoyl-1H-benzimidazol-2-yl)-carbamate

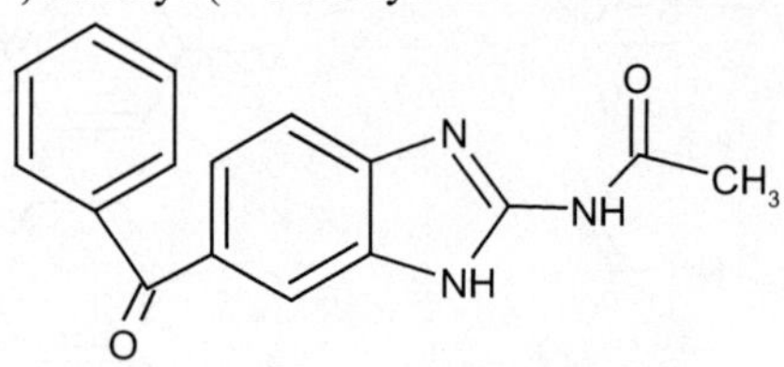

Fig. 22.4 Mebendazole

Synthesis:

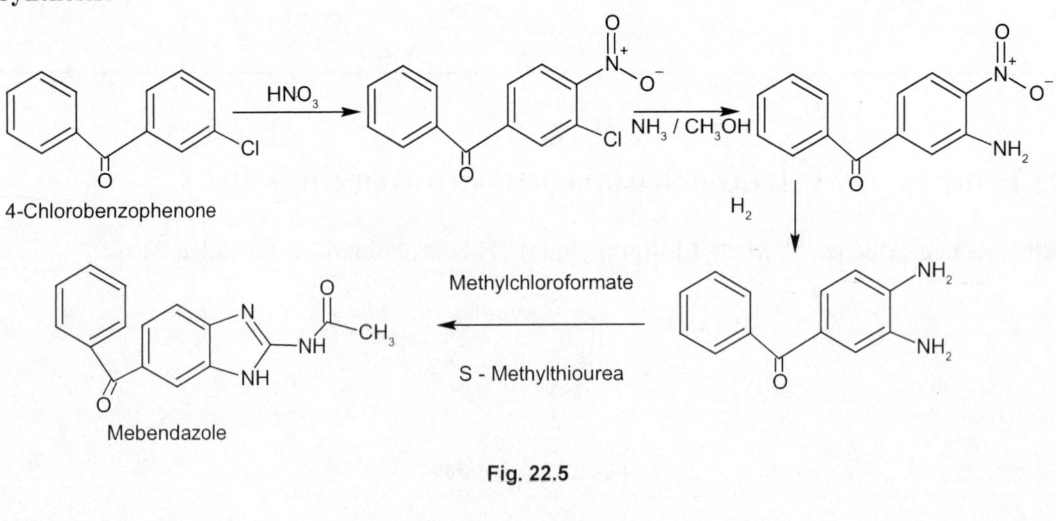

Fig. 22.5

Mebendazole is used to treat serious parasitic worm infestations. This includes ascariasis, pinworm infection, hookworm infections, guinea worm infections, hydatid disease, and giardiasis, among others.

Thiabendazole: Mintezol$^{(R)}$, 4-(1H-benzimidazol-2-yl)-1,3-thiazole

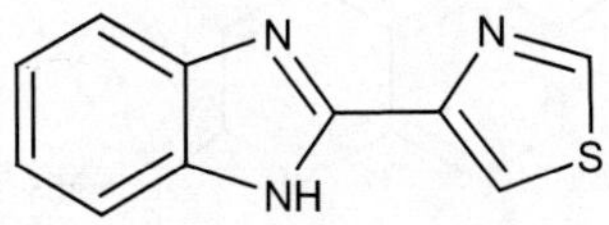

Fig. 22.6 Thiabendazole

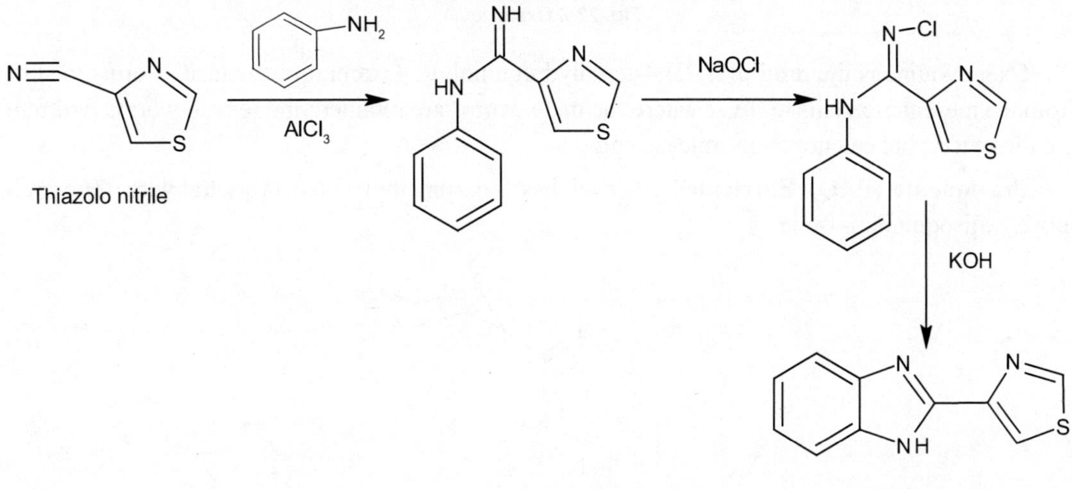

Fig. 22.7

Thiabendazole is used as an antiparasitic agent. Initially, it was used as antifungal and food additives.

22.5 QUINOLINE AND ISOQUINES DERIVATIVES

Quinoline and Isoquines is the class of anthelmintic agents. Oxamquinine, Praziquental, Niclosemide are the active drugs of this category.

Mode of Action: The active members of quinoline include Oxamniquine associated with an irreversible inhibition of the nucleic acid metabolism of the parasites. On the other side Praziquantel works by causing severe spasms and paralysis of the worms' muscles caused by a rapid Ca^{2+} influx inside the schistosome.

22.6 CLINICALLY USEFUL DERIVATIVES

Oxamniquine: Vansil$^{(R)}$, {2-[(isopropylamino)-methyl]-7-nitro-1,2,3,4-tetrahydroquinolin-6-yl} methanol

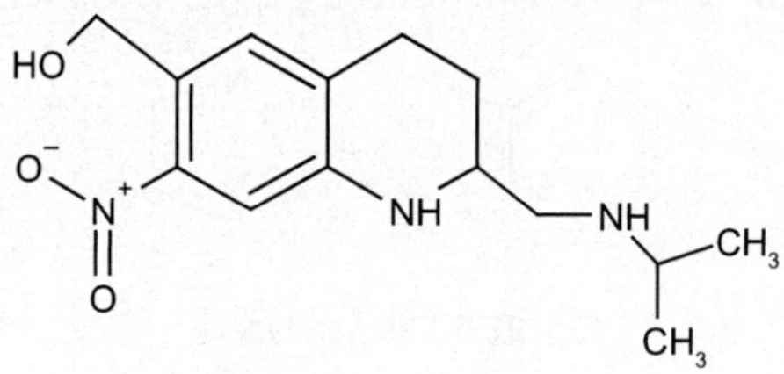

Fig. 22.8 Oxamniquin

Oxamniquine is the drug of 1,2,3,4-tetrahydroquinoline. Oxamniquine causes worms to shift from the mesenteric veins to liver where the male worms are retained; the female worms return to the mesentery, but can no longer release egg.

Praziquentel (PZQ): Biltricide$^{(R)}$, 2-[cyclohexyl(oxo)methyl]-3,6,7,11b-tetrahydro-1*H*-pyrazino[2,1-a]isoquinolin-4-one.

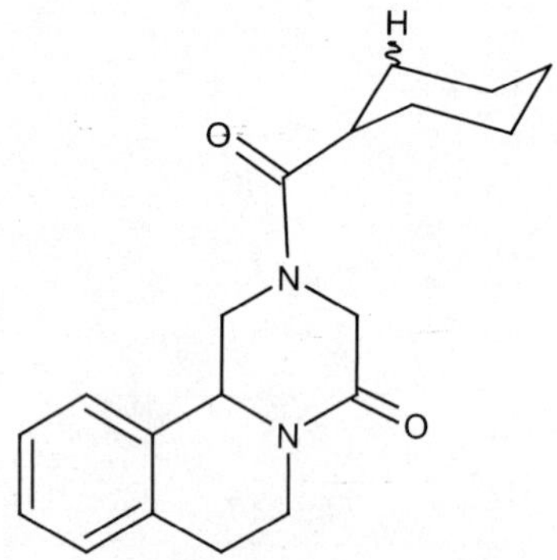

Fig. 22.9 PZQ

Praziquentel is isoquinoline derivative. Its levo-enentiomer is found to be most active. It is a potent drug for cestodes and trematodes infections. PZQ is medicinal agent of choice for treatment of schistosomiasis and liver flukes. It is rapidly absorbed and undergoes first pass metabolism and less active serum metabolites contains hydroxyl group. This hydroxylation catalysed by CYP2B6 and CYP3A4.

Pyrentel: Combantrin$^{(R)}$, 1-Methyl-2-[(E)-2-thiophen-2-ylethenyl]-5,6-dihydro-4H-pyrimidine

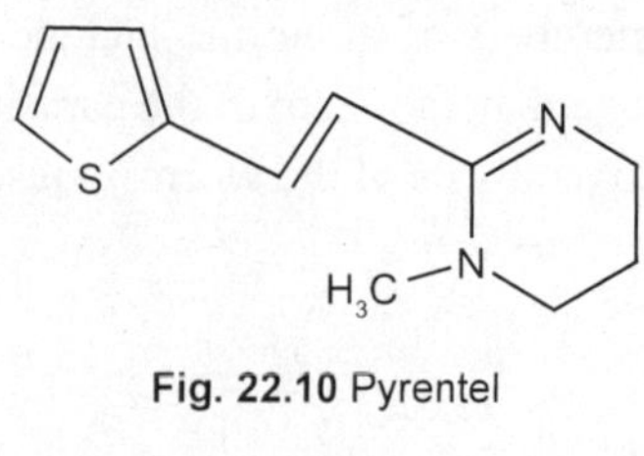

Fig. 22.10 Pyrentel

Pyrentel was initially discovered for its anthelmintic activity. It is active angaist intestinal roundworms infections and pinworms. This is available in insoluble acid, pamoate salt that improves the usefulness of drug in the treatment of intestinal helminthis. It acts as depolarising neuromuscular blocking agents to worm paralysis.

Niclosemide: Niclocide[(R)], 5-Chloro-N-(2-chloro-4-nitrophenyl)-2-hydroxybenzamide

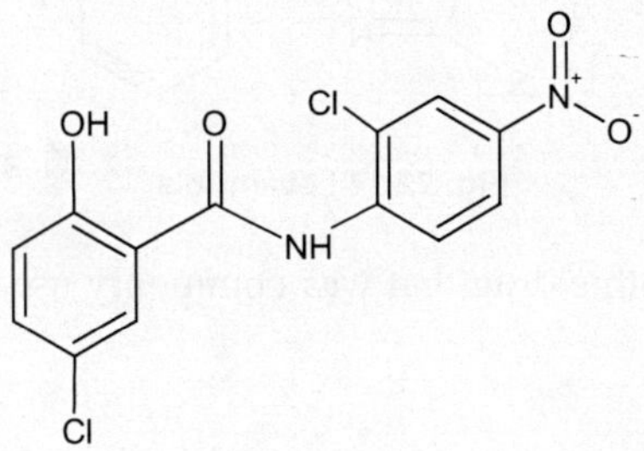

Fig. 22.11 Niclosemide

Niclosamide is suggested to treat broad or fish tapeworm, dwarf tapeworm, and beef tapeworm infections. Niclosamide is also used for other tapeworm infections. It will not work for other types of worm infections including pinworms or roundworms.

22.7 MISCELLANEOUS AGENTS

Diethylcarbamazine: *N,N*-Diethyl-4-methyl-1-piperazinecarboxamide

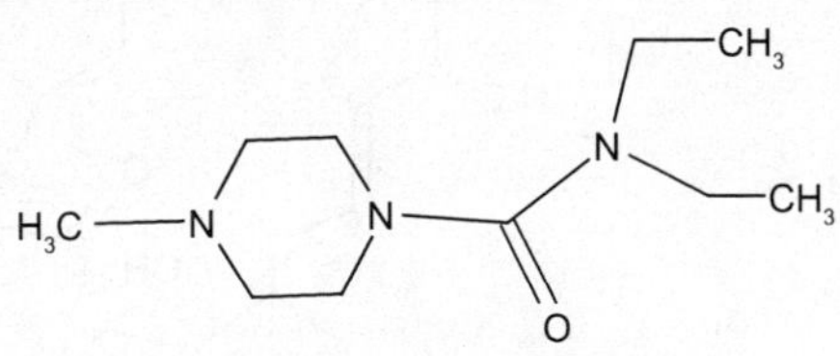

Fig. 22.12 Diethylcarbamazine

Synthesis:

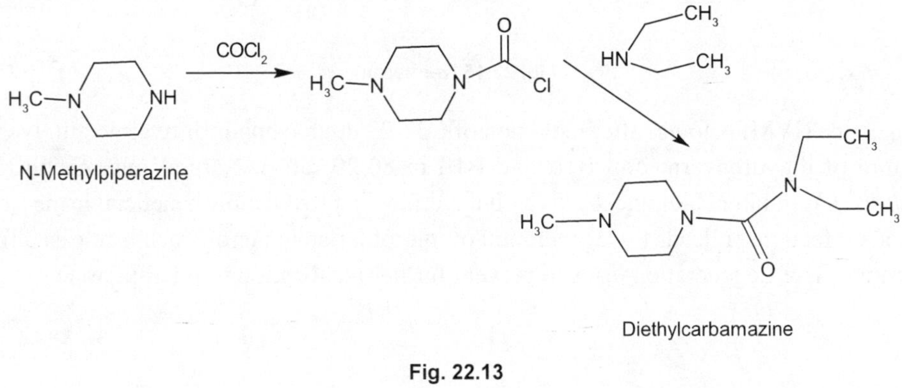

Fig. 22.13

Diethylcarbamazine is a piperazine derivative. It is used in the treatment of filariasis including lymphatic filariasis, tropical pulmonary eosinophilia, and loiasis.

Levamisole: Ergamisol(R), (S)-6-Phenyl-2,3,5,6-tetrahydroimidazo[2,1-b][1,3]-thiazole

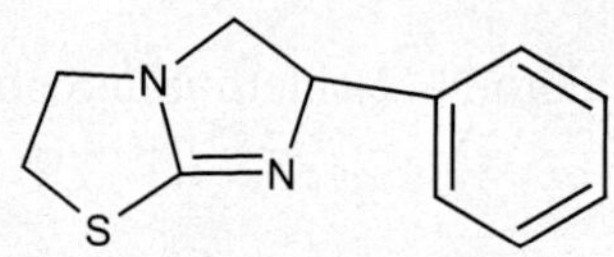

Fig. 22.14 Levamisole

Levamisole is an antihelminthic drug that was commonly used to treat parasitic, viral as well as a bacterial infections.

22.8 NEWER AGENTS

Ivermectin: Stromectol(R)

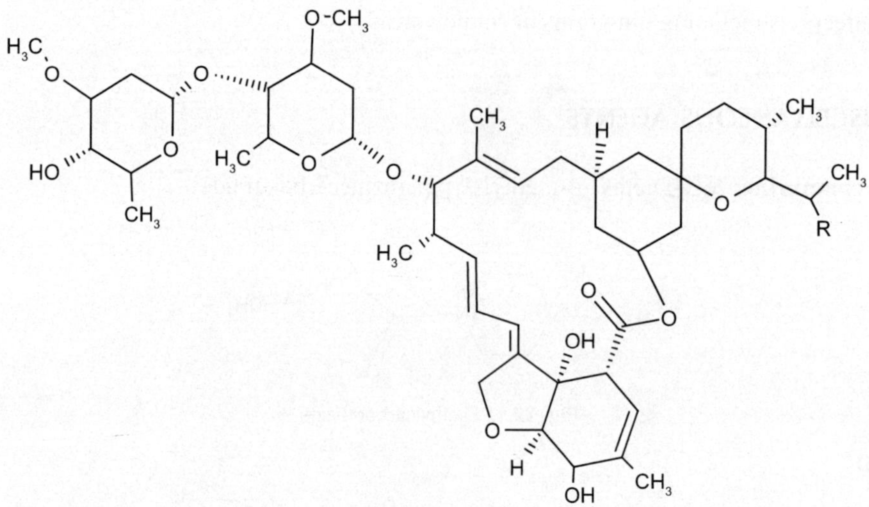

Fig. 22.15 Ivermectin

Ivermectine (IVM) is found after reduction of C22-23 double bonds of Avermectin. Ivermectin is a mixture of dihydroavermectin B1α and B1β in 80:20 ratios. Natural Avermectin (source: *S. avermitilis*) has minimal biological activity but ivermectin proved quite beneficial in the treatment of nematode infections. It leads to degeneration of microfilariae in utero which is released from the female worms. This degeneration leads to prevent further fertilization of microfillariae.

Moxidectin:

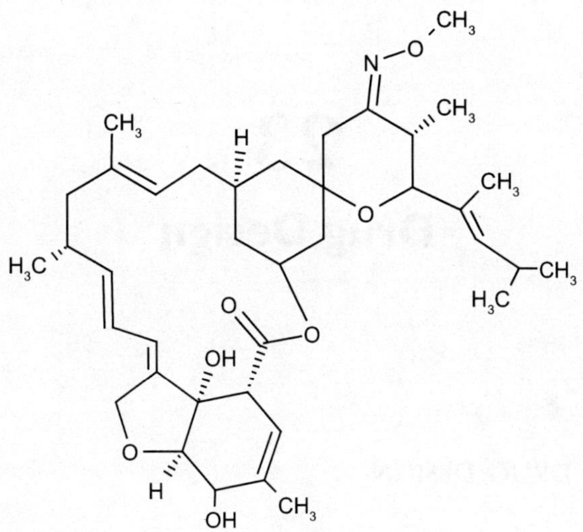

Fig. 22.16

Moxidectin is a recently approved drug for young and older patient of onchocerciasis. This drug is effective against microfilaria. It has similar adverse effect like ivermectine. It is also used for infected dogs, cats and sheep for the treatment of worms.

EXERCISE

1. Classify Benzimidazole derivatives as anthelmintic agents.
2. Explain the synthesis, MOA and uses of Piperazine derivative as anthelmintic agents.
3. Write synthesis, MOA and uses of the following drugs:
   (a) Albendazole
   (b) Mebendazole
   (c) Thiabendazole
4. Write an explanatory note on quinoline and isoquinolins derivatives.
5. Write a note on the following drugs.
   (a) Niclosemide
   (b) Ivermectin
   (c) Diethylcarbamazine

23
Drug Design

23.1 CONCEPTS OF DRUG DESIGN

The discovery of new drugs needs a design process along with the synthetic techniques, development of tests, and further procedures to establish pharmacodynamics and their toxicological tools. The diversity in synthetic chemistry is ever increasing as a result of the increase in the number of chemical scaffolds, which requires a sequence of screening modules with appropriate time consumption and achieve little success. These consequences made it necessary to search the possibilities for the development of new, logical, and molecular approaches in the discovery of new scaffolds or drugs and are known as drug designing.

Drug design is an integrated advancing discipline, in which a biologically active molecule is produced by chemical synthetic approaches followed by an evaluation of pharmacological and toxicological studies with the limitation of trial-and-error screening. In the exploration, it implies evaluation (randomly) of congeners or derivatives produced from the lead molecule either by execution of tailor-made techniques or by applying the basic concepts of physicochemical properties to produce pharmacological active molecule.

1. Design of Analogues and Prodrugs

Drug design involves the development of analogues and prodrugs by some chemical modifications from the lead molecule or from a parent compound by modifying the carbon skeletal transformation or by the synthesis of compounds of the same nucleus with various substitutions. Analogues can also be synthesized by changing the position of substitution group. For example, synthesis of *trans*-diethylstilbestrol by the modification of oestradiol produced better oestrogenic activity than the latter one.

The term prodrug implies an appropriate derivative of a drug. Prodrugs are discovered by screening of active metabolites after in vivo biotransformation that is formed from parent compounds. For example, phenylbutazone is transformed into oxyphenbutazone by hydroxylation reaction mediated by phase I (non-synthetic) metabolic reaction.

2. Design of Lead and Lead Discovery

The concept of lead discovery envisages two investigational processes. They are:

1. Search of new lead molecules.
2. Assessment, molecular modelling, and extension of scaffold.

A drug discovery is based on lead molecule. The most prominent examples include penicillium and librium. Librium is the first benzodiazepine tranquillizer, synthesized by Leo Sternbach at Roche in a development programme of newer tranquillizer agents. None of the molecules showed reliable pharmacological activity. One of the molecules that had been ignoring was found to possess probable activity and chemically it was benzodiazepine-4-oxide. It produced chloromethyl quinazoline-3-oxide with methylamine. This identification was further exploited to develop potent analogues such as diazepam, which was found to be 10 times more potent than the lead molecule.

3. Lead Identification

The lead identifications require a series of biological evaluations of the lead molecules. After the identification, it can be structurally modified for the potency and the biological activity are improved.

4. Random Screening

The synthesized compounds or any chemical constituents obtained from natural products are evaluated in a series for their biologically active components. Thus, random screening may produce active medicines with unwanted biological activity. Earlier some antibiotics, such as streptomycin, tetramycins, and fungal metabolites, such as lovastatin and cyclosporins, were discovered through this method. This approach is expensive and time consuming and the success rate is considerably low.

5. Non-random Screening

Non-random screening processed the compounds with similar structural skeletons were evaluated from their particular properties.

6. Pharmacokinetic Studies

Biotransformation occurs as the fate by metabolizing enzymes. In order to develop new leads, the metabolites or biotransformed compounds are studied for their properties, and such studies are expected to evaluate the activity from a differentiation with the parent molecule. The discovery of sulphanilamide is reported through the metabolic studies of prontosil is an unique example of pharmacokinetic studies.

7. Pharmacodynamic Studies

The effects, apart from the therapeutic actions, that is, side effects may lead to the finding out of a new molecule with some appreciable structural modification. For example, sulphonamide used specifically for the treatment of typhoid, lowered the blood sugar levels drastically. This exerted action led to the finding of aryl sulphonyl thiourea moiety that is responsible for the lowering of blood glucose level. Another example is amino alkyl derivatives of iminodibenzyl, were synthesized as analgesic, sedative, and antihistamines that was found to possess anti-depressive and anti-psychotic actions. Many tricyclic antidepressants are synthesized from amino alkyl derivatives of iminodibenzyl moiety.

8. Drug Design through Disjunction

Disjunction comes into the systemic formulation of analogues of a prototype agent, generally, towards structurally simpler products, which may be carried as quasi-replicas or partials of the prototype agent. This is employed by various methods, that is:

- Unjoining of certain bonds.
- Substitution of aromatic cyclic system for saturated bonds.
- Elimination of the size of hydrocarbon portion of the parent molecule.
- Decyclization of ring system.
- Cyclization of hydrocarbon chains.

23.2 QSAR (QUANTITATIVE STRUCTURE ACTIVITY RELATIONSHIP)

QSAR methodology has progressed remarkably in 19th century. In early 1868, Crum-Brown and Fraser expressed an equation of physiological action and chemical compositions. Richet discovered that diversity of the cytotoxicity of organic compounds such as Aldehyde, alcohol, ethers, ketone, etc., were inversely related to their water solubilities. At the turn of 20th century Meyer and Overton have worked independently and observed the linear relationships parallel to their olive oil/water partition coefficient and narcotic activities. Further, Fuhner has given the first evidence of additivity of groups contributing to biological activity values in a homologous series of compounds for their narcotic activity. Ferguson, 1939, postulated that the thermodynamic principles could be related to drug activities.

23.2.1 Hansch Analysis

QSAR paradigm is based on Hammett's relationship utilize electronic properties as the descriptors of structures. Problems were encountered when investigators attempted to apply Hammett-type relationships to biological systems, indicating that other structural descriptors were necessary. In 1962, Hansch et al., published a study on structure activity relationship with the numerical information on lipophilicity, electronic, and steric effect on the model development. The general form of Hansch equation is as follows:

$$\text{Log BA} = a \log p + b\,\sigma + c\,Es + \text{constant (linear)}$$

$$\text{Log BA} = a \log p + b\,(\log p)^2 + c\,\sigma + d\,Es + \text{constant (nonlinear)}$$

Hansch's model correlates biological activity with physicochemical properties. The coefficients (a, b, c, d, and constant) are determined by multiple regression analysis. BA represents the biological activity while log p represents the partition coefficient.

23.2.2 Free-Wilson Analysis

Free-Wilson analysis is also known as the additivity model or de novo approach which addresses structure activity studies in congeneric series. This method is based on the facts that the introduction of a particular substituent at a particular molecular position always contributes in the same way to the biological potency of the whole molecule, as expressed by the equation:

Biological activity (log BA) is contribution of unsubstituted parent compound and contribution of corresponding substituents in the molecule.

$$\text{Log BA} = \mu + \Sigma i \; ai \; aj$$

where, ai = average number of substituted positions

aj = average number of substituents at that position

μ = overall average.

$X_1 = 1$ (presence) and $X_1 = 0$ (absence), denoted of the different substituents as independent parameters, while the measured activity serves as dependent variable.

23.2.3 Mixed Approach

Kubinyi has presented the combination of Hansch and Free-Wilson approach as mixed approach.

$$\text{Log BA} = k1\,\pi + k2\,\sigma + k3\,Es + k \text{ (Hansch equation)}$$

$$\text{Log BA} = \mu + \Sigma\; ai\; aj \text{ (Free-Wilson equation)}$$

Therefore, mixed approach is

$$\text{Log BA} = \Sigma\; ai\; aj + \Sigma\; ki\; \phi j + k$$

where, Σ (ai aj) is the Free-Wilson part for the substituents

$\phi j = \sigma$, π and Es contribution of the parent skeleton.

Among the above-mentioned approaches, Hansch approach became the most popular approach in QSAR. The high-dimensional QSAR analyses (3D, 4D, and 5D) are developed to avoid pitfalls of classical method and to create the hypothetical drug receptor model.

23.2.4 Primacy of QSAR

- Quantifying the relationship between structure and activity with their physiochemical property basis.
- Predictions of designed possible compounds before the chemical synthesis of novel analogues.
- Understand the interactions between functional group of designed molecules and their activity of targeted enzyme or protein.

23.2.5 Limitations of QSAR

- False correlations of biological data may occur due to experimental error.
- Inefficient set of molecule give unsatisfactory descriptive reflections of molecular property which is unable to predict the most active compounds.

- In some 3D QSAR study ligands binding receptor or protein may not be available thus QSAR evaluation would be unsuccessful.
- Sometimes common approach results may not show the valid data score.

A model perfectly predicts that the training data may not be good or even useless for prediction. The problem of QSAR is to find coefficients C_0, C_1 , ... , C_n such that

$$\text{Biological activity} = C_0 + (C_1 \times P_1) + ... + (C_n \times P_n)$$

Partial least squares (PLSs) test is statically used for computation of the coefficients of descriptors for the prediction error is minimized for a list of given compounds.

23.2.6 Basic Requirements for QSAR Analysis

Basic requirements for best model development are the following:

- All analogues belong to a congeneric series, i.e., classical QSAR studies, exerting the same mechanism of action. This is a series of compounds with a similar basic structure, but with varying substituents. Non-congeneric series are widely used for higher dimensional (3D and 4D) studies.
- The set of bioactive compounds with same biological of action is essential for QSAR study.
- Exploration of biological response should be distributed to large number of derivatives.
- Observed biological activity should be in in molar units or IC_{30} or percentage inhibition.
- Total number of compounds in the training set divided by the number of variables in the final model should be greater than approximately five or six.

This will assure that a dataset will not be 'over predicted' and that the model will have a better chance to retain the predictive value.

23.2.7 Steps Involved in QSAR Studies

The QSAR methodology enables the development of mathematical models, which can be used to predict the biological activity of newly designed compounds. There are three steps involved in this procedure.
Creation of a database in which calculation of various physicochemical and structural parameters of a congeneric series takes place.
Regression analyses leading to model development between biological activities versus derived physiochemical descriptors.
Validation of the models and prediction of the biological activity of the designed compounds.

23.2.8 Parameters used in QSAR

1. **Electronic parameters:** Extensive studies using electronic parameter revealed that electronic attribution of molecules are related to their chemical reactivities and biological activities. The reaction responds to electronic perturbation which measures the electronic demand determined by its mechanism. Introduction of substituent's groups and substituent alteration of reaction rate helps to describe overall mechanism of reaction. The role of electronic substituents on reaction rates was established by Hammett, as a model reaction, ionization in water of substitueted benzoic acid and determined the equilibrium constant, $K_{a,.}$ where σ

is the substituent constant which measures the electronic effect of substituent and represent the electronic charge distribution in benzene nucleus.

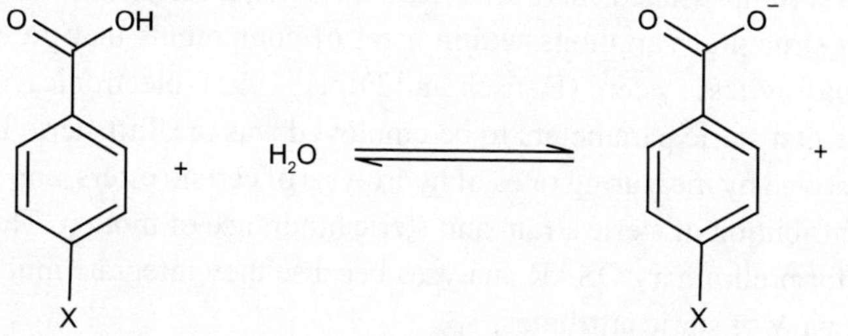

Fig. 23.1

The basic equation is

$$\sigma_X = \log K_X - \log K_H$$

Hence,

$$\text{Log}\ (K_X/K_H) = -pK_X + pK_H$$

Thus,

$$\text{Log}\ (K_X/K_H) = \rho\ \sigma$$

Electronic withdrawing substituent is characterized by positive charge, whereas electronic donating ones are characterized by negative charge. In the extension of this approach, ionization is the substitueted phenylacetic acid was measured.

2. **Hydrophobicity parameter:** The most practical method for assigning a hydrophobic parameter to a given solute is to dissolve it in a standard nonpolar solvent, which is immiscible in water, then determine how much of it can be extracted by water. If the chemical is more soluble in water, it could be dissolved in that phase and extracted by the nonpolar liquid. No nonpolar liquid could possess all the properties that would make it completely ideal; but-1-octanol is firmly established as the primary choice.

$$\log P\ (\text{oct/water})$$

Hydrogen bonding strength has been shown to be an important property in the design of some pharmaceuticals, and thus should not be neglected in pure enzyme studies.

Meyer and Overtone discovered correlation between oil and water partition coefficient, Ferguson extended correlation between depressant and hydrophobicity in thermodynamic context. Hansch and his colleguges suggested using multi-parameter approach that includes both electronic and hydrophobic terms. Hydrophilicity (lower log P values) in the octanol/water system is much more sensitive to solute H-acceptor strength (β) than it is to solute H-donor strength (α). These parameters can be developed independently of each other from the formation constants of 1: 1 H-bond complexes in CCl_4 but only for simpler monofunctional solutes. The difference in log P_{oet} values and those from other solvent systems have been examined in terms of evaluating the H-bonding ability of solutes. For example, cyclohexane as the nonpolar solvent would respond differently to solutes strong in either α or β, but it is also more sensitive to any purely hydrophobic enhancement in solute structure, such as a methylene fragment. It is for this reason that propylene glycol dipelargonate (PGDP) has been proposed as a more suitable "complementary" solvent system to octanol.

3. **Steric parameters:** Simply stated, there are at least three major factors that seem to be involved in understanding structural variations within a set of compounds that produce a common response in a bioassay test system (Hansch and Fujita, 1995): electronic, hydrophobic and steric. One of the first steric parameters to be employed was the Taft steric E_S in 1952. This parameter was derived by measuring rates of hydrolysis of certain esters, and it is a composite reflecting the contribution of steric strain and steric hindrance of motion. The E_S parameters are often useful for preliminary QSAR analyses because they integrate multiple factors and provide a global view of steric attributes.

Taft equation

$$E_S = \log (k_X/k_H)\ A$$

where, E_S is the Taft constant and k_X and k_H are rates of acid hydrolysis of esters, XCH_2COOR and CH_2COOR

Unlike most of these steric parameters that are experimentally derived, Verloop (1976) described several computationally derived parameters to define steric constraints of substituent groups: substituent length, minimal substituent width and maximal substituent width. Substituent volume is another computationally derived parameter that can be used to define substituent groups. Hence, the Verloop and volume parameters permit deconstruction of steric bulk into subcomponents that might contribute differently to structure activity relationships (SAR). Accordingly, in the event that steric factors are implicated by an initial QSAR analysis using E_S, subsequent analysis with computationally derived parameters permits a more refined evaluation of steric factors that might generate added precision.

23.3 STATISTICAL METHODS USED IN QSAR ANALYSIS

Statistical methods are an essential component of QSAR work. They help to build models, estimate a model's predictive abilities, validate an already existing model and find the relationships and co-relationship among the variables and the activities. Data analysis methods are used to recombine data into forms and groups and observations into hierarchies.

1. Regression Methods

It is a mathematical procedure, which correlates dependent (X) variable with the independent (Y) variables. There can be different forms of regression analysis:

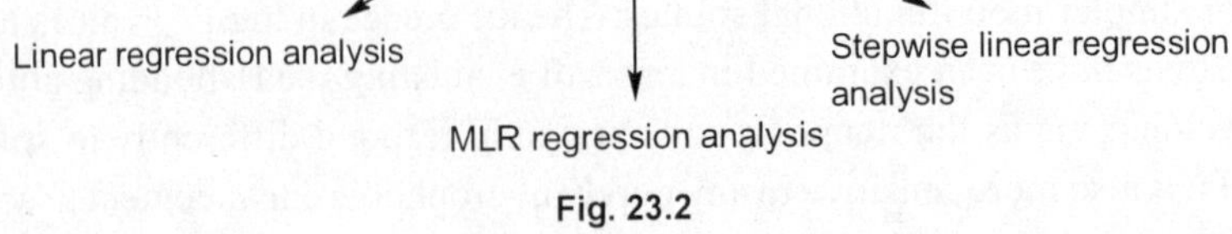

Fig. 23.2

Linear regression analysis: An independent variable is correlated with a dependent variable and produces a linear one-term equation of the most important descriptors.

MLR analysis: More than one independent variable is correlated with a dependent variable and a single multi-term equation is formed.

Stepwise linear regression analysis: This is useful when the number of independent variables is very high and is thus correlated in a stepwise manner with the dependent variable producing a multi-term linear equation.

2. Partial Least Square (PLS)

PLS is used when X data contain co-linearity or when N is less than 5M. It is the method of choice in 3D QSAR method.

where, **N** → Number of compounds

M → Number of independent variables.

Hundreds or even thousands of independent variables (X-block) can be correlated with one or several dependent variables (Y-block). Often perfect correlations are obtained in PLS analysis, due to the usually large number of X variables and cross-validation procedure.

23.4 EVOLUTION PROCEDURES OF QSAR ANALYSIS

1. Classical or 2D QSAR Analysis

2D descriptors are usually developed by using the atoms and connective information of the molecule. In 2D techniques, physicochemical parameters such as hydrophobic (π), steric (molar refractivity or MR), hydrogen acceptor (HA), hydrogen donor (HD), and electronic (field effect or F, resonance or R, Hammett's constant or σ) are normally used. In addition to these parameters, de novo constants or indicator variables with 0 or 1 values denoting the absence or presence of certain features (cis/trans ring atom and bridge atom or chain, different test model, etc.) are also used to adequately parameterize the compounds. In all this, many topological indices are also considered parameters for analysis.

Drug distribution and binding processes are equilibrium processes described by free energy differences equation use logarithmic scale.

$$K = e–\Delta G/RT = e–\Delta H–T\Delta S)/RT$$

2. 3D-QSAR Analysis

Three-dimensional quantitative structure-activity relationships (3D-QSARs) are known as quantitative models that relate the biological activity of small molecules with their properties calculated in 3D space. Hence, 3D properties of a molecule are considered rather than that of the individual substituents. The 3D structures are usually generated from 2D with configurational information or 3D-structure database or X-ray crystallographic analysis either 2D NMR study. This structure is optimized to refine the geometry based on the size of the molecule such as molecular mechanics (large systems; thousands of atoms) or semi-empirical (medium size systems; hundreds of atoms) or *ab initio* (small systems; tens of atoms), in order to obtain one lowest energy structure per molecule. There are many 3D-QSAR techniques used for various purposes. Comparative molecular field analysis (CoMFA), Comparative molecular similarity indices analysis (CoMSIA), Molecular shape analysis (MSA), distance geometry approach, binding site model approach, COMPASS, the hypothetical active lattice methods are used in QSAR analysis.

3. Comparative Molecular Field Analysis (COMFA)

CoMFA is a 3D-QSAR technique employing both interactive graphics and statistical techniques are used for correlating the shapes and the biological properties of the molecules. It was proposed and developed by R.D. Cramer in 1988. The principle underlying CoMFA is that differences in a target property related to differences in the shapes of the noncovalent fields of tested molecules. The molecular shape of tested moelcule field into a QSAR table and the magnitude of steric and electrostatic fields are sampled at regular intervals throughout a defined region of rigid box.

CoMFA tries to identify the quantitative influence of the specific chemical features of the molecules on their potencies. In 3D space the contour plots result are showing that biological activity important regions of designed or active molecules. CoMFA technique includes the prediction of activity of new compounds and representation of QSAR models in the form of contour maps.

There are many important aspects that need to be considered for developing an ideal CoMFA model. They include the following factors:

- Biological data, selection of compounds, and series design, generation of 3D structure of ligand molecules.
- Conformational analysis of each and every molecule.
- Establishment of bioactive conformation of each molecule, binding mode and superimposition of the molecules.
- Position of lattice points, choice of force fields and calculation of interaction energies.
- Statistical analysis of the data and selection of the 3D QSAR model.
- Display of results in contour plots and interpretation, design and forecasting the activity of unknown scaffolds.

23.5 MOLECULAR MODELLING

Computer-aided drug design (CADD) has been recognized as a powerful technology in the drug discovery worldwide. In accelerating drug discovery, molecular modelling techniques have experienced advances in computational capabilities over the last decade. Pharmaceutical companies and academic research organizations are currently using various computational modelling techniques to lower the cost and time required for the discovery of an effective drug. There are three components of molecular modelling are molecular docking, molecular dynamics, and ADMET modelling, their applications, and limitations in small-molecule drug discovery. In molecular dynamics and docking, the algorithms are used to develop the docking software, and the models explored by these algorithms coupled with their scoring functions.

1. Molecular Docking (In *silico* method)

Molecular docking is an *in silico* method for recognizing the correct binding pose of a protein-ligand complex and evaluating its strength using various scoring functions for selecting the best pose generated by each molecule to a rank-order by using most well software DOCK program. The docking techniques attempt to fit a ligand into the binding site of a target protein by combining and optimizing variables such as hydrophobic, steric, and electrostatic complementarily and thereby estimating their binding free energy. Docking process requires the high resolution of X-ray crystallographic structures, NMR or homology-modelled structures with a known binding site.

There are thousands of experimentally determined 3D-structures of proteins available in the protein data bank (PDB). X-ray and NMR machines allow scientists to experimentally solve the detailed 3D structures of molecules in their interaction forms. the development of sophisticated bioinformatics techniques, the 3-dimensional structure of a target protein that is unavailable in the protein data bank could be modelled using homology modelling servers.

Homology modelling is used to predict the tertiary structure of the target based on a template with similar residue. The theoretical evidence behind this is based on the fact that proteins that share similar residues have similar structures and the same ancestral origin.

Molecular docking encompasses three major objectives: virtual screening, pose prediction, and binding affinity estimation. A reliable docking technique must be able to differentiate binding sites from non-binding sites as well as their molecular interactions. In addition, when dealing with large compound libraries, it is expected that the method used must successfully distinguish binding molecules from non-binding and rate the binding molecules among the best compounds in the database.

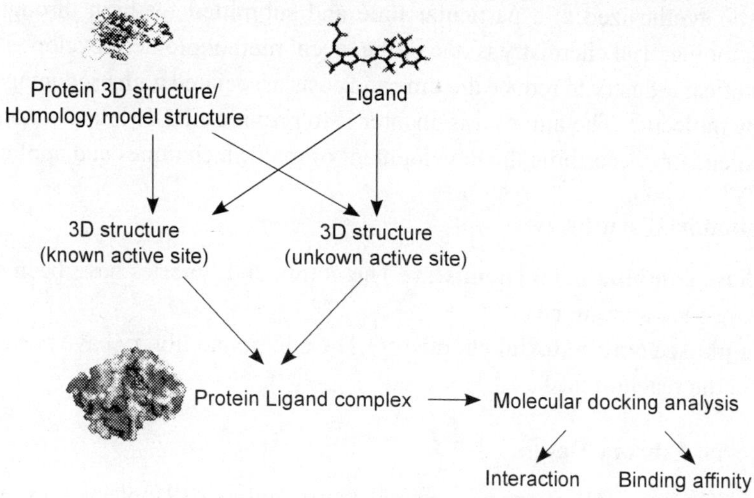

Fig. 23.3

2. De Novo Design

De Novo design includes the novel chemical structures that are capable of interacting a receptor of known structure uses the methodologies of thermodynamically stable. This is well promising method to interact with well known structure. In this method, molecule acts simply as a scaffold to align the appropriate groups in 3D arrangement that is crucial to molecular recognition. Various software are already under development in design objective. An alternative tool is BRIDGE which is based on geometrical generation of possible cyclic compounds as scaffold. Monte Carlo approach also came in BRIDGE to generating a novel structure that fit in to receptor cavity. CAVEAT is the program used for correct vectorial arrangement of appended groups. One more program, ALADIN, is used for the design or recognition of compounds that meet geometrical, steric and substrctural criteria.

23.6 COMBINATORIAL CHEMISTRY

Combinatorial chemistry is a useful tool for rapidly optimizing molecular properties, particularly ones that are difficult to design. The combinatorial approach is used to generate a large number of respective variables in a format that allows the selection of an optimal subset of variables. Naturally occurring combinatorial approach is exist to generate diverse functional macromolecules, such as the large number of antibodies that recognize non-self molecules. Combinatorial chemistry develops on this transformative approach by generating diversity in a controlled setting and then applying it to different problems such as drug discovery or catalysis development. Nowadays, combinatorial chemistry is a fundamental approach to the identification of novel molecular properties.

1. Principles of Combinatorial Chemistry

Combinatorial chemistry is a technique through which large numbers of structurally distinct molecules can be synthesized at a particular time and submitted for high throughput screening (HTS) assay. Combinatorial chemistry is one of the recent methodologies developed by researchers in the pharmaceutical industry to reduce the time and costs associated with producing successful and competitive new molecule. The aim of this chapter is to provide a basic introduction to the field of combinatorial chemistry describing the development of major techniques and applications.

Types of combinatorial synthesis

- **(a) Solid phase combinatorial chemistry:** The compound libraries have been synthesized on solid phase such as resin bead.
- **(b) Solution phase combinatorial chemistry:** The compound library have been synthesized in solvent in the reaction flask.

2. Combinatorial Library Design

Combinatorial libraries are collections of chemical compounds, small molecules or macromolecules such as proteins, synthesized by combinatorial chemistry, in which multiple different combinations of related chemical moieties are reacted together in similar chemical reaction approach.

Peptoid combinatorial libraries

Combinatorial libraries are known for collections of large number of synthetic, chemical or biological compounds that are synthesized and stored in mixtures for all possible combinations of the building blocks. This feature permits the synthesis of a large number of compounds in a relatively short time. The utility of having such libraries is that they can be screened to identify useful components, such as ligands and drug candidates.

Due to facilitate the synthesis and structural similarity to natural peptides, peptoids have been used as the basis for the combinatorial drug discovery process, gene delivery and for identification of new compound scaffold. From a synthetic perspective, combinatorial strategies through submonomer are of three types.

1. Peptoid synthesis protocol.
2. Split-and-mix format
3. One-bead two-compound strategies or the creation of encoded combinatorial libraries for cyclic peptoid microarrays.

The ability to screen and analyze peptoids rapidly and inexpensively from the libraries has been a primary tool in modern peptoid research. For instance, scientist Miller and his co- workers in 1994, designed, synthesized and screened a diverse peptoid library of 5000 compounds and discovered potent ligands that bind opiate and α1-adrenergic receptors. The library was synthesized using the submonomer peptoid synthesis protocol.

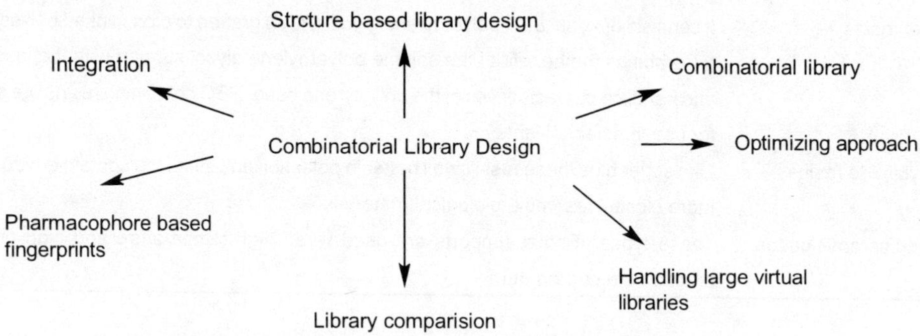

Fig. 23.4 Schemetic diagram of combinatorial library design

3. Combinatorial Synthesis on Solid Phase

Merrifield discovered the solid phase synthesis (SPS) in 1962, which earned him a Nobel prize. Merrifield's SPS concept was first applied for a developed biopolymer, recently it has extensively used in organic synthesis. The impressive outbreak of combinatorial chemistry profoundly changed the approach to new drugs, new catalyst, or new natural discovery.

The utilization of solid support for the organic synthesis relies on three interconnected following requirements.

1. A cross-linked, insoluble polymeric material should be inert to the condition of synthesis.
2. The linking substrate (linker) to the solid phase that permits selective cleavage of some or all the products from the solid support during synthesis for analysis of the extent of reaction (s) and ultimately to give the final product of interest.
3. The chemical protection strategy must allow selective protection and de-protection of reactive groups.

3.1 (Solid support)

Most solid state combinatorial chemistry is conducted by using polymer beads ranging from 10 to 750 μm in diameter.

The solid support must have the following characteristics:

1. Physical stability
2. Chemical inertness
3. Permeation of solvents and reagents to the reactive sites within the resin
4. Derivatization with functional groups to allow for the covalent attachment of an appropriate linker or first monomeric unit.

Table 23.1: Types of solid phase

Polystyrene resins	Polystyrene is cross-linked with divinyl benzene (about 1% cross linking). Polystyrene resin is suitable for non-polar solvents.
Tenta gel resins	It consists of about 80% polyethylene glycol (PEG) grafted to cross-linked polystyrene1. It Combines the benefit of the soluble polyethylene glycol support with the insolubility and handling characteristics of the polystyrene bead. PEG containing resins are suitable for use in polar solvents.
Polyacrylamide resins	Like super blue these resin swell better in polar solvent, since the contain amide bonds, more closely resemble biological materials.
Glass and ceramic beads	These types of solid supports are used when high temperature and high pressure reaction are carried out.

Table 23.2: Types of commercially available resins used in solid phase synthesis

Types of resins	Strctures
Merrifield resin: For -COOH	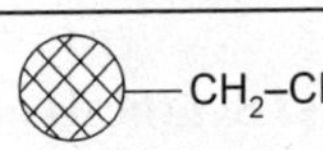 Merrifield resin
Tritylchloride resin: For –COOH, -OH, C_6H_5OH, R-NH_2, R-SH attachment	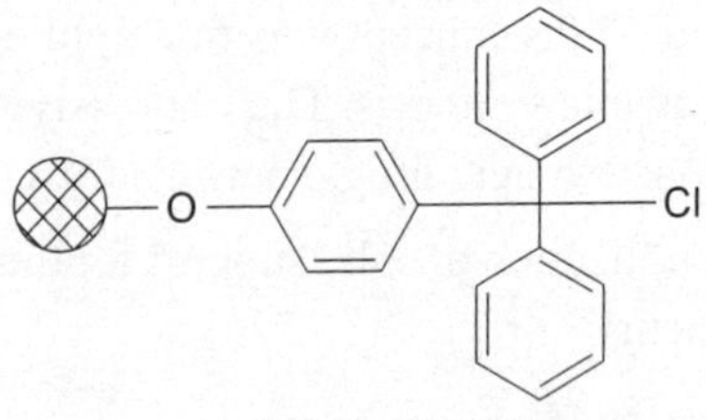 Tritylchloride resin

Wang resin: For -COOH attachment

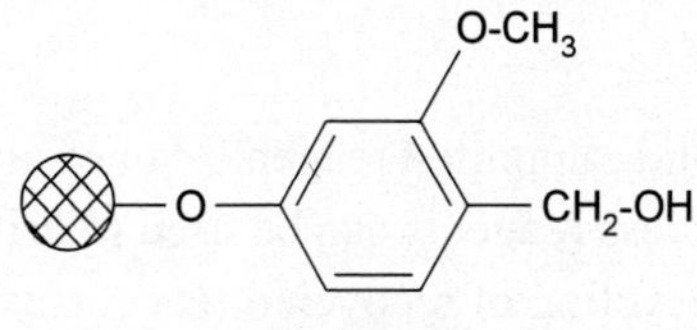

Wang resin

Hydroxymethyl resin: For -COOH attachment

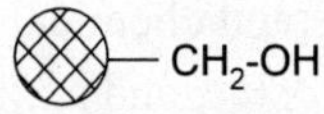

Hydroxymethyl resin

Aminomethyl resin: For -COOH attachment

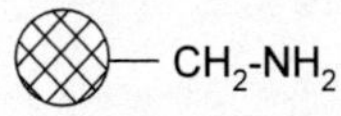

Aminomethyl resin

Rink amide resin: : For -COOH attachment

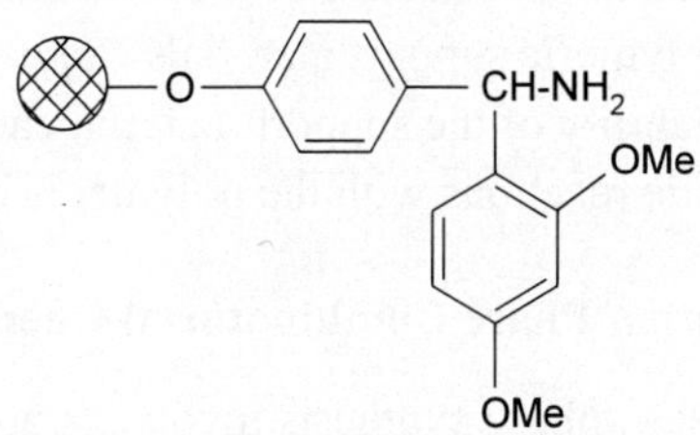

Rink amide resin

3.2 Protecting groups used in solid phase synthesis

A protecting group is reversibly attached to the functional group to convert it to a less reactive form. When the protection is no longer needed, the protecting group is cleaved and the original functionality is restored. A large number of protecting groups were developed for use in peptide synthesis since amino acids are multifunctional compounds. It is an important requirement for a protecting group to be stable under the expected reaction conditions and to be cleavable, if possible, at mild reaction conditions. The stability/cleavage conditions of a protecting group are considered relative to those of the others. Two protecting groups are said to be orthogonal if either of them can be removed without affecting the stability of the other one. Some of the protecting groups most widely used in peptide synthesis are described below.

(a) Benzyl carbonyl (Z) group
(b) t-butoxy carbonyl (Boc) group
(c) 9-fluorenyl methoxy carbonyl (Fmoc) group

3.3 Pros and cons of combinatorial chemistry

Primacy

- Solid-supported reagents can easily remove from reactions by filtration method.
- Excess reagents can be used to drive reactions to completion with easy in purification.
- Recycling of recovered reagents is economical, environment-friendly, and efficient.
- Ease to handling is especially important when dealing with expensive or time-intensive catalysts, which can be incorporated into reactors and automated processes.
- Finely accepted chemical properties by altering choice of support and its preparation.
- Toxic, explosive, and noxious reagents can easily and safely handle when contained on solid support.
- Reagents on solid-support react differently, mostly more selectively, than their unbound counterparts.

Drawbacks

- Some selective reagents may not interact well with solid-support media.
- Ability to recycle reagents on solid support is not confirmed.
- Reactions time may be slow due to diffusion constraints.
- Polymeric support materials can be very expensive to be developed.
- Stability of the support material can be poor under harsh reaction conditions.
- Side reactions with the polymer support itself may occur.

4. Solution Phase Combinatorial Chemistry

The solution-phase synthesis involves conducting chemical reaction simultaneously, preferably in well-ordered sets (arrays) of reaction vessels in solution. Most ordinary synthetic chemistry takes place in solution phase. The use of solution phase techniques has been explored as an alternative to solid-phase chemistry approaches for the preparation of arrays of compounds in the drug discovery process. Solution phase work is free from some of the constraints of solid-phase approaches but has disadvantages with respect to purification.

In solution phase synthesis soluble polymer is used as support for the product. PEG is a common vehicle which is used in solution-phase synthesis it can be liquid or solid at room temperature and show varying degrees of solubility in aqueous and organic solvent. By converting one OH group of PEG to methyl ether (MeO-PEG-OH) it is possible to attach a carboxylic acid to the free OH and use in solution-phase combinatorial synthesis. Another common support which is used in solution-phase synthesis is liquid Teflon consisting mainly of long chain of (-CF2 -) groups attached to a silicon atom. When these phases are used as a soluble support for synthesis the resulting product can be easily separated from any organic solvent.

The main disadvantage of this method is when number of reagents is taken together in a solution; it can result in several side reactions and may lead to polymerization giving a tarry mass. Therefore, to avoid this, the new approach is developed in which all chemical structure combinations are prepared separately, in parallel on a giving building block using an automated robotic apparatus.

For example, hundreds and thousands of vials are used to perform the reactions and laboratory robots are programmed to deliver specific reagents to each vial.

EXERCISE

1. Explain QSAR in detail.
2. Explain the following terms with examples:
   (a) Hansch equation
   (b) Free Wilson analysis
   (c) Hamette equation
3. Write an explanatory approach of molecular docking.
4. Write a note on combinatorial chemistry in drug design.
5. Explain electronic parameters used in QSAR.
6. Write a note on combinatorial synthesis on solid phase.

24

General Considerations of Heterocyclic Chemistry Concept

24.1 INTRODUCTION

Concept of heterocyclic chemistry in medicinal compounds is that which possesses the cyclic ring structure, having at least two different kinds of atoms in the ring. The most common type of atoms include nitrogen, oxygen and sulphur heteroatoms that contains largely carbon atoms. Apart from that, many atom elements including even bromine also is part of heterocyclic chemistry. Heterocyclic compounds are widely arranged in nature and are essential for life in various ways. They can be aliphatic or aromatic in nature depending upon the distribution of electronic constitution.

In this chapter a brief introduction is detailed to the simple types of three-, five- and six- membered heterocyclic system and their naming.

24.2 NOMENCLATURE OF HETEROCYCLIC COMPOUNDS

According to Systemic Nomenclature System (Hantzsch-Widman System)

According to this system, heterocyclic compounds are recognized by their trivial names.

Examples:

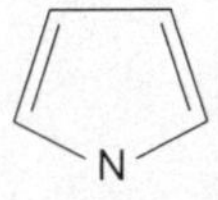

Pyrroleⓐ

Furan

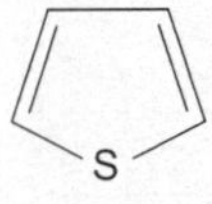

Thiophene

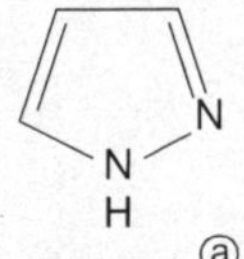

Pyrazoleⓐ

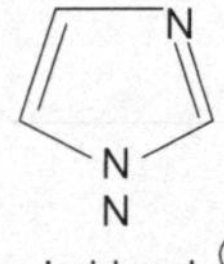

Imidazoleⓐ

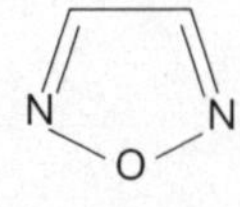

Furazan

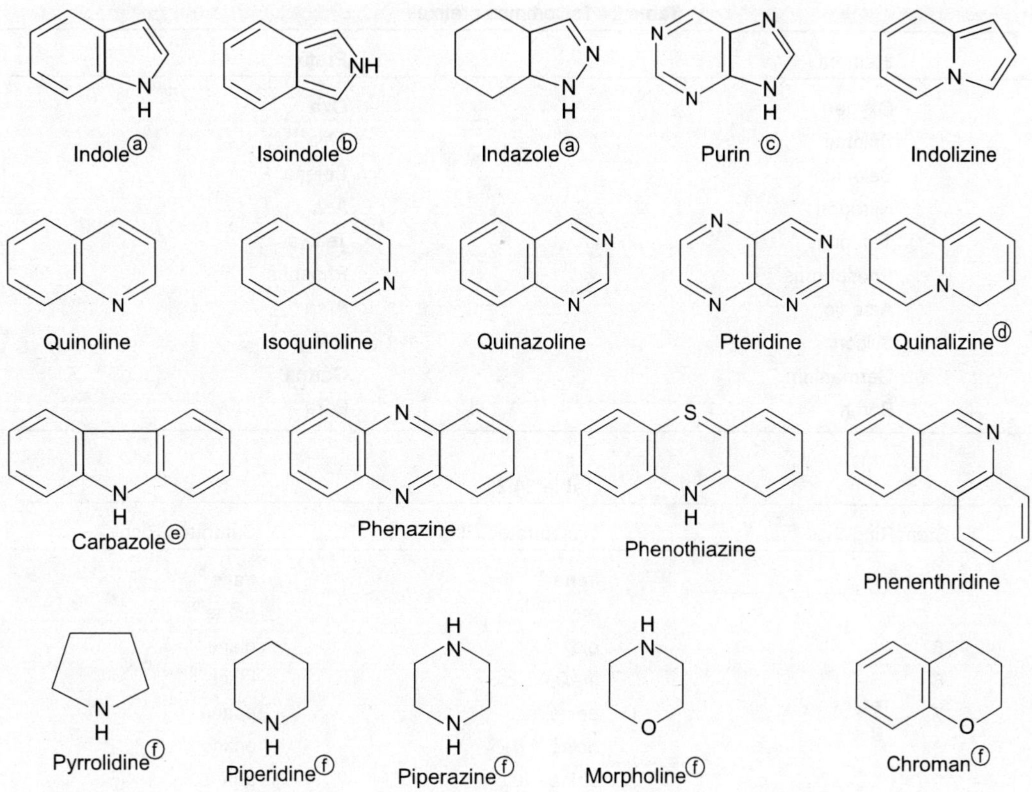

Symbols represents

a → 1H Tautomer

b → 2H Tautomer

c → 9H Tautomer

d → 4H Tautomer

e → 10H Tautomer

f → Saturated heterocyclic rings

Fig. 24.1

24.3 NAMING OF HETEROCYCLIC RINGS

Fig. 24.2

Table 24.1: Common prefixes

Element	Prefix
Oxygen	Oxa
Sulphur	Thia
Selenium	Selena
Nitrogen	Aza
Tellurium	Tellura
Phosphorus	Phospha
Arsenic	Arsa
Silicon	Sila
Germanium	Germa
Boron	Bora

Table 24.2

Stem Ring Size	Unsaturated Ring	Saturated Ring
3	irene[a]	irane[b]
4	ete	etane[b]
5	ole	olane[b]
6	ine©	inane[d]
7	epine	epane
8	ocine	ocane
9	onine	onane
10	ecine	ecane

SYMBOLS REPRESENT:-

a → Three-member ring contain nitrogen

b → Saturated ring containing nitrogen

c → 'inine' used with some phosphorous, arsenic and boron heterocycles

d → 'ane' is used if immediately preceded by the Prefixes 'ox', 'thi', selen.

Examples:

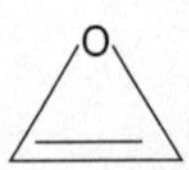
Thirene

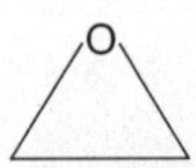
Oxirane

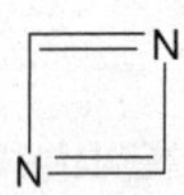
1, 3-diazete

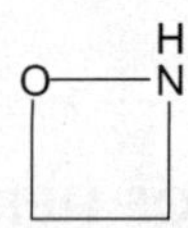
1, 2-Oxazetidine

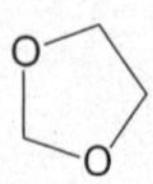
1, 3-Dioxolane

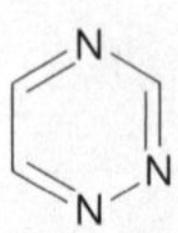
1,2,4 Triazine

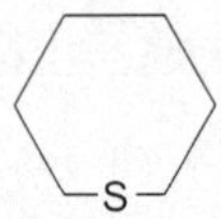
Thiepene

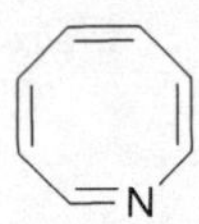
Azocine

Fig. 24.3

24.4 NOMENCLATURE OF FUSED RING SYSTEM

Fused ring system involving the benzene ring may be fused on to the carbon atoms of each of these three heterocyclic compounds in two different positions. Six possible structures are illustrated below:

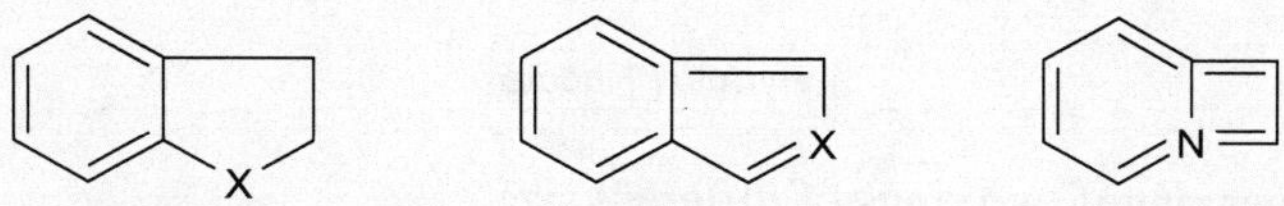

Indole X = NH
Benzofuran X = O
Benzo[b] thiophene X= S

Isoindole
Isobenzofuran
Benzo[c]thiophene

Indolizine

Fig. 24.4

24.5 RULES OF NOMENCLATURE OF FUSED RING SYSTEM

(1) Select prefix (O)
(2) Select base component
Example

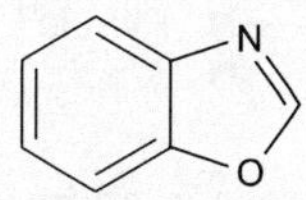

Benzoxazole

Fig. 24.5

Base Component:

1. Choose, one ring containing 'N'
Example

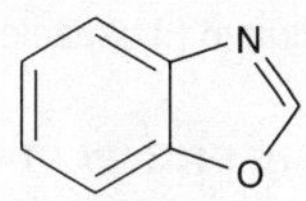

Benzoxazole

Fig. 24.6

2. Prefer than oxa, thia, aza

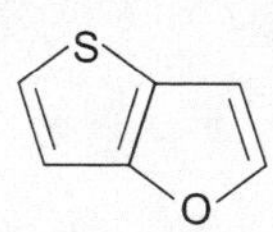

Thieno [ ] furane

Fig. 24.7

3. Select one consisting two or more rings

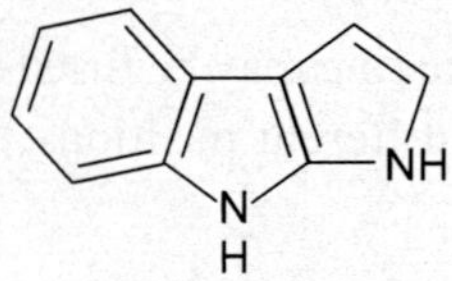

Pyrrolo [ ] indole

4. Select the larger ring if two rings of different size

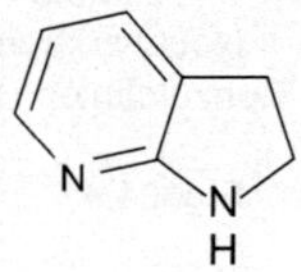

Pyrrolo [ ] Pyridine

5. Select the most heteroatom containing ring

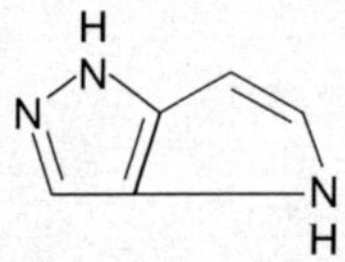

Pyrrolo [ ] pyrrazole

6. If same number of heteroatoms are present, choose with priority oxa > thia > aza with higher electronegativity

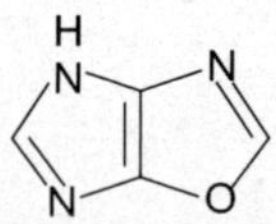

Imidazo [ ] oxazole

7. If the same number of heteroatoms are present of oxa, thia, aza then select with lower numbering

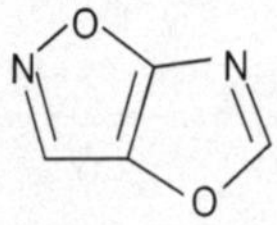

Oxazola [ ] isoxazole

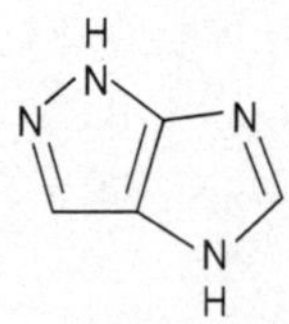

Imidazo [ ] Pyrrazola

8. Another aspect to indicate the fusion by the given letter symbols to the base components and numbers to the prefix (in same direction)
   Example

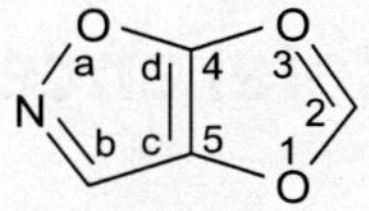

Oxazolo[5,4-d]isoxazole

EXERCISE

1. Give a brief note on heterocyclic chemistry
2. Discuss nomenclature of heterocyclic compounds.
3. Write about Hantzsch-Widman system.
4. Write the name of tricyclic heterocyclic compounds
5. Discuss with examples of nomenclature of fused ring system.

References

1. Abraham. D. J. (Editor), Burger's Medicinal Chemistry. Vol. 6, Nervous System Agents, Wiley, New York, 6th edition, 2003.
2. Wilson and Gisvold's Textbook of Organic Medicinal and Pharmaceuticals Chemistry, Block, J. H. and Beal J. M. Jr (Editors), 11th edition, 2004.
3. Foye's Principles of Medicinal Chemistry, Williams, D. A. and Lemke, L. T. (Editors), 5th edition, 2002.
4. Elsevier's Synthesis of Essential Drugs, Vardanyan R. and Hruby V-(Editors), 1st edition, 2006.
5. The Principles of Heterocyclic Chemistry, Katrifzky. A. R. and Langowski. M. (Editors), Pharma-Med Press, 2009.
6. An Introduction to the Chemistry of Heterocyclic Compounds, Acheson. R. M. (Editor), 3rd edition, (Wiley Student Edition), 2008.
7. Essential of Medicinal Chemistry, Korolkovas A. (Editor), 2nd edition (Wiley India Edition), 2008.
8. Elsevier's Comprehensive Medicinal Chemistry, Hansch. C, Sanmes, P.G. and Taylor. J. B. (Editors), Volume 1 (Pergamon Press), 2005.
9. Name Reaction: A Collection of Detailed Reaction Mechanism, Li. J. J. (Editor), 2nd edition (Springer International Edition), 2004.
10. Medicinal Chemistry, A Molecular and Biochemical Approach, Nogrady. N, Weaver D. F. (Editors), 3rd edition (Oxford University Press), 2005.
11. The Textbook of Pharmaceutical Medicine, Grijjin J. P and Grady J. O. (Editors), 5th edition (Blackwell Publishing), New Age International Publishers), 2006.
12. Elsevier's The Practice of Medicinal Chemistry, Wermuth C.G. (Editor), 2nd edition (Academic Press), 2004.
13. Principles of Medicinal Chemistry including Proteomics, Ranganathan S. and Suresh, A. J., CBS Publishers and Distributors, 2011.
14. Chemistry for Pharmacy Student's General, Organic and Natural Product Chemistry, Sarker, S.D. and Nahar, L. (Editors), John Wiley and Sons Ltd., 2007.
15. Harrison's Principles of Internal Medicine, Braunwald E., Fauci. A.S., Kasper, L. D. (Editors), 2nd Volume, 17th edition (McGraw-Hill), 2008.

Index

R

S